Animal Manure:
Production, Characteristics, Environmental Concerns, and Management

H.M. Waldrip, P.H. Pagliari, and Z. He, editors

Books and Multimedia Editorial Board Chair: Shuyu Liu
ASA Editor-in-Chief: Daniel Sweeney
SSSA Editor-in-Chief: David Myrold
Managing Editor: Danielle Lynch

American Society of Agronomy
Soil Science Society of America

5585 Guilford Road, Madison, WI 53711-58011 USA
agronomy.org · soils.org

ASA Special Publication 67
doi:10.2134/asaspecpub67
ISSN: 2165-9710
ISBN: 978-0-89118-370-9 [print]
ISBN: 978-0-89118-371-6 [online]

Library of Congress Control Number: 2019947063

FSC
MIX
Paper from responsible sources
www.fsc.org FSC® C013604

Contents

Foreword

Animal manures have been recognized for centuries as a critical resource for promoting crop production. Modern animal production systems provide a concentrated source of these manures. These systems require science-based management to effectively use them as valuable resources for recycling and replenishing soil nutrients and carbon, while mitigating their potential impacts on the environment. I am honored to introduce this book, while taking the opportunity to thank Heidi Waldrip, Paulo Pagliari, and Zhongqi He, the editors, and all of the 48 contributing authors for lending their extensive expertise and efforts towards this valuable, comprehensive compilation.

William L. Pan, President, Soil Science Society of America

Preface

The ever-increasing global human population is putting an enormous amount of pressure on current food production systems. This increase includes demands for animal products, such as milk, meat and eggs, the production of which occurs simultaneously with enormous quantities of animal manure. As a result, environmental concern (i.e., air, water and soil) due to non-point source pollutants is increasing while producers struggle to find economically and environmentally sustainable solutions for both animal farming and manure disposal management. Frequently, animal producers are blamed for ongoing environmental issues because of negative public perception of concentrated animal feeding operations (CAFOs) and the large amounts of manure produced by these CAFOs. Animal manure, when used according to best management practices (BMPs), is considered one of the most economically profitable source of nutrients required for crop production. By recycling manure back to the land, farmers reduce the cost of purchased inorganic fertilizers for feed and forage production, while adding essential organic matter to the soil, which improves soil health via various mechanisms for long-term cropland productivity. In general, animal manure contains all the macro- and micro-nutrients required for optimum plant growth and maximum yield, although supplemental fertilizers may be required to improve the manure nutrient balance. Even though, inappropriately managed animal manure may present significant environmental risks, including eutrophication/contamination of surface and subsurface water sources, emissions of greenhouse gases and air pollutants, contamination of ecosystems with antibiotic residues, antibiotic-resistant microbial genes and pathogens, as well as other health and quality-of-life-related issues. The goal of this monograph is to address the environmental concerns of modern animal manure systems while simultaneously pinpointing areas where manure management can be improved. Topics covered in the book include, but are not limited to, characteristics of different manure production practices and corresponding manure characteristics associated with traditional and newer management strategies; manure-derived emissions of greenhouse gases and other components that impair air quality; transport of nutrients and other contaminants from animal housing and field-applied manure. This work also includes timely information on potential fate and transport of pharmaceutical residues, microbial pathogens and antibiotic resistance from manure. In addition, beneficial manure utilization practices are reviewed as a step towards improved manure management. The overall objective is the production of a timely, comprehensive reference that is applicable and useful for the scientific community, regulatory agencies, environmental advocacy groups, and animal farm managers. The majority of the content was derived from peer-reviewed scientific publications; however, the editors feel that there is also value in information garnered from the popular press and non-traditional sources. Ideally, this approach results in the publication of a highly accessible, relatively informal, and reader-friendly text for use by anyone interested in optimal management of manure for both environmental quality and soil fertility.

The monograph is divided into five parts. Part I comprises two introductory chapters. The first chapter reviews the concept of CAFOs and provides estimations for the amount of manure produced in the USA. The second chapter uses data from two service laboratory to analyze the temporal changes in manure chemical composition in the southern Great Plains. Part II delves into the vast diversity of animal production practices and their impact on manure characteristics. Seven chapters in this part review common animal production systems and manure storage methods; nutrient characteristics of manures from beef, dairy, poultry, swine, small ruminants and cervids, horses, and other equids. In addition, Part II contains a chapter comparing manure produced under organic and conventional management systems. Part III covers environmental aspects of animal manures and also presents mitigation technologies that are being considered to help alleviate potential negative impacts of animal manure. Five chapters in Part III report on how to manage manure to minimize phosphorus losses; one chapter investigates the use of whole farm modeling as an approach to improve manure management for greenhouse gas mitigation and improved environmental quality; the last three chapters in Part III handle the fate and transport of estrogen, antibiotics, and antibiotic resistance to pharmaceuticals. Part IV presents current knowledge on numerous beneficial uses of animal manure. Six chapters in Part IV review the potential for improved manure use as a stable composted product, manure digestion for energy production, production of bioenergy and biochar from manure in thermochemical reactors, and nutrient removal and reutilization from manure. The last two chapters in Part IV provide an in-depth look at pelletization and production of organomineral fertilizers from manure. Concluding remarks and visions for the future are presented in Part V of the monograph.

Chapter contribution to this monograph was by invitation only. Each chapter was designed to cover a specific topic and the authors for each topic were selected after extensive communication between the editors. It is possible that overlap in literature review is present among the different chapters, as each chapter was intended to serve as a stand-alone document. All 21 chapters were written by accomplished experts in their relevant fields and were subject to peer review and revision processes prior to being accepted. Constructive criticisms from reviewers were required before a chapter was finally accepted for publication. The editors thank all reviewers for their comments and suggestions which improved the quality of the work reported in this monograph.

This editorial duty is a team work. It was initiated by Zhongqi He, while serving in the Tri-Societies' Book and Multimedia Publishing Committee and encouraged by the committee managing editor, Lisa Al-Amoodi. Heidi Waldrip led the draft of the book proposal and outline of the tentative Table of Contents. All three editors were actively involved in invitation of chapter contributions. Zhongqi He and Paulo Pagliari have taken the main responsibility in the peer review and revision processes and made acceptances of these chapter contributions. Finally, we like to thank the Tri-Societies' Book Managing Editor Danielle Lynch for her editorial coordination and technical support throughout the process of executing this book project.

Paulo H. Pagliari, Zhongqi He, Heidi M. Waldrip, Editors

Animal Manure Production and Utilization: Impact of Modern Concentrated Animal Feeding Operations

Paulo Pagliari,* Melissa Wilson, and Zhongqi He

Abstract

The total number of livestock and poultry animals being raised in the United States has shown some fluctuation over time. From 2012 to 2018, however, there was a consistent increase of approximately 96 million head annually. This trend was fueled by strong growth in global demand over the past decade and net exports of meat are forecast to continue to grow rapidly. Swine and poultry tend to be raised in confinement, while beef cattle tend to be raised on pastures, but then are finished in confinement, while dairy cattle are often fed and milked in confinement but are allowed on pasture periodically. The number of small animal farms have continually decreased since 1982, while the number of confined animal feeding operations (CAFOs) have increased. In the United States, as much as 1.4 billion tons of manure is produced by the 9.8 billion heads of livestock and poultry produced yearly. These manures produced are primarily used as a nutrient source for crop production. However, because of the shift toward CAFOs producing most of the meat consumed, the mismatch of manure produced to area manure is applied is leading to negative environmental impacts.

By the traditional definition, animal manure is animal excreta (urine and feces) and bedding materials, usually applied to soils as a fertilizer for agricultural production (He, 2012). In fact, before the extensive application of synthetic fertilizer, the majority of outsourced crop nutrient inputs were from animal manure. Indeed, since the beginning of human agricultural activities, animal manure has been an integral part of sustainable crop production. For example, stable isotope analysis of charred cereals and pulses from 13 Neolithic sites across Europe (dating approximately 5900–2400 B.C.) has shown that early farmers used livestock manure and water management to enhance crop yields (Bogaard et al., 2013). However, the development of concentrated animal feeding operations (CAFOs) has produced a large amount of animal manure that far exceeds the needs of regional soils and crops. This practice toward fewer but larger operations of animal production has created environmental concerns in recycling and disposing of surplus animal manure (He, 2011; He et al., 2016). This introductory chapter first compares and discusses several terms used in animal manure management and research communities, then highlights the modern CAFO practices and their impacts on animal manure production and its utilization in the United States.

Abbreviations: AFO, animal feeding operation; AU, animal units; CAFO, concentrated animal feeding operation.

P. Pagliari, Department of Soil, Water, and Climate, University of Minnesota. Southwest Research and Outreach Center. 23669 130th St. Lamberton, MN 56152; M. Wilson, Department of Soil, Water, and Climate, University of Minnesota, 439 Borlaug Hall, 1991 Upper Buford Circle, Saint Paul, MN 55108; Z. He, USDA-ARS Southern Regional Research Center, 1100 Robert E. Lee Blvd. New Orleans, LA 70124. *Corresponding author (pagli005@umn.edu)

doi:10.2134/asaspecpub67.c1

Animal Manure: Production, Characteristics, Environmental Concerns and Management. ASA Special Publication 67. Heidi M. Waldrip, Paulo H. Pagliari, and Zhongqi He, editors.
© 2019. ASA and SSSA, 5585 Guilford Rd., Madison, WI 53711, USA.

Field Operation Terms Used Alternatively and/or Inappropriately

Animal Versus Livestock

In the scientific literature, there are publications that misuse terms "livestock" and "animal" by assuming poultry production to be a livestock operation. Per the U.S. Code of Federal Regulations (2019a, 2019b), the term "livestock" includes the following animals, among others: cattle (both dairy and beef cattle), sheep, swine, horses, mules, donkeys, and goats. However, turkeys or domesticated fowl are considered poultry and not livestock within this definition. The USDA and EPA statistics and literature listed livestock and poultry as two separate categories (e. g. USDA, 2015; USEPA, 2013). Based on these definitions, livestock manure is not equivalent to animal manure as the former does not include poultry manure. Thus, extra effort should be encouraged to apply these terms correctly.

Animal Head Versus Animal Unit

There are multiple ways of discussing how many animals are raised in the U.S. farms. Often times we see either animal head counts or "animal units". Animal head counts are intuitive as they represent the number of animal bodies being produced. On the other hand, an animal unit (AU) is 1000 pounds of live animal weight and is often the term used by agricultural engineers, as well as regulatory or conservation agencies in the United States (National Research Council, 2003). This allows for a common measurement between livestock and poultry species, which is particularly important when considering manure production. For example 1000 broiler chickens produce significantly less manure than 1000 dairy cows, but 1000 AU of broiler chickens (about 125,000 head of chickens) produce about the same amount of manure as 1000 AU of dairy cows (about 700 head of dairy cows). The average number of animals per AU, the amount of manure, and the amount of nutrients typically excreted by various livestock and poultry types can be found in Table 1. It should be noted that when converting animal head counts to AU, there are certain assumptions made that may differ across the United States (Gollehon et al., 2016) and thus the values shown in Table 1 are only estimates.

Animal Manure Versus Animal Waste

Animal manure was traditionally a valuable resource for agricultural production, and is still an essential fertilizer for organic farming today (He, 2019). On the other hand, the impact of manure generation and disposal by CAFOs is far greater than the role of organic fertilizers. Thus, in the last three decades or so, the objectives of animal manure management research have been focused on the basic knowledge of manure's impacts on the environment and relevant technologies for the best management of animal manure (He, 2011; He et al., 2016), leading to increased usage of the term "animal waste" (Food Print, 2019). While "animal waste" could be more than "animal manure", the two terms are equivalent in many cases. For example, under WAC 16–250–010 (WAC, 2003), Washington State Legislature defines "animal wastes" as a material composed of excreta, with or without bedding materials and/or animal drugs, collected from poultry, ruminants, or other animals except humans. Thus, the two terms are sometimes used alternatively, such as in the book title "Animal Waste Utilization: Effective Use of Manure as a Soil Resource" (Hatfield and Stewart, 1998). It is apparent that the term "animal waste" is negative, but "animal manure" is neutral if not positive. Thus, while both terms are acceptable for management, the neutral term "animal manure" seems more appropriate for use in research-relevant work and publications (He, 2012; He and Zhang, 2014; Sommer et al., 2013).

As-Excreted Versus As-Stored Manure

The amount of manure produced and nutrients present in raw manure as it comes out of the animal is often preceded by the term "as-excreted". These values can be found in the literature (ASABE, 2014; Lorimor et al., 2004) and are for planning purposes when designing manure storage facilities. The actual values can vary by up to 30% due to individual farm management as well as genetics and animal performance (Lorimor et al., 2004). It is important to distinguish between the characteristics of "as-excreted" manure from manure that has been stored (sometimes referred to "as-stored" manure) because housing, handling, and storage can significantly impact manure characteristics (ASABE, 2014; Lorimor et al., 2004). For example, approximately 25%, 10%, 26%, and 53% of nitrogen can be lost as ammonia before manure is land applied in dairy, beef, poultry, and swine operations, respectively (USEPA, 2004). In some cases, researchers do not identify what characteristics they are using in the literature (as-excreted versus as-stored) and caution must be taken to use the appropriate terminology in future.

Composition of Manure: Liquid Versus Slurry, Semi-Solid, and Solid

The composition of manure varies with the amount of solids that it contains. Liquid manures have less than 4% solids, slurries have between 4 and 10% solids, semi-solids have between 10 and 20% solids and may or may not be stackable, while manures with greater than 20% solids are stackable and considered to be solid manures (Lorimor et al., 2004). The amount of solids in the manure dictates how it can be handled, stored, and land applied. For instance, stackability, or the ability to pile manure, is determined by whether manure can be piled in

Table 1. Average number of animals in one animal unit (AU), plus the manure produced and nitrogen and phosphorus excreted per animal unit by livestock and poultry in the United States. (Source: Kellogg et al., 2014; USDA-NRCS, 1992; Wheeler and Zajaczkowski, 2009; Pagliari and Laboski, 2012, 2013).

Livestock Type	Number of animals in one Animal Unit	Manure produced	Nitrogen produced	Phosphorus produced
	(AU)	——— lb d^{-1} AU^{-1}———		
Beef Cattle	1.0	59.1	0.31	0.11
Dairy	0.7	80.0	0.45	0.07
Swine	2.5	63.1	0.42	0.16
Chicken (layer)	82.0	60.5	0.83	0.31
Chicken (broiler)	125.0	80.0	1.10	0.34
Turkey	56.0	43.6	0.74	0.28
Horse	0.9	51.0	0.28	0.20

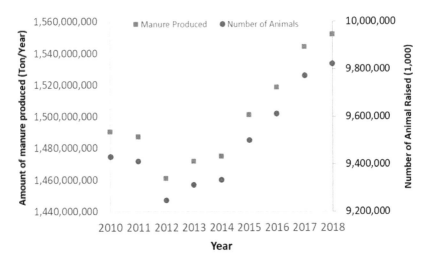

Fig. 1. Total number of animals produced and manure produced from 2010 to 2018. (Source: USDA-APHIS: www.aphis.usda.gov)

a 1:3 vertical to horizontal ratio or greater (USDA-NRCS, 2017). Waste from animal barns with bedding is typically treated as semi-solid or solid manure, while more intensive livestock rearing systems collect the manure as a liquid or slurry because concrete or slats are used to minimize the need for bedding (He et al., 2012; Janni and Cortus, 2019). Another term that is used in the literature outside of the United States includes "farmyard manure," which typically refers to semi-solid or solid manure from beef or dairy operations, but may also include manure from other species in operations that have diverse livestock and poultry.

Poultry Manure Versus Poultry Litter

Both terms of poultry manure (Dail et al., 2007; Waldrip-Dail et al., 2009) and poultry litter (Tewolde et al., 2018; Waldrip et al., 2015) can be found in literature. However, there are some differences in their compositions. In the poultry industry, poultry manure commonly refers to the bird excreta (feces and urine) collected from egg-laying and breeder facilities and can be treated as a solid, slurry, or liquid manure, depending on the storage and handling of the farm (USDA-NRCS, 2009). Poultry litter comes from chicken broiler and turkey grow-out facilities that produce meat and refers to the mixture of bird excreta and bedding materials removed from poultry houses (USDA-NRCS, 2009). Both manure and litter may include wasted feed and feathers as well. Of the poultry waste generated in 1990, approximately 68% was poultry litter and greater than 90% of both litter and manure was land applied (Moore et al., 1995). Despite the distinct differences between poultry manure and litter, it should be noted that the terms are occasionally interchanged in the literature (Bitzer and Sims, 1988; Billen et al., 2015) which can be confusing for those working in the industry and for researchers.

Unit Conversions

English units are used throughout the chapter as it is relevant to the livestock industry in the United States. Additionally, manure analyses often report various forms of nutrients which can be converted as needed (i.e., reports may show total phosphorus or total phosphorus as P_2O_5). Table 2 shows common conversions that can be used throughout this book.

Livestock and Poultry Production

The total number of livestock and poultry animals being raised in the United States has fluctuated over time. There was a decrease from 2010 to 2012 from around 9.4 billion head to 9.2 billion head, respectively (Fig. 1). From 2012 to 2018, however, there was a consistent increase in every year by approximately 96 million head annually, primarily for beef, chicken, and swine (Fig. 1). This trend was fueled by strong growth in global demand over the past decade and net exports of meat are forecast to continue to grow rapidly (Jones et al., 2018).

Swine and poultry tend to be raised in confinement while beef cattle tend to be raised on pastures, but then are often finished in confinement (Gillespie, 2019). Dairy cattle are often fed and milked in confinement but are allowed on pasture periodically (Gillespie, 2019). These systems are known as animal feeding operations (AFOs) and are defined by the USEPA as agricultural enterprises in which livestock (cattle, swine, and so forth) and poultry are kept and raised in confined conditions until they are transported to processing plants for slaughter, for a minimum of 45 d in a 12-mo period (USEPA, 2019). In these operations, food is typically harvested from the surrounding land and brought to the confinement area, as opposed to having animals directly graze forages. In some cases, feed can originate from areas further away and even other countries, depending on price.

The number of AFOs in the United States has been reported to be around 450,000 (USEPA, 2004). The number of farms considered small confined operations (with less than 300 animals) decreased from 435,000 in 1982 to 213,000 in 1997 (Ribaudo et al., 2003). In 2010, the number of operations with more than 200

Table 2. Conversion table.

To convert from	To	Multiply by
Pound	Kilogram	0.454
Ton (short)	Ton (metric)	0.907
Acre	Hectare	0.405
Gallons	Liters	3.79
%	Milligrams per liter	1.0
K	K_2O	1.2
P	P_2O_5	2.3
Pounds per acre	Kilograms per hectare	1.12
Pounds per acre	Megagrams per hectare	0.00112
Pounds per 1000 gal	Milligrams per liter	119.8
Pounds per ton	Milligrams per kilogram	500
Gallons per acre	Liters per hectare	9.354
Tons (short) per acre	Metric tons per hectare	2.24

Confined Animal Feeding Operations and Human Health: Case Studies

Wing and Wolf (2000) surveyed residents of three rural communities in North Carolina totaling 155 respondents. The communities could be distributed as one with residents living in the vicinity of a 6,000-hog operation, one in the vicinity of beef cattle operations, and the last one in an area with no livestock operation nearby. The results of the survey showed that communities living in close proximity to hog farms have increased respiratory related problems, gastrointestinal problems, and mucous membrane irritation, in addition to lower quality of life compared with the other two communities.

Hooiveld et al., (2016) surveyed a community to understand the prevalence of respiratory and gastrointestinal conditions of people (119,036 participants) living in close proximity (about 10 km) to swine, poultry, cattle and goat CAFOs. In this report the authors found that poultry and cattle had no effect on health conditions, but swine and goat CAFOs significantly increased the odds of unspecified infectious disease and pneumonia.

Radon et al. (2006) surveyed and performed medical tests in approximately 7,000 residents from four rural towns in northwest Germany between 2002 and 2004. Survey participants ranged from 18 to 44 years of age. The results showed that high density of CAFOs near residential areas significantly affected respiratory health. Among all potential environmental concerns, manure management will be the focus of this chapter.

Fig. 2. Case-studies of the human health impacts of confined animal feeding operations (CAFOs) on local communities.

heads of milking cows in the United States constituted 69% and those with 199 or less were 31%; in comparison, for swine, the number of operations with 2000 of more heads constituted 85% and those with 1999 or less made up 15% (USDA-APHIS, 2011). In contrast, 87% of all beef cattle were being raised in small-scale operations (USDA-APHIS, 2011). The USDA-APHIS considers small-scale operations the farms which report annual gross sales of agricultural goods ranging from $10,000 to $499,999, and in 2011 there were approximately 350,000 small farms in the United States (USDA-APHIS, 2011). Horses and other equids are the second largest group of animals to be raised by small-scale farms with 38% of total animals grown in small farms in the United States. In contrast, dairy cattle, swine, and poultry grown in small-scale farms represent 8.5%, 5.1%, and 16.9%, respectively, of total animals grown in the United States (USDA-APHIS, 2011).

Concentrated animal feeding operations, or CAFOs, are AFOs with more than 1000 animal units (AU). The use of CAFOs to raise livestock and poultry have been shown to improve farm income, increase animal density, and lower the amount of land required to raise large amounts of animals. On the other hand, many environmental concerns have been brought forward since the inception of CAFOs, such as large amounts of waste produced in a relatively small area, lagoon management, odor, pathogens, presence of significant amounts of pharmaceuticals in the waste, and nutrient management (Cole et al., 2000; Wing and Wolf, 2000). Confined animal feeding operations also pose significant risks to human health (Ribaudo et al., 2003; Hooiveld et al., 2016). See Fig. 2 for some cases studies exploring the risks. Due to the potential environmental and human health concerns, manure management will be the focus of this chapter.

Animal Manure Produced

The amount of manure produced on a daily basis by different livestock and poultry, as well as the daily amount of nitrogen (N) and phosphorus (P) excreted, can be found in Table 1. These values are estimates based on literature data and reflect the amount of nutrients in the manure without bedding. The amount of manure produced per animal species varies substantially, even when adjusted to an animal unit (AU) as observed in Table 1. Chicken broiler and dairy are the biggest producers generating about 80 lb manure d^{-1} AU^{-1}; while turkey produces the least amount, 43.6 lb manure d^{-1} AU^{-1} (Table 1). Similarly, the amount of N and P in the manures varies and is highest in poultry and turkey (range 0.74 to 1.10 lb N and 0.28 to 0.34 lb P d^{-1} AU^{-1}) than in the other animal species considered (range 0.28 to 0.45 lb N and 0.07 to 0.20 lb P d^{-1} AU^{-1}) (Table 1).

It is estimated that in 2017, about 1.4 billion tons of manure was produced from over 9.8 billion heads of livestock including beef cattle, dairy cows, swine, poultry, goat, sheep, horse, and others (USDA-APHIS, 2017). Table 3 presents the number of animals raised in 2017, potential amount of manure produced by each animal species, and the amount of N and P in the animal manures. Figure 1 shows the total amount of manure produced from 2010 to 2018 (2018 being estimated based on limited information available at the time the report was generated). The amount of manure produced decreased from 2010 to 2012 following the decrease in the number of total head produced during that same time, and then showed a consistent increase of 15 million tons of manure yr^{-1} from 2012 to 2018 (Fig. 1). Beef cattle (1.2 billion ton yr^{-1}) generated about 78% of the total amount of manure produced in 2017, followed by dairy and horse (296 and 46 million ton yr^{-1}), while turkey and chicken layers produced the least amount of manure among all species considered, at 35,000 and 49,000 ton yr^{-1}, respectively (Table 3). It is nearly impossible to estimate the amount of bedding that is used by livestock and poultry operations, which end up being mixed in with the manure. In most published reports, manure is regarded as the feces, urine,

Table 3. Number of poultry and livestock animals (source: USDA-APHIS, 2017) in 2017 in the United States, along with the estimated amount of manure produced and nitrogen and phosphorus excreted that year.

Livestock Type	Number of animals	Manure produced‡	Nitrogen produced‡	Phosphorus produced‡
	— 1000 head—		ton yr^{-1}	
Beef Cattle	108,405	1200,000,000	6294,416	2233,503
Dairy	14,181	145,000,000	1663,735	258,803
Swine	129,388	596,000	3967	1511
Chicken (layer)	375,845	49,000	672	251
Chicken (broiler)	8913,000	1050,000	14,438	4463
Turkey	242,500	35,000	594	225
Horse	3914†	36,000,000	197,647	141,176
Total	9,822,460	1,382,730,000	8,175,469	2,639,932

† Data available is from 2012 only.

‡ Data estimated using values reported in Table 1.

and bedding all combined, but in some cases, the term feces or feces and urine is used to specify that bedding is not included. In this chapter, manure is referred to feces and urine and manure plus bedding will be regarded as litter.

In 2017, 8.1 million tons of manure N and 2.6 million tons of manure P were excreted by livestock and poultry (Table 3). More than 75% of the total manure N and P was excreted by beef cattle, with dairy, horse, and broilers following behind (Table 3). The lowest amount of manure N and P was found in manure from turkeys and layers (Table 3). In comparison, the amount of N and P excreted in 1997 was 1.23 million and 0.66 million tons, respectively (Gollehon et al., 2001). When bedding is added, the amount of nutrients in the manure may change, or it may stay the same. For example, Long et al. (2018) reported that dairy manure mixed with bedding had a composition of 5.8 lb P_2O_5 ton^{-1} and 8.0 lb N ton^{-1}, while as-excreted book values for the region would be 4.4 lb P_2O_5 ton^{-1} and 9.9 lb N ton^{-1}. For swine manure the values for manure plus bedding and as-excreted book values for the region were 19.3 lb P_2O_5 and 43.1 lb N 1000 gal^{-1} and 37.9 lb P_2O_5 and 45.5 lb N 1000 gal^{-1}, respectively.

Most of the estimates reported in Table 3 refer to the amount of manure and nutrients excreted, but the amount that is land applied may be different. For example, the USDA-NRCS has reported that under the best circumstances only 90% to 95% of the total manure produced can be recovered and reutilized (USDA-NRCS, 1992). This indicates that as much as 0.15 billion tons of manure is lost or remains in storage every year. Additionally, manure nutrients can be lost through handling and storage. For example, N is constantly being volatilized (Cole et al., 2005; Selbie et al., 2015) and N and P, in addition to all other nutrients found in manure, can run-off from litter piles any time there is a rain event and water runs through the litter pile (Pagliari 2014). These losses make manure nutrient availability unpredictable unless a sample is collected and sent to a lab immediately prior to land application.

Land Application of Manure

Land application is the most economical way to reuse animal manure, although other possibilities are currently being investigated for manure reuse, including the use of manure for biochar and energy production (Pagliari et al., 2010; Lundgren and Pettersson, 2009; Ro, 2012). Unfortunately, many livestock and poultry producers are not using best management practices for manure management, which is exceedingly problematic (Jackson et al., 2000; Ribaudo et al., 2003; Long et al., 2018). For example, the repeated application of manure to meet the N needs of the crop will bring soil test P (STP) to levels that are excessive (Lehmann et al., 2005; Waldrip et al., 2015). Excessive nutrients in the soil will increase the potential for nonpoint-source pollution to reach ground and surface waters by leaching and runoff. However, the potential for environmental problems depends on many factors, such as the amount of nutrient applied in excess, nutrient management practices, soil type, cropping systems used, and others.

On the other hand, many producers see manure as a useful soil amendment and utilize nutrient management plans to help them apply the manure properly (Dagna and Mallarino, 2014). When used as recommended, manure can result in similar or higher crop yields when compared with inorganic fertilizer, thereby minimizing reliance on synthetic chemical inputs for food production (Carter et al., 2010; Ayinla et al., 2018). Manure can be applied to fields in a few different methods: as a liquid using sprinkler irrigation; liquids and slurries can be injected underneath the soil surface; and liquids, slurries, semi-solids, or solids can be broadcasted and either incorporated or not incorporated after application. Incorporation of manure soon after application is vital to assure N is kept in the field and does not volatilize contributing to nonpoint source–pollution. Jackson et al. (2000) reported that sprinkler irrigation of liquid swine manure resulted in 88% of manure N being transferred to the atmosphere which was much higher than when manure was injected into the soil (34% of total N lost to the atmosphere). Nitrogen losses from liquid dairy manure are also much higher when manure was broadcast (60% of total ammonium N) than when manure was injected (40% of total ammonium N) into a poorly drained clay soil (Pfluke et al., 2011). In general, ammonia emission reduction can vary from 0 to 100% when manure is injected using open slot injection and from 58 to 100% when using closed slot injection (Dell et al., 2011).

Regardless of whether farmers are utilizing their manure as a waste or as a resource, overapplications may still occur due to the significant variability in nutrient content which could lead to erroneous calculations. Nutrient values presented as reference values (also known as "book" values) are usually a generalization of many samples collected over a wide area and different practices. It is well known that nutrient content changes based on feed, animal age, health status, and climatic conditions. For example, collecting manure samples for chemical analyses after rain periods when manure is stored outdoors could dilute manure nutrients due to the addition of water. Similarly, collecting manure after a prolonged dry season could increase the amount of nutrients in manure due to the loss of moisture. This reiterates the need to test manure for nutrient availability as close to field application as possible.

Common book values for the nutrient content of manure can be found through the MidWest Plan Service (MWPS) and American Society Agricultural and Biological Engineers (ASABE). Figure 3 reports data for beef and dairy manure book values (liquid and solid) as indicated by the MWPS and ASABE as well as data that were obtained from commercial laboratories around the upper Midwest United States for chemical tests. The commercial laboratory data represents real analyses from livestock and poultry operations in the region. For beef manure, the MWPS reports, in most cases, lower nutrient for Total N, P, and K than all other samples, while ammonia-N is usually higher in the MWPS, with exception for liquid beef manure (Fig. 3). For dairy manures, values are more similar between the MWPS and ASABE, and both are slightly higher than the values observed in the commercial laboratories (Fig. 3). It is also interesting to note how variable the samples are. Total N in liquid beef manure ranges from 0 to 45 lb 1000 gal^{-1}, while total K in liquid dairy manure ranges from 0 to near 70 lb 1000 gal^{-1} (Fig. 3). Most importantly, data from Fig. 3 demonstrate how nutrient concentration changes dramatically across farms and species, and thus regular manure sampling is the best strategy to apply correct amounts of nutrients to fields. Ribaudo et al. (2003) reported that in 1998 only 19% of animal feeding operations with less than 1000 AU were testing manure for total N and P; while the number of CAFOs testing manure for total N and P prior to application was 73%. However, the data reported by Ribaudo et al. (2003) are 20 yr old and new, more reliable information is warranted.

Special Considerations for CAFO Manure

While AFOs and CAFOs have increased in number, the amount of land for manure application has decreased, which is becoming problematic. In 1982, the available cropland managed by farmers with livestock and poultry operations available for manure

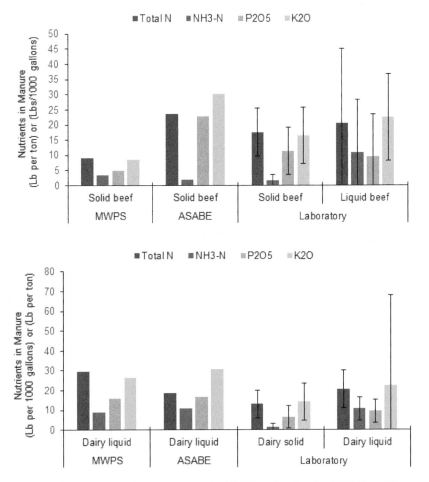

Fig. 3. Nutrient concentration as reported by MidWest Plan Service (MWPS) and American Society Agricultural and Biological Engineers (ASABE), and from samples submitted to various laboratories throughout the United States.

application was 3.6 acres per AU, which declined to 2.2 acres per AU by 1997 (Ribaudo et al., 2003). This is likely due to the decrease in the number of small farms with lower number of livestock (less than 1000 heads) and an increase in the number of livestock (1000 or more heads) in larger operations (USDA-APHIS, 2017). Nowadays, we are starting to see more livestock operations with no crop acreage. The percent of farms producing livestock with no harvested crop acres have increased from 21% to 36% from 1996 to 2015, respectively, with beef cattle and swine being the biggest contributors (MacDonald et al., 2018). This represents an increase of 71% in the number of farms with no land for manure application. The reduction in acreage per AU suggests that more manure will be applied on land that does not necessarily need manure.

Some reports have suggested that the amount of N and P in manure being produced by CAFOs is in excess of more than 50% of what would be safe to apply in land in a way that remains economically and environmentally viable (Jackson et al., 2000; Ribaudo et al., 2003). In fact, Long et al. (2018) assessed how manure and its associated nutrients produced by CAFOs in Michigan are handled on a farm-basis scenario. The authors reported that 42% of all manure application between 2013 and 2015 happened in soils which already had excessive amounts of soil available P (soil test *P* > 50 ppm). The excess nutrients applied to the field with excessive soil test P levels would be enough to supply almost 10,000 acres with manure P for a soybean crop (Long et al., 2018). Similarly, Jackson et al. (2000) estimated that if manure applications were based on supplying the P needs of the crop (not N), then the 10 CAFOs used in the study would need ten times more land to safely use manure than the land they were currently using for manure application, increasing from 2449 ac to 23,109 ac. In a different study, Ribaudo et al. (2003) used data from the 1998 Hog Agricultural Resource Management Survey to generate a detailed report on the acreage used by confined hog operations and whether more acreage is needed for proper manure disposal. The authors found that only 37% of operations with less 1000 AU had the sufficient amount of land needed to safely apply manure using N rate applications (Ribaudo et al., 2003). In contrast, the authors reported that only 3% of operations with more than 1000 AU had been applying manure on agronomic rates (Ribaudo et al., 2003). Fertilizer recommendations are usually based on yield goal, and farmers are responsible for setting their own yield goal based on soil type and yield potential for a given crop and soil. In many cases, CAFO operators and farmers will overestimate yield potential of a certain field to increase the amount of nutrients that are needed for such field (Jackson et al., 2000; Long et al., 2018). Other ways to maximize how much manure will be applied to a field is by basing the manure application on the N requirement of the crop without much consideration for the soil test P levels (Ribaudo et al., 2003). These practices force overapplication of manure on land that should not receive the manure, or perhaps, should not receive as much as what ends up being applied. Furthermore, in many cases farmers will add inorganic fertilizer, N and P, in addition to the manure already applied in excess (Long et al., 2018). In cases where manure nutrients are mishandled, farms tend to apply 100% more P than farms that do not use manure (91 lb of P_2O_5 acre^{-1} compared with 45 lb of P_2O_5 acre^{-1}) (Long et al., 2018).

Final Considerations

Production of livestock and poultry under CAFOs has proven to be an effective way to raise a very large number of animals in a small area. However, the practice is also proving to be detrimental to the environment. When manure is applied at agronomic rates, it can be considered a valuable nutrient source and a soil amendment that improves soil physical, chemical, and biological properties. However, many farmers are either not using management practices that are ideal or simply lack information to maximize the use of manure for food production because of the variability inherently found in manure nutrient content. When misused, manure can be a source of nonpoint-source pollution to aquatic ecosystems. Repeated application of high rates of manure to the same land is causing nutrients to build up to levels considered excessive. Future work should focus on better predicting nutrient availability across soil types and climates, as well as best alternative options for manure reuse in farms that no longer have land where manure can safely be applied. Education and outreach will be critical so that livestock and poultry farm operators have the best information for making site-specific manure management decisions. Future chapters outline how mitigation strategies are being developed for sustainable use of these important nutrient sources.

References

ASABE. 2014. ASABE Standard: Manure production and characteristics. ASAE D384.2 MAR2005 (R2014). American Society of Agricultural and Biological Engineers, St. Joseph, MI.

Ayinla, A., I.A. Alagbe, B.U. Olayinka, A.R. Lawal, O.O. Aboyeji, and E.O. Etejere. 2018. Effects of organic , inorganic and organo-mineral fertilizer on the growth, yield and nutrient composition of *Corchorus Olitorious* (L.). Ceylon J. Sci. 47(1): 13–19.

Billen, P., J. Costa, L. Van der Aa, J. Van Caneghem, and C. Vandecasteele. 2015. Electricity from poultry manure: A cleaner alternative to direct land application. J. Clean. Prod. 96:467–475. doi:10.1016/j.jclepro.2014.04.016

Bitzer, C.C., and J.T. Sims. 1988. Estimating the availability of nitrogen in poultry manure through laboratory and field studies. J. Environ. Qual. 17:47–54. doi:10.2134/jeq1988.00472425001700010007x

Bogaard, A., R. Fraser, T.H. Heaton, M. Wallace, P. Vaiglova, M. Charles, G. Jones, R.P. Evershed, A.K. Styring, and N.H. Andersen. 2013. Crop manuring and intensive land management by Europe's first farmers. Proc. Natl. Acad. Sci. USA 110:12589–12594. doi:10.1073/pnas.1305918110

Carter, J.E., W.E. Jokela and S.C. Bosworth. 2010. Grass forage response to broadcast or surface-banded liquid dairy manure and nitrogen fertilizer. Agron. J. 102: 1123-1131.

Cole, D., L. Todd, and S. Wing. 2000. Concentrated swine feeding operations and public health: A review of occupational and community health effects. Environ. Health Perspect. 108:685–699. doi:10.1289/ehp.00108685

Cole, N., R. Clark, R. Todd, C. Richardson, A. Gueye, L. Greene, and K. McBride. 2005. Influence of dietary crude protein concentration and source on potential ammonia emissions from beef cattle manure. J. Anim. Sci. 83:722–731. doi:10.2527/2005.833722x

Dagna, N.E., and A.P. Mallarino. 2014. Beef cattle manure survey and assessment of crop availability of phosphorus by soil testing. Soil Sci. Soc. Am. J. 78:1035–1050. doi:10.2136/sssaj2013.06.0223

Dail, H.W., Z. He, M.S. Erich, and C.W. Honeycutt. 2007. Effect of drying on phosphorus distribution in poultry manure. Commun. Soil Sci. Plant Anal. 38:1879–1895. doi:10.1080/00103620701435639

Dell, C.J., J.J. Meisinger, and D.B. Beegle. 2011. Subsurface application of manures slurries for conservation tillage and pasture soils and their impact on the nitrogen balance. J. Environ. Qual. 40:352–361. doi:10.2134/jeq2010.0069

FoodPrint. 2019. What happens to animal waste? https://foodprint.org/issues/what-happens-to-animal-waste/ FoodPrint. (Accessed 26 Apr. 2019).

Gillespie, J. 2019. Animal policy & regulatory issues. USDA-Economic Research Service, Washington, D.C. https://www.ers.usda.gov/topics/animal-products/animal-policy-regulatory-issues/ (Accessed on 1 May 2019).

Gollehon, N.R., M. Caswell, M. Ribaudo, R.L. Kellogg, C. Lander and D. Letson. 2001. Confined animal production and manure nutrients. Agricultural Information Bulletin No. (AIB-771). USDA-ERS, Washington, D.C. p. 39.

Gollehon, N.R., R.L. Kellogg, and D.C. Moffitt. 2016. Estimates of recoverable and non-recoverable manure nutrients based on the Census of Agriculture-2012 Results. USDA-NRCS. Washington, D.C. https://www.nrcs.usda.gov/wps/portal/nrcs/detail/national/technical/nra/rca/?cid=nrcseprd1360819. (Accessed 29 Apr 2019).

Hatfield, J.L., and B.A. Stewart, editors. 1998. Animal waste utilization: Effective use of manure as a soil resource. Ann Arbor Press, Chelsea, MI.

He, Z., editor. 2011. Environmental chemistry of animal manure. Nova Science Publishers, NY. p. 1–459.

He, Z., editor. 2012. Applied research of animal manure: Challenges and opportunities beyond the adverse environmental concerns. Nova Science Publishers, New York. p. 1–325.

He, Z. 2019. Organic animal farming and comparative studies of conventional and organic manures. In: H.M. Waldrip, P. Pagliari, and Z. He, editors, Animal manure: Production, characteristics, environmental concerns and management. ASA Spec. Publ. 67. ASA and SSSA, Madison, WI. p. 139–156.

He, Z., and H. Zhang, editors. 2014. Applied manure and nutrient chemistry for sustainable agriculture and environment. Springer, Amsterdam, the Netherlands. p. 1–379. doi:10.1007/978-94-017-8807-6

He, Z., P.H. Pagliari, and H.M. Waldrip. 2016. Applied and environmental chemistry of animal manure: A review. Pedosphere 26:779–816. doi:10.1016/S1002-0160(15)60087-X

He, Z., M. Guo, N. Lovanh, and K.A. Spokas. 2012. Applied manure research-Looking forward to the benign roles of animal manure in agriculture and the environment. In: Z. He, editor, Applied research of animal manure: Challenges and opportunities beyond the adverse environmental concerns. Nova Science Publishers, New York. p. 299–309.

Hooiveld, M., L.A. Smit, F. van der Sman-de Beer, I.M. Wouters, C.E. van Dijk, P. Spreeuwenberg, D.J.J. Heederik, and C.J. Yzermans. 2016. Doctor-diagnosed health problems in a region with a high density of concentrated animal feeding operations: A cross-sectional study. Environ. Health 15:24. doi:10.1186/s12940-016-0123-2

Jackson, L.L., D.R. Keeney, and E.M. Gilbert. 2000. Swine manure management plans in north-central Iowa: Nutrient loading and policy implications. J. Soil Water Conserv. 55:205–212.

Janni, K., and E. Cortus. 2019. Common animal production systems and manure storage methods. In: H.M. Waldrip, P. Pagliari, and Z. He, editors, Animal manure: Production, characteristics, environmental concerns and management, Vol. ASA Special Pub. 67. ASA, Madison, WI. p. 27–44.

Jones, K., M. Haley, and A. Melton. 2018. Per capita red meat and poultry disappearance: Insights into its steady growth. USDA-Economic Research Service, Washington, D.C. https://www.ers.usda.gov/amber-waves/2018/june/per-capita-red-meat-and-poultry-disappearance-insights-into-its-steady-growth/ (Accessed 1 May 2019).

Kellogg, R.L., D.C. Moffitt, and N.R. Gollehon. 2014. Estimates of recoverable and non-recoverable manure nutrients based on the census of agriculture. USDA-NRCS, Washington, D.C. https://www.nrcs.usda.gov/wps/portal/nrcs/detail/national/technical/nra/rca/?cid=nrcseprd1360819. (Accessed 29 Apr 2019).

Lehmann, J., Z. Lan, C. Hyland, S. Sato, D. Solomon, and Q.M. Ketterings. 2005. Long-term dynamics of phosphorus forms and retention in manure-amended soils. Environ. Sci. Technol. 39:6672–6680. doi:10.1021/es047997g

Long, C.M., R.L. Muenich, M.M. Kalcic, and D. Scavia. 2018. Use of manure nutrients from concentrated animal feeding operations. J. Great Lakes Res. 44:245–252. doi:10.1016/j.jglr.2018.01.006

Lorimor, J., W. Powers, and A. Sutton. 2004. Midwest planner service: Manure characteristics. Iowa State University, 2nd ed. Ames, IA.

Lundgren, J., and E. Pettersson. 2009. Combustion of horse manure for heat production. Bioresour. Technol. 100:3121–3126. doi:10.1016/j.biortech.2009.01.050

MacDonald, J.M., R.A. Hoppe and D. Newton. 2018. Three decades of consolidation in US agriculture. Economic Information Bulletin Number 189. USDA-ERS, Washington, D.C.

Moore, P.A., T.C. Daniel, A.N. Sharpley, and C.W. Wood. 1995. Poultry manure management: Environmentally sound options. J. Soil Water Conserv. 50(3):321–327.

National Research Council. 2003. Air emissions from animal feeding operations: Current knowledge, future needs. The National Academies Press, Washington, D.C. doi:10.17226/10586.

Pagliari, P.H. 2014. Variety and solubility of phosphorus forms in animal manure and their effects on soil test phosphorus. In: Z. He, Applied manure and nutrient chemistry for sustainable agriculture and environment. Springer, Dordrecht, The Netherlands. p. 141-161.

Pagliari, P.H., and C.A.M. Laboski. 2012. Investigation of the inorganic and organic phosphorus forms in animal manure. J. Environ. Qual. 41:901–910. doi:10.2134/jeq2011.0451

Pagliari, P.H., and C.A.M. Laboski. 2013. Dairy manure treatment effects on manure phosphorus fractionation and changes in soil test phosphorus. Biol. Fertil. Soils 49:987–999. doi:10.1007/s00374-013-0798-2

Pagliari, P.H., J.S. Strock, and C.J. Rosen. 2010. Changes in soil pH and extractable phosphorus following application of turkey manure incinerator ash and triple superphosphate. Commun. Soil Sci. Plant Anal. 41:1502–1512. doi:10.1080/00103624.2010.482172

Pfluke, P.D., W.E. Jokela, and S.C. Bosworth. 2011. Ammonia volatilization from surface-banded and broadcast application of liquid dairy manure on grass forage. J. Environ. Qual. 40:374–382. doi:10.2134/jeq2010.0102

Ribaudo, M., J.D. Kaplan, L.A. Christensen, N. Gollehon, R. Johansson, V.E. Breneman, M. Aillery, J. Agapoff, and M. Peters. 2003. Manure management for water quality costs to animal feeding operations of applying manure nutrients to land. doi:10.2139/ssrn.757884

Ro, K. 2012. Thermochemical conversion technologies for production of renewable energy and value-added char from animal manures. In: Z. He, Applied research of animal manure: Challenges and opportunities beyond the adverse environmental concerns. Nova Science Publishers, Inc. New York. p. 63-81.

Selbie, D.R., L.E. Buckthought, and M.A. Shepherd. 2015. The challenge of the urine patch for managing nitrogen in grazed pasture systems. Advances in agronomy. Elsevier, New York. p. 229-292.

Sommer, S.G., M.L. Christensen, T. Schmidt, and L.S. Jensen, editors. 2013. Animal manure recycling: Treatment and management. John Wiley & Sons, West Sussex, UK. p. 1–384. doi:10.1002/9781118676677

Tewolde, H., M.W. Shankle, T.R. Way, D.H. Pote, K.R. Sistani, and Z. He. 2018. Poultry litter band placement affects accessibility and conservation of nutrients and cotton yield. Agron. J. 110:675–684. doi:10.2134/agronj2017.07.0387

USDA. 2015. Organic farming-Results from the 2014 Organic Survey. ACH12-29. www.agcensus. usda.gov (Accessed 22 July 2019).

U.S. Code of Federal Regulations. 2019a. U.S. Code of Federal Regulations § 780.120- Raising of "livestock." United States Government, Washington, D.C.

U.S. Code of Federal Regulaions. 2019b. U.S. Code of Federal Regulations § 780.328- Meaning of livestock. United States Government, Washington, D.C.

USDA-APHIS, 2011. Overview of U.S. Livestock, Poultry, and Aquaculture Production in 2011 and Statistics on Major Commodities. USDA-APHIS, Washington, D.C. https://www.aphis.usda.gov/ animal_health/nahms/downloads/Demographics2011.pdf (Accessed 24 July 2019).

USDA-APHIS, 2017. Overview of U.S. Livestock, Poultry, and Aquaculture Production in 2011 and Statistics on Major Commodities. USDA-APHIS, Washington, D.C. https://www.aphis.usda.gov/ animal_health/nahms/downloads/Demographics2017.pdf (Accessed 24 July 2019).

USDA-NRCS. 1992. Agricultural waste management field handbook. USDA-Natural Resources Conservation Service, Washington, D.C.

USDA-NRCS. 2009. Agricultural waste management field handbook. 210–VI–AWMFH, Amend. 31. USDA-Natural Resources Conservation Service, Washington, D.C.

USDA-NRCS. 2017. NRCS Minnesota Agronomy Technical Note 32: Nutrient loss risk assessments, sensitive features and definitions for Minnesota. USDA-Natural Resources Conservation Service, Saint Paul, MN.

USEPA. 2004. National emission inventory—Ammonia emissions from animal husbandry operations. US Environmental Protection Agency, Washington, D.C.

USEPA. 2013. Literature review of contaminants in livestock and poultry manure and implications for water quality. EPA 820-R-13-002. Office of Water, Washington, D.C.

USEPA. 2019. Animal feeding operations (AFOs). US Environmental Protection Agency, Washington, D.C. https://www.epa.gov/npdes/animal-feeding-operations-afos. (Accessed 1 May 2019).

WAC. 2003. "Animal waste" and "processed" defined. WAC 16-256-010. https://app.leg.wa.gov/wac/ default.aspx?cite=16-256-010, Washington State Legislature, Olympia, WA. (Accessed 26 Apr 2019).

Waldrip-Dail, H., Z. He, M.S. Erich, and C.W. Honeycutt. 2009. Soil phosphorus dynamics in response to poultry manure amendment. Soil Sci. 174:195–201. doi:10.1097/SS.0b013e31819cd25d

Waldrip, H.M., P.H. Pagliari, Z. He, R.D. Harmel, N.A. Cole, and M. Zhang. 2015. Legacy phosphorus in calcareous soils: Effects of long-term poultry litter application. Soil Sci. Soc. Am. J. 79:1601–1614. doi:10.2136/sssaj2015.03.0090

Wheeler, E., and J.S. Zajaczkowski. 2009. Horse stable manure management. Publication code UB035. The Pennsylvania State University Extension Services, State College, PA.

Wing, S., and S. Wolf. 2000. Intensive livestock operations, health, and quality of life among eastern North Carolina residents. Environ. Health Perspect. 108:233–238. doi:10.1289/ehp.00108233

Temporal Changes of Manure Chemical Compositions and Environmental Awareness in the Southern Great Plains

Hailin Zhang*, Fred Vocasek, Joao Antonangelo, and Christopher Gillespie

Abstract

Knowing the nutrient contents of animal manure is important in nutrient management plan development. Nutrient contents of manure may have been changed over time due to improvement of breeding, feeding, and manure handling. Therefore, the major characteristics of beef feedlot manure, dairy manure, poultry litter, and swine effluent were summarized using the data from two service laboratories in Kansas and Oklahoma. In general, dry matter contents, pH, and macro- and micronutrient contents of the manures had little changes over time in the last 5 to 20 yr. Only a trend of phosphorus decrease over time in swine effluent was observed. The nutrient contents of various manures largely depend on the dry matter contents. The nitrogen (N), phosphorus (P) and potassium (K) contents are in the following order: broiler litter > beef feedlot manure > dairy manure > swine effluent. Various environmental regulations related to animal manure management have been established and implemented in most parts of the world. The awareness of sustainable manure application to cropland has greatly improved in the last twenty years. More efforts need to be made to further improve nutrient use efficiency of animal manure, protect soil health, and environmental quality.

Animal production is a large segment of the U.S. economy and of many other countries in the world (Zhang and Schroder, 2014). Confined animal feeding operations (CAFOs) produce large quantities of manure that demand proper management; cattle are the greatest of manure producer, followed by pigs, poultry, sheep, and goats worldwide (Sommer and Christensen, 2013). The United States Department of Agriculture (USDA) estimated that in 2007 there were over 2.2 billion head of livestock and poultry in the U.S. (US-EPA, 2012), which produced over 1.1 billion tons of manure. Manure storage and disposal is a major expense and potential liability for animal operations.

Animal manures have been used by ancient and modern farmers to enhance crop production and to improve soil health (Bogaard et al., 2013; He, 2011; He, 2012). Besides providing valuable macro- and micronutrients to the soil, manure also supplies organic matter (OM) to improve soil tilth, enhance water infiltration, increase nutrient retention, reduce wind and water erosion, and promote growth of beneficial soil organisms. Manure application to croplands succeeds in both recycling nutrients and sustaining crop production (Fig. 1). Animal manure can be an asset rather than a liability for producers when effectively managed and properly used on field crops (Richards et al., 2011; Tang et al., 2007; Zhang et al., 1998).

Key Words: Animal manure, feedlot manure, dairy manure, poultry litter, swine effluent, nutrient content, nutrient loss, water quality, best management practices, manure regulations

H. Zhang, J. Antonangelo, and C. Gillespie, Oklahoma State University, Department of Plant and Soil Sciences, Stillwater, OK; F. Vocasek, Servi-Tech Laboratories, Dodge City, KS. *Corresponding author (hailin.zhang@okstate.edu)

doi:10.2134/asaspecpub67.c2

Animal Manure: Production, Characteristics, Environmental Concerns and Management. ASA Special Publication 67. Heidi M. Waldrip, Paulo H. Pagliari, and Zhongqi He, editors. © 2019. ASA and SSSA, 5585 Guilford Rd., Madison, WI 53711, USA.

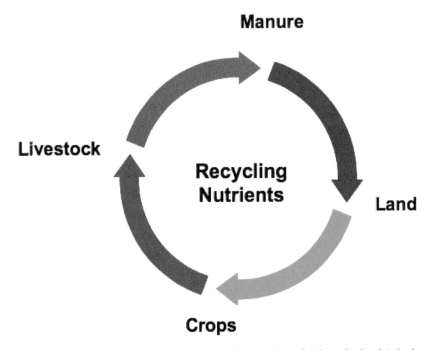

Fig. 1. Land application of animal manure recycles nutrients back to the land. It is the most economical and environmentally sound method to handle byproducts generated from meat and milk production if managed properly.

Mismanaged manure applications can cause both surface water and ground water pollution (He, 2011). Surface runoff from manured land usually contains plant nutrients and organic materials. Excess nutrients and organic materials that reach surface water often cause algal blooms that increase the turbidity and biochemical oxygen demand (BOD). The polluted water may generate disagreeable odors and the pollutants can cause fish kills if the dissolved oxygen falls below critical thresholds. Excess manure that remains on the application field may also cause nitrate–nitrogen (NO_3–N) and phosphorus (P) to accumulate in the treated soil. The excess NO_3–N can also reach surface water via drainage pathways or can leach into underlying ground water (He and Zhang, 2014).

To minimize the impact of animal manure land application on air and water quality, new regulations on manure management have been established and implemented in many countries and in most states of the United States due to pollution potential from improper manure applications (Sommer et al., 2013). The key to successful manure management is to develop and follow a nutrient management plan by applying the right amount at the right time. This requires one to know the nutrient needs of the intended crop through soil testing and the nutrient contents of the manure by manure analysis. When manure test is not available to some farmers, book values of nutrient contents of manure have been used. However, manure nutrient contents can change with time due to changes in animal feeding practices, manure management systems, nutrient use efficiency by animal due to breeding efforts, and other factors. It is essential to know the manure nutrient content to use it properly as a nutrient source and a soil amendment.

Average nutrient contents of manure are available in many publications and manuals (Midwest Plan Service, 1993; Christensen and Sommer, 2013; Hatfield and Stewart, 1998), but few highlighted the changes with time. This chapter presents a summary of inorganic characteristics of feedlot manure, poultry litter, dairy manure and swine effluent analyzed within the last two decades by two service laboratories in the Southern Great Plains and highlights environmental awareness by evaluating certain policies related to manure management.

Manure Chemical Composition

Animal manure contains valuable nutrients that can support crop production and can enhance soil chemical, physical, and biological properties. Manure can thus be an asset to a livestock production operation if the nutrient value is maximized. The nutrient composition of livestock manure varies widely between operations even for the same animal species. Some have raised concerns about whether manure nutrients or other constituents have changed over time due to genetic improvement of animals or due to updated management practices. Analysis results of four major manure types were obtained from Oklahoma State University (OSU) Soil, Water and Forage Analytical Laboratory- SWFAL (http://www.soiltesting.okstate.edu) and Servi-Tech Laboratories (https://servitechlabs.com) to identify possible temporal variability of the manure properties. Both laboratories use standardized methods for manure analyses (Peters, 2003). The analysis values listed in the Tables are averaged from actual samples submitted by the respective laboratory clients over a long period of time. These averages are valuable data, but represent wide ranges in the sample population. Manure should always be analyzed before application to determine the actual nutrient values and characteristics.

Poultry Litter

"Poultry litter" is a general term for chicken and turkey manure mixed with spilled feed and poultry house bedding materials. Litter may include wood shavings, saw dust, rice hulls and peanut hulls (Harsch, 1995). Moisture and nutrient contents can vary widely, depending on the type and age of the poultry. However,

Table 1. Characteristics of poultry litter tested by the Oklahoma State University Soil, Water and Forage Analytical Laboratory (results are expressed in "fresh weight"). §

Sample Period		Solids	pH	TDS†	TN	NO_3-N	NH_4-N	OC	P_2O_5	Ca	K_2O	Mg	Na	S	Fe	Zn	Cu
		%		ppm	%	—ppm—					—%—					—ppm—	
2001-04	Average	74.0	8.4	7114	3.12	605	4396	NA	3.13	2.92	2.70	0.53	0.66	0.68	394	452	574
(1250)‡	Median	75.8	8.3	7110	3.14	211	4398	NA	3.18	2.46	2.81	0.55	0.70	0.69	270	434	548
2005-09	Average	73.4	8.2	7673	3.19	580	3714	27.1	3.02	2.39	2.84	0.52	0.74	0.70	300	404	338
(1694)‡	Median	74.4	8.3	7845	3.27	755	2920	27.9	2.98	2.19	2.65	0.52	0.75	0.67	242	398	338
2010-13	Average	74.7	8.2	6826	3.23	71	2784	28.5	2.72	2.43	2.90	0.48	0.64	0.74	460	359	213
(1401)‡	Median	75.6	8.2	6726	3.29	19	2423	29.2	2.68	2.03	2.93	0.48	0.62	0.73	355	344	197
2014-18	Average	74.9	8.3	8019	3.16	NA	NA	29.4	3.00	2.63	2.52	0.52	0.71	0.86	554	468	217
(1294)‡	Median	75.7	8.3	8314	3.23	NA	NA	29.8	2.96	2.37	2.53	0.52	0.69	0.86	461	431	200

† TDS; total dissolved solids.

‡ Number of samples included in this group.

§ Na; not analyzed.

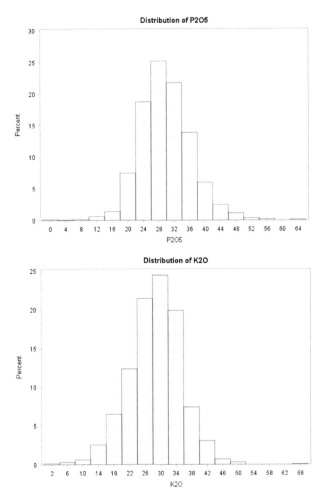

Fig. 2. Distribution of P_2O_5 (top) and K_2O (bottom) (kg Mg^{-1}) in poultry litter from 2014 to 2018 (n = 1294).

most poultry litter samples that are submitted for analysis are submitted by broiler operations and are managed like "dry manure" due to their low moisture content. The characteristics of poultry litter tested by OSU-SWFAL from 2001 to 2018 are presented in Table 1. The results were grouped into 4 to 5 year segments to show any temporal changes in nutrient content. The results are shown on an "as is" or "wet" basis (not corrected for moisture).

Poultry litter is alkaline with an average median pH of 8.3. This may be benefi-cial for maintaining soil pH or to help neutralize soil acidity when it is applied in acid soils. The pH values of poultry litter samples were not changed since 2001, indicating no or little feeding changes. The average solids content was 74% with no significant changes with time. Poultry litter contains significant amounts of both macronutri-ents and micronutrients, as well as some salts (as shown by "TDS" total dissolved solids). There were no significant temporal changes in the nutrients analyzed, except for copper (Cu), which showed a gradual decreasing trend over time, but the cause for this decline is unknown. Because those were actual farmer samples instead of replicated sampling, no statistical analyses were applied. The averages are very close to the median of all analytes, which suggests they were normally distributed. This is confirmed by the P and K distributions of the 2014 to 2018 data (Fig. 2).

Swine Lagoon Effluent

Swine manure is typically treated in two-stage, anaerobic digestion lagoon systems, common in the Central and Southern regions of the United States. Manure produced in the swine housing unit is initially directed into the primary lagoon. The manure solids settle to the bottom of this lagoon and slowly undergo decomposition by acid-forming and methane-forming microorganisms. Liquids from the primary lagoon flow into the secondary lagoon, where further settling and decomposition may occur. Manure liquids from the surface of the secondary lagoon are removed by pumping. Liquids from the upper two feet or so are then applied to nearby fields as a nutrient and water source. The samples sent for lab analysis typically represent the liquids used for irrigation. The effluent sample results are found in Table 2. Results of other swine manure types were not included.

The solid contents from Table 2 results are less than 0.50%, because they represent effluent samples from the irrigation lagoon surface. The low solid contents resulted in low concentrations of all nutrients analyzed. However, significant amounts of soluble solids (i.e., "soluble salts") were present in the effluent. Irrigators should use caution when using effluent on salt sensitive crops. The phosphorus (P) concentration in the effluent is much lower than nitrogen (N) or potassium (K) because most of the phosphorus is found in the settled sludge solids on the lagoon bottoms. Nitrogen and potassium are more soluble, so tend to remain in the liquid fractions.

There were no obvious temporal trends among analytes, except for phosphorus. Phosphorus concentrations gradually decreased from 141 ppm P to 90 ppm P over the past 14 yr (Fig. 3). The declining phosphorus content could be attributed to improved feeding management and to supplementing swine feeds with phytase (Dr. Scott Carter, Oklahoma State University Animal Nutritionist, personal communication). Monogastric animals such as swine are unable to utilize the phytate phosphorus in the feed. It is therefore necessary to supplement some swine feeds with inorganic phosphorus. Adding phytase to a swine ration will

Table 2. Characteristics of swine effluent tested by Oklahoma State University Soil, Water and Forage Analytical Laboratory (results are expressed in "fresh weight").

Sample Period		Solids	pH	TDS†	TN	NO_3-N	NH_4-N	OC	P_2O_5	Ca	K_2O	Mg	Na	S	Fe	Zn	Cu
		%									―――ppm―――						
2005-09	Average	0.43	7.9	5398	626	0.20	426	1572	324	72.2	989	45.4	350	47.4	6.64	1.80	0.48
(115)‡	Median	0.36	7.9	5244	516	0.14	238	1270	246	49	968	15.7	320	40.2	2.52	2.52	0.18
2010-13	Average	0.48	7.8	4404	666	1.38	420	2000	288	128	626	60.4	240	61.4	7.46	4.88	0.80
(192)‡	Median	0.45	7.9	3996	588	0.84	334	1784	222	97.6	674	45.0	210	42.4	5.04	3.32	0.62
2014-18	Average	0.48	8.0	4766	606	5.18	114	1610	146	72.2	884	36.8	392	57.4	4.66	1.88	0.46
(161)‡	Median	0.40	8.0	2490	608	1.74	104	1484	145	58.4	874	21.2	324	37.6	2.44	1.06	0.18

† TDS; total dissolved solids.

‡ Number of samples included in this group.

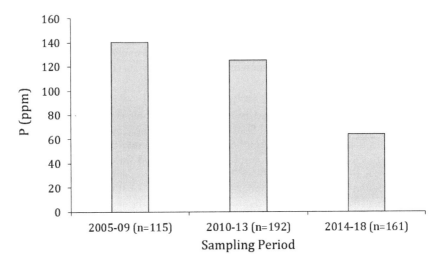

Fig. 3. Phosphorus (P) in swine effluent has been declining gradually over the last 14 yr as observed from samples analyzed by Oklahoma State University Soil, Water and Forage Analytical Laboratory.

increase phosphorus use efficiency and reduce the required amount of inorganic phosphorus supplement. Reducing the phosphorus in the manure will result in less phosphorus enrichment in soil receiving swine effluent and thus less potential phosphorus loss to surface waters.

Almost all the treated effluent generated by U.S. swine producers is land applied using high-volume sprinklers or center-pivot irrigation systems. Potential concerns for sprinkler systems include application timing, water losses, odors, and ammonia emissions. These become more of a problem under hot and dry conditions (Stone et al., 2008).

Application problems mentioned above can be minimized with subsurface drip irrigation (SDI) systems (Stone et al., 2008). Subsurface drip irrigation has been used to increase water and nutrient use efficiency, reduce nutrient losses from runoff, and to mitigate odor. In general, SDI systems conserve water and achieve higher crop or forage yields when compared with sprinkler or center-pivot irrigation systems (Stone et al., 2008).

Beef Feedlot Manure

Feedlot manure samples were submitted by farmers or consultants to Sevi-Tech Lab; only data from those two periods were available. The samples were typically collected directly from cattle feedlot pens or from manure storage piles. The test results of feedlot manure samples from two time periods are shown in Table 3. The average median solid content was about 73%, and total N, P_2O_5 and K_2O were 1.23%, 1.04%, and 1.32%, respectively. There were few differences found for all analytes between the two sampling periods. Therefore, we conclude that the characteristics of beef feedlot manure have not changed significantly in the last two decades.

Dairy Manure

The characteristics of dairy manure tested by Servi-Tech Laboratories from the middle of the last decade and the middle of this decade showed few changes (Table 4) for all analytes. Dairy manure had higher moisture contents than poultry litter or beef feedlot manure and accordingly had comparatively lower nutrient contents. We also conclude that the characteristics of dairy manure have not changed significantly in the last two decades.

Manure Nutrient Content

Manure nutrient contents are typically expressed on test reports as "pounds per ton" of solid manure or as "pounds per 1000 gallons" of liquid manure in the United States. Phosphorus and potassium are expressed with the fertilizer industry conventions of "phosphate, P_2O_5" and "potash, K_2O". This makes it easier for farmers to calculate the equivalent amount of fertilizer nutrients and to develop nutrient management plans.

The most recent average nutrient contents of the four types of manure presented above are shown in Table 5. All manures contain significant amounts of N, P, and K making them valuable resources for crop production. Poultry litter had the highest wet weight based nutrient content followed by feedlot, dairy manure, and swine effluent. The difference may reflect the respective moisture contents and the amount and types of bedding materials in different manures. Both the nutrient content and manure weight or volume must be used to calculate the final nutrient application rate. Although the nutrient contents of swine effluent are very low, nutrient credits can be quite high if a large amount is applied. For example, swine effluent can contribute 137 lb (62 kg) N, 46 lb (21 kg) P_2O_5, and 200 lb (91 kg) K_2O for each acre-inch of effluent (27,000 gallons) that is applied as irrigation.

Table 3. Characteristics of Beef Feedlot Manure Tested by Servi-Tech Laboratories (results are expressed in "fresh weight").

Sample Period		Solids	OC	Total N	NH$_4$-N	NO$_3$-N	P$_2$O$_5$	K$_2$O	Ca	Mg	S	Na	Zn	Fe	Mn	Cu	B
					%—————————										———ppm———		
2006-08	Average	70.8	15.7	1.28	0.17	0.01	1.05	1.38	1.98	0.50	0.30	0.23	191	4420	181	36	13
(157)†	Median	72.7	14.3	1.24	0.150	0.002	0.97	1.29	1.48	0.47	0.28	0.21	163	3227	167	35	10
2014-17	Average	70.1	14.2	1.22	0.15	0.01	1.22	1.40	2.08	0.47	0.31	0.23	188	4907	211	35	14
(1336)†	Median	71.3	13.7	1.21	0.130	0.001	1.10	1.34	1.77	0.44	0.28	0.21	168	4601	187	32	12

† Number of samples included in this group.

Table 4. Characteristics of Dairy Manure Tested by Servi-Tech Laboratories (results are expressed in "fresh weight").

Sample Period		Solids	OC	Total N	NH$_4$-N	NO$_3$-N	P$_2$O$_5$	K$_2$O	Ca	Mg	S	Na	Zn	Fe	Mn	Cu	B
					%—————————										———ppm———		
2006-08	Average	53.7	13.4	0.94	0.09	0.01	0.58	1.09	0.82	0.38	0.23	0.23	77	2251	92	39	12
(202)†	Median	54.7	12.5	0.86	0.069	0.001	0.51	0.97	0.52	0.33	0.25	0.15	61	1504	77	15	10
2014-17	Average	55.5	12.0	0.84	0.06	0.01	0.59	1.02	1.38	0.36	0.19	0.21	101	3432	122	38	14
(560)†	Median	60.4	10.7	0.76	0.044	0.001	0.52	0.83	1.18	0.31	0.16	0.14	76	3030	114	19	11

† Number of samples included in this group.

Table 5. Range and average of nutrient contents in "fresh weight" of major manure types analyzed in the last 4 to 5 years.

Manure Type	Dry Matter	Total N	P_2O_5†	K_2O†	Total N	P_2O_5	K_2O
	%	————————lb ton⁻¹————————			————————kg Mg⁻¹————————		
Broiler Litter	42-92 ‡	5.2-153	0-128	0-135	2.6-77	0-64	0-68
	75±7.3 §	63±27.1	59±13.8	57±14	31.5±14	30±7	29±6.8
Feedlot Manure	41-90	0.59-60	0.39-302.8	1.38-63.9	21-45	0.3-30	0.20-151.4
	70±11.4	24±9.2	24±14.4	28±11.8	35±5.7	12±4.6	12±7.2
Dairy Manure	41-90	2.4-50.2	0.79-47.6	0.59-76.3	1.2-25.1	0.40-23.8	0.30-38.2
	56±22.3	17±8.9	12±7.7	20±15.3	28±11.2	6±3.9	10±7.7
	————————lb 1000 gal⁻¹————————			————————kg m⁻³————————			
Swine Effluent	0.05-1.4	0.1-12.8	0.03-15.2	0.72-14.6	0.01-1.54	0.00-1.83	0.09-1.8
	0.48±0.26	5.1±2.6	1.2±1.9	7.4±2.6	0.6±0.3	0.14±0.23	0.89±0.31

† P_2O_5 and K_2O are commonly used expressions for fertilizer ingredients instead of P and K. Laboratories may report in elemental P and K content. To convert, use the following equations: $K = K_2O \times 0.83$ or $P = P_2O_5 \times 0.44$

‡ Range from minimum to maximum value observed

§ Average values ± Standard deviation

The average N, P, and K contents in the four types of manure summarized in Table 5 are similar to averages published in extension fact sheets (Mitchell, 2008; Zhang, 2002). Although the average values can serve as a reference for planning purposes, it is important to have individual manure tested before land application because the ranges of nutrient content are very large (Table 5), for example, the P_2O_5 of dairy manure ranged from 0.4 to 24 kg Mg⁻¹.

Environmental Awareness and Regulations Related to Manure Management

Environmental Concerns

There are numerous benefits resulting from manure land application, including: building organic matter content, supplying nutrients, buffering soil pH, enhancing biological activities, and improving soil physical properties (Zhang and Schroder, 2014). There are also many concerns involving manure application impacts on environmental quality (Moore, 1998; Sharpley et al., 1998). The top concerns include soil phosphorus accumulations leading to increased offsite loss potential to water bodies, elevated soil concentrations of certain metals (Moore, 1998; Richards et al., 2011; Sweeten, 1998; Sharpley et al., 1998), and potential pathogens.

Manure application rates were historically based on the agronomic nitrogen requirement. The soil phosphorus content or the crop phosphorus requirement was typically not considered in the past. The typical N/P ratio of manure is usually lower than that required by plants (Pote et al., 1996; Gotcher et al., 2014). Fertilizing crops with manure based on the agronomic nitrogen needs alone will result in over applying phosphorus. This practice has resulted in significant accumulations of soil phosphorus where manure has been repeatedly applied (Moore, 1998). Over time, failure to consider the manure phosphorus contribution may saturate the soil's phosphorus sorption capacity near the soil surface and also in deeper layers, which often increases the dissolved phosphorus in runoff (Moore, 1998; Sharpley et al., 1994; Sharpley et al., 1998; Wang et al., 2011).

Significant quantities of micronutrients such as copper (Cu), iron, (Fe), manganese (Mn), or zinc (Zn) can also accumulate in the soil if large amounts of manure are applied (Penha et al., 2015). Long-term applications could cause some of these elements to reach toxic levels within or near the application zone.

Environmental Awareness and Regulations

New regulations on manure management have been established and implemented in most states due to pollution potential from improper manure applications. The U.S. Environmental Protection Agency (US-EPA) considers eutrophication through organic enrichment the most widespread water quality impairment in the United States (US-EPA, 2017). The greatest potential for eutrophication of surface waters usually occurs in watersheds with intensive animal production (CAST, 1996; He et al., 2016).

In 1998, the USDA and EPA announced a joint strategy to implement "comprehensive nutrient management plans (CNMPs)" for animal feeding operations (AFOs) by 2008. A CNMP is a conservation farm plan that is specific to AFOs. The CNMP incorporates practices to utilize animal manure and organic by-products as beneficial resources. It documents the management strategies adopted by the AFO to address natural resource concerns related to soil erosion, animal manure management, and disposal of organic by-products. Nutrient management plans emphasize a balance between nutrient supply and crop utilization and may include nutrient removal.

The CNMP normally contains six different elements: (i) manure and wastewater handling and storage, (ii) land treatment practices, (iii) the nutrient management plan, (iv) record keeping, (v) feed management considerations, and (vi) other waste utilization options.

The NRCS 590 planning standard that deals with nutrient management has been revised several times to include a phosphorus-based planning standard as a consequence of the USDA/EPA joint strategy (Sharpley et al., 2003; USDA-NRCS, 2013). The strategy made three phosphorus-based choices available for states to use in developing nutrient management planning policies. These approaches were to use: (i) agronomic soil test phosphorus recommendations, (ii) environmental soil test phosphorus thresholds, or (iii) a phosphorus loss index (P-index) that ranked fields according to their vulnerability for potential phosphorus loss (Sharpley et al., 2003).

By the year 2003, 48 states had adopted the phosphorus index approach (Sharpley et al., 2003). Overall, 23 states have adopted the original P-index format or have modified the index for local conditions. There are 25 states that use the P-index and/or an environmental phosphorus threshold. Two states use agronomic soil test phosphorus recommendations (Sharpley et al., 2003). The P-index policy is thus an integral part of most CNMPs and is widely used to determine manure application rates.

The original P-index was developed by Lemunyon and Gilbert (1993) to identify the vulnerability of agricultural fields to phosphorus loss. The P-index accounts for and ranks certain "source" factors (soil test P, fertilizer and manure rates, application methods, etc.) and certain "transport" factors" (erosion, runoff or leaching potential, landscape and soil characteristics, connectivity to surface water, etc.) that could control offsite phosphorus loss potential. These individual site characteristics were weighted differently. It was assumed that different site characteristics have different effects on phosphorus loss. Each site characteristic was assigned a phosphorus loss rating value: negligible (0), low (1), medium (2), high (4), and very high (8). The phosphorus loss rating value was multiplied by the weighting coefficient to obtain a P-index value for the individual source or transport factor. The individual P-index values were then added together to obtain the final P-index value.

Most states use the P-index approach, but many of them have modified the original P-index format. The changes included: multiplication instead of addition of source and transport factors; consideration of distance to water; and quantification of erosion, soil test P, and phosphorus application rate. Other states have derived predictive loading models that calculate edge-of-the-field phosphorus loss in kg P ha^{-1} yr^{-1} (Osmond et al., 2006). Additional states have moved to more quantitative predictions since then (Osmond et al., 2017). Those P-indices have played a major rule in reducing phosphorus loss and protecting water quality. However, agronomic phosphorus recommendations still differ widely between states (Osmond et al., 2006, 2012, and 2017). The Europe Union and other regions of the world have also established and implemented various environmental regulations related to animal manure management (Sommer et al., 2013)`

Conclusion

The major characteristics of beef feedlot manure, dairy manure, poultry litter, and swine effluent were summarized using the data from two service laboratories in Kansas and Oklahoma. In general, dry matter contents, pH, and macro- and micronutrient contents of the manure had little changes over time in the last 5 to 20 yr. Only a trend of phosphorus decrease over time in swine was observed. Various environmental regulations related to animal manure management have been established and implemented in most parts of the world. The awareness of sustainable manure application to cropland has greatly improved in the last twenty years. More efforts need to be made to further improve nutrient use efficiency, soil health, and environmental quality.

References

Bogaard, A., R. Fraser, T.H. Heaton, M. Wallace, P. Vaiglova, M. Charles, G. Jones, R.P. Evershed, A.K. Styring, N.H. Anderson, R.M. Arbogst, L. Gardelsen, M. Karnstrup, U. Maier, E. Marinova, L. Ninov, M. Schafer, and E. Stephan. 2013. Crop manuring and intensive land management by Europe's first farmers. Proc. Natl. Acad. Sci. USA 110:12589–12594. doi:10.1073/pnas.1305918110

CAST. 1996. Integrated animal waste management. Council for Ag Sci and Tech, Ames, IA.

Christensen, M.L., and S.G. Sommer. 2013. Manure characterization and inorganic chemistry. In: S.G. Sommer, M.L. Christensen, T. Schmidt, and L.S. Jensen, editors, Animal manure recycling: Treatment and management. Wiley, West Sussex, UK. doi:10.1002/9781118676677.ch4

Gotcher, M.J., H. Zhang, J.L. Schroder, and M.E. Payton. 2014. Phytoremediation of soil phosphorus with crabgrass. Agron. J. 106:528–536. doi:10.2134/agronj2013.0287

Harsch, J. 1995. Poultry litter marketing and utilization project: A case study. Winrock International, Little Rock, AR.

Hatfield, J.L., and B.A. Stewart. 1998. Animal waste utilization: Effective use of manure as a soil resource. Sleeping Bear Press, Ann Arbor, MI.

He, Z. 2011. Environmental chemistry of animal manure. Nova Science Publishers, New York.

He, Z. 2012. Applied research in animal manure: Challenges and opportunities beyond the adverse environmental concerns. Nova Science Publishers, New York.

He, Z., P.H. Pagliari, and H.M. Waldrip. 2016. Applied and environmental chemistry of animal manure: A review. Pedosphere 26:779–816. doi:10.1016/S1002-0160(15)60087-X

He, Z., and H. Zhang, editors. 2014. Applied manure and nutrient chemistry for sustainable agriculture and environment. Springer, Amsterdam, The Netherlands. doi:10.1007/978-94-017-8807-6

Lemunyon, J.L., and R.G. Gilbert. 1993. The concept and need for a phosphorus assessment tool. J. Prod. Agric. 6:483–486. doi:10.2134/jpa1993.0483

Mitchell, C.C. 2008. Nutrient content of fertilizer materials. ANR-174. Alabama Cooperative Extension System, Tuskegee, AL. http://www.aces.edu/pubs/docs/A/ANR-0174/ANR-0174.pdf

MidWest Plan Service. 1993. Livestock waste facilities handbook. Iowa State Univ. Ames, IA.

Moore, P.A., Jr. 1998. Best management practices for poultry manure utilization that enhance agricultural productivity and reduce pollution. In: J.L. Hatfield and B.A. Stewart, editors, Animal waste utilization: Effective use of manure as a soil resource. Sleeping Bear Press, Ann Arbor, MI.

Osmond, D.L., M. Cabrera, S. Feagley, G. Hardy, C. Mitchell, R. Mylavarapu, P. Moore, L. Oldham, B. Thom, J. Stevens, F. Walker, and H. Zhang. 2006. Comparing P-indices for the southern region. J. Soil Water Conserv. 61:325–337.

Osmond, D., A. Sharpley, C. Bolster, M. Cabrera, S. Feagley, B. Lee, C. Mitchell, R. Mylavarapu, L. Oldham, F. Walker, and H. Zhang. 2012. Comparing phosphorus indices from twelve Southern U.S. States against monitored phosphorus loads from six prior southern studies. J. Environ. Qual. 41:1741–1749. doi:10.2134/jeq2012.0013

Osmond, D., C. Bolster, A. Sharpley, M. Cabrera, S. Feagley, A. Forsberg, C. Mitchell, R. Mylavarapu, J.L. Oldham, D.E. Radcliffe, J.J. Ramirez-Avila, D.E. Storm, F. Walker, and H. Zhang. 2017. Southern P indices, water quality data, and modeling (APEX, APLE, and TBET) results: A comparison. J. Environ. Qual. 46:1296–1305. doi:10.2134/jeq2016.05.0200

Penha, H.G.V., J.F.S. Menezes, C.A. Silva, G. Lopes, C.A. Carvalho, S.J. Ramos, and L.R.G. Guilherme. 2015. Nutrient accumulation and availability and crop yields following long-term application of pig slurry in a Brazilian Cerrado soil. Nutr. Cycl. Agroecosyst. 101:259–269. doi:10.1007/s10705-015-9677-6

Peters, J. 2003. Recommended methods of manure analysis. Univ. of Wisconsin-Madison, Madison, WI. http://learningstore.uwex.edu/assets/pdfs/A3769.PDF (Accessed 22 Feb. 2019).

Pote, D.H., T.C. Daniel, A.N. Sharpley, P.A. Moore, D.R. Edwards, and D.J. Nichols. 1996. Relating extractable soil phosphorus to phosphorus losses in runoff. Soil Sci. Soc. Am. J. 60:855–859. doi:10.2136/sssaj1996.03615995006000030025x

Richards, J., H. Zhang, J.L. Schroder, J.A. Hattey, W.R. Raun, and M.E. Payton. 2011. Micronutrient availability as affected by the long-term application of phosphorus fertilizer and organic amendments. Soil Sci. Soc. Am. J. 75:927–939. doi:10.2136/sssaj2010.0269

Sweeten, J.M. 1998. Cattle feedlot manure and wastewater management practices. In: J.L. Hatfield and B.A. Stewart, editors, Animal waste utilization: Effective use of manure as a soil resource. Sleeping Bear Press, Ann Arbor, MI.

Sharpley, A.N., S.C. Chapra, R. Wedepohl, J.T. Sims, T.C. Daniel, and K.R. Reddy. 1994. Managing agricultural phosphorus for protection of surface waters-Issues and options. J. Environ. Qual. 23:437–451. doi:10.2134/jeq1994.00472425002300030006x

Sharpley, A., J.J. Meisinger, A. Breeuwsma, J.T. Sims, T.C. Daniel, and J.S. Schepers. 1998. Impacts of animal manure management on ground and surface water quality. In: J.L. Hatfield and B.A. Stewart, editors, Animal waste utilization: Effective use of manure as a soil resource. Sleeping Bear Press, Ann Arbor, MI.

Sharpley, A.N., J.L. Weld, D. Beegle, P. Kleinman, W.J. Gburek, P.A. Moore, Jr., and G. Mullins. 2003. Development of phosphorus indices for nutrient management planning strategies in the United States. J. Soil Water Conserv. 58:137–152.

Sommer, S.G., and M.L. Christensen. 2013. Animal production and animal manure management. In: S.G. Sommer, M.L. Christensen, T. Schmidt, and L.S. Jensen, editors, Animal manure recycling: Treatment and management. Wiley, West Sussex, UK. doi:10.1002/9781118676677.ch2

Sommer, S.G., O. Oenema, T. Matsunaka, and L.S. Jensen. 2013. Regulations on manure management. In: S.G. Sommer, M.L. Christensen, T. Schmidt, and L.S. Jensen, editors, Animal manure recycling: Treatment and management. Wiley, West Sussex, UK. doi:10.1002/9781118676677.ch3

Stone, K.C., P.G. Hunt, J.A. Miller, and M.H. Johnson. 2008. Forage subsurface drip irrigation using treated swine wastewater. Trans. ASABE 51:433–440. doi:10.13031/2013.24385

Tang, Y., H. Zhang, J.L. Schroder, M.E. Payton, and D. Zhou. 2007. Animal manure reduces aluminum toxicity in an acid soil. Soil Sci. Soc. Am. J. 71:1699–1707. doi:10.2136/sssaj2007.0008

USDA-NRCS. 2013. Conservation practice standard code 590: Nutrient management. USDA-NRCS, Washington, D.C. https://www.nrcs.usda.gov/Internet/FSE_DOCUMENTS/stelprdb1192371.pdf (Accessed 22 Feb. 2019).

U.S. EPA. 2017. National Water Quality Inventory: Report to Congress. EPA 841-R-16-011, Office of Water, Washington, D.C.

U.S. EPA. 2012. NPDES permit writers' manual for CAFOs. EPA 833-F-12-001, Office of Water, Washington, D.C.

Wang, J., H. Zhang, J.L. Schroder, T.K. Udeigwe, Z. Zhang, S.K. Dodla, and M.S. Stietiya. 2011. Reducing potential leaching of phosphorus, heavy metals, and fecal Coliform from animal wastes using bauxite residues. Water Air Soil Pollut. 214:241–252.

Zhang, H. 2002. Fertilizer nutrients in animal manure. PSS-2228. Oklahoma Cooperative Extension Service, Oklahoma City, OK. http://pods.dasnr.okstate.edu/docushare/dsweb/Get/Rendition-5067/unknown (Accessed 22 Feb. 2019).

Zhang, H., D. Smeal, and J. Tomko. 1998. Nitrogen fertilization value of feedlot manure for irrigated corn production. J. Plant Nutr. 21:287–296. doi:10.1080/01904169809365403

Zhang, H., and J. Schroder. 2014. Animal manure production and utilization in the US. In: Z. He and H. Zhang, editors, Applied manure and nutrient chemistry for sustainable agriculture and environment. Springer, Dordrecht. p. 1–21. doi:10.1007/978-94-017-8807-6_1

Common Animal Production Systems and Manure Storage Methods

Kevin Janni and Erin Cortus*

Abstract

Livestock and poultry production systems include diverse housing and manure storage practices depending on species, animal age, stage of production, products produced and climate. Swine housing systems typically segregate animals into different facilities based on the pig life cycle phase. Most swine systems handle manure as a slurry or liquid. Manure collected below slotted flooring in swine barns can either be stored in barn briefly before being transferred to outdoor storage, or stored for longer periods (> six months) in a deep pit before removal for land application. Broiler chickens and turkeys are typically raised in barns with organic litter covered floors. Used litter may be stacked prior to land application. Layer hen production systems include cage, enriched colony, and cage free-housing systems. In most cases layer hen manure is handled as a solid and removed from the bird area to a stacking area. Dairy calves are usually raised on straw in various types of housing. Most lactating dairy cow housing includes bedded freestalls and alleys that are cleaned multiple times a day. Sand is commonly used as freestall bedding but recycled solids can be used. Bedding selection impacts outdoor manure storage options and equipment used for manure removal and land application. Some dairy cattle are raised in compost barns that use practices that keep the bedded pack aerobic. Milking center washwater is usually added to the farm's manure storage system. Beef cattle facilities depend on production stage. Most cow–calf operations are pasture based. As calves grow and enter backgrounding or finishing phases, they tend to transition to feedlot-based systems. Feedlots may be open lots with runoff containment, under roof with manure storage in bedded packs or under-floor storage, or combinations thereof. Responsible animal care and environmental considerations are paramount for housing and manure storage practices across all livestock and poultry production systems.

Introduction

Livestock and poultry are raised using production practices that depend on species, animal age, stage of production, products produced and climate. Manure collection, handling and storage are also done in different ways. The result is a myriad of livestock and poultry production systems that incorporate manure storage.

Livestock and poultry can be raised in buildings (indoors) or outdoors. Most animals raised indoors are housed in buildings specially designed for the animals and managing them in environments that suit their needs (Fig. 1). The buildings are commonly managed to modify or control the building thermal environment and ventilation air exchange. Buildings also have systems and equipment for supplying feed, water, and lighting. Animal buildings also have methods for handling voided animal feces and urine and spilled feed and water. Some animals are raised using bedding that adds to the manure that must be collected. While some animal buildings incorporate manure storage for varying lengths of time, other barns use various options of manure removal to separate manure storage units (Fig. 2). This chapter is focused on animal feeding operations that incorporate some type of manure storage.

Department of Bioproducts and Biosystems Engineering, University of Minnesota, 1390 Eckles Avenue, St Paul, MN 55108. *Corresponding author (ecortus@umn.edu).

doi:10.2134/asaspecpub67.c3

Animal Manure: Production, Characteristics, Environmental Concerns and Management. ASA Special Publication 67. Heidi M. Waldrip, Paulo H. Pagliari, and Zhongqi He, editors.

Fig. 1. Curtain sided swine finishing barn with deep pit manure storage and pit fans (NPB, 2016a).

Manure storage is one part of the manure collection, handling, and storage system that livestock and poultry producers use to manage the nutrients in the manure (USDA NRCS, 1992). Manure includes voided feces and urine, spilled feed and water, wash water, bedding, and other wastes (Fulhage et al., 2001). Some systems include manure treatment such as solid–liquid separation or anaerobic digestion. Treatment can generate energy (i.e., anaerobic digestion to produce methane run through a genset or burned as fuel) or make the manure easier to handle. In most cases, manure storage is used so that nutrients can be land applied when the soil and plants can best utilize them and avoid having to land apply fresh manure on a daily basis. Well-designed and managed manure systems with storage can help farmers better manage manure nutrients for enhanced crop production.

Numerous federal, state, and local regulations may apply to livestock and poultry operations depending on location and size. The regulations may impact siting decisions, manure handling practices, storage structure design and construction, and manure land application amounts and timing. Many manure storage structures are required to be designed and inspected by a professional engineer (Fulhage et al., 2001).

Poorly designed and managed manure systems and storages can create environmental and community problems. Manure systems and storages can cause odor complaints, contribute to fly problems, and pollute ground and surface waters if not designed and managed properly. Manure nutrients represent a valuable resource for crop production. Well-designed and managed manure systems more effectively use the manure as a nutrient source and soil amendment.

Manures are normally handled either as a solid (greater than 15% dry matter), slurry (5 to 10% dry matter), or liquid (less than 5% dry matter) (Fulhage et al., 2001).

Manure solids content and consistency impact the manure handling and storage options and the equipment that is used to collect, handle, and land-apply the manure.

This chapter describes common systems for swine, poultry, dairy, and beef production. Each section describes common production phases, waste sources, manure form (i.e., solid, slurry or liquid), collection, and storage options. There are other unique livestock and poultry productions systems that exist but are beyond the scope of this chapter.

Livestock and Poultry Productions Systems

Livestock and poultry are commonly raised in specialized buildings that separate animals by age and type of production. This is commonly done to match thermal environmental needs and to reduce disease exposure of younger animals. For example, day-old turkey poults are commonly placed in brooder barns and raised to about six weeks of age before being moved to a grower barn where they are raised until they reach market weight (FAD PReP Poultry Industry Manual, 2013). Brooder barns provide very warm conditions for the poults during the first few weeks. New born calves are fed colostrum within hours of birth to give calves some immunity protection against common infectious organisms until their immature immune system develops over several weeks (Godden, 2008). Calves are also commonly moved away from older animals that shed organisms that can make the calves sick. Weaned piglets are commonly raised in nursery rooms until they are large enough to do well in a finishing barn (Jacobson et al., 1997). All of these specialized barns have manure generated in them that needs to be collected, handled, and stored in ways that enhance the animals' wellbeing.

Swine Production Systems

Swine facilities typically segregate animals into different barns based on the life cycle phase of the pig. The life-cycle phases are commonly referred to as: Breeding and Gestation (or simply Gestation), Farrowing, Nursery, and Growing–Finishing (NPB, 2009; 2016b). Breeding and gestation facilities house female pigs after their

Fig. 2. Above ground concrete manure storage.

piglets were weaned, through impregnation (by either artificial insemination or a boar), and the gestation period until they are moved to the farrowing facilities for farrowing. Farrowing facilities are where pregnant female pigs give birth and they nurse the piglets until they are weaned. Breeding, gestation and farrowing rooms are commonly built connected together to facilitate pig movement from one facility to another through covered aisles. Nursery facilities and growing–finishing facilities focus on the feeding and growth of groups of pigs. The nursery phase requires close attention to piglet health as the piglet transitions from its mother's milk to solid food.

Most farrowing, nursery, and finishing facilities are managed as all-in and all-out operations– with the exception of mortalities or sick animals, all animals enter and leave the facility at the same time and the rooms are cleaned and disinfected thoroughly between groups to reduce the spread of disease (NPB, 2009). The washwater used during the cleaning operation is added to the manure storage facilities. Gestation facilities are considered continuous flow, because small groups of sows will exit (to farrowing) and return (for breeding), with the majority of sows remaining in the gestation barn at various stages of gestation.

Within a gestation barn, sows may be housed individually in stalls, in groups, or in a combination of stalls and group-housing. Farrowing sows are typically in individual stalls, although group farrowing is sometimes practiced. Nursery and growing–finishing animals are housed in groups.

Swine Manure Storage

At least 85% of the swine production systems in the United States handle manure as a semisolid (slurry) or liquid (Hatfield et al., 1998), and is the focus of this section. Additional information on solid manure systems is presented in Hatfield et al. (1998) and USDA-NRCS (2009).

Manure collection in most swine production facilities is through slotted flooring that allows voided feces and urine to drop through to a storage area below the floor (Fig. 3). This below-floor storage can be designed for short (< 1 d) or long-term (> 6 mo) accumulation of manure and wastewater (USDA-NRCS, 2009). Short-term under-floor storages are referred to as shallow pits or gutters, and manure is subsequently transferred to longer-term storage in another barn or an outdoor storage system. The manure from farrowing barns is often transferred to a neighboring gestation barn to facilitate frequent cleaning between groups.

Gravity drained shallow pits accumulate manure for several days. When sufficient manure builds up for sufficient flow, a manually operated plug in the shallow pit floor is removed and the pit is drained. After reinstalling the plug, one to two inches of water are added to the shallow pit. Scraped gutters have mechanical scrapers that push the manure to storage or channels or pumps that move the manure to storage or treatment facilities. Scrapers can be designed and managed to remove manure at different schedules ranging from two to four times a day to once every 36 h. Flush gutters use either fresh or recycled lagoon water to flush the gutter once or twice a day.

Longer-term under-floor storage is called a deep pit. In a deep pit system, the manure is stored for several months to up to one year, before the manure is agitated, removed, and land-applied according to nutrient management plans (Fig. 4). Long-term outdoor manure storages include above ground tanks and lined earthen basins. These storage options simply store the manure from the time it is collected until it is agitated and removed for land application. Manure agitation and mixing helps to create a more uniform slurry for pumping and land applying. Deep pits, above ground tanks and lined earthen basins all need to be well designed and constructed to store the accumulated manure and avoid leaks to ground or surface waters.

Anaerobic lagoons are a treatment system that both treat and store the accumulated manure. Anaerobic lagoons are used only in warm climates where the anaerobic microorganisms treating the manure are sufficiently active enough all year to breakdown the accumulated manure in a reasonably sized lagoon (Fig. 5). Insufficient lagoon size, excess daily manure loading, and cold temperatures can overwhelm the microbial capacity of the lagoon and lead to odor problems and poor

Fig. 3. Pigs on slatted flooring (NPB, 2015).

Fig. 4. Equipment for agitation and pumping of a deep pit manure storage (NPB, 2018).

Fig. 5. Mechanically ventilated pig barn with adjacent lined manure storage (NPB, 2008).

anaerobic treatment. Anaerobic lagoons are commonly used with flushed shallow pit manure collection, or with irrigation distribution systems (Hatfield et al., 1998).

Poultry Production Systems
Broiler Chicken and Turkey Production

Broiler, meat bird, production is organized in three general operations. Breeder operations focus on genetic improvement to improve growth rate, feed efficiency, and breast meat quantity and quality. Parent farms produce the eggs of the commercial broilers that are hatched and delivered as day-old chicks to broiler farms. More detailed information about broiler production can be found in the Poultry Industry Manual (FAD PReP Poultry Industry Manual, 2013). Over 8.9 billion broilers were produced in the United States in 2017 (USDA NASS, 2018a).

Commercial broiler chickens are raised in environmentally-controlled barns with up to 50,000 birds per barn on litter covered floors (Fig. 6). The litter is commonly an organic material that varies depending on what is locally available. Bedding materials include: wood shavings, and rice, peanut or sunflower hulls, and cotton gin trash. The birds in broiler barns are raised from day-old chicks to market weight, commonly five pounds, in six to seven weeks depending on the bird size wanted (FAD PReP Poultry Industry Manual, 2013).

The vast majority of turkeys raised in the United States are broad-breasted white turkeys. There are breeder flocks that focus on genetic improvement and produce parent stock that produce the eggs of the commercial turkeys grown for meat.

Breeder flocks are typically raised on separate farms with the toms on "stud" farms and hens on breeder replacement farms (FAD PReP Poultry Industry Manual, 2013). The hens are artificially inseminated with semen from a stud farm. Hens lay fertile eggs in nest boxes. After collection and processing, the eggs are hatched at a hatchery. The day old poults are delivered to commercial turkey grower farms. Over 242 million turkeys were produced in the United States in 2017 (USDA NASS, 2018b).

Turkeys grown for meat are raised in two phases and types of environmentally-controlled barns, brooder and grower, on litter-covered floors. The brooder phase is for day-old poults through about six weeks of age. The grower phase is for birds from six weeks of age to the market weight desired (Fig. 7). Hens and toms are raised separately. Hens can be raised to about 20 pounds by 14 wk of age and toms can reach nearly 50 pounds by 20 wk of age. Turkeys are commonly raised on organic bedding materials such as wood shavings, and rice or sunflower hulls (FAD PReP Poultry Industry Manual, 2013).

Litter management is very important for good broiler and turkey health and growth. Proper ventilation is required to manage litter moisture content to avoid caking (i.e., wet litter compacting) and feet problems if the litter becomes too wet. Litter amendments and ventilation are important for ammonia management too. Broiler and turkey barn litter can be reused for multiple flocks for up to one year before being removed during clean-out. Between flocks the litter can be de-caked, composted in windrows, tilled or have litter amendments applied. Litter treatment between flocks is usually done to reduce microbe numbers and dry the litter (Tabler and Wells, 2018). Broiler and turkey mortalities during production are commonly composted on site in a composting shed with bins that can be filled and stirred as needed to effective mortality composting (FAD PReP Poultry Industry Manual, 2013).

Fig. 6. Broiler chickens raised on litter (Courtesy of the Chicken and Egg Association of Minnesota).

Fig. 7. Turkeys raised on litter (Courtesy of the Minnesota Turkey Growers Association).

Broiler Chicken and Turkey Litter Storage

State regulations usually specify poultry litter storage requirements before the litter is land-applied, according to a farm nutrient management plan. A stacking shed for storing litter has a concrete floor and walls, short or full, and a roof. The purpose of the stacking shed is to keep the litter dry and avoid nutrient seepage if the litter becomes wet. Some states allow field stacks for short-term storage during periods of wet weather during barn clean-out. Rules for location, surface preparation, and covering of the stack for short-term storage vary. Runoff and seepage from uncovered solid manure stacks must be managed to prevent movement into surface or ground water. More information on solid manure storage structures is available in USDA NRCS (2009).

Layer Hen Production

Layer hens are raised for egg production. Genetics operations produce day old chicks for commercial egg production companies. Commercial egg production operations purchase day-old chicks from the hatchery and raise them on pullet growing farms. Pullets are raised from day-old chicks to about 18 wk of age in pullet houses before being moved to a laying house. Laying flocks begin laying eggs at around 20 wk of age and continue for about 60 to 65 wk. At this time the hens are either marketed as spent hens or molted before being brought back into egg production. Laying hens can be kept for one to three laying cycles before being replaced with pullets. Approximately 80% of laying hens in the United States have molted at least once (FAD PReP Poultry Industry Manual, 2013).

Production practices for layers are changing (UEP, 2018). Many layer hens are housed in cages in environmentally-controlled chicken houses. Older houses are commonly called high-rise houses. The cages on the top floor of two-story houses hold small numbers of birds in each cage. The stacked cages are in an A shape where manure from the top cages is deflected into a slot in the floor to a manure storage on the bottom floor. Many newer barns have manure belts below each level of cages that catch the manure. The belts carry the manure away for storage outside of the layer house. Some newer barns have enriched colony systems that provide more space for larger groups of hens. Some new barns are cage free. Cage free housing systems have a litter floor area and may have a manure belt below the feeders and waterers (UEP, 2018).

Layer Hen Manure Storage

Layer hen manure in older high-rise houses is stored on the bottom level of two-story houses (LPELC, 2018). Manure in layer hen houses is commonly removed at least once a year and land applied. Some farms remove the manure more often and compost it. States may allow field stacks for short-term storage during clean out (LPELC, 2018). Rules for locating, surface preparation and covering the stack for short-term storage vary.

Manure from layer houses with manure belts is commonly removed from the layer house every day to every third day (Fig. 8). The removed manure is stored in a separate structure. The manure storage shed has a concrete floor and walls, short or full, and a roof (USDA NRCS, 2009). The manure can be stored for up to one year before being land-applied to available cropland.

There are a small percentage of layer houses that have flush systems that handle the manure as a liquid. These operations store the manure in lined earthen basins, or steel or concrete structures for a few months to a year before land application. Design information for liquid manure storage facilities is available in USDA NRCS (2009) handbook.

Manure from cage-free houses can be stored in a separate manure storage shed similar to that used with layer houses with manure belts (USDA NRCS, 2009). The litter in cage-free laying houses can be added to the manure from the belt.

Dairy Production Systems

Dairy production systems are very diverse in size and layout. Dairy animals are commonly raised in facilities for different groups including: calves, heifers, lactating cows, and dry cows. Many dairy operations have space for special needs cows that need more observation and care due to recent calving or health problems. Dairy animals do very well in unheated facilities if they have clean and

Fig. 8. Layer hens over manure belt (Courtesy of the Chicken and Egg Association of Minnesota).

Fig. 9. Dairy calves on straw bedding.

Fig. 10. Naturally ventilated freestall dairy barn.

dry places to lie down, plenty of fresh draft-free ventilation air, and easy access to nutritious feed and clean water (Holmes et al., 2013).

Calves are born without many protective antibodies. Within hours of birth calves are fed colostrum, which contains colostral antibodies that the calves are able to absorb for protection against common infectious organisms until their immature immune systems develop during the first weeks and months of life (Godden, 208). For this reason, calves are commonly housed in facilities away from older animals that carry and shed disease organisms. Facilities for calves range from well-bedded individual calf hutches, pens, or stalls to group pens with automated feeding systems (Fig. 9). Straw is commonly used for bedding.

After weaning, heifers are raised in group pens in barns that can be open front sheds, naturally ventilated or mechanically ventilated barns (Holmes et al., 2013). Most of these sheds and barns use organic bedding and have a bedded pack. Heifers can also be raised in barns with freestalls. Some barns have exercise lots or access to pastures. Heifers can also be raised on pasture or open lots if given access to water and feed.

Lactating cows are housed in facilities that depend in part on climate. All lactating cow facilities will have space for the cows to lie down, a feeding area, waterers, and a milking center where the cows are milked two to three times a day (Holmes et al., 2013). Lactating cow facilities range from tie stall barns, freestall barns to open lots with barns with feed mangers. The most common housing system for lactating cows are freestall barns which have rows of stalls for the cows to lie in, one or more feed mangers and waterers (Fig. 10). The cows are free to get up and go to eat or drink whenever they wish. Cows in freestall barns are typically milked in a milking parlor twice a day, once every 12 h, or three times a day, every 8 h. Some dairy farms have replaced the milking parlor with milking robots in the freestall barn (Devir et al., 1999; Salfer et al., 2018). The robots allow cows to be milked more often every day. Freestall barns can be naturally ventilated, tunnel ventilated, or cross-ventilated.

Freestalls are commonly bedded with sand, which makes a very comfortable surface for cows to lie on (Fig. 11). When sand is used for bedding the manure becomes sand-laden which can make the manure handling hard on equipment. Some manure handling systems incorporate sand separating equipment or

Fig. 11. Freestall dairy barn with recycled sand bedding and scraped manure alleys.

Fig. 12. Naturally ventilated compost dairy barn.

sand-settling lanes to separate the sand from the sand-laden manure to be able to recycle the sand and use it in the sand-bedded freestalls.

Freestalls can also have mattresses covered with a small amount of organic bedding. The organic bedding from the freestalls will increase the solids content of the manure. Some dairy operations use solid–liquid separation to collect manure and organic bedding solids. These solids can be used to bed freestalls too. The liquid portion is collected and stored in a liquid manure storage facility.

The alleys where the cows walk or stand for eating in freestall barns can be cleaned by scraping with a skid loader or a vacuum manure wagon while the cows are being milked at the milking center. The alleys can also be cleaned with a mechanical scraper that slowly scrapes the alleys several times a day with cows present. Alleys can also be flushed with fresh or recycled water, recycled is more common.

Lactating cow facilities in hot and dry climates can be open lots with pole barns without sidewalls or freestalls that have mangers for feeding the cows and waterers. In these dry climates the manure and urine dry quickly and the manure is handled as a dry solid scrapped together for land application.

In tie-stall barns cows are kept in bedded stalls most of the day. Feed and water is provided to each stall. In some situations the cows are milked while in the stalls or they may be taken to a milking parlor. Manure and urine are collected in gutters behind the stalls. The gutter can be emptied daily with a gutter cleaner.

Compost barns are another lactating cow housing option (Barberg et al., 2007; Janni et al., 2007). Compost barns are similar to the freestall barns except that the freestalls are replaced by a bedded pack that is stirred every time that the cows are milked (Fig. 12). The pack is commonly bedded with wood shavings. The pack is stirred to refresh the surface and incorporate oxygen to encourage organic biological activity similar to composting. Compost barns have mangers similar to freestall barns and the alley where the cows stand to eat is scraped each time the cows are milked.

Dry-cows are pregnant cows that are nearing the time for them to give birth to a calf. They are dry, not milked, for 45 to 60 d before they are expected to give birth. Dry cows are commonly housed in a group pen that provides more space and either larger well-bedded stalls or a bedded pack. There may be a far-off pen for cows recently dried off and a close-up pen for cows expected to calve within a week. Some dairy operations will have maternity pens for cows to calve that have clean organic bedding.

Many dairy operations have a special needs area or pen for cows that need special care or veterinary care. Cows that recently calved may be held in a special needs pen for a few days before being moved into a pen with other lactating cows.

The milking center has milking center wastewater (manure and urine) and washwater that are commonly added to the manure storage. Depending on the size of the herd and the number of milkings per day, the milking center may be cleaned once, twice, or three times a day.

Dairy Manure Storage

Dairy operations can generate manure and wastewater that can be handled as a solid, slurry, or a liquid (Fulhage et al., 2001; Holmes et al., 2013). The manure from bedded calf facilities, heifer barns, compost barns, dry-lots in dry climates, dry-cow pens, and special needs pens is handled as a solid. Solid manure storage may be needed to store the solid manure until cropland, weather, and regulations allow land application. Rules and guidelines for locating the storage, containing seepage, and preventing surface water and rain from entering the storage vary. Concrete floors and walls are commonly used with a roof to store solid manure.

Liquid and slurry manure can often be stored in similar structures including below-ground tanks, lined-earthen basins and above ground tanks (Fig. 2 and 13). These structures commonly require engineered designs to meet local regulations designed to prevent pollution and storage failure. These storage systems may include multiple basins that provide some solid separation and allow for wastewater to be used for flushing manure alleys or manure flumes. Most liquid and slurry manure

Fig. 13. Lined dairy manure storage with natural crust.

Fig. 14. Outdoor beef feedlot (Courtesy of the U.S. Meat Animal Research Laboratory).

storage systems are designed to store manure for at least a year to facilitate land application after fall crop harvest and before the ground freezes in winter.

There are some dairy barns that have mattresses in the freestalls, use organic bedding on the mattresses, and have slotted floors and deep pit manure storage under the barn. Sand bedding is not used in barns with deep pit manure storage. Deep pit manure storages are commonly sized to provide storage for six months to a year.

Beef Production Systems

Beef production is commonly done in two stages. Cow–calf operations breed cows to produce calves that are sold after several months to feedlots where they are raised to market weight. Most cow–calf operations raise the cows and calves on pasture and do not collect or store manure. However, some cow–calf operations use facilities similar to those described below for raising cattle to market weight.

Beef cattle feeding operations vary widely depending on operation size, climate, manure handling, runoff handling, and regulations. Beef feeding can be done in outdoor feedlots with concrete or earthen lots, in barns with solid floors and bedded packs, or in barns with slotted flooring and under-floor manure storage (Lawrence et al., 2007). Outdoor feedlots may have a shed or roofed area covering part of the pen. Feedlot cattle are fed a mixed ration one or two times a day.

Outdoor beef feedlots are very common, with cattle fed via fence line bunks. Open-front sheds or windbreaks may provide cattle with some protection in adverse weather. Outdoor lots are laid out and sloped to move runoff from rain and snow melt out of the cattle area and should have a system for handling and storing the runoff at the bottom of the slope (Fig. 14). Solid manure accumulates on the pen surface over time. Concrete lots require approximately 4.6 m² per head, whereas earthen lots are call for 23 to 37 m² per head, depending on the level of drainage (MWPS, 1987). Earthen lots commonly have mounds to improve drainage and provide dry areas for cattle to rest.

Bedded pack beef barns are typically naturally ventilated pole barns with feed bunks along one or more sides of the pens (Fig. 15). Stocking densities are 3.9 to 4.6 m² per head. The area were the cows stand to eat is scraped one to two times per week and the manure removed is placed in a storage facility. This scraped manure can account for around 60% of all manure generated. A large portion of the barn is a bedded pack where the beef animals lie down. Organic bedding types include, but are not limited to corn stover, soybean residue, and wheat straw. The type and amount of bedding added and weather influence pack surface conditions. Approximately 2 to 3 kg of bedding per head per day is typical (Jones et al., 2013). The bedded pack accumulates manure and bedding for one week to over a year when it is removed and either land applied or stored.

Slotted floor beef barns are typically naturally-ventilated pole barns that have deep pit manure storage under the slotted flooring (Fig. 16). The manure may be stored for six months to a year. Feed bunks are along one or more sides of the pen, similar to bedded pack barns. Stocking densities are 1.9 to 2.3 m² per head. No bedding and no scraping are required on a regular basis. Mats are often installed to lessen leg and hoof injuries compared to the concrete surface (Euken, 2013).

Beef Manure Storage

Regulations and best management practices dictate that outdoor feedlots have facilities to handle lot runoff to prevent surface water pollution. This liquid handling system is in addition to the solid manure collected on the feedlot pen surface. Feedlot runoff is often handled using a two-stage system. The first stage is a smaller settling basin designed to separate solids from the liquid. The liquid portion proceeds to a larger, longer-term storage in a larger basin. Solids need to be periodically removed from the settling basin. The basin requires adequate storage to accommodate normal and storm-event precipitation over the entire feedlot area between periods when the liquid can be land applied. Evaporation from the basin may be significant in dry climates. The basin may be referred to as a detention basin, holding pond, or evaporation pond in different areas and for different modes of management. The low solids and nutrient content of the basin liquid is amenable to distribution by irrigation. Vegetative treatment systems are alternative systems for handling and utilizing feedlot runoff for vegetative growth (Andersen et al., 2013).

Solid manure removed from open feedlots or barns with solid floors (with or without bedding) requires storage if it cannot be land-applied immediately. Solid manure can be stored in uncovered stacks on impervious surfaces that prevent nutrient leaching into the soil. Uncovered stacks usually are required to have runoff

Fig. 15. Naturally ventilated beef barn with bedded manure pack.

Fig. 16. Naturally ventilated beef barn with slatted flooring and deep pit manure storage. A section of slats has been temporarily removed for manure agitation and removal.

collection systems and storage to prevent surface water pollution. Solid manure from beef operations can also be stored in stacks in covered sheds that reduce runoff.

Manure collected in deep pits requires agitation before removal. Additional water may be required to reduce the solids content and enable removal by pumps. The manure is directly land applied in accordance to nutrient management plans.

Conclusions

Livestock production, housing, and manure collection, handling, and storage practices are intricately linked. Many factors affect the design options considered and choices made by owners and managers as they balance the animal, labor, and environmental needs and applicable regulations. Well planned and managed animal facilities with manure storage can help farmers use manure nutrients effectively to grow needed feed and fiber while protecting their land, water, and air resources.

References

Andersen, D.S., R.T. Burns, L.B. Moody, M.J. Helmers, B. Bond, I. Khanijo, and C. Pederson. 2013. Impact of system management on vegetative treatment system effluent concentrations. J. Environ. Manage. 125:55–67. doi:10.1016/j.jenvman.2013.03.046

Barberg, A.E., M.I. Endres, and K.A. Janni. 2007. Compost dairy barns in Minnesota: A descriptive study. Appl. Eng. Agric. 23(2):231–238. doi:10.13031/2013.22606

Devir, S., C.C. Ketelaar-deLauwere, and J.P.T.M. Noordhuizen. 1999. The milking robot dairy farm management: Operational performance characteristics and consequences. Trans. ASABE 42(1):201–213. doi:10.13031/2013.13197

Euken, R. 2013. Evaluation of rubber mats over concrete slats in cattle confinement facilities. In: Proceedings of the Beef Facilities Conference, Sioux Falls, SD. 21 Nov. 2013. Iowa State Extension and Outreach, Ames, IA.

FAD PReP Poultry Industry Manual. 2013. Poultry industry Manual. Foreign Animal Disease Preparedness & Response Plan (FAD PReP). Center for Food Security and Public Health, Ames, IA. www.cfsph.iastate.edu/pdf/fad-prep-nahems-poultry-industry-manual (Accessed 19 Apr. 2019).

Fulhage, C., J. Hoehne, D. Jones, and R. Koelsch. 2001. Manure storages. Section 2, MWPS-18 Manure Management System Series. MidWest Plan Sevice, Iowa State University, Ames, IA.

Godden, S. 2008. Colostrum management for dairy calves. Vet Clin. North Amer.: Food Anim. Pract. 24(1):19-39. doi:10.1016/j.cvfa.2007.10.005

Hatfield, J.L., M.C. Brumm, and S.W. Melvin. 1998. Swine manure management. In: Agricultural uses of municipal, animal, and industrial byproducts (pp. 78-90). USDA ARS Conservation Research Report 44. USDA-ARS, Washington, D.C.

Holmes, B., N. Cook, T. Funk, R. Graves, D. Kammel, D. Reinemann, and J. Zulovich. 2013. Dairy freestall housing and equipment (MWPS-7, 8th ed). MidWest Plan Sevice, Iowa State University, Ames, IA.

Jacobson, L.D., H.L. Person, and S.H. Pohl. 1997. Swine nursery facilities handbook. MidWest Plan Sevice, Iowa State University, Ames, IA.

Janni, K.A., M.I. Endres, J.K. Reneau, and W.W. Schoper. 2007. Compost dairy barn layout and management recommendations. Appl. Eng. Agric. 23(1):97–102. doi:10.13031/2013.22333

Jones, D., R. Lemenager, K. Foster, B. Doran, R. Euken, and S. Shouse. 2013. Cattle feeding in monoslope and gable roof buildings (No AED60). Midwest Plan Service, Ames, IA.

Lawrence, J.D., W.M. Edwards, S. Shouse, D. Loy, and J.J. Lally. 2007. Beef feedlot systems manual (No. 12797). Iowa State University, Department of Economics, Ames, IA.

LPELC. 2018. Layer hen housing and manure management. Livestock Poultry Environmental Learning Center, Lincoln, NE. https://lpelc.exposure.co/layer-chicken-housing-and-manure-management (Accessed 18 Apr. 2019).

MWPS. 1987. Structures and environment handbook (No MWPS-1). Midwest Plan Service, Ames, IA.

NPB. 2008. Water reflection of pig barn. National Pork Board, Des Moines, IA. https://library.pork. org/media/?mediaId=91778D15-20A9-4012-9206DFFD3DCE1534 (Accessed 18 Apr. 2019).

NPB. 2009. Quick Facts: The Pork Industry at a Glance. National Pork Board, Ames, IA. http:// porkgateway.org/wp-content/uploads/2015/07/quick-facts-book1.pdf (Accessed 19 Apr. 2019).

NPB. 2015. Pigs. National Pork Board, Des Moines, IA. https://library.pork.org/ media/?mediaId=78107B75-3879-4B06-93F953608B36A0AD (Accessed 19 Apr. 2019).

NPB. 2016a. Andersen Farm. National Pork Board, Des Moines, IA. https://library.pork.org/ media/?mediaId=A3CFC18C-CDA9-4294-AF0602020A1A74BE (Accessed 19 Apr. 2019).

NPB. 2016b. Life cycle of a market pig. Natl. Pork Board- Pork Checkoff, Des Moines, IA. https:// www.pork.org/facts/pig-farming/life-cycle-of-a-market-pig (Accessed 19 Apr. 2019).

NPB. 2018. Nutrient management with tractor. National Pork Board, Des Moines, IA. https://library. pork.org/media/?mediaId=9F7CE927-D834-4D3A-90A8F27727E989F0 (Accessed 19 Apr. 2019).

Salfer, J.A., J.M. Siewert, and M.I. Endres. 2018. Housing, management characteristics, and factors associated with lameness, hock lesion, and hygiene of lactating dairy cattle on Upper Midwest United States dairy farms using automatic milking systems. J. Dairy Sci. 101:8586–8594. doi:10.3168/jds.2017-13925

Tabler, T., and J. Wells. 2018. Poultry litter management. Mississippi State University Extension, Starkville, MS. https://extension.msstate.edu/sites/default/files/publications/publications/ p2738.pdf (Accessed 18 Apr. 2019).

UEP. 2018. Choices in hen housing. United Egg Producers, Johns Creek, GA. https://uepcertified. com/choices-in-hen-housing/ (Accessed 19 Apr. 2019).

USDA NASS. 2018a. Broiler production by state. USDA-NASS, Washington, D.C. www.nass.usda. gov/Charts_and_Maps/Poultry/brlmap.php (Accessed 19 Apr. 2019).

USDA NASS. 2018b. Turkey production by state. USDA-NASS, Washington, D.C. https://www.nass. usda.gov/Charts_and_Maps/Poultry/tkymap.php (Accessed 19 Apr. 2019).

USDA-NRCS. 2009. Agricultural waste management system component design, Chapter 10. In: USDA-NRCS, editor, Agricultural waste management field handbook, Part 651. USDA-NRCS, Washington, D.C. https://www.nrcs.usda.gov/wps/portal/nrcs/detailfull/national/ water/?&cid=stelprdb1045935 (Accessed 19 Apr. 2019).

USDA-NRCS. 1992. Agricultural waste management systems, Chapter 9. In: USDA-NRCS, editor, Agricultural Waste Management Field Handbook, Part 651. USDA-NRCS, Washington, D.C. https://www.nrcs.usda.gov/wps/portal/nrcs/detailfull/national/water/?&cid=stelprdb1045935 (Accessed 19 Apr. 2019).

Nitrogen and Phosphorus Characteristics of Beef and Dairy Manure

Paulo H. Pagliari,* Melissa Wilson, Heidi M. Waldrip, and Zhongqi He

Abstract

The ever-growing global population puts enormous pressure on the food production systems by always demanding increased food production. In the last four years, the amount of milk produced in the US has increased on average by 1100 metric ton yr[-1], with the number of milking cows (*Bos taurus*) increasing by about 41,000 animals over the same timeframe. Managing the manure produced by livestock operations is one of the major challenges in food production and the one that has the highest potential negative impact on the environment. Most of the N required by animals is provided in the form of protein present in the feed. In general, animals excrete a large quantity of the N consumed (as much as 95%) either in the urine or in the feces. Changing the total amount of protein in feed is an option to reduce the amount of N in manure. For P, feeding too low P content can be dangerous for the lactating cows; feeding rations with less than 0.31% of P leads to P translocation from bone deposits. The amount of N and P present in manure is directly related to the amount of N and P in the feed. Increasing feed N and P as a percentage of feed dry matter causes a significant increase in the amount of N and P in the manure. Understanding manure nutrient composition is key for developing sound strategies for manure reuse in the farm and minimizing the negative impacts of manure on the environment.

As of 2019, the human global population was 7.7 billion individuals, a number that is increasing annually by 1.5 to 5.8% in most developed and emerging nations. This ever-increasing global population puts enormous pressure on current food production systems, as it coincides with a desire for higher quantities of affordable animal products, including milk and beef. In the U.S., there were 95 million cattle (*Bos taurus* and *Bos indicus*) in 2019 (USDA-NASS, 2018). The amount of U.S. milk produced within the last 4 yr increased by approximately 1100 metric ton yr[-1], with the number of milking cows increasing by about 41,000 animals over the same timeframe (USDA-ERS, 2019). From 2013 to 2018, the number of beef cattle and calves increased from 95 to 103 million animals (USDA-NASS, 2018). As a result, the amount of manure (i.e., a mixture of urine, feces, scurf, dropped feed and water, and any bedding) generated by livestock operations also increased annually. When managed improperly, manure use can increase environmental risks to water, air, and soil from rogue carbon (C) and nutrients (nitrogen [N], phosphorus [P]). Although the volume of manure generated per animal varies with breed, body size, diet, performance, and other variables, it is estimated to range between 27 and 36 kg d[-1] for a beef steer and 52 kg d[-1] for a larger milking cow. In 2013 there were about 89.3 million cattle animals raised in the United States, with an estimated manure production of 1.17 billion Mg (He et al., 2016).

P.H. Pagliari, Department of Soil, Water, and Climate, University of Minnesota. Southwest Research and Outreach Center. 23669 130th St. Lamberton, MN 56152; M. Wilson, Department of Soil, Water, and Climate, University of Minnesota. 439 Borlaug Hall, 1991 Upper Buford Circle, Saint Paul, MN 55108; H.M. Waldrip, USDA-ARS Conservation and Production Research Laboratory, Bushland, TX 79012; Z. He, USDA-ARS Southern Regional Research Center, New Orleans, LA 70124. *Corresponding author (pagli005@umn.edu)

doi:10.2134/asaspecpub67.c4

Animal Manure: Production, Characteristics, Environmental Concerns and Management. ASA Special Publication 67. Heidi M. Waldrip, Paulo H. Pagliari, and Zhongqi He, editors.
© 2019. ASA and SSSA, 5585 Guilford Rd., Madison, WI 53711, USA.

Managing livestock manure, and its N, P, and C contents, is a major challenge for producers, but it is also an area with high mitigation potential against the negative environmental impact of livestock production (He, 2012; He and Zhang, 2014).

Animal manure can impact both agriculture and the environment in multiple complicated ways. Many producers have developed or employed nutrient management plans to recycle manure for maximum fertilizer value and minimum environmental loss. However, other producers view manure as a waste product requiring disposal, which often leads to soil application at rates exceeding crop needs (Dagna and Mallarino, 2014; Long et al., 2018). Inappropriate use of animal manure can result in water quality degradation and has been linked to eutrophication of aquatic systems (Biagini and Lazzaroni, 2018). Climate change has caused increased amounts of localized precipitation, which presents further risks, as rainfall increases the amount of dissolved P and runoff from manured fields into lakes, rivers, and streams (Pagliari, 2014; Motew et al., 2018). The high N and P contents of manure, in addition to enteric methane production, are one major source of environmental degradation from livestock production (Pagliari and Laboski, 2012; Dagna and Mallarino, 2014). In addition, ammonia (NH_3) and other nitrogenous emissions (i.e., nitrous oxide [N_2O], nitric oxide [NO]) from cattle operations have significant impacts on air quality and global climate (Ndegwa et al., 2011; Parker et al., 2017). Thus, knowledge of N and P characteristics in animal manure is crucial for the development of best management practices for sustainable manure recycling. This chapter will review and synthesize information on relevant properties of N and P in dairy and beef manure, and how they are affected by diet, climate, and management.

Nitrogen Intake and Excretion by Beef and Dairy Cattle

Nitrogen is an essential element for proteins, amino acids, nucleic acids, and energy transfer in vertebrate animals; thus, large amounts of dietary N are required to produce profitable quantities of meat and milk and ensure successful stock procreation. Most required N is provided in the form of feed or supplemental protein (e.g., soybean meal, cottonseed meal, blood meal, urea). When dairy rations are adequately balanced for energy, crude protein (CP), and minerals, dietary N is metabolized and transferred to milk, urine, and feces in approximately equal proportions (Broderick, 2003; Powell, 2011). However, feeding excess dietary N to cattle can lead to increased concentrations of urine-N, and to a lesser extent fecal-N; as much as 75 to 95% of N fed can be excreted (Cole et al., 2006; Selbie et al., 2015).

The nitrogen use efficiency (NUE) of beef cattle is generally low, as 80 to 90% of the dietary feed N is excreted in manure (Cole et al., 2006). The majority of excess N is excreted in urine when high dietary CP concentrations are fed to beef cattle (Waldrip et al., 2013). Cole et al. (2006) found that urinary N content increased from 84 to 94 g d^{-1}, while fecal N increased from 39 to 65 g d^{-1}, when dietary CP was increased from 11.5 to 13.0%, which is within National Research Council (NRC) recommended CP levels for finishing cattle. Levels of dietary CP can easily reach 20% of dietary dry matter (DM) when distiller's dried grains plus solubles (DDGS) replace a portion of the grain in a finishing diet. Cole et al. (2005) reported that decreasing CP intake from 13.0 to 11.5% led to a 44% reduction in manure-derived NH_3

emissions under laboratory conditions and up to 28% lower cumulative daily emissions under field conditions. Similarly, Cole et al. (2005) reported 140% increased NH_3 emissions when beef cattle were fed 13.0% CP, versus 11.5% CP. For this reason, cattle NUEs, and effects of diet composition on N use have remained hot topics for livestock managers (Table 1). It was estimated that cattle NUE could potentially be 30 to 54% if diets were properly managed at beef feedlots. One example would be phase feeding, where dietary CP is adjusted to meet, but not exceed, animal CP requirements at differing stages of growth and production.

For dairy cows, N requirements are about two to three times higher than beef animals due to the N requirement for milk production (Table 1). It was estimated that a diet containing 23% CP produced maximum milk production (NASEM, 2001). However, overfeeding protein N can have negative effects on milk N content. For example, Groff and Wu (2005) reported that increasing CP intake from 15 to 19% led to a decrease in milk N from 0.27 to 0.22 g d^{-1}, an 18% decrease. In another study, Tomlinson et al. (1996) reported that increasing milking cow dietary CP from 12 to 18% caused a linear increase in the amount of total N excreted in urine and feces. The authors also reported that adding bypass protein (i.e., ruminally undegraded protein [RUP]), such as canola meal, increased fecal N concentrations but decreased urinary N concentrations. This decrease in urine-N would result in reduced NH_3 losses and increased manure fertilizer value, as fecal N is mineralized slowly over time. Castillo et al. (2001) increased the dietary N content of lactating cows by 100 g N d^{-1} and found no significant changes in milk- or fecal-N outputs. In contrast, there was a significant nonlinear increase in urine-N when dietary protein intake increased, especially when total N consumption was > 400 g N d^{-1} (Castillo et al., 2001). Figure 1 summarizes relevant data from the literature and shows clearly how urine-N increases as dietary CP increases. It can also be seen that fecal-N (as a percent of manure DM) decreased as CP intake increased (Fig. 1). The decrease in fecal-N was accompanied by a nonlinear increase in urine-N. Over the last two decades, corn silage production in the top five U.S. dairy states (California, Wisconsin, Idaho, New York, and Pennsylvania) has increased; whereas, alfalfa acreage and hay production has declined (USDA–NASS, 2014). The continuing shift toward feeding more high-silage diets has stimulated NUE research on dairy farms (Table 1). As N transformation on dairy farms are interrelated and complex, Barros et al. (2017) and Powell et al. (2017) applied stable ^{15}N isotope analyses to measure relative partitioning of N at the whole farm scale and from alfalfa silage, corn silage, corn grain, and soybean meal into milk, urine, and feces. Based on these ^{15}N recovery data, the authors reported differences in undigested dietary N in feces, indicating that differences in dietary components could impact long-term contributions to stable organic matter (OM) in manure-amended soils. In other words, feeding more corn silage and grain may have profound impacts on the N cycle of dairy farms, where more N is lost from urine as NH_3; thus, producers must purchase additional fertilizer N and also face increased N loss per unit land area and milk produced.

P Intake and Excretion by Beef and Dairy Cattle

Phosphorus is an essential component of cattle diets due to its importance in bone formation, cell growth and differentiation (i.e., DNA and RNA), energy use and transfer (i.e., ATP, ADP, AMP), phospholipid formation, and maintenance of acid-base and osmotic balances (NASEM, 2001). Fewer studies are available that focused on the effects of diet on P, rather than N, excretion by dairy and beef

Table 1. Selected studies of nitrogen (N) intake and excretion by beef and dairy cattle conducted within the last decade.

Reference	Farm type, region	Average N input	Study subject and time frame	Average N excreted	Feces-N	Urine-N	NUE †
Aarons et al. (2017)	Grazed dairy, Australia	545 g head^{-1} d^{-1} (feed and grazing)	43 farms, 1 yr	433 g head^{-1} d^{-1} - ‡	–		21% (11–39%) †
Aguerre et al. (2011)	Dairy, WI, USA	507 g head^{-1} d^{-1} (feed)	16 animals, 21 d	297 g head^{-1} d^{-1} –	–		31% (30–32%)
Barros et al. (2017)	Dairy, WI, USA	638 g head^{-1} d^{-1} (^{15}N-labeled diets)	12 animals, 35 d	381 g head^{-1} d^{-1}	166 g head^{-1} d^{-1}	241 g head^{-1} d^{-1}	23.5% (21.5-25.2%)
Bernier et al. (2014)	Beef, Canada	122 g head^{-1} d^{-1} (low quality forage with supplements)	24 animals, 5 mo	104 g head^{-1} d^{-1}	54 g head^{-1} d^{-1}	50 g head^{-1} d^{-1}	15% (1–24%)
Cole and Todd (2009)	Beef, TX, USA	165 g head^{-1} d^{-1} (various diets)	> 1000 animals, 1 yr	141 g head^{-1} d^{-1}	50.9 g head^{-1} d^{-1}	90.5 g head^{-1} d^{-1}	15%
Gandra et al. (2011)	Beef, Brazil	217 g head^{-1} d^{-1} (corn silage and concentrates)	30 animals, 21 d	100 g head^{-1} d^{-1}	58 g head^{-1} d^{-1}	42 g head^{-1} d^{-1}	54% (43–63%)
Gourley et al. (2012)	Dairy, Australia	52.5 kg ha^{-1} yr^{-1} (Concentrates and grain)	84 farms, 1 yr	–	–	–	25%
Hassanat et al. (2013)	Dairy, Quebec, Canada	584 g head^{-1} d^{-1} (total mixed rations)	9 animals, 21 d	369 g head^{-1} d^{-1}	171 g head^{-1} d^{-1}	196 g head^{-1} d^{-1}	29% (26–31%)
Hünerberg et al. (2013b)	Growing beef, Canada	280 g head^{-1} d^{-1} (diets with dried distiller grains with solubles)	16 animals, 21 d	235 g head^{-1} d^{-1}	85 g head^{-1} d^{-1}	149 g head^{-1} d^{-1}	16% (13–18%)
Hünerberg et al. (2013a)	Finishing beef, Canada	278 g head^{-1} d^{-1} (diets with dried distiller grains with solubles)	16 animals, 28 d	220 g head^{-1} d^{-1}	63 g head^{-1} d^{-1}	157 g head^{-1} d^{-1}	21% (18–27%)
Kobayashi et al. (2010)	Dairy, Japan	503 kg ha^{-1} yr^{-1} (home-grown and purchased feed)	1 farms, 5 yr	297 kg ha^{-1} yr^{-1}	100 g head^{-1} d^{-1}	100 g head^{-1} d^{-1}	59% (51–59%)
Koenig and Beauchemin (2013)	Beef, Canada	166 g head^{-1} d^{-1} (barley-based diets)	16 animals, 28 d	112 g head^{-1} d^{-1}	46 g head^{-1} d^{-1}	72 g head^{-1} d^{-1}	28% (26–31%)
Koenig and Beauchemin, (2018)	Beef, Canada	290 g head^{-1} d^{-1} (finishing diets with tannins)	32 animals, 35 d	205 g head^{-1} d^{-1}	34 g head^{-1} d^{-1}	136 g head^{-1} d^{-1}	30% (24–34%)
Luebbe et al. (2012)	Beef, TX, USA	232 g head^{-1} d^{-1} (diets with wet distillers grain plus solubles)	37 animals, 21 d	153 g head^{-1} d^{-1}	55 g head^{-1} d^{-1}	98 g head^{-1} d^{-1}	34% (30–38%)
Morris et al. (2018)	Dairy, OH, USA	698 g head^{-1} d^{-1} (diets with reduced-fat distillers grain)	36 animals, 63 d	475 g head^{-1} d^{-1}	191 g head^{-1} d^{-1}	284 g head^{-1} d^{-1}	32% (31–34%)
Nair et al. (2016)	Beef, Canada	248 g head^{-1} d^{-1} (Canola-meal-based)	25 animals, 160 d	191 g head^{-1} d^{-1}	51 g head^{-1} d^{-1}	140 g head^{-1} d^{-1}	23% (20–27%)
Powell and Rotz (2015)	Dairy, WI, USA	288 kg ha^{-1} yr^{-1} (feed)	2 farms, 1 yr	209 kg ha^{-1} yr^{-1} –	–		27.2% (27.1-27.5%)
Spiehs and Varel (2009)	Beef, NE, USA	201 g head^{-1} d^{-1} (diets with corn wet distillers grains with solubles)	24 animals, 96 h	139 g head^{-1} d^{-1}	58 g head^{-1} d^{-1}	81 g head^{-1} d^{-1}	31% (28–35%)
Vasconcelos et al. (2009)	Beef, TX, USA	131 g head^{-1} d^{-1} (crude protein formulas with urea)	2 farms, 159 d	93 g head^{-1} d^{-1}	38 g head^{-1} d^{-1}	51 g head^{-1} d^{-1}	30% (29–32%)
Zenobi et al. (2015)	Beef, Canada	281 g head^{-1} d^{-1} (blended feed pellets)	25 animals, 160 d	209 g head^{-1} d^{-1}	78 g head^{-1} d^{-1}	131 g head^{-1} d^{-1}	25% (19–31%)

† NUE, N use efficiency. Data represent the mean NUE, with the range in parentheses.

‡ No data available

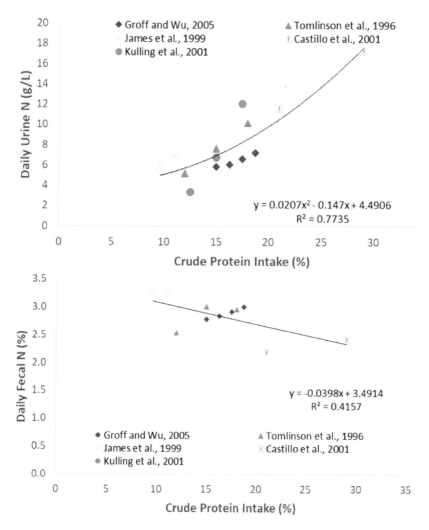

Fig. 1. a. Amount of nitrogen (N) in urine as a function of crude protein (CP) intake by lactating dairy cows. b. Amount of nitrogen (N) in feces as a function of crude protein (CP) intake by lactating dairy cows.

cattle. In contrast to N excretion, which can be partitioned differently depending on diet and animal characteristics, most excess P is excreted in cattle feces, rather than urine (Table 2). The dietary P requirement for a feedlot steer is 0.30% of the diet DM (NRC, 1996), but could be as low as 0.16% of dietary DM (Erickson et al., 2007). More recently, Geisert et al. (2010) fed steers diets ranging from 0.10 to 0.38% P and concluded that the P requirement for finishing steers was between 0.10 to 0.17% of dietary DM, or 7.2 to 12.1 g d^{-1}. For dairy cows, the dietary P requirement has been reported to range from 0.35 to 0.40% of DM (Reid et al., 2015). Reid et al. (2015) investigated the results of feeding diets with P content ranging from 0.22 to 0.37% of DM. The authors observed that feeding the lowest amount of P

Table 2. Selected studies on phosphorus (P) intake and secretion by beef and dairy cattle conducted within the last decade.

Reference	Farm type, region	Average _P input	Study subject and time frame	Average P excreted	P in feces	P in urine	PUE†
Bernier et al. (2014)	Beef, Canada	17.9 g head^{-1} d^{-1} (low quality forage with supplements)	24 animals, 5 mo	22.2 g head^{-1} d^{-1}	18.3 g head^{-1} d^{-1}	3.9 g head^{-1} d^{-1}	0.0% (0.0%)
Cole and Todd (2009)	Beef, TX, USA	28.9 g head^{-1} d^{-1} (various diets)	> 1000 animals, 1 yr	23.1 g head^{-1} d^{-1}	14.5 g head^{-1} d^{-1}	8.6 g head^{-1} d^{-1}	20.1%
Geisert et al. (2010)	Beef, NE, USA	28.6 g head^{-1} d^{-1} (diets with different P contents)	5 animals, 21 d	17.2 g head^{-1} d^{-1}	15.5 g head^{-1} d^{-1}	3.0 g head^{-1} d^{-1}	36.8% (11.8–51.6%)
Morris et al. (2018)	Dairy, OH, USA	111.6 g head^{-1} d^{-1} (diets with reduced-fat distillers grain)	36 animals, 63 d	64.2 g head^{-1} d^{-1}	60.2 g head^{-1} d^{-1}	3.9 g head^{-1} d^{-1}	39.9% (31–34%)
Souza et al. (2016)	Beef, Brazil	31.4 g head^{-1} d^{-1} (high grain diest with or without P supplements)	50 animals, 116 d	18.3 g head^{-1} d^{-1}	17.2 g head^{-1} d^{-1}	1.2g head^{-1} d^{-1}	41.4% (40.3–43.5%)
Spiehs and Varel (2009)	Beef, NE, USA	31.1 g head^{-1} d^{-1} (diets with corn wet distillers grains with solubles)	24 animals, 96 h	19.8 g head^{-1} d^{-1}	10.9 g head^{-1} d^{-1}	8.9 g head^{-1} d^{-1}	36.5% (29.2–49.5%)
Vasconcelos et al. (2009)	Beef, TX, USA	131 g head^{-1} d^{-1} (crude protein formulas with urea)_	2 farms, 159 d	20.0 g head^{-1} d^{-1}	8.5 g head^{-1} d^{-1}	3.4 g head^{-1} d^{-1}	40.8% (39.7–42.4%)
Zanetti et al. (2017)	Beef, Brazil	13.0 g head^{-1} d^{-1} (different P diets)	28 heads, 63 d	8.4 g head^{-1} d^{-1}	8.2 g head^{-1} d^{-1}	0.2 g head^{-1} d^{-1}	31.4% (5.6–43.9%)

†PUE, P use efficiency. Data represent the mean PUE, with the range in parentheses.

decreased blood P concentration but had no effect on milk yield, likely due to the short nature of the study, eight weeks only.

In practice, it is not uncommon for dietary P to exceed requirements (Spiehs and Varel, 2009). Phosphorus excretion has been reported to increase from 12.6 to 18.6 g d^{-1} when CP increased from 11.5 to 13.0%, respectively (Cole et al., 2006). The P use efficiency of lactating cows is relatively low, and ~67% of total consumed P is excreted (Wu et al., 2003). Reducing manure P is primarily conducted by decreasing dietary P levels (Feng et al., 2015), either by manipulating the amount of supplemental P or changing the ratio of silage to forage being fed. Cereal grains, silage, and other feedstuffs contain higher P levels than forages (Singh et al., 2018). For example, steam-flaked corn contains approximately 0.25% P, while meadow hay contains approximately 0.18% P (NASEM, 2016). In general, grains contain 3.5 to 4.5 g of P kg^{-1} DM, while straw contains about less than 1 g kg^{-1} DM (Singh et al., 2018). Feeding dairy cows feather meal, which contains ~0.44% P, has been reported to change the P balance in dairy cows, with subsequent increases in soluble and total P in urine (though the amount of P in urine is lower than in feces) from milking cows (Tomlinson et al., 1996). However, feeding P below maintenance requirements (e.g., < 0.34% P as dietary DM) can lead to negative P balance, which results in mobilization of P in bones and other body reserves (Knowlton and Herbein, 2002). A negative P balance occurred when Reid et al. (2015) and Wu et al. (2001) fed diets containing 0.22 and 0.31% P, respectively. Milk production can also be affected when not enough dietary P is provided in the feed, and Wu et al. (2000) reported decreased milk production during late lactation by feeding feed with 0.31% P. The P content of milk is generally 0.083 to 0.100%, and is distributed as casein, colloidal inorganic calcium phosphate, phosphate ions, and a small portion of phospholipids (NASEM, 2001).

Research has shown that decreasing dietary P content from 0.67 to 0.34% DM decreased daily P excretion from 113 g $d^{-1}cow^{-1}$ to 43 g $d^{-1}cow^{-1}$ and caused changes in blood mineral concentrations (Knowlton and Herbein, 2002). Feng et al. (2015) studied the effects of dietary P content (ranging from 0.15 to 0.45% P of DM) on beef cattle performance and found a linear increase in excreted P as the amount of dietary P increased. Wu et al. (2003) reported that decreasing dietary P from 0.42 to 0.33% DM resulted in 25% less P excreted in the feces of lactating cows. Wu et al. (2000) reported a positive linear relationship between the amount of P in the diet and feces-P, where increasing diet P content by 0.01% increased feces-P by an average of 0.02% (note that feces has higher moisture content than feed, hence the relative higher percentage). Similarly, in a different study Wu (2005) reported that reducing dietary P content from 0.44 to 0.37% reduced feces-P by about 12%: a reduction of about 0.017% in feces-P for each 0.01% increase in dietary P. The source of P, whether added as calcium diphosphate or other mineral form or as a component of dietary constituents, also may impact the amount of P in feces. Knowlton et al. (2001) reported that lactating cows fed wheat bran (1.04% P) showed lower P retention than cows fed supplemental mineral P. Although diet has a profound effect on the nutrient concentration of animal manure, the actual nutrient concentration in the manure can be estimated fairly accurately if the feed nutrient concentration is known (Nennich et al., 2005; Rotz et al., 2005). Selected studies of P intake and excretion are listed in Table 2.

Characteristics of N Excreted by Cattle

Both organic and inorganic N is excreted by cattle. These manure N compounds can be valuable plant nutrients or have adverse environmental effects on air and water quality (He and Olk, 2011; Honeycutt et al., 2011; Miller and Varel, 2011; Ndegwa et al., 2011). Inorganic N is mainly present in urine in the form of ammonium (NH_4^+) or NH_3 (collectively called NH_x) (Bernier et al., 2014; Jardstedt et al., 2017). Urea and other minor compounds (such as hippuric acid, creatinine, allantoin, and uric acid) are some organic forms of urinary N (Jardstedt et al., 2017; Ndegwa et al., 2011). Spek et al. (2013) reported that urea accounted for 82% and 79% of total urinary N from lactating cows in Northwestern Europe and North America, respectively. Thus, urea N content has been regarded as an excellent indicator of the efficiency of N turnover in ruminants, where high values indicated elevated, and therefore excessive, feed N content (Schuba et al., 2017). Urea and NH_x are both valuable sources of fertilizer N and are readily utilized by plants. Studies on manure NH_x and N_2O from cattle operations of have focused primarily on potential adverse environmental impacts rather than manure fertilizer value (Bonifacio et al., 2015; He et al., 2016). Urea ($[NH_2]_2CO$) is rapidly hydrolyzed into two NH_4^+

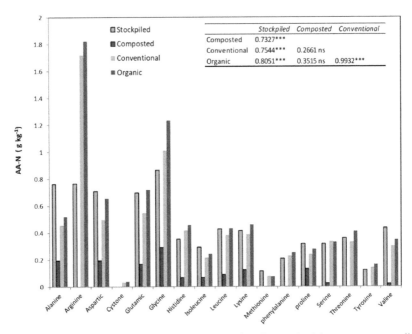

Fig. 2. Amino acid (AA)-N contents in conventional and organic dairy manure as well as water extracts of stockpiled and composted dairy manure. The table insert shows the correlation coefficents between the two samples with statistically significant difference at p = 0.001 (***) and no significant difference at p = 0.05 (ns). Data for the conventional and organic manure are the averages of 4 and 14 samples, respectively, and adapted from He and Olk (2011). Data of stockpiled and composted samples are adapted from Liang et al. (1996).

molecules by the urease enzyme that is ubiquitous in manure and soil. This NH_4^+ is in equilibrium with free NH_3. Under appropriate conditions (e.g., high pH, high temperatures, sufficient wind speed and moisture content), NH_3 can be transported to the manure surface by diffusive mass transfer and then emitted to the atmosphere. Urea is the primary source of NH_3 volatilized from cattle manure. In a review on beef feedlot sustainability, Waldrip et al. (2015) reported that up to 90% of the NH_3 volatilized from beef cattle feedlots originated from urine deposited in cattle pens. In contrast, N volatilization from feces is considerably low, ranging from 1 to 13% due to slow organic matter mineralization (Ndegwa et al., 2011).

The chemical composition of fecal N is more complicated than that of urine. Feces-N is primarily in organic forms and can be divided into two general pools: (i) endogenous N consisting of microbial N, sloughed cells from the rumen, intestine, and the hindgut, and undigested enzymatic secretions; and (ii) undigested dietary N consisting mainly of undigested, lignified cell wall components derived from plant biomass (Powell et al., 2017). Protein and amino acid components are the major manure organic N forms, as they are crucial dietary components for cattle growth and milk production (Lean et al., 2018). However, there are only limited studies on manure amino acid N (AA-N) profiles (Hacking et al., 1977; He et al., 2014).

The amino acid profiles of four dairy manure samples are presented in Fig. 2. In water extracts of stockpiled and composted dairy manures, total amounts of AA-N in the water extracts of the stockpiled manures were five times greater than those extracted from the composted manure, with 7.16 and 1.37 g AA-N kg^{-1} DM, respectively (Liang et al., 1996). The most abundant AA compounds were neutral and acidic compounds. The water extract of the composted dairy manure was relatively richer in aspartic acid, glutamic acid, lysine, proline, glycine, alanine, isoleucine, and leucine, but depleted in serine and valine compared with the water extracts of the stockpiled manure. On the other hand, arginine, threonine, tyrosine, phenylalanine, and methionine were not present in the water extract of the composted manure. These changes in AA-N contents in the water extracts of stockpiled and composted manures reflect decomposition of proteinaceous materials during dairy manure composting. Two samples that were analyzed were manures collected from conventional and organic dairy farms. The management of organic dairy farms differ from their conventional counterparts by using less imported protein and energy from purchased feed and supplements, including a higher proportion of forage in the diet and increased reliance on manure and compost as nutrient sources for growing crops, rather than chemical fertilizers. These differences have shown some impacts on the availability, utilization, and cycling of manure nutrients (He, 2019; He et al., 2015). The AA distribution patterns of the two types of dairy manure are similar (Fig. 2). In addition, the distribution pattern of AA in these dairy manures is similar to that of water extracts of dairy manure (Liang et al., 1996) and other manure, such as poultry litter (He et al., 2015). Figure 4 also shows that the concentration of each AA compound is always equal or greater (up to 22% higher with glycine) in organic dairy manure than in conventional dairy manure. This observation implies either a protein-rich diet for cows on organic dairies or less digestibility of AA or proteins in organic diets compared to conventional dairy management. Correlation analysis showed a strong relationship between the AA–N distribution among stockpiled, conventional, and organic samples; however, this same relationship was not observed among manure compost and any of the other manure samples (stockpiled, organic, and conventional dairy manure). Total AA–N content was similar for stockpiled, conventional, and organic manures, with values of 7.17, 7.33, and 8.44 g kg^{-1} DM, respectively. Total AA–N content of the composted dairy manure is much lower, with the value of 1.37 g kg^{-1} DM. This observation implies that composting decomposes manure-bound AA compounds, but individual amino compounds degrade at different rates. In addition to mineralization, some amino acids may also affect the AA profile of soil and directly promote plant growth (Ma et al., 2017; Perez et al., 2015).

As previously discussed, fecal organic N mineralization is a much slower process than urea hydrolysis; thus, the contribution of fecal N to NH$_3$ emissions is generally low (He and Waldrip, 2016f). Most research on fecal N (or more generally manure organic N) have focused on characterization and mineralization rates for N cycling (Agomoh et al., 2018; He, 2012; He and Zhang, 2014; Wang et al., 2018). Generally, fecal endogenous N (i.e., microbial cells and sloughed gastrointestinal cells) is rapidly mineralized and available to crops, while undigested dietary N in feces is not readily available to microbes and could be a stable component in soil organic matter (Powell et al., 2009; Powell et al., 2017). In lactating dairy cows, endogenous N represented an average of 17% of feces-N (Ouellet et al., 2002)

Another gaseous component from beef and dairy manure is N$_2$O, which is a potent GHG implicated in global climate change (Külling et al., 2001). While the volume of N$_2$O lost from cattle production systems is generally small compared to

Table 3. Comparison of specific phosphorus (P) fractions of cattle manures from selected studies.

	Sample	Total P (g kg⁻¹)	H₂O-P†	NaHCO₃–P	NaOH-P	HCl-P	Residual-P
Ajiboye et al. (2007)	Beef manure	2.2	53%	35%	8%	4%	1%
Ajiboye et al. (2007)	Dairy manure	13.4	18%	35%	10%	8%	29%
Dou et al. (2000)	Dairy manure	4.9	70%	14%	6%	5%	5%
He and Honeycutt (2001)	Unspecified cattle manure	3.5	51%	29%	13%	4%	1%
He et al. (2004)	Dairy manure	9.2	42%	41%	5%	6%	8%
Hong et al. (2018)	Dairy manure	7.0	37%	43%	16%	5%	–
Li et al. (2014)	Dairy manure	6.8	38%	48%	8%	3%	2%
Pagliari and Laboski (2012)	Beef manure	NA	36%	31%	18%	6%	9%
Pagliari and Laboski. (2013)	Dairy manure	NA	46%	37%	11%	7%	9%
Yan et al. (2015)	Beef and Dairy	9.6	27%	34%	18%	16%	5%

†H₂O-P, water extractable P; NaHCO₃–P, sodium bicarbonate extractable P; NaOH-P, sodium hydroxide extractable P; HCl-P, hydrochloric acid extractable P; residual-P, total P minus the sum of extractable fractions.

NH_3, N_2O emissions from livestock housing can be significant under certain conditions. Nitrous oxide can be produced enterically in small quantities; however, most N_2O is derived from excreted manure. In addition, the manure or slurry accumulated in lagoons and retention ponds are often anaerobic and serve as a source of N_2O from feedlots and dairies (Külling et al., 2003).

Characteristics of P Excreted by Cattle

Unlike the diversity found in the chemical structures of excreted N compounds, most, if not all, P in manure is present in the form of inorganic phosphate. Inorganic P (P_i) (HPO_4^-, $H_2PO_4^{2-}$ and PO_4^{3-}) may be soluble or insoluble, depending on manure pH and specific metal species present in the manure (Akinremi et al., 2011; He et al., 2009). Organic P (P_o) is present as phosphate esters (i.e., organophosphates with the general formula $O = P(OR)_3$] that contain a diverse range of forms, including DNA, RNA, and ATP (He and Honeycutt, 2011). To date, characterization of manure P has focused largely on P_i solubility and specific ester forms of P_o. The former is done by extraction with different solvents, separately or sequentially (Bernier et al., 2014; Toth et al., 2011). In contrast, P_o can be analyzed by enzymatic hydrolysis (He and Honeycutt, 2001; Pagliari and Laboski, 2012) and/or nuclear magnetic resonance (NMR) and other spectroscopic techniques (Akinremi et al., 2011; Cade-Menun, 2011). Urine-P is considered to be labile because it is excreted in solution. Thus, the forms and lability of feces-P have been the focus of most research. Bernier et al. (2014) examined water extractable P (H_2O-P), sodium bicarbonate extractable P (NaHCO₃–P), and residual P (Res-P) in fresh feces from mature beef cattle consuming either low-quality forage or low-quality forage supplemented with DDGS. These researchers found that labile P (H_2O-P and NaHCO₃–P) accounted for 74 to 81% of total feces-P. In another study, Cade-Menun et al. (2015) used solution ³¹P NMR analysis to compare P forms in dairy feces extracted with (1) H_2O, (2) sodium acetate buffer (pH 5.0) with fresh sodium dithionite (NaAc-SD), or (3) 0.25

M NaOH–0.05 M ethylenediaminetetraacetic acid (NaOH-EDTA). They reported that H_2O extracted 35% of the total P present, whereas NaAc-SD and NaOH-EDTA extracted 80 to 100% of total manure P. In addition to differences in the amounts of P extracted with the various solvents, there were differences in the forms of P extracted: H_2O extracted a greater proportion of P_o, particularly orthophosphate digesters, but a lower proportion of P_i relative to NaOH-EDTA and NaAc-SD. Thus, results from studies differ depending on the extractant used.

While such fecal P research is helpful in manure P chemistry, the characteristics of P in accumulated cattle manure (a mixture of urine and feces), as found in livestock housing, is often studied to provide data aligned with real-world practices. Table 3 lists the distribution of cattle manure P in the five fractions extracted with a typical modified Hedley sequential fractionation strategy (H_2O, $NaHCO_3$, NaOH, HCl, and total P). These data were collected over two decades; however, the P distribution pattern was not significantly changed during this time frame. In most samples, the first two P fractions (i.e., H_2O-P and $NaHCO_3$–P) were the most abundant. These two P fractions are generally considered labile P fractions and shed be easily available for plant growth (Pagliari, 2014; Toth et al., 2011). In addition, H_2O-P content is a critical indicator of the P runoff potential (Kleinman et al., 2019; Liu et al., 2018). The high portion of H_2O-P and $NaHCO_3$–P in cattle manure implies that P management should focus on immediate impacts on water quality and plant response. This is in contrast with other types of manure with more stable P fractions (such as poultry litter) (He et al., 2008; Waldrip et al., 2015).

Cattle production systems generally capture and store manure runoff in settling basins, retention ponds, and lagoons. Research has shown that these storage systems are effective at reducing runoff potential (He and Waldrip, 2015; Jannin and Cortus, 2019). Hong et al. (2018) characterized P in the sediment sludge and crust formed on retention ponds of a dairy manure management system. In this study, P in pond sludge and crust were characterized with a sequential fractionation method, as previously described. Pond sludge and crust contained significant amounts of labile P (H_2O-P and $NaHCO_3$–P). There were lower H_2O-P levels in sludge and crust than found in raw manure, implying that the use of sludge and crust instead of raw manure as a soil amendment could reduce the possibility of P loss by surface runoff and leaching, as well as providing soluble P for immediate plant utilization.

Enzymatic hydrolysis and solution ^{31}P NMR spectroscopic analysis have identified major P_o species in cattle manure as general monoesters, general diesters, specific diesters DNA-P, specific monoester phytate, or phytate-like P (Cade-Menun, 2011; He and Honeycutt, 2011). There were some differences in P_o distribution patterns measured by the two methods (enzyme hydrolysis vs. NMR), but the overall variation was slight. In dairy manure with 6.88 g total P kg^{-1}, He et al. (2007) reported the content of monoester-P was 1.49 g P kg^{-1}, phytate was 0.41 g P kg^{-1}, and DNA-P was 0.09 g P kg^{-1} when measured with ^{31}P NMR spectroscopy. At the same time, enzyme hydrolysis showed lower amounts of monoester-P (0.61 g P kg^{-1}), but increased concentrations of phytate (0.69 g P kg^{-1}) and DNA-P (0.43 g P kg^{-1}).

Phytate (i.e., inositol hexakisphosphate) is present in large quantities in the manure of monogastric animals, such as poultry, which can cause issues with legacy P due to binding of the 6 phosphate groups with soil OM and metal components (He et al., 2006; He et al., 2007). However, the digestive tract of ruminants contains microorganisms that can effectively degrade phytate before it is excreted; thus, phytate and total P_o contents of cattle manure are much lower than swine, poultry, or other monogastric animals

(He et al., 2007; Pagliari, 2014). Pagliari and Laboski (2012) reported that phytate (0.47g P kg^{-1}) was the major P$_o$ form in beef manure (average total P of 8.5 g P kg^{-1}), followed by monoester-P (0.3 g P kg^{-1}), and DNA-P (0.18 g P kg^{-1}). In contrast, Turner and Leytem (2004) used ^{31}P NMR to identify the forms of P$_o$ in beef manure (total P of 3.9 g P kg^{-1}), where monoester-P (1.1 g P kg^{-1}) was the major form of P$_o$, with small amounts of DNA-P (0.11 g P kg^{-1}). For dairy manure (total P of 3.5 g P kg^{-1}), He and Honeycutt (2001) reported that most of the hydrolyzable P$_o$ was DNA-P (3.9% of total P), followed by phytate (3.5% of total P), organic pyrophosphate (3.4% of total P), and small amounts of monoester-P (1.4% of total P). In studying 13 dairy manure samples (average total P was 9.1 g P kg^{-1}), He et al. (2004) reported the phytate content was 5.0% of total P. Pagliari and Laboski (2012) reported similar total P amounts and percentages of monoester-P (3.4% of total P), phytate (3.0% of total P), and DNA-P (1.2% of total P) in 18 dairy manure samples (total P average 8.6 g P kg^{-1}). Li et al. (2014) reported that the P$_o$ distribution was phytate (9%), other monesters (4%) and diesters (1%) in a dairy manure with 7.1 g total P kg^{-1}.

The different P forms impact soils and water bodies differently (Pagliari, 2014). Inorganic P is readily available in any ecosystem, soil or water; while P$_o$ must first be hydrolyzed before it can become available for use (Pagliari and Laboski, 2012, 2013). Manure P$_i$ has been reported to behave like inorganic fertilizer P when manure is added to soil and increases available soil P rather quickly soon after manure is applied to soils (Pagliari and Laboski, 2012, 2013). Organic P behavior is more complex. Many of the organic compounds containing P in manure have high affinity for soil clay particles and they will adsorb to soil clay particles when the conditions are right (Pagliari and Laboski, 2013 and 2014). When P$_o$ binds to soil particles they become more resistant to hydrolysis and therefore are not bioavailable (Celi et al., 1999; Karathanasis and Shumaker, 2009). Therefore, understanding the forms of P in manure can help develop best management practices, which maximizes the beneficial reuse of manure as a nutrient source in crop production.

Conclusions

The ever-increasing demand to produce food for the exponentially increasing global population requires improved waste management strategies. Although some approaches to reduce nutrients in animal manure are being investigated, there is still a very large amount of manure produced globally. The amount of nutrients in animal manure varies greatly with breed, body size, diet, performance, and other variables. However, knowing the total amount of manure being produced in an operation along with the nutrients present in the manure is key in optimizing manure reuse. Greenhouse gas emissions from livestock operations, primarily beef and dairy cattle, are also a significant concern, and a substantial amount of the GHG emissions from agriculture come from cattle production. It is possible that manure will become a primary resource in the near future. Strategies that investigate the use of manure for biogas production, nutrient removal with further development of chemical fertilizers such as struvite, and the production of manure for bacterial biomass production are techniques that are currently being investigated and are promising.

References

Aarons, S.R., C.J. Gourley, J.M. Powell, and M.C. Hannah. 2017. Estimating nitrogen excretion and deposition by lactating cows in grazed dairy systems. Soil Res. 55:489–499. doi:10.1071/SR17033

Agomoh, I., F. Zvomuya, X. Hao, O.O. Akinremi, and T.A. McAllister. 2018. Nitrogen mineralization in Chernozemic soils amended with manure from cattle fed dried distillers grains with solubles. Soil Sci. Soc. Am. J. 82:167–175. doi:10.2136/sssaj2017.08.0282

Aguerre, M.J., M.A. Wattiaux, J. Powell, G.A. Broderick, and C. Arndt. 2011. Effect of forage-to-concentrate ratio in dairy cow diets on emission of methane, carbon dioxide, and ammonia, lactation performance, and manure excretion. J. Dairy Sci. 94:3081–3093. doi:10.3168/jds.2010-4011

Ajiboye, B., O.O. Akinremi, Y. Hu, and D.N. Flaten. 2007. Phosphorus speciation of sequential extracts of organic amendments using nuclear magnetic resonance and X-ray absorption near-edge structure spectroscopies. J. Environ. Qual. 36:1563–1576. doi:10.2134/jeq2006.0541

Akinremi, O.O., B. Ajiboye, and Z. He. 2011. Metal speciation of phosphorus derived from solid state spectroscopic analysis. In: Z. He, editor, Environmental chemistry of animal manure. Nova Science Publishers, NY. p. 301–324.

Barros, T., J. Powell, M. Danes, M. Aguerre, and M. Wattiaux. 2017. Relative partitioning of N from alfalfa silage, corn silage, corn grain and soybean meal into milk, urine, and feces, using stable 15N isotope. Anim. Feed Sci. Technol. 229:91–96. doi:10.1016/j.anifeedsci.2017.05.009

Bernier, J.N., M. Undi, K.H. Ominski, G. Donohoe, M. Tenuta, D. Flaten, J.C. Plaizier, and K.M. Wittenberg. 2014. Nitrogen and phosphorus utilization and excretion by beef cows fed a low quality forage diet supplemented with dried distillers grains with solubles under thermal neutral and prolonged cold conditions. Anim. Feed Sci. Technol. 193:9–20. doi:10.1016/j.anifeedsci.2014.03.010

Biagini, D., and C. Lazzaroni. 2018. Eutrophication risk arising from intensive dairy cattle rearing systems and assessment of the potential effect of mitigation strategies. Agric. Ecosyst. Environ. 266:76–83. doi:10.1016/j.agee.2018.07.026

Bonifacio, H.F., C.A. Rotz, A.B. Leytem, H.M. Waldrip, and R.W. Todd. 2015. Process-based modeling of ammonia and nitrous oxide emissions from open-lot beef and dairy facilities. Trans. ASABE 58:827–846.

Broderick, G. 2003. Effects of varying dietary protein and energy levels on the production of lactating dairy cows. Journal of dairy science 86: 1370-1381.

Cade-Menun, B.J. 2011. Characterizing phosphorus in animal waste with solution 31P NMR spectroscopy. In: Z. He, editor, Environmental chemistry of animal manure. Nova Science Publishers, New York, N.Y. p. 275–299.

Cade-Menun, B.J., Z. He, and Z. Dou. 2015. Comparison of phosphorus forms in three extracts of dairy feces by solution 31P NMR analysis. Commun. Soil Sci. Plant Anal. 46:1698–1712. doi:10.1080/00103624.2015.1047512

Castillo, A., E. Kebreab, D. Beever, J. Barbi, J. Sutton, H. Kirby, and J. France. 2001. The effect of protein supplementation on nitrogen utilization in lactating dairy cows fed grass silage diets. J. Anim. Sci. 79:247–253. doi:10.2527/2001.791247x

Celi, L., S. Lamacchia, F.A. Marsan and E. Barberis. 1999. Interaction of inositol hexaphosphate on clays: adsorption and charging phenomena. Soil Science 164: 574.

Cole, N., R. Clark, R. Todd, C. Richardson, A. Gueye, L. Greene, et al. 2005. Influence of dietary crude protein concentration and source on potential ammonia emissions from beef cattle manure. Journal of Animal Science 83: 722-731.

Cole, N.A., and R.W. Todd. 2009. Nitrogen and phosphorus balance of beef cattle feedyards, p. 17-24 Proceedings of the Texas animal manure management issues conference. Round Rock, TX.

Cole, N., R. Clark, R. Todd, C. Richardson, A. Gueye, L. Greene, and K. Bride. 2005. Influence of dietary crude protein concentration and source on potential ammonia emissions from beef cattle manure. J. Anim. Sci. 83:722–731. doi:10.2527/2005.833722x

Cole, N., P. Defoor, M. Galyean, G. Duff, and J. Gleghorn. 2006. Effects of phase-feeding of crude protein on performance, carcass characteristics, serum urea nitrogen concentrations, and manure nitrogen of finishing beef steers. science. Anim. Sci. 84:3421–3432. doi:10.2527/jas.2006-150

Dagna, N.E., and A.P. Mallarino. 2014. Beef cattle manure survey and assessment of crop availability of phosphorus by soil testing. Soil Sci. Soc. Am. J. 78:1035–1050. doi:10.2136/sssaj2013.06.0223

Dou, Z., J.D. Toth, D.T. Galligan, C.F. Ramberg, Jr., and J.D. Ferguson. 2000. Laboratory procedures for characterizing manure phosphorus. J. Environ. Qual. 29:508–514. doi:10.2134/jeq2000.00472425002900020019x

Erickson, G.E., V.R. Bremer, T.J. Klopfenstein, A. Stalker, and R. Rasby. 2007. Utilization of corn co-products in the beef industry, 2nd ed. University of Nebraska – Lincoln, Lincoln, NE. http://beef.unl.edu/byprodfeeds/07CORN-048_BeefCoProducts.pdf (Accessed 23 May 2019).

Feng, X., E. Ronk, M. Hanigan, K. Knowlton, H. Schramm and M. McCann. 2015. Effect of dietary phosphorus on intestinal phosphorus absorption in growing Holstein steers. Journal of dairy science 98: 3410-3416.

Gandra, J.R., J. Freitas Jr, R.V. Barletta, M. Maturana Filho, L. Gimenes, F. Vilela, P.S. Baruselli, and F.P. Rennó. 2011. Productive performance, nutrient digestion and metabolism of Holstein (Bos taurus) and Nellore (Bos taurus indicus) cattle and Mediterranean Buffaloes (Bubalis bubalis) fed with corn-silage based diets. Livest. Sci. 140:283-291. doi:10.1016/j.livsci.2011.04.005

Geisert, B.G., G.E. Erickson, T.J. Klopfenstein, C.N. Macken, M.K. Luebbe, and J.C. MacDonald. 2010. Phosphorus requirement and excretion of finishing beef cattle fed different concentrations of phosphorus. J. Anim. Sci. 88:2393-2402. doi:10.2527/jas.2008-1435

Gourley, C.J., W.J. Dougherty, D.M. Weaver, S.R. Aarons, I.M. Awty, D.M. Gibson, M.C. Hannah, A.P. Smith, and K.I. Peverill. 2012. Farm-scale nitrogen, phosphorus, potassium and sulfur balances and use efficiencies on Australian dairy farms. Anim. Prod. Sci. 52:929-944. doi:10.1071/AN11337

Groff, E., and Z. Wu. 2005. Milk production and nitrogen excretion of dairy cows fed different amounts of protein and varying proportions of alfalfa and corn silage. J. Dairy Sci. 88:3619-3632. doi:10.3168/jds.S0022-0302(05)73047-2

Hacking, A., M.T. Dervish, and W.R. Rosser. 1977. Available amino acid content and microbiological condition of dried poultry litter. Br. Poult. Sci. 18:443-448. doi:10.1080/00071667708416384

Hassanat, F., R. Gervais, C. Julien, D.I. Massé, A. Lettat, P.Y. Chouinard, H.V. Petit, and C. Benchaar. 2013. Replacing alfalfa silage with corn silage in dairy cow diets: Effects on enteric methane production, ruminal fermentation, digestion, N balance, and milk production. J. Dairy Sci. 96:4553-4567. doi:10.3168/jds.2012-6480

He, Z., editor. 2012. Applied research of animal manure: Challenges and opportunities beyond the adverse environmental concerns. Nova Science Publishers, New York. p. 1-325.

He, Z. 2019. Organic animal farming and comparative studies of conventional and organic manures. In: H.M. Waldrip, P.H. Pagliari, and Z. He, editors, Animal manure: Production, characteristics, environmental concerns and management. ASA Spec. Publ. 67. ASA and SSSA, Madison, WI. p. 139-156.

He, Z., and C.W. Honeycutt. 2001. Enzymatic characterization of organic phosphorus in animal manure. J. Environ. Qual. 30:1685-1692. doi:10.2134/jeq2001.3051685x

He, Z., and D.C. Olk. 2011. Manure amino compounds and their bioavailability. In: Z. He, editor, Environmental chemistry of animal manure. Nova Science Publishers, Inc., N.Y. p. 179-199.

He, Z., and C.W. Honeycutt. 2011. Enzymatic hydrolysis of organic phosphorus. In: Z. He, editor, Environmental chemistry of animal manure. Nova Science Publishers, N.Y. p. 253-274.

He, Z., and H. Zhang, editors. 2014. Applied manure and nutrient chemistry for sustainable agriculture and environment. Springer, Amsterdam, the Netherlands. p. 1-379. doi:10.1007/978-94-017-8807-6

He, Z., and H.W. Waldrip. 2015. Composition of whole and water-extractable organic matter of cattle manure affected by management practices. In: Z. He and F. Wu, editors, Labile organic matter-Chemical compositions, function, and significance in soil and the environment. SSSA Spec. Publ. 62. Soil Science Society of America, Madison, WI. p. 41-60. doi:10.2136/sssaspecpub62.2014.0034

He, Z., T.S. Griffin, and C.W. Honeycutt. 2004. Phosphorus distribution in dairy manures. J. Environ. Qual. 33:1528-1534. doi:10.2134/jeq2004.1528

He, Z., P.H. Pagliari, and H.M. Waldrip. 2016. Applied and environmental chemistry of animal manure: A review. Pedosphere 26:779-816. doi:10.1016/S1002-0160(15)60087-X

He, Z., Z.N. Senwo, R.N. Mankolo, and C.W. Honeycutt. 2006. Phosphorus fractions in poultry litter characterized by sequential fractionation coupled with phosphatase hydrolysis. J. Food Agric. Environ. 4(1):304-312.

He, Z., C.W. Honeycutt, B.J. Cade-Menun, Z.N. Senwo, and I.A. Tazisong. 2008. Phosphorus in poultry litter and soil: Enzymatic and nuclear magnetic resonance characterization. Soil Sci. Soc. Am. J. 72:1425-1433. doi:10.2136/sssaj2007.0407

He, Z., Z.N. Senwo, H. Zou, I.A. Tazisong, and D.A. Martens. 2014. Amino compounds in poultry litter, litter-amended pasture soils and grass shoots. Pedosphere 24:178-185. doi:10.1016/S1002-0160(14)60004-7

He, Z., B.J. Cade-Menun, G.S. Toor, A. Fortuna, C.W. Honeycutt, and J.T. Sims. 2007. Comparison of phosphorus forms in wet and dried animal manures by solution phosphorus-31 nuclear magnetic resonance spectroscopy and enzymatic hydrolysis. J. Environ. Qual. 36:1086-1095. doi:10.2134/jeq2006.0549

He, Z., C.W. Honeycutt, T.S. Griffin, B.J. Cade-Menun, P.J. Pellechia, and Z. Dou. 2009. Phosphorus forms in conventional and organic dairy manure identified by solution and solid state P-31 NMR spectroscopy. J. Environ. Qual. 38:1909–1918. doi:10.2134/jeq2008.0445

He, Z., M. Zhang, X. Cao, Y. Li, J. Mao, and H.M. Waldrip. 2015. Potential traceable markers of organic matter in organic and conventional dairy manure using ultraviolet-visible and solid-state 13C nuclear magnetic resonance spectroscopy. Organic Agriculture 5:113–122. doi:10.1007/s13165-014-0092-0

Honeycutt, C.W., J.F. Hunt, T.S. Griffin, Z. He, and R.P. Larkin. 2011. Determinants and processes of manure nitrogen availability. In: Z. He, editor, Environmental chemistry of animal manure. Nova Science Publishers, NY. p. 201–224.

Hong, W.T., D. Hagare, M. Khan, and J. Fyfe. 2018. Phosphorus characterisation of sludge and crust produced by stabilisation ponds in a dairy manure management system. Water Air Soil Pollut. 229:276. doi:10.1007/s11270-018-3912-0

Hünerberg, M., S. McGinn, K. Beauchemin, E. Okine, O. Harstad, and T. McAllister. 2013a. Effect of dried distillers' grains with solubles on enteric methane emissions and nitrogen excretion from finishing beef cattle. Can. J. Anim. Sci. 93:373–385. doi:10.4141/cjas2012-151

Hünerberg, M., S. McGinn, K. Beauchemin, E. Okine, O. Harstad, and T. McAllister. 2013b. Effect of dried distillers grains plus solubles on enteric methane emissions and nitrogen excretion from growing beef cattle. J. Anim. Sci. 91:2846–2857. doi:10.2527/jas.2012-5564

Jannin, K., and E. Cortus. 2019. Common animal production systems and manure storage methods, p. In: press. In: H.M. Waldrip, et al., editors, Aminal manure: Production, characteristics, environmental concerns and management, Vol. ASA Special Pub. 67. ASA, Madison, WI.

Jardstedt, M., A. Hessle, P. Nørgaard, W. Richardt, and E. Nadeau. 2017. Feed intake and urinary excretion of nitrogen and purine derivatives in pregnant suckler cows fed alternative roughage-based diets. Livest. Sci. 202:82–88. doi:10.1016/j.livsci.2017.05.026

Karathanasis, A.D. and P.D. Shumaker. 2009. Preferential sorption and desorption of organic and inorganic phospates by soil hydroxyinterlayered minerals. Soil science 174: 417-423.

Kleinman, P.J.A., S. Spiegal, J. Liu, M. Holly, C. Church, and J. Ramirez-Avila. 2019. Managing animal manure to minimize phosphorus losses from land to water. In: H.M. Waldrip, et al., editors, Animal manure: Production, characteristics, environmental concerns and management, Vol. ASA Special Pub. 67. ASA, Madison, WI.

Knowlton, K., and J. Herbein. 2002. Phosphorus partitioning during early lactation in dairy cows fed diets varying in phosphorus content. J. Dairy Sci. 85:1227–1236. doi:10.3168/jds.S0022-0302(02)74186-6

Knowlton, K., J. Herbein, M. Meister-Weisbarth, and W. Wark. 2001. Nitrogen and phosphorus partitioning in lactating Holstein cows fed different sources of dietary protein and phosphorus. J. Dairy Sci. 84:1210–1217. doi:10.3168/jds.S0022-0302(01)74582-1

Kobayashi, R., A. Yamada, H. Hirooka, Y. Tabata, J. Zhang, K. Nonaka, M. Kamo, K. Hayasaka, Y. Aoki, and H. Kawamoto. 2010. Changes in the cycling of nitrogen, phosphorus, and potassium in a dairy farming system. Nutr. Cycling Agroecosyst. 87:295–306.

Koenig, K.M., and K.A. Beauchemin. 2013. Nitrogen metabolism and route of excretion in beef feedlot cattle fed barley-based backgrounding diets varying in protein concentration and rumen degradability. J. Anim. Sci. 91:2295–2309. doi:10.2527/jas.2012-5652

Koenig, K.M., and K.A. Beauchemin. 2018. Effect of feeding condensed tannins in high protein finishing diets containing corn distillers grains on ruminal fermentation, nutrient digestibility, and route of nitrogen excretion in beef cattle. J. Anim. Sci. 96:4398–4413. doi:10.1093/jas/sky273

Külling, D., H. Menzi, T. Kröber, A. Neftel, F. Sutter, P. Lischer, and M. Kreuzer. 2001. Emissions of ammonia, nitrous oxide and methane from different types of dairy manure during storage as affected by dietary protein content. J. Agric. Sci. 137:235–250. doi:10.1017/S0021859601001186

Külling, D., H. Menzi, F. Sutter, P. Lischer, and M. Kreuzer. 2003. Ammonia, nitrous oxide and methane emissions from differently stored dairy manure derived from grass-and hay-based rations. Nutr. Cycling Agroecosyst. 65:13–22. doi:10.1023/A:1021857122265

Lean, J., M.B.d. Ondarza, C.J. Sniffen, J.E.P. Santos, and K.E. Griswold. 2018. Meta-analysis to predict the effects of metabolizable amino acids on dairy cattle performance. J. Dairy Sci. 101:340–364. doi:10.3168/jds.2016-12493

Lee, M., C. Jung, E. Shevliakova, S. Malyshev, H. Han, S. Kim, and P.R. Jaffé. 2018. Control of nitrogen exports from river basins to the coastal ocean: Evaluation of basin management strategies for reducing coastal hypoxia. J. Geophys. Res.: Biogeosciences 123(10):3111–3123.

Li, G., H. Li, P.A. Leffelaar, J. Shen, and F. Zhang. 2014. Characterization of phosphorus in animal manures collected from three (dairy, swine, and broiler) farms in China. PLoS One 9:e102698. doi:10.1371/journal.pone.0102698

Liang, B.C., E.C. Gregorich, M. Schnitzer, and H. Schulten. 1996. Characterization of water extracts of two manures and their adsorption on soils. Soil Sci. Soc. Am. J. 60:1758–1763. doi:10.2136/sssaj1996.03615995006000060021x

Liu, J., J.T. Spargo, P.J.A. Kleinman, R. Meinen, J.P.A. Moore, and D.B. Beegle. 2018. Water extractable phosphorus in livestock manures and composts: Quantities, characteristics and temporal changes. J. Environ. Qual. 47:471–479. doi:10.2134/jeq2017.12.0467

Long, C.M., R.L. Muenich, M.M. Kalcic and D. Scavia. 2018. Use of manure nutrients from concentrated animal feeding operations. Journal of Great Lakes Research 44: 245-252.

Luebbe, M.K., J.M. Patterson, K.H. Jenkins, E.K. Buttrey, T.C. Davis, B.E. Clark, F.T. McCollum, N.A. Cole, and J.C. MacDonald. 2012. Wet distillers grains plus solubles concentration in steam-flaked-corn-based diets: Effects on feedlot cattle performance, carcass characteristics, nutrient digestibility, and ruminal fermentation characteristics. J. Anim. Sci. 90:1589–1602. doi:10.2527/jas.2011-4567

Ma, H., G. Pei, R. Gao, and Y. Yin. 2017. Mineralization of amino acids and its signs in nitrogen cycling of forest soil. Acta Ecol. Sin. 37:60–63. doi:10.1016/j.chnaes.2016.09.001

Miller, D.N., and V.H. Varel. 2011. Origins and identities of key manore ordor components. In: Z. He, editor, Environmental chemistry of animal manure. Nova Science Publishers, N.Y. p. 153–177.

Morris, D.L., S.H. Kim, and C. Lee. 2018. Effects of corn feeding reduced-fat distillers grains with or without monensin on nitrogen, phosphorus, and sulfur utilization and excretion in dairy cows. J. Dairy Sci. 101:7106–7116. doi:10.3168/jds.2018-14528

Motew, M., E.G. Booth, S.R. Carpenter, X. Chen, and C.J. Kucharik. 2018. The synergistic effect of manure supply and extreme precipitation on surface water quality. Environ. Res. Lett. 13:044016. doi:10.1088/1748-9326/aaade6 [erratum: 13(7): 044016].

Nair, J., G.B. Penner, P. Yu, H.A. Lardner, T.A. McAllister, D. Damiran, and J.J. McKinnon. 2016. Evaluation of canola meal derived from Brassica juncea and Brassica napus on rumen fermentation and nutrient digestibility by feedlot heifers fed finishing diets. Can. J. Anim. Sci. 96:342–353. doi:10.1139/cjas-2015-0184

NASEM. 2001. Nutrient requirements of dairy cattle. Seventh Revised ed. Subcommittee on Dairy Cattle Nutrition, Committee on Animal Nutrition, Board on Agriculture and Natural Resources, National Research Council. Washington, D.C. p. 43-104. https://profsite.um.ac.ir/~kalidari/software/NRC/HELP/NRC%202001.pdf (Accessed 23 May 2019).

Ndegwa, P.M., A.N. Hristov, and J.A. Ogejo. 2011. Ammonia emission from animal manure: Mechanisms and mitigation techniques. In: Z. He, editor, Environmental chemistry of animal manure. Nova Science Publishers, N.Y. p. 107–151.

Nennich, T., J.H. Harrison, L.M. Vanwieringen, D. Meyer, A. Heinrichs, W.P. Weiss, N.R. St. Pierre, et al. 2005. Prediction of manure and nutrient excretion from dairy cattle. J. Dairy Sci. 88:3721–3733. doi:10.3168/jds.S0022-0302(05)73058-7

NRC. 1996. Nutrient requirements of beef cattle. 7th rev. ed. Natl. Acad. Press, Washington, D.C.

Ouellet, D.R., M. Demers, G. Zuur, E. Lobley, J.R. Seoane, J.V. Nolan, and H. Lapierre. 2002. Effect of dietary fiber on endogenous nitrogen flows in lactating dairy cows. J. Dairy Sci. 85:3013–3025. doi:10.3168/jds.S0022-0302(02)74387-7

Pagliari, P.H. 2014. Variety and solubility of phosphorus forms in animal manure and their effects on soil test phosphorus. In: Z. He and H. Zhang, editors, Applied manure and nutrient chemistry for sustainable agriculture and environment. Springer, Amsterdam, the Netherland. p. 141–161. doi:10.1007/978-94-017-8807-6_8

Pagliari, P.H., and C.A.M. Laboski. 2012. Investigation of the inorganic and organic phosphorus forms in animal manure. J. Environ. Qual. 41:901–910. doi:10.2134/jeq2011.0451

Pagliari, P.H., and C.A.M. Laboski. 2013. Dairy manure treatment effects on manure phosphorus fractionation and changes in soil test phosphorus. Biol. Fertil. Soils 49:987–999. doi:10.1007/s00374-013-0798-2

Parker, D.B., H.M. Waldrip, K.D. Casey, R.W. Todd, W.M. Willis and K. Webb. 2017. Temporal nitrous oxide emissions from beef cattle feedlot manure after a simulated rainfall event. Journal of environmental quality 46: 733-740.

Perez, P.G., R. Zhang, X. Wang, J. Ye, and D. Huang. 2015. Characterization of the amino acid composition of soils under organic and conventional management after addition of different fertilizers. J. Soils Sediments 15:890–901. doi:10.1007/s11368-014-1049-3

Powell, J., and C. Rotz. 2015. Measures of nitrogen use efficiency and nitrogen loss from dairy production systems. J. Environ. Qual. 44:336–344. doi:10.2134/jeq2014.07.0299

Powell, J.M., G.A. Broderick, J.H. Grabber, and U.C. Hymes-Fecht. 2009. Effects of forage protein-binding polyphenols on chemistry of dairy excreta. J. Dairy Sci. 92:1765–1769. doi:10.3168/jds.2008-1738

Powell, J., M. Aguerre and M. Wattiaux. 2011. Dietary crude protein and tannin impact dairy manure chemistry and ammonia emissions from incubated soils. Journal of environmental quality 40: 1767-1774.

Powell, J.M., T. Barros, M. Danes, M. Aguerre, M. Wattiaux, and K. Reed. 2017. Nitrogen use efficiencies to grow, feed, and recycle manure from the major diet components fed to dairy cows in the USA. Agric. Ecosyst. Environ. 239:274–282. doi:10.1016/j.agee.2017.01.023

Reid, M., M. O'Donovan, C. Elliott, J. Bailey, C. Watson, S. Lalor, et al. 2015. The effect of dietary crude protein and phosphorus on grass-fed dairy cow production, nutrient status, and milk heat stability. Journal of dairy science 98: 517-531.

Rotz, C., D. Buckmaster, and J. Comerford. 2005. A beef herd model for simulating feed intake, animal performance, and manure excretion in farm systems. J. Anim. Sci. 83:231–242. doi:10.2527/2005.831231x

Schuba, J., K.-H. Südekum, E. Pfeffer, and A. Jayanegara. 2017. Excretion of faecal, urinary urea and urinary non-urea nitrogen by four ruminant species as influenced by dietary nitrogen intake: A meta-analysis. Livest. Sci. 198:82–88. doi:10.1016/j.livsci.2017.01.017

Selbie, D.R., L.E. Buckthought, and M.A. Shepherd. 2015. The challenge of the urine patch for managing nitrogen in grazed pasture systems. Advances in Agronomy. Elsevier, New York. p. 229-292.

Singh, J., J. Hundal, A. Sharma, U. Singh, A. Sethi and P. Singh. 2018. Phosphorus Nutrition in Dairy Animals: A Review. Int. J. Curr. Microbiol. App. Sci 7: 3518-3530.

Souza, V.C., P. Malafaia, B.R. Vieira, Y.T. Granja-Salcedo, and T.T. Berchielli. 2016. Phosphorus supplementation with or without other minerals, ionophore and antibiotic did not affect performance of Nellore bulls receiving high-grain diets, but increased phosphorus excretion and dietary costs. Anim. Prod. Sci. 58:871–877. doi:10.1071/AN16420

Spek, J.W., J. Dijkstra, G.V. Duinkerken, W.H. Hendriks, and A. Bannink. 2013. Prediction of urinary nitrogen and urinary urea nitrogen excretion by lactating dairy cattle in northwestern Europe and North America: A meta-analysis. J. Dairy Sci. 96:4310–4322. doi:10.3168/jds.2012-6265

Spiehs, M.J., and V.H. Varel. 2009. Nutrient excretion and odorant production in manure from cattle fed corn wet distillers grains with solubles. J. Anim. Sci. 87:2977–2984. doi:10.2527/jas.2008-1584

Tomlinson, P., W.J. Powers, H.H. Van Horn, R.A. Nordstedt, and C.J. Wilcox. 1996. Dietary protein effects on nitrogen excretion and manure characteristics of lactating cows. Trans. ASAE 39: 1441-1448. doi:10.13031/2013.27637

Toth, J.D., Z. Dou, and Z. He. 2011. Solubility of manure phosphorus characterized by selective and sequential extractions. In: Z. He, editor, Environmental chemistry of animal manure. Nova Science Publishers, NY. p. 227–251.

Turner, B.L., and A.B. Leytem. 2004. Phosphorus compounds in sequential extracts of animal manure: Chemical speciation and a novel fractionation procedure. Environ. Sci. Technol. 38:6101–6108. doi:10.1021/es0493042

USDA-ERS. 2019. Dairy data. United States Department of Agriculture, Economic Research Service, Washington, D.C. https://www.ers.usda.gov/data-products/dairy-data/dairy-data/ (Accessed 23 May 2019).

USDA–NASS. 2014. Crop production historical track records. United States Department of Agriculture, National Agriculture Statistics Service, Washington, D.C. http://www.nass.usda.gov/Publications/Todays_Reports/reports/croptr14.pdf (Accessed 23 May 2019).

USDA-NASS. 2018. Cattle. United States Department of Agriculture, National Agricultural Statistics Service, Washington, D.C. https://www.nass.usda.gov/Publications/Todays_Reports/reports/catl0718.pdf (Accessed 23 May 2019).

Vasconcelos, J.T., N.A. Cole, K.W. McBride, A. Gueye, M.L. Galyean, C.R. Richardson, and L.W. Greene. 2009. Effects of dietary crude protein and supplemental urea levels on nitrogen and phosphorus utilization by feedlot cattle. J. Anim. Sci. 87:1174–1183.

Waldrip, H.M., R.W. Todd, and N.A. Cole. 2013. Prediction of nitrogen excretion by beef cattle: A meta-analysis. J. Anim. Sci. 91:4290–4302. doi:10.2527/jas.2012-5818

Waldrip, H.M., P.H. Pagliari, Z. He, R.D. Harmel, N.A. Cole, and M. Zhang. 2015. Legacy phosphorus in calcareous soils: Effects of long-term poultry litter application. Soil Sci. Soc. Am. J. 79:1601–1614. doi:10.2136/sssaj2015.03.0090

Wang, J., D. Wang, C. Li, T.R. Seastedt, C. Liang, L. Wang, W. Sun, M. Liang, and Y. Li. 2018. Feces nitrogen release induced by different large herbivores in a dry grassland. Ecol. Appl. 28:201–211. doi:10.1002/eap.1640

Wu, Z. 2005. Utilization of phosphorus in lactating cows fed varying amounts of phosphorus and sources of fiber. J. Dairy Sci. 88:2850–2859. doi:10.3168/jds.S0022-0302(05)72966-0

Wu, Z., L. Satter, A. Blohowiak, R. Stauffacher, and J. Wilson. 2001. Milk production, estimated phosphorus excretion, and bone characteristics of dairy cows fed different amounts of phosphorus for two or three Years. J. Dairy Sci. 84:1738–1748. doi:10.3168/jds.S0022-0302(01)74609-7

Wu, Z., L. Satter, and R. Sojo. 2000. Milk production, reproductive performance, and fecal excretion of phosphorus by dairy cows fed three amounts of phosphorus. J. Dairy Sci. 83:1028–1041. doi:10.3168/jds.S0022-0302(00)74967-8

Wu, Z., S. Tallam, V. Ishler, and D. Archibald. 2003. Utilization of phosphorus in lactating cows fed varying amounts of phosphorus and forage. J. Dairy Sci. 86:3300–3308. doi:10.3168/jds.S0022-0302(03)73931-9

Yan, Z.-J., C. Shuo, M.-F. Wang, M.-F. Wei, and C. Qing. 2015. Characteristics and availability of different forms of phosphorus in animal manures. Nongye Ziyuan Yu Huanjing Xuebao 32:31-39.

Zanetti, D., S.C. Valadares Filho, L.F. Prados, E. Detmann, M.V.C. Pacheco, L.A. Godoi, L.N. Rennó, and T.E. Engle. 2017. Impacts of reduction of phosphorus in finishing diets for Holstein× Zebu steers. Livest. Sci. 198:45–51. doi:10.1016/j.livsci.2017.02.001

Zenobi, M.G., H.A. Lardner, P.G. Jefferson, and J.J. McKinnon. 2015. Effect of feeding strategically blended feed pellets on rumen fermentation and nutrient digestion. Can. J. Anim. Sci. 95:243–254. doi:10.4141/cjas-2014-131

Nutrient Characteristics of Poultry Manure and Litter

A.J. Ashworth*, J.P. Chastain, and P.A. Moore, Jr.

Abstract

Animal manures have been used to supply nutrients for plant growth and production since the inception of agriculture. Concurrent with chemical fertilizer cost increases, farmers are developing a renewed interest in poultry (layers, pullets, broilers, turkeys) litter for fertilizer in pasturelands and croplands, as well as organic crop production systems. Therefore, given the increasing acreage of organic or local usage of fertility sources for forage and crop production, and the increased demand for protein-based diets, there is a growing need for improved poultry manure and litter-based nutrient management for sustained water quality and plant and animal productivity. This chapter aims to provide an overview of poultry manure and litter characteristics in U.S. production systems so that producers and scientists can develop nutrient management plans for minimizing excess nutrients in groundwater systems and optimize targeted nutrient applications.

Overview and Scale of Poultry Operations in the United States

The Census of Agriculture reported there were 233,770 poultry farms in the United States in 2015 (NASS, 2015). In 2016, the U.S. poultry industry produced 8.78 billion broilers, 101 billion eggs, and 244 million turkeys (Table 1). The combined value of production from broilers, eggs, turkeys, and sales from chickens in 2016 was $38.7 billion, down 20% from $48.1 billion in 2015 (Table 1; NASS, 2016). Of that combined total, 67% was from broilers, 17% from eggs, and 16% from turkeys. In addition to these trends, over the last 50 yr, the U.S. poultry industry has evolved to a vertically integrated and highly efficient industry, which is large in part due to improved disease management (e.g., in 1965, 112 rearing days would produce a 1.13 kg chicken with a 4.7 feed conversion ratio [weight/feed intake]; whereas current rearing periods are 42 d for a 2.7 kg chicken with a feed conversion ratio of 1.8) (National Chicken Council, 2015). Concurrent with production efficiency increases is consumption, as the average American now consumes 41 kg of broiler meat per year (National Chicken Council, 2018).

Amount of Litter and Manure Produced During a Poultry Lifecycle

Overall, the amount of poultry manure produced, or excreted, can be approximately estimated by the diet, whereas the amount of total poultry litter (excreta plus bedding and dead birds) is more difficult to ascertain as this varies based on management practices. However, if we take broiler chickens for example, they

A.J. Ashworth and P.A. Moore, Jr., USDA-ARS, Poultry Production and Product Safety Research Unit, 1260 W. Maple St. Fayetteville, AR; J.P. Chastain, Clemson University, Clemson, SC 29657. *Corresponding author (Amanda.ashworth@ars.usda.gov)

USDA is an equal opportunity provider and employer. Mention of trade names or commercial products in this article is solely for the purpose of providing specific information and does not imply recommendation or endorsement by the USDA.

doi:10.2134/asaspecpub67.c5

Animal Manure: Production, Characteristics, Environmental Concerns and Management. ASA Special Publication 67. Heidi M. Waldrip, Paulo H. Pagliari, and Zhongqi He, editors. © 2019. ASA and SSSA, 5585 Guilford Rd., Madison, WI 53711, USA.

Table 1. Broiler and turkey production and value in 2016 for the United States (NASS, 2016).

	Broiler Production and Value		
State[†]	Number produced	Pounds produced	Value of production
	(1000 head)	(1000 pounds)	(1000 dollars)
Alabama	1,070,100	5,992,600	2,864,463
Arkansas	1,009,400	6,561,100	3,136,206
Delaware	252,500	1,843,300	881,097
Florida	63,200	366,600	175,235
Georgia	1,367,100	8,065,900	3,855,500
Kentucky	300,300	1831,800	875,600
Maryland	303,500	1851,400	884,969
Minnesota	48,400	295,200	141,106
Mississippi	7,39,400	4,658,200	2,226,620
Missouri	2,93,000	1,435,700	686,265
North Carolina	8,18,700	6,467,700	3,091,561
Ohio	88,000	475,200	227,146
Oklahoma	209,700	1,363,100	651,562
Pennsylvania	185,700	1,039,900	497,072
South Carolina	244,700	1,810,800	865,562
Tennessee	175,200	928,600	443,871
Texas	629,500	3,840,000	1,835,520
Virginia	269,100	1,533,900	733,204
West Virginia	90,300	352,200	168,352
Wisconsin	54,100	227,200	108,602
Other States	564,800	3,318,700	1,586,339
United States	8,776,700	54,259,100	25,935,852

	Turkey Production and Value		
State †	Number raised	Pounds produced	Value of production
	(1000 head)	(1000 pounds)	(1000 dollars)
Arkansas	26,000	525,200	433,815
California	11,400	323,760	267,426
Indiana	19,500	758,550	626,562
Iowa	11,700	460,980	380,769
Michigan	5400	216,540	178,862
Minnesota	44,500	1,103,600	911,574
Missouri	19,200	625,920	517,010
North Carolina	33,500	1,202,650	993,389
Ohio	5800	236,640	195,465
Pennsylvania	7300	189,800	156,775
South Dakota	4200	180,180	148,829
Utah	4700	121,730	100,549
Virginia	17,200	467,840	386,436
West Virginia	3700	111,000	91,686
Other States ‡	29,900	962,588	795,100
United States	244,000	7,486,978	6,184,247

† California, Illinois, Indiana, Iowa, Louisiana, Michigan, Nebraska, New Jersey, New York, Oregon, and Washington combined to avoid disclosing individual operations.

‡ Includes State estimates not shown and states withheld to avoid disclosing data for individual operations.

roughly consume 2.5 to 3.0 kg of feed (dry matter basis) for the first 35 d (representative first clean-out), with approximately 5 to 6 kg of total solids being consumed after 49 d (typical final clean-out; Bolan et al., 2010). As for feed digestibility, chickens are able to digest approximately 85 to 90% of feed (dry matter total solid basis; NRC, 1994; Bolan et al., 2010). Therefore, it is estimated (based on 87.5% dry matter digestibility of the diet) that nearly 0.34 and 0.63 kg of solid excreta is produced by a 35 and 49 d old bird, respectively (FSA, 2007). Consequently, assuming a wet basis of 90%, total manure production of a 35 and 49 d old bird will be 4 and 6 kg, respectively (ASABE, 2005). Therefore, approximately 6.9 kg per 1000 kg live weight per day is produced for a typical broiler operation, or 0.6 to 1.8 Mg per 1000 broilers per flock (dry weight basis). In one study, Moore et al. (2011) measured the amount of litter produced from five flocks of 50 d old birds in four commercial poultry houses throughout the year. There were 504,702 birds marketed with an average weight of 2.58 kg (total live weight of 1,303,124 kg). Over the course of the year, 596,937 kg of litter and "cake" (or clumping due to excess moisture) were produced. In this study, birds produced an average of 1.18 kg of litter; 11% of which was cake and 89% was collected during the full clean-out.

The quantity of annual litter produced per house varies bases on externalities such as house and flock size, but ranges between 106 and 137 Mg (Sharpley et al., 2009b). In addition, roughly 82.9 kg of litter per square meter of floor space is produced per year (Tabler et al., 2009; Locatelli et al., 2017). A current 25,000-bird poultry house will produce approximately 5.5 flocks of birds a year, resulting in about 127 Mg of litter (Tabler et al., 2009). Therefore, by some estimates, nearly 13 million Mg (14 million tons) of broiler litter is produced on U.S. poultry farms (Moore et al., 1995b; Gollehon et al., 2001; Locatelli et al., 2017).

Commercial turkey production differs from broiler production in that two types of farms are used. Newly hatched chicks are initially placed on a poultry farm where they are brooded in a manner similar to chickens and are raised to a weight of about 2.3 to 3.6 kg (5 to 8 lb). The poults are then moved to a grow-out farm where the poults are reared to the desired market weight that can range from 5.4 to 22.7 kg (12 to 50 lb). As a result, the amount of litter removed from the houses can vary greatly depending on clean-out frequency, the removal of caked manure between flocks, amount of bedding used per flock, and final bird weight. Turkey litter production can range from 1 Mg per 1000 birds yr^{-1} for poultry houses to 14 Mg per 1000 birds yr^{-1} for heavy tom turkeys (Collins et al., 1999; Chastain et al., 2001). The largest turkeys are on breeder farms where litter production can be 50 Mg per 1000 birds year $^{-1}$ (Collins et al., 1999). Variation in litter moisture content between farms will result in differences in bedding management, which further confounds litter quantity estimates per year on a regional basis.

Factors that Influence Poultry Manure and Litter Composition

Poultry litter is not just feces from a bird, but is a combination of bedding material, wasted water, feathers, soil, spilt feed, and total excrement (feces and urine; Locatelli et al., 2017; Tasistro et al., 2004). For example, nitrogen (N) forms and micronutrients vary for layer, broiler, and turkey as excreted (Table 2). Furthermore, as with other manure sources, the moisture content, pH, soluble salt level, and plant availability of poultry litter has been shown to vary widely as a function of dietary supplement, feed ratios, and the litter storage and handling conditions (Bolan et al., 2010; He et al., 2016).

The variability in average bird weights, amount of bedding used, litter removal frequency, use of windrows to treat litter between flocks, and number of flocks are other litter management factors that dictate composition of litter. For example, 5 to 9 broiler flocks prior to sampling results in 6.78%, 1.8, and 0.84% C, C/N, and K differences, respectively (Table 3). Similar variations are also seen for different types of turkey facilities (Table 4). In addition, the nutrient composition of poultry manure as excreted (Table 2) is greatly different from litter removed from a commercial poultry production facility (Tables 3 and 4).

Table 2. Composition of chicken and turkey manure, as excreted (Collins et al., 1999).

	Layer Chicken		Broiler Chicken		Turkey	
Moisture (%)	75.0		74.0		75.0	
TS (%)	25.0		26.0		25.0	
VS/TS (%) †	76.0		73.1		76.0	
COD (ppm) ‡	176,000		197,000		236,000	
Density (kg m⁻³)	993.1		1025.2		1009.2	
	kg 1000 kg⁻¹	% d.b.	kg 1000 kg⁻¹	% d.b.	kg 1000 kg⁻¹	% d.b.
TKN §	13.5	5.40	13.0	5.00	14.0	5.60
TAN ¶	3.3	1.32	3.35	1.29	4.05	1.62
Organic-N	10.2	4.08	9.65	3.71	9.95	3.98
C #	–	42.2	–	40.6	–	42.2
C/N	–	7.8	–	8.12	–	7.54
P_2O_5 ††	10.5	4.20	8.0	3.08	12.0	4.80
K_2O ‡‡	6.0	2.40	6.0	2.31	6.0	2.40
Ca	20.5	8.20	5.0	1.92	13.5	5.40
Mg	2.15	0.86	1.75	0.67	1.55	0.62
S	2.15	0.86	1.0	0.38	1.65	0.66
Na	1.80	0.72	1.75	0.67	1.40	0.56
Cl	10.0	4.00	9.0	3.46	9.0	3.60
Fe	1.0	0.40	0.95	0.37	1.60	0.64
	g 1000 kg⁻¹	ppm, d.b.	g 1000 kg⁻¹	ppm, d.b.	g 1000 kg⁻¹	ppm, d.b.
Zn	70	280	40	162	310	1240
Cu	10	40	10	38	15	60
Mn	80	320	100	385	50	200
B	25	100	30	115	0.06	30

† VS/TS, the fraction of the total solids consumed by burning in a furnace.

‡ COD, chemical oxygen demand on a wet basis.

§ TKN, Total Kjeldahl Nitrogen = NH_4^+-N + Organic-N.

¶ TAN = total ammonical nitrogen = (NH_4^+-N + NH_3-N), the fraction of TAN that is in ammonia form is dependent on pH. The fraction of TAN in ammonia form is typically in the range of 8% to 10%.

Carbon estimated from the VS. content as: % Cd.b. = (VS/TS, %)/1.8 (Rynk et al., 1992).

†† Total P expressed as P_2O_5. To convert to elemental P multiply by 0.44.

‡‡ Total K expressed as K_2O. To convert to elemental K multiply by 0.83.

Management varies across production systems and influences plant nutrients in litter, such as N, P, and K, trace elements such as Cu, Zn, and As (Schroder et al., 2011), pesticides, veterinarian pharmaceuticals (Zheng et al., 2019), endocrine disruptors, and pathogens and microorganisms (Chen et al., 2019). Therefore, the next sections of this chapter aim to describe the differences in composition and plant availability based on factors affecting these important and abundant fertilizer sources.

Macronutrient Availability of Poultry Litter Sources

Nitrogen Forms and Fractions in Litter

Poultry litter contains both organic and inorganic forms of nutrients. Inorganic forms of N, such as nitrate and ammonium, are readily available for plant uptake and use, regardless of source, whereas organic forms of nutrients are less immediately available for plant adsorption. Because mineralization of the organic-N fraction in poultry litter is also mediated by externalities, such as climate, it is generally thought that 50 to 60% of the Organic-N in poultry litter becomes available through mineralization the first year, with about 20% the second year, and 10% the following year. This general rule can be used to adjust litter application rates based on N. Although, within that first year of application, the release of Organic-N fraction is quite rapid, with the majority being usable within two to four weeks after application (assuming a spring

Table 3. Composition of whole broiler litter as-removed from the house as influenced by litter management. Bedding used was predominately pine shavings and sawdust (NR = not reported).

Average Bird Weight	4.5 lb†	8.0 lb‡	8.5 lb§	7.5 lb§	8.8 lb§	8.8 lb§
Litter Removal by Year	Complete	Complete	Partial	Partial	Partial	Partial
No. of Flocks at Sample	6	5	9	5.5	7	8
Bedding Use	High	High	None	Low	Low	Medium
Windrow?	No	No	No	No	Yes	Yes
Litter Treatment?	No	No	Yes	Yes	Yes	Yes
Moisture (%)	22.3	24.0	29.2	27.8	29.5	35.5
TS (%)	77.7	76.0	70.9	72.20	70.6	64.5
VS/TS (%)	75.5	77.6	70.3	65.1	68.7	71.4
pH	8.7	NR	7.85	8.19	7.91	7.95
Density (kg m⁻³)	512.6	464.5	NR	NR	NR	NR
TAN (%, d.b.) #	0.71	1.05	0.96	0.80	1.19	0.94
Organic-N (%, d.b.)	3.48	3.49	3.33	2.68	3.23	2.62
NO₃–N (%, d.b.)	0.47	NR	0.15	0.19	0.12	0.01
Total-N (%, d.b.)	4.66	4.54	4.44	3.67	4.53	3.57
C (%, d.b.)	41.9 ¶	43.1 [4]	36.32	34.04	35.79	36.55
C/N	9	10	8.2	9.3	7.9	10.2
P₂O₅ (%, d.b.)	4.37	4.61	4.28	4.16	4.09	3.93
K₂O (%, d.b.)	3.23	3.09	4.03	3.85	3.78	4.72
Ca (%, d.b.)	2.80	2.70	2.77	2.63	2.67	2.76
Mg (%, d.b.)	0.56	0.55	0.69	0.79	0.70	0.81
S (%, d.b.)	0.74	0.92	1.16	0.92	1.33	1.43
Na (%, d.b.)	0.78	0.86	0.95	0.97	0.825	0.99
Fe (%, d.b.)	0.08	0.11	0.18	0.49	0.19	0.26
Al (%, d.b.)	NR	NR	0.26	0.75	0.45	0.69
Zn (ppm, d.b.)	420	450	600	645	555	560
Cu (ppm, d.b.)	360	320	365	225	485	590
Mn (ppm, d.b.)	450	500	550	645	570	570
B (ppm, d.b.)	40	30	NR	NR	NR	NR

† Means based on data from Collins et al. (1999), Chastain et al. (2001), and Chastain et al. (2012)

‡ Means from Collins et al. (1999)

§ Means from Chastain and Smith (2018)

¶ Carbon estimated from the VS. content as: % C d.b. = (VS/TS, %)/1.8 (Rynk et al., 1992). All other total carbon concentrations were measured using the Elementar procedure (ASL, 2018).

TAN = total ammonical nitrogen (NH₄⁺–N + NH₃–N)

Table 4. Composition of turkey litter, as removed from the house (NR = not reported).¶

	Grow-Out Litter †	Grow-Out Cake†	Grow-Out Litter‡	Poult Litter†	Poult Litter 1 Flock§	Poult Litter 2 Flocks‡	Breeder Litter†
Moisture (%)	27.0	45.0	26.5	20.0	14.5	24.7	22.0
TS (%)	73.0	55.0	73.5	80.0	85.5	75.3	78.0
VS/TS (%)	72.6	80.0	NR	77.5	92.4	83.3	43.6
Density (kg m⁻³)	512.6	560.6	NR	368.4	357.2	NR	800.9
TAN (%, d.b.) #	0.82	1.82	1.37	0.60	0.15	0.68	0.49
Organic-N (%, d.b.)	2.95	2.27	4.04	1.90	2.18	3.29	1.76
NO₃–N (%, d.b.)	NR	NR	0.06	NR	0.03	0.07	NR
Total-N (%, d.b.)	3.77	4.09	5.46	2.50	2.36	4.03	2.24
C (%, d.b.)	40.3 [4]	44.4 [4]	NR	43.1 [4]	51.3 [4]	43.30	24.2 [4]
C/N	10.7	10.9	NR	17.2	21.8	10.7	10.8
P₂O₅ (%, d.b.)	4.32	4.27	6.02	2.69	1.72	2.73	3.01
K₂O (%, d.b.)	2.74	2.73	3.19	1.69	1.19	1.96	1.15
Ca (%, d.b.)	2.60	2.36	3.84	1.63	1.24	1.76	4.62
Mg (%, d.b.)	0.51	0.49	0.53	0.32	0.21	0.32	0.30
S (%, d.b.)	0.58	0.57	0.67	0.38	0.25	0.44	0.47
Na (%, d.b.)	0.52	0.50	0.52	0.29	0.10	0.17	0.28
Fe (%, d.b.)	0.10	0.11	NR	0.13	NR	0.15	0.06
Zn (ppm, d.b.)	450	430	610	290	280	420	320
Cu (ppm, d.b.)	410	440	510	240	160	290	260
Mn (ppm, d.b.)	550	510	540	330	280	420	280
B (ppm, d.b.)	40	30	NR	20	NR	NR	20

† Means based on data from Collins et al. (1999)

‡ Means from Chastain and Smith (2018)

§ Chastain et al. (2013)

¶ Carbon estimated from the VS. content as: % C d.b. = (VS/TS, %)/1.8 (Rynk et al., 1992). All other total carbon concentrations were measured using the Elementar procedure (ASL, 2018).

TAN = total ammonical nitrogen.

application). Therefore, to maximize plant uptake and use of N in poultry litter, and to minimize external losses, applications should be synchronized with crop uptake and growth as much as is practical.

Organic N is not available to plants until it has been mineralized to ammonium N. Most N-based litter is from uric acid, with other forms of N coming from ammonia salts and fecal matter (organic). Uric acid is easily transformed to ammonium, which is highly volatile and susceptible to losses. However, if uric acid is thoroughly mixed in soil, it can be converted to ammonium (NH_4^+), which can be held on clay particles and organic matter; thus minimizing losses and improving plant available-N uptake (Zublena et al., 1993).

Proteins, amino acids, and amino sugars are also important forms of organic N in poultry litter. The amino acid fraction in poultry manure is largely composed of proteins from feed and urinary compounds (He et al., 2014). In addition, N content in poultry manure is reportedly much higher today compared with historical levels. For example, amino N represents approximately 75% of the total N in current poultry manure, compared with only 39% several decades ago (Blair, 1974; He and

Olk, 2011). This trend is assumedly due to over-supplementation of proteins and amino acids in modern poultry diets (He et al., 2014; Perez et al., 2015).

Ammonium N content is important in poultry litter, particularly when considering its use as a fertilizer source. Many laboratories report a single value for the ammonium N content of manure, which includes both ammoniacal forms of N (NH_4^+ and NH_3^-). The combination of NH_4^+ and NH_3^- N is often referred to as the total ammoniacal N (TAN). From laboratory analyses, one can obtain TAN and total Kjeldahl N (TKN); organic N is obtain by subtracting TAN from total Kjeldahl N (e.g., TAN + Organic N = TKN). Having knowledge of both TAN and organic N allows producers to estimate the amount of N available to plants immediately and pools that will become more slowly available over time. Nitrate N is usually only present in poultry litter in trace amounts, unless more aerobic conditions exist for manure.

Ammonium (NH_4^+) and ammonia (NH_3^-) can interchange rapidly depending on pH. Most litter has a pH close to 8.0. As a result, the percentage of TAN in the ammonia form is on the order of 10% (Meisinger and Jokela, 2000). If litter pH is reduced to 7.0 the fraction of TAN in the ammonia form is about 1%. At a litter pH 6.5 or below, the ammonia fraction is zero (Meisinger and Jokela, 2000). The plant available N (PAN) is the sum of the available ammonium N, the available organic N, and the nitrate N. The estimate of PAN is used to calculate the amount of manure that is needed to satisfy the crop N needs. Equation 1 can be used to estimate the plant available N content:

$$[PAN]=A_f[TAN]+m_f[Organic-N]+[NO_3-N] \qquad [1]$$

Where,

A_f = the ammonium availability factor (decimal), and

m_f = the organic-N mineralization factor (decimal).

Moore et al. (2011) conducted a N mass balance for a typical broiler farm with four broiler houses over the course of one year (Table 5). They reported that the N inputs from feed, chick placement, and bedding were 70, 435, 602, and 303 kg N yr[-1], respectively, which was equivalent to 98.74, 0.84, and 0.42% of the inputs. This corresponded to inputs from feed, chick placement, and bedding on a per bird basis of 139.56, 1.19 and 0.06 g N bird[-1] marketed. The annual N outputs reported in this study were birds marketed, mortality, ammonia emissions, nitrous oxide emissions, and litter plus cake, which were 39,485; 635; 15,571; 241; and, 14,464 kg N yr[-1], respectively, which corresponds to 56.09, 0.90, 22.12, 0.34, and 20.55% of the outputs. The N mass balance recovery (outputs/inputs) reported by Moore et al. (2011) was 98.8%. Surprisingly, more N was lost via ammonia emissions in this study than was removed from the barns as litter plus cake; however, it should be noted that there were no amendments utilized to control ammonia volatilization in that study. These data are similar to values obtained in a pen trial by Coufal et al. (2006) who estimated that N partitioning as a percentage of inputs averaged 15.3, 6.8, 55.5, and 21.1% for litter, caked litter, broiler carcasses, and N loss, respectively. These data indicate that in-house ammonia losses are a much bigger problem than losses that occur during storage or following land application (Moore et al., 2011).

Table 5. Annual mass balance of N for a typical broiler farm with four houses (data represent 20 flocks; five flocks from four houses). Taken from Moore et al. (2011).

Nitrogen inputs	Weight	Inputs	N per bird marketed
	kg N yr^{-1}	%	g
Bedding	303	0.42	0.06
Chick placement	602	0.84	1.19
Feed	70,435	98.74	139.56
Total	71,340	100.00	141.35
Nitrogen Outputs	Weight	Outputs	N per bird marketed
	kg N yr^{-1}	%	g
Birds marketed	39,485	56.09	78.23
Mortality	635	0.90	1.26
Ammonia emissions	15,571	22.12	30.85

Phosphorus and Potassium Sources and Forms in Poultry Litter

Phosphorus (P) is generally imported into poultry systems by mineral-P supplements in feed, including enzymes integrated into diets to improve nutrient adsorption by poultry, as well as the use of crops that contain lower levels of indigestible phytate-P (Sharpley et al., 2007). In general, one third or less of the P in feed is actually utilized during the bird life cycle, with the remaining P being excreted in manure (Patterson et al., 2005). The majority of organic P in poultry litter is in the form of phytic acid salts (Turner and Leytem, 2004).

Researchers suggest that, in general, P and potassium (K) in poultry litter are as labile as those in chemical fertilizers (Pagliari and Laboski, 2014). Further, it is widely assumed that at least 90% of P and K in poultry litter are readily available for plant-use during the season of application (Sharpley and Moyer, 2000), and is more equivalent to inorganic fertilizer forms (P_2O_5 and K_2O). However, some work suggests P bound in litter may be less bio-available than mentioned above or about two thirds in the solid phase (organic) and one-third inorganic (Edwards and Daniel, 1992). Nonetheless, total phosphorus has two fractions, orthophosphorus and organic phosphorus, both of which are essentially immediately labile (excluding phytate). Pagliari and Laboski (2014) showed that the solid phase is not entirely organic but a combination of minerals and organic compounds. Specifically, the inorganic phosphate fraction includes calcium phosphate, dibasic calcium phosphate, and weakly bound water-soluble P (Bolan et al., 2010; He et al., 2016). Potassium has no organic fraction and is thus immediately plant available.

Feed Additive Impact on Poultry Litter

One management practice to reduce P runoff from poultry litter is to reduce the amount of P in feed. In the past, heavy P supplementation was required in poultry rations, because most grains store between 80 and 90% of their P as phytate (Kornegay, 1996; Turner et al., 2002). Phytate-P is relatively stable and not readily digested by poultry. However, phytase enzymes can be used in the diet to cleave the P from the phytate molecule, making it available. Although the advantages of using phytase in poultry diets have been recognized for decades (Nelson et al.,

1968, 1971), the technology to produce phytase cheaply on an industrial scale was not available until the late 1990s. Another diet modification technology that can be used to reduce the amount of dicalcium phosphate added in the feed is the use of high available P (HAP) corn, which are varieties that have been selected for their ability to store P in forms more bioavailable than phytate (Raboy, 2002). Miles et al. (2003) showed that poultry diets developed with phytase and HAP corn resulted in lower P in manure. Smith et al. (2004) found that these diets also resulted in lower P runoff. Currently, the great majority of poultry diets use phytase enzymes; however, the use of HAP corn has never been performed beyond research.

Poultry diets are supplemented with enzymes, nutrients, and antibiotics in efforts to enhance and accelerate growth, which may dictate the quantity of nutrients such as P in excrement (Sharpley et al., 2007). For example, enzymes such as phytase, can degrade digestible fibers from indigestible grain fractions and reduce the need for P supplement, consequently minimizing the P level in poultry manure. In addition, adding amino acids has shown to reduce P excretion by nearly 50%, although this increases production costs (Keshavarz and Austic, 2004).

Optimizing feed efficiency minimizes flock disease outbreaks, although because many of the growth promoting supplements are metals (such as Cu and Zn) this results in greater metals in poultry excrement (He et al., 2014). Essential nutrients are supplemented to conduct enzymatic process involved in many physiological processes, considering they act as catalysts for essential growth functions (Bolan et al., 2010). In addition, the use of coccidiostats as growth promoters rather than arsenical compounds can greatly reduce arsenic levels in litter, which is now commonplace for the industry (Bolan et al., 2010). Feed additives may have the added benefit of decreasing P and N contents of litter, as well as substantially reducing the odor of chicken manure (Nahm, 2002).

Variation in Nutrient Content from Various Poultry Sources

The quantity and nutritive composition can vary from farm-to farm within a poultry type (i.e., broiler, layer, or turkey), as well as across regions. Information of the nutrient content of poultry manure is an essential component in the design of a poultry manure management plan. In sections below, we will describe poultry manure and litter compositional differences from the three major poultry (broiler, layer, and turkey) production systems.

Broiler Manure and Litter Characteristics

Broiler litter is produced in the greatest quantity in the U.S. and is consequently the most common land applied poultry litter source. Inorganic forms of N in broiler litter account for approximately 14% of the total N in litter and is labile, and consequently vulnerable to losses. The majority of P is dominated by the inorganic fraction (35 to 41%), with only about 6% being water extractable P (WEP; Turner and Leytem, 2004). Although, more recent literature illustrates that the actual amount of WEP in manure depends heavily on the water: manure ratio used in the extraction (Pagliari and Laboski, 2014). Nonetheless, WEP is important for determining plant uptake and the P fraction that is susceptible to runoff. Amounts of K and Ca average about 2.5%, with K ranging from 1.1 to 3.4% (Table 3). Broiler litter has

Table 6. Composition of other types of chicken litter (non-broiler litter), as removed from the house (NR, not reported). §

	Broiler cake †	Pullet house ‡	Cornish hen litter †	Cornish hen cake †	Breeder mixed–litter & manure ‡	Breeder litter-center alley ‡	Breeder manure-under slats ‡
Moisture (%)	40.0	37.8	32	46	33.5	18.6	54.1
TS (%)	60.0	62.2	68	54	66.5	81.4	45.9
VS/TS (%)	78.3	NR	79.4	77.8	42.0	66.2	39.0
pH	NR	NR	NR	NR	NR	8.47	7.60
Density (kg m⁻³)	544.6	NR	480.6	544.6	800.9	403.7	581.5
TAN (%, d.b.)	1.00	0.68	0.88	1.57	0.57	0.20	0.47
Organic-N (%, d.b.)	2.83	2.18	4.19	5.28	1.99	2.12	0.68
NO₃-N (%, d.b.)	-	-	-	-	NR	0.02	1.57
Total-N (%, d.b.)	3.83	2.85	4.34	5.74	2.55	2.34	2.72
C (%, d.b.)	43.5 [3]	NR	44.1 [3]	43.2 [3]	23.3 [3]	35.42	22.12
C/N	11.4	NR	10.2	7.5	9.0	15.2	8.2
P₂O₅ (%, d.b.)	4.42	3.64	4.19	3.61	4.21	3.59	7.24
K₂O (%, d.b.)	3.00	2.08	4.34	3.61	2.48	2.28	4.18
Ca (%, d.b.)	2.83	2.47	3.01	2.78	6.68	9.31	13.34
Mg (%, d.b.)	0.58	0.52	1.62	1.30	0.56	0.45	0.84
S (%, d.b.)	0.77	0.44	NR	NR	0.61	0.45	0.85
Na (%, d.b.)	0.83	0.47	NR	NR	0.64	0.42	0.69
Fe (%, d.b.)	0.10	NR	NR	NR	0.09	0.14	0.47
Al (%, d.b.)	NR	NR	NR	NR	NR	0.10	0.52
Zn (ppm, d.b.)	500	402	676	463	430	350	700
Cu (ppm, d.b.)	342	201	449	426	165	150	330
Mn (ppm, d.b.)	575	498	809	620	475	440	740
B (ppm, d.b.)	37	NR	NR	NR	20	NR	NR

† Means based on data from Collins et al. (1999).

‡ Means from Chastain and Smith (2018)

§ Carbon estimated from the VS. content as: % C d.b. = (VS/TS, %)/1.8 (Rynk et al., 1992). All other total carbon concentrations were measured using the Elementar procedure (ASL, 2018).

a very high carbon content, averaging 25.2%, indicating great potential for soil organic carbon formation (Sharpley et al., 2009b; Ashworth et al., 2014).

Most of the broiler litter is removed as whole litter (Table 3); however, it is a common practice for broiler producers to remove small amounts of wet manure from near the feed and water lines between flocks. This material is often called caked litter and has macronutrient concentrations that are lower than whole litter (Table 6).

While most broiler farms produce meat birds with finish bird weights, ranging from 4.5 to 8.8 lb, each poultry production company also has farms for raising replacement breeder hens and roosters (pullet farms), and breeder farms that produce fertilized eggs for the hatchery (Table 6). Pullet farms generally raise two flocks per year and remove litter once a year. The stocking density (birds per unit area) is much lower compared to a broiler farm. In addition, the feeding program and rations are designed to enhance sexual development while preventing rapid weight gain. As a result, the composition of the litter can differ greatly as compared to other broiler litter sources. Once the chickens (hens and roosters) raised on a pullet farm have reached maturity they are moved to a broiler breeder farm. A broiler breeder barn differs from all other types of broiler barns in that a heavily bedded center alley is provided to allow birds to roam freely and breed. Raised nesting boxes and slotted flooring are also provided on either side of the building. Manure, without bedding, is allowed to accumulate in a shallow pit below the nesting boxes and slotted flooring. Manure is removed as dry litter from the center alley, and as thick solid manure from below the slotted area once a year. In many cases, the alley litter and solid manure are mixed prior to land application. Data on the composition of these types of broiler chicken manures are provided in Table 6.

Layer Manure Composition

The majority of commercial table egg farms use high-rise buildings that are designed to store solid manure below cages inside the building in an area similar to a basement. No bedding is used and the ventilation system is designed and operated so as to dry the manure as it accumulates for one year in a long conical piles. The manure is typically removed once a year with a front-end loader and applied to cropland. The next most popular manure handling system for layer farms involves removing manure from below cages every one to three days. The manure is either scraped from a belt that conveys the manure out of the barn or it is scraped from an ally. There are a few laying operations that use flush systems to daily remove layer manure from below the cages. These types of facilities handle and apply layer manure as a liquid and may also provide storage and treatment in a lagoon. Examples of manure composition for layer manure handled as a solid or liquid are provided in Tables 7 and 8.

Table 7. Composition of commercial layer manure handled as a solid (Collins et al., 1999).

	Under-cage, alley-scraped †		High-rise, stored manure ‡	
Moisture (%)	65.0		47.0	
TS (%)	35.0		53.0	
VS/TS (%)	71.4		60.4	
Density (kg m⁻³)	993.1		816.9	
	kg 1000 kg⁻¹	% d.b.	kg 1000 kg⁻¹	% d.b.
TKN	14.00	4.00	17.00	3.21
TAN	7.00	2.00	6.00	1.13
Organic-N	7.00	2.00	11.00	2.08
C §		39.7		33.5
C:N		9.9		10.5
P_2O_5	16.00	4.57	25.50	4.81
K_2O	10.00	2.86	13.00	2.45
Ca	20.50	5.86	38.00	7.17
Mg	2.75	0.79	2.85	0.54
S	3.55	1.01	2.40	0.45
Na	2.00	0.57	3.00	0.56
Cl	1.20	0.34	1.40	0.264
Fe	0.15	0.04	0.22	0.0
	g 1000 kg⁻¹	ppm, d.b.	g 1000 kg⁻¹	ppm, d.b.
Zn	155.0	440	175.0	330
Cu	17.0	50	29.0	50
Mn	1400	4000	1650	3110
B	11.0	30	18.0	30

† Manure scraped from paved alley every 2 d.

‡ Annual accumulation of manure, wasted feed, and wasted water below cages on an unpaved floor·

§ Carbon estimated from the VS. content as: % C d.b. = (VS/TS, %)/1.8 (Rynk et al., 1992).

Turkey Litter Chemical Composition

The majority of turkey litter produced is whole litter and litter cake removed from grow-out barns. Grow-out farms receive turkey poults from a brooder farm, which are then raised until they reach the desired market weight. Like broiler litter, the composition of grow-out litter can vary depending on litter clean-out frequency, bedding practices, and the market bird weight (5.4 to 22.7 kg; Table 4). Poult or brooder farms represent about 20% of the farms in a turkey production system. They raise hatchlings to a weight similar to a broiler chicken prior to being moved to a grow-out farm. Litter in poult houses is typically replaced after every one or two flocks to help prevent disease. As a result, poult litter is typically much lower in plant nutrients and has a high C to N ratio as compared to grow-out or breeder litter (see Table 4). Poult litter has been shown to be an excellent material for composting, and may be preferred to land application due to the potential for nitrogen immobilization due to its high C to N ratio (Chastain et al., 2013). Turkey breeder farms, similar to broiler breeder farms, are an essential component of turkey production, and since the bird density (birds/unit floor area) tends to be lower as compared to grow-out farms the litter contains less plant nutrients (Table 4).

Other Types of Litter

There are many other types of poultry farms that handle and store manure as litter. These include farms that produce Cornish hens, and ducks. Manure composition for these specialty poultry farms are provided in Tables 6 and 9.

Impact of Flock and Litter Management on Manure Characteristics

While the value of poultry litter as fertilizer is well recognized, exact management factors affecting its nutritive quality are less known (VanDevender et al., 2000). Previous work has estimated that as a general rule, from their diet and during their lifecycle, poultry excrete 55% of N, about 70% of P, and 80% of K (Bolan et al., 2010). To estimate nutrients in excreted manure, one can subtract levels in the feed by the amount assimilated. Feed spilt during feeding is another important factor when determining total solids and nutrient levels in litter (Leytem et al., 2007). In general, a broiler will drink approximately 0.9 kg of water per 0.45 kg feed (2 pounds of water per pound of feed) consumed per day, which translates into approximately 189,270 L (50,000 gallons) consumed per 20,000 birds (Payne, 2012; Tabler et al., 2009). Only about 20% of this total water consumption is actually used for growth, the rest reaches the litter through manure.

Reducing poultry litter moisture is critical to flock and nutrient management, as moisture levels affect bird performance, which effects on-farm economic viability and nutrient volatility. Briefly (see Chapter 10 on Ammonia Emissions), moisture levels influence ammonia levels and disease and injury incidence (e.g., footpad burns). If inadequate poultry house ventilation exists, caking occurs. Over-ventilation often occurs to rectify caking, which results in excessive gas use (especially during winter months). In general, all poultry are most sensitive to elevated ammonia levels during early brooding (up to 21 d old), although feed supplements, litter amendments, and proper ventilation helps control NH_3 losses. Handheld ammonia sensors can assess in-house ammonia levels, though they can be somewhat expensive ($300–500) and may underestimate levels (Payne, 2012). However, if excess levels occur, brooding poultry are susceptible to blindness, which affects feed consumption and subsequent weight gain. Therefore, caking can have negative influences on broiler welfare, animal performance, and litter quality through N losses.

Litter pH is another important factor influencing poultry litter composition and fertilizer value. In house poultry litter needs to be under pH 7.0 to reduce ammonia losses; consequently, many producers will add acidifying agents before new flock placement. However, it is generally difficult to manage flock-level litter pH. Overall, these acidifying agents can influence terminal litter pH, which influences the labile fraction of nutrients, microbial community structure, and litter liming-agent potential.

Treating Poultry Litter to Reduce Nutrient Losses and Improve Production

Atmospheric emissions of compounds like ammonia from poultry manure not only cause production problems for poultry producers, but cause environmental issues as well (He et al., 2016). Gaseous ammonia concentrations can reach very high levels in poultry houses, sometimes exceeding 100 ppm during winter months. Moore

Table 8. Composition of commercial layer manure handled as a liquid (Collins et al., 1999).

	Storage – Slurry[†]		Anaerobic Lagoon – Liquid[‡]		Anaerobic Lagoon – Sludge[‡]	
Moisture (%)	89.0		99.5		83.0	
TS (%)	11.0		0.5		17.0	
VS/TS (%) [§]	67.3		44.9		42.9	
Density (kg m⁻³)	929.1		993.1		993.1	
	kg 1000 kg⁻¹	kg 1000 L⁻¹	kg 1000 kg⁻¹	kg 1000 L⁻¹	kg 1000 kg⁻¹	kg 1000 L⁻¹
TKN	28.50	26.48	3.30	3.28	10.50	10.43
TAN [¶]	18.50	17.19	2.80	2.78	3.25	3.23
Organic-N	10.00	9.29	0.50	0.50	7.25	7.20
C #	41.1	38.2	1.2	1.2	40.6	40.3
C/N	1.4		0.37		3.9	
P_2O_5	26.00	24.16	0.85	0.84	38.50	38.24
K_2O	16.50	15.33	5.15	5.12	4.90	4.87
Ca	16.50	15.33	0.55	0.55	23.50	23.34
Mg	2.00	1.86	0.17	0.17	6.00	5.96
S	2.00	1.86	0.31	0.30	3.55	3.53
Na	3.30	3.07	1.70	1.69	1.20	1.19
	g 1000 kg⁻¹	g 1000 L⁻¹	g 1000 kg⁻¹	g 1000 L⁻¹	g 1000 kg⁻¹	g 1000 L⁻¹
Cl	850.0	789.7	30.0	29.8	2400	2383
Fe	190.0	176.1	3.5	3.5	800.0	794.4
Zn	195.0	180.9	8.0	7.9	550.0	546.4
Cu	36.5	33.6	2.0	2.0	70.0	69.5
Mn	2400	2230	900.0	893.9	1650	1639

† Includes all manure, wasted feed, wasted water, and net rainfall.

‡ Lagoon designed to provide volume for manure storage, rainfall, anaerobic treatment, and sludge storage. No solids removal prior to lagoon treatment.

§ VS/TS = the fraction of the total solids consumed by burning in a furnace (volatile solids fraction).

¶ TAN = total ammonical nitrogen = (NH_4^+-N + NH_3^--N), the fraction of TAN that is in ammonia form is dependent on pH.

Carbon estimated from the VS content as: % C d.b. = (VS/TS, %) / 1.8 (Rynk et al., 1992).

Table 9. Composition turkey litter as removed from storages (NR, Not Reported). §

	Uncovered pile 6 Months[+†]	Uncovered pile 3 Months[‡]	Covered pile (Tarp) 1 to 2 mo[‡]	Covered litter 6 Months[+‡]
Moisture (%)	37.5	47.1	26.9	40.5
TS (%)	62.5	53.0	73.1	59.5
VS/TS (%)	72.1	NR	NR	NR
TAN (%, d.b.)	0.46	0.96	1.07	1.61
Organic-N (%, d.b.)	2.22	2.56	2.56	3.43
NO_3–N (%, d.b.)	0.062	0.08	0.14	0.17
Total-N (%, d.b.)	2.75	3.60	3.78	5.22
C (%, d.b.) [3]	40.1	NR	NR	NR
C:N	14.6	NR	NR	NR
P_2O_5 (%, d.b.)	5.99	4.78	3.91	2.54
K_2O (%, d.b.)	2.56	3.48	2.19	1.62
Ca (%, d.b.)	3.95	2.70	2.29	1.79
Mg (%, d.b.)	0.60	0.39	0.33	0.29
S (%, d.b.)	0.77	0.58	0.39	0.41
Na (%, d.b.)	0.48	0.52	0.35	0.30
Fe (%, d.b.)	0.17	NR	NR	NR
Zn (ppm, d.b.)	470	550	400	240
Cu (ppm, d.b.)	280	510	390	280
Mn (ppm, d.b.)	540	550	510	240
B (ppm, d.b.)	30	NR	NR	NR

† Means based on data from Collins et al. (1999)
‡ Means from Chastain and Smith (2018)
§ Carbon estimated from the VS. content as: % C d.b. = (VS/TS, %)/1.8 (Rynk et al., 1992).

et al. (2011) found the overall average ammonia concentrations in broiler houses was 25 ppm. Anderson et al. (1964) showed that ammonia concentrations as low as 20 ppm compromised the immune system of chickens, making them more susceptible to viral diseases such as New Castle disease. Poor weight gains and feed conversion and ocular damage to broilers has also been reported when ammonia levels are high (Miles et al., 2004; Miles et al., 2006).

Atmospheric ammonia can also cause enviro%nmental issues, such as soil acidification, formation of fine particulate matter, and excessive N deposition into aquatic and terrestrial ecosystems (Hutchinson and Viets, 1969; Van Breemen et al., 1982; Behera and Sharma, 2010). When atmospheric ammonia is deposited onto soil it is converted to nitrate via nitrification, which is an acid-generating process (Van Breemen et al., 1982). At present, the only technology used by the poultry industry to reduce ammonia volatilization from manure is acidification with acid salts, such as aluminum sulfate (alum), sold under the tradename of Al+Clear (Moore et al., 1999, 2000; Moore, 2011; Eugene et al., 2015) or sodium bisulfate, sold under the tradename PLT (Blake and Hess, 2001). When litter is acidified, it shifts the ammonia/ammonium equilibrium toward ammonium, which is not volatile.

Moore et al. (1995a, 1996) conducted experiments to evaluate effects of poultry litter amendments on ammonia volatilization. These studies showed that alum could reduce ammonia losses by up to 99% if adequate amounts of alum is applied relative to the manure in the barn. Reductions in ammonia resulted in significantly higher N content in litter, as well as a higher N/P ratios than litter treated with other products (Moore et al., 1995a; Moore et al., 1996; Anderson et al., 2018). In an additional study on commercial broiler farms, Moore et al. (1999, 2000) found alum significantly reduced litter pH for the first 3 to 4 wk after the beginning of each flock, but after 5 wk, pH leveled off around 7.5, while the litter pH of the control houses remained relatively constant at pH 8. The reduction in litter pH is due to Eq. [2]:

$$Al_2(SO_4)3.14H_2O + 6H2O + 2Al(OH)^3 + 3SO_4^{2-} + 6H + 14H_2O \ [2]$$

Poultry Litter and Manure Handling Considerations

Manure disposal and management are among the most important issues facing producers, as improper manure management can lead to excessive odor, fly breeding, and nonpoint-source pollution. Depending on flock disease issues, some producers will reuse bedding material for multiple years; however, an annual clean

Table 10. Composition broiler and turkey litter as removed from storage type (NR is Not Reported). §

	Broiler litter			Turkey litter			
	Uncovered pile[†]	Covered pile (Tarp)[‡]	Open stacking shed[‡]	Uncovered pile 6 Months[+†]	Uncovered pile 3 Months[‡]	Covered pile (Tarp) 1 to 2 mo[‡]	Covered litter 6 Months[+‡]
Moisture (%)	39.0	17.4	10.9	37.5	47.1	26.9	40.5
TS (%)	61.0	82.6	89.1	62.5	53.0	73.1	59.5
VS/TS (%)	70.5	NR	NR	72.1	NR	NR	NR
TAN (%, d.b.) [¶]	0.59	0.60	0.62	0.46	0.96	1.07	1.61
Organic-N (%, d.b.)	2.22	3.26	2.62	2.22	2.56	2.56	3.43
NO$_3$–N (%, d.b.)	0.04	NR	0.21	0.062	0.08	0.14	0.17
Total-N (%, d.b.)	2.85	3.86	3.45	2.75	3.60	3.78	5.22
C (%, d.b.) [3]	39.2	NR	NR	40.1	NR	NR	NR
C:N	13.8	NR	NR	14.6	NR	NR	NR
P$_2$O$_5$ (%, d.b.)	6.43	3.97	2.20	5.99	4.78	3.91	2.54
K$_2$O (%, d.b.)	2.70	3.15	2.52	2.56	3.48	2.19	1.62
Ca (%, d.b.)	4.80	2.56	1.55	3.95	2.70	2.29	1.79
Mg (%, d.b.)	0.66	0.49	0.35	0.60	0.39	0.33	0.29
S (%, d.b.)	0.90	0.70	0.44	0.77	0.58	0.39	0.41
Na (%, d.b.)	0.53	0.72	0.41	0.48	0.52	0.35	0.30
Fe (%, d.b.)	0.15	NR	NR	0.17	NR	NR	NR
Zn (ppm, d.b.)	480	460	260	470	550	400	240
Cu (ppm, d.b.)	230	680	140	280	510	390	280
Mn (ppm, d.b.)	530	520	330	540	550	510	240
B (ppm, d.b.)	30	NR	NR	30	NR	NR	NR

† Means based on data from Collins et al., (1999)

‡ Means from Chastain and Smith (2018)

§ Carbon estimated from the VS content as: % C d.b. = (VS/TS, %)/1.8 (Rynk et al., 1992).

¶ TAN, total ammonical nitrogen.

Fig. 1. Typical poultry litter storage operation prior to land application. Photo credit: Steven Haller, USDA-ARS.

out may also be practiced. Generally, either in-house windrowing or partial house cleanout are practiced to reuse litter for multiple production cycles (Tabler et al., 2009). In-house windrowing is practiced by placing litter into windrows (46 to 61 cm high) between flocks and turning windrows about once every 5 or 6 d. In doing so, high enough temperatures are reached (> 130 °F) to kill pathogens and thus minimize disease outbreaks. In this practice, windrows are leveled before chick placement to minimize ammonia volatilization and allow litter to cool. Ro et al. (2017) also showed that windrowing, while popular with many growers, resulted in significantly higher ammonia and nitrous oxide emissions during the period between flocks than houses that were not windrowed. Ro et al. (2017) also measured pathogen levels in litter; however, the results were inconclusive since few pathogens were present. In the latter practice, or partial house cleanout, a limited amount of poultry litter is removed from the center of the house, with the remaining being redistributed (Tabler et al., 2009).

Ideally, poultry litter should be stored for a short period of time and in usable quantities. Longer storage periods will result in reduced poultry litter nutrient concentrations, particularly if the poultry litter is exposed to excess moisture. See Fig. 1 for a typical stacking operation after cleanout.

Storage of poultry litter is among the most influential factors influencing composition, which is largely driven by moisture (Table 10) and can vary based on farm location (e.g., warm wet climates versus cold climates). Considering this, moisture differences range from 40.5 to 10.9% under uncovered and open staking shed conditions, which in turn influences poultry litter nutrient content such as NO_3–N (0.04–0.21%), P_2O_5 (2.20–6.43%), and K_2O (1.62–3.48%), respectively (Table 10). Laboraory analyses may be reported on a wet basis (as-is) or on a dry matter basis, or both. Therefore, nutrients reported on a wet basis refer to the nutrient per unit or volume of nutrient per mass of wet manure (as-is). Conversely, poultry manure nutrients reported on a dry weight-basis signify the weight of the nutrient without moisture; consequently, dry weight basis is usually converted to wet weight, as poultry litter is handled and applied on the wet or as-is basis. Another important attribute of poultry litter is the total solids levels, as they also affect handing qualities and bulk density characteristics.

If manure is stored in a predominantly anaerobic condition, then very little nitrate nitrogen will be present. Substantial amounts of total N can be lost depending on storage method (Chastain et al., 2001). Considering, N losses averaged 30% for uncovered stored poultry litter piles, with 26% losses occurring in a stacking shed due to periodic turning of the litter (Table 11). Therefore, manure that receives a significant amount of aeration should also be analyzed to determine the nitrate N content prior to land application.

Due to increased fertilizer costs, economic analyses suggest greater poultry litter transportation distances are viable. In general, poultry litter hauling is estimated at $0.24 per km Mg[-1] (assuming a one-time poultry house clean out, loading, spreading cost totaling $28.90 Mg[-1], and at 2008 fertilizer prices [$108 Mg[-1]]), which translates into litter being able to be transported in up to a 933-km radius and still be considered economical (Doye et al., 1992; Mullen et al., 2011). This distance could potentially increase if more nutrient-dense litters were transported (pelletized poultry litter). Although recently, the USDA-NRCS Environmental Quality Incentives Program now offers to incentivize poultry litter nutrient export in nutrient dense watersheds by paying $4.1 Mg[-1] up to 40 km and over 120 miles, $10.8 Mg[-1] (USDA-NRCS, 2006), which may increase the transportation radius considered to be economical.

Considerations for Land Application of Manure

Not surprisingly, a multitude of studies have documented substantial yield increases (Gilley and Risse, 2000; Ashworth et al., 2016; Chastain et al., 2007) and soil health improvements (Ashworth et al., 2017a, 2017b,) from poultry litter applications. In addition, other ancillary benefits to poultry litter applications are increases in soil organic carbon formation and consequently greater soil water-holding capacity (Ashworth et al., 2014). Although, land application, particularly under non-tillage and high slope, has been scrutinized for eutrophication from excess P applications and subsequent runoff into neighboring waterbodies (Moore et al., 1998). Therefore, poultry litter management is an increasing important consideration for land managers, federal and state agencies, industry practitioners, and consumers. For example, roughly 45% of broiler litter is produced in approximately four Mid-south states (Arkansas, North Carolina, Georgia, and Alabama) and consequently concentrated manure application areas in these places occur, which may result in nutrient imbalances in these watersheds (Moore et al., 1995b; Gollehon et al., 2001). This has resulted in poultry litter application rates and locations being determined through litigated nutrient export programs (Sharpley et al., 2009a) and requirements for producers to develop certified nutrient management plans (Goodwin et al., 2003).

Since poultry litter typically consists of approximately 3% N, P (P_2O_5), and K (K_2O) each; applications targeting N needs for forage (primarily tall fescue (*Lolium arundinaceum* [Schreb.] Darbysh) and bermudagrass (*Cynodon dactylon* [L.] Pers.) and field crops often exceed the amount of usable P (see Chapter 13 on Phosphorus Runoff and Water Quality). As a result, when poultry litter is applied annually to meet the N needs of the crop, the amount of P applied exceeds plant P removal, resulting in an increase in soil test P (Pote et al., 1996). Consequently, most state extension programs suggest sampling and analyzing litter, as well as

Table 11. Nitrogen losses from poultry litter storage methods (calculated on a dry matter basis, Chastain et al., 2001).

Type of storage	Moisture content	Total ammonical N lost	Total-N lost
		———%———	
Uncovered Pile	39- 47	21	30
Covered Pile	16- 19	13	17
Stacking Shed	7- 15	11	26

Table 12. Recommended estimates of ammonium nitrogen availability for poultry manure (Chastain et al., 2001, Meisinger and Jokela, 2000).

Application method (all manure types)	Fraction of TAN available †
Surface application without incorporation	0.5
Surface application with incorporation within 1 d	0.8
Direct injection below the soil surface or immediate incorporation	1.0

† TAN, total ammonia nitrogen.

Fig. 2. A prototype, tractor-drawn implement for subsurface applications of dry poultry litter in conservation tillage systems developed by research team at USDA's Agricultural Research Service in Booneville, AR. Photo credit: Dr. Dan Pote, USDA-ARS.

applying poultry litter based on the crop P requirements to prevent overapplication and subsequent P buildup in soils (Espinoza et al., 2007). Soluble P losses are greatest in areas with high rainfall and courser textured soils, as these soils have greater water infiltration and lower cation exchange capacity.

Rainfall simulation studies have often shown a positive relationship between soil test P and P concentrations in runoff water (Pote et al., 1999a, b; Sharpley, 1995). Research has also shown that P concentrations in runoff water from fields fertilized with poultry litter can be high, even when litter is applied at a moderate rate (Edwards and Daniel, 1993a; Shreve et al., 1995; DeLaune et al., 2004a). Edwards and Daniel (1993b) found that more than 90% of the P in surface runoff from pastures fertilized with animal manure was in the soluble form. Soluble P has a greater effect on water quality since it is directly bioavailable to algae and macrophytes, whereas particulate P is only available once it has been converted to inorganic P (Sonzogni et al., 1982).

Several studies illustrated that the WEP content of land-applied manure was a key water quality indicator for P runoff (DeLaune et al., 2004b; Haggard et al., 2005; Kleinman et al., 2002). The method used by DeLaune et al. (2004a,b) to determine soluble P concentration in poultry litter was a 1:10 manure:water extraction according to the method described by Self-Davis and Moore (2000), developed for dry poultry litter. Although this method is still used by many researchers, Kleinman et al. (2002, 2007) conducted a study comparing it to a method using a 1:200 manure:water ratio. Since many manures, such as dairy manure, are not suitable for the 1:10 method of Self-Davis and Moore (2000), Kleinman et al. (2007) suggested this method should be used to determine soluble P. For further explanation why different extraction ratios extract various amounts of P see Pagliari and Laboski (2014).

Alternative Methods to Poultry Litter Land Application

Conservation management practices continue to evolve, however, surface application continues to be the primary practice for agronomic poultry litter application. Poultry litter is generally applied in the spring and if left on the soil surface without incorporation for an extended period is highly susceptible to losses by water (erosion, runoff, or leaching), volatilization, and wind movement. Poultry litter is a alkaline material (pH > 7), therefore, the N available in the litter can continue to be lost via ammonia volatilization, particularly if left unincorporated. Consequently, innovative subsurface tractor drawn implements have been developed, which can reduce ammonia losses by over 75% (Fig. 2; Pote et al., 2009; 2011).

The amount of ammonium nitrogen that is lost from manure depends on the method of land application. Immediate incorporation of manure by disking has been shown to reduce volatilization losses by 85 to 90% (Brunke et al., 1988). An additional incorporation method available is tillage following application, or irrigation following application. Incorporation of manure conserves valuable N and increases the precision of manure fertilizer applications. Application of at least 6.4 mm (0.25 inches) of irrigation will carry most of the ammonium nitrogen into the soil and will reduce ammonia losses (Chastain et al., 2001). Recommended estimates of ammonium N availability are provided in Table 12.

Primary factors that affect ammonia losses are manure moisture content, application method, application rate, temperature, rainfall, and wind speed. Therefore, it is not surprising that a great deal of variation in volatilization losses has been observed (Meisinger and Jokela, 2000; Table 13). The magnitude of volatilization losses from surface applied poultry manure range from 15% to 46% of the TAN applied (Table 13).

One method of poultry manure management that has received increased attention in recent years is composting, another is fabrication of organomineral fertilizer. Composting results in a more uniform product and 30 to 50% reduction in mass (Dao, 1999). The biggest advantage of composting is it reduces pathogen populations and odors (Sweeten, 1988). However, there are two big disadvantages of composting poultry litter: (i) approximately half of the total N in poultry litter is lost due to ammonia volatilization during the composting process, resulting in a poor fertilizer compared to fresh litter (DeLaune et al., 2004b, 2006); (ii)

Table 13. Magnitude of volatilization losses from surface applied poultry manure expressed as percentage of total ammoniacal nitrogen (TAN). Adapted from Meisinger and Jokela (2000).

Description	Range of Loss % of TAN†	Source
Surface application of poultry litter in Europe	15 to 45%	Jarvis and Pain (1990), Moss et al. (1995), Chambers et al. (1997)
Surface application of poultry litter on fescue pasture in the Southeastern U.S.	28 to 46%	Marshall et al. (1998)
Surface applied poultry litter in fall at a temperature of 64 °F (18 °C) in Maryland	30%	Meisinger and Jokela (2000)
Surface applied poultry litter in spring at a temperature of 48 °F (9 °C) in Maryland	15%	Meisinger and Jokela (2000)

† TAN, total ammonical nitrogen (NH_4^+-N + NH_3-N), the fraction of TAN that is in ammonia form is dependent on pH.

soluble and total P levels are significantly increased, resulting in greater P runoff than fresh litter (DeLaune et al., 2006); and, (iii) copper and zinc runoff is higher in composted litter compared with fresh litter (DeLaune and Moore, 2016). See Chapters of Hao and He (2019) and Modderman (2019) in this volume for additional information for poultry litter pelletization and composting.

Summary

Authors hope that the factors outlined in this chapter help readers understand the complex network of production system practices and externalities that drive poultry litter compositional and nutritive differences throughout U.S. poultry farms. Knowledge of the nutritive content of this important fertilizer source is essential in the nutrient management planning of poultry operations and for the reduction of nonpoint-source pollution.

Poultry manure contains all of the 13 essential plant macro- and micro-nutrients that are required for plant growth and development. These elements originate from supplements, feed, and enzymes. Some generalities about poultry litter nutrient characteristics and composition include: i) a multitude of factors influence elemental composition and form, which is not limited to bedding material, supplement and feed rates and sources, number of flocks between poultry house clean-out, litter storage and handling, and poultry operation type; ii) compared with other manure sources, poultry litter and manure is exceptionally high in total N and P, although some fractions of N are not immediately available to plants during the first year; iii) to prevent excess P loading, poultry litter and manure should be applied based on crop P requirements; and, iv) poultry litter and manure soil amendments have the ability to improve yield and soil quality, and is an important fertility source in agriculture.

References

Anderson, D.P., C.W. Beard, and R.P. Hanson. 1964. The adverse effects of ammonia on chickens including resistance to infection with Newcastle Disease virus. Avian Dis. 8:369–379. doi:10.2307/1587967

Anderson, K.R., P.A. Moore, Jr., D.M. Miller, P.B. DeLaune, D.R. Edwards, P.J.A. Kleinman, and B.J. Cade-Menum. 2018. Phosphorus leaching from soil cores from a twenty-year study evaluating alum treatment of poultry litter. J. Environ. Qual. 47:530–537. doi:10.2134/jeq2017.11.0447

ASABE. 2005. Manure production characteristics. ASABE Standard D384.2. American Society of Agricultural and Biological Engineers. St. Joseph, MI.

Ashworth, A.J., F.L. Allen, J. Wight, A. Saxton, and D. Tyler. 2014. Soil organic carbon sequestration rates under crop sequence diversity, bio-covers, and no-tillage. Soil Sci. Soc. Am. J. 78:1726–1733. doi:10.2136/sssaj2013.09.0422

Ashworth, A.J., F.L. Allen, A.M. Saxton, and D.D. Tyler. 2016. Long-term corn yield impacted by cropping rotations and bio-covers under no-tillage. Agron. J. 108:1495–1502. doi:10.2134/agronj2015.0453

Ashworth, A.J., F.L. Allen, D.D. Tyler, D. Pote, and M.J. Shipitalo. 2017a. Earthworm populations are affected from long-term crop sequences and bio-covers under no-tillage. Pedobiologia–International Journal of Soil Ecology 60:27–33. doi:10.1016/j.pedobi.2017.01.001

Ashworth, A.J., J. DeBruyn, F.L. Allen, M.A. Radiosevich, and P.R. Owens. 2017b. Microbial community structure is affected by cropping sequences and poultry litter under long-term no-tillage. Soil Biol. Biochem. 114:210–219. doi:10.1016/j.soilbio.2017.07.019

ASL. 2018. Compost Procedures, Agricultural service laboratory, Clemson Cooperative Extension, Clemson University, Clemson, SC. https://www.clemson.edu/public/regulatory/ag-srvc-lab/compost/procedures/index.html.

Behera, S.N., and M. Sharma. 2010. Investigating the potential role of ammonia in ion chemistry of fine particulate matter formation for an urban environment. Sci. Total Environ. 408: 3569–3575. doi:10.1016/j.scitotenv.2010.04.017

Blake, J.P., and J.B. Hess. 2001. Sodium bisulfate (PLT) as a litter amendment. ANR-1208. ACES Publication. Auburn, AL.

Blair, B. 1974. Evaluation of dehydrated poultry waste as a feed ingredient for poultry. Fed. Proc. 33:1934–1936.

Bolan, N.S., A.A. Szogi, T. Chuasavathi, B. Seshadri, M.J. Rothrock, Jr., and P. Panneerselvam. 2010. Uses and management of poultry litter. Worlds Poult. Sci. J. 66:673–698. doi:10.1017/S0043933910000656

Brunke, R., P. Alvo, P. Schuepp, and R. Gordon. 1988. Effect of meteorological parameters on ammonia loss from manure in the field. J. Environ. Qual. 17:431-436. doi:10.2134/jeq1988.00472425001700030014x

Chambers, B.J., K.A. Smith, and T.J. van der Weerden. 1997. Ammonia emissions following the land spreading of solid manures. In: S.C. Jarvis and B.F. Pain, editors, Gaseous nitrogen emissions from grasslands. CAB Internat. Oxon, UK. p. 275–280.

Chastain, J.P., and W.B. Smith. 2018. Unpublished manure nutrient data compiled by the Confined Animal Manure Managers Program. Clemson Cooperative Extension, Department of Agricultural Sciences, Clemson University, Clemson, SC.

Chastain, J.P., P.A. Rollins, and K.P. Moore. 2013. Composting of Turkey Brooder Litter in South Carolina: An On-Farm Demonstration Project. Journal of Agricultural Systems, Technology, and Management 24:36–50.

Chastain, J.P., A. Coloma-del Valle, and K.P. Moore. 2012. Using broiler litter as an energy source: Energy content and ash composition. Appl. Eng. Agric. 28(4):513–522. doi:10.13031/2013.42081

Chastain, J.P., P.A. Rollins, and M. Riek. 2007. Using poultry litter to fertilize longleaf pine plantations for enhanced straw production. 2007 ASABE Annual International Meeting, Minneapolis, MN. 17–20 June 2007. ASABE, St. Joseph, MI 49085. doi:10.13031/2013.23431

Chastain, J.P., J.J. Camberato, and P. Skewes. 2001. Poultry manure production and nutrient content. In: Clemson Cooperative Extension, editor, Confined animal manure managers certification program manual: Poultry version. Clemson University Extension, Clemson, SC. p. 3b-1–3b-17.

Chen, C., S. Hilaire, and K. Xia. 2019. Veterinary pharmaceuticals, pathogens and antibiotic resistance. In: H.M. Waldrip, P. H. Pagliari, and Z. He, editors, Animal manure: Production, characteristics, environmental concerns and management. ASA and SSSA, Madison, WI, Spec. Publ. 67. p. 359–382.

Collins, J.C., Jr., L.E. Barker, L.E. Carr, H.L. Brodie, and J.H. Martin, Jr. 1999. Poultry waste management handbook (NRAES-132). Natural Resource, Agriculture, and Engineering Service, Cooperative Extension, Ithaca, NY.

Coufal, C.D., C. Chavez, P.R. Niemeyer, and J.B. Carey. 2006. Nitrogen emissions from broilers measured by mass balance over 18 consecutive flocks. Poult. Sci. 85:384–391. doi:10.1093/ps/85.3.384

Dao, T.H. 1999. Coamendments to modify phosphorus extractability and nitrogen/phosphorus ratio in feedlot manure and composted manure. J. Environ. Qual. 28:1114–1121. doi:10.2134/jeq1999.00472425002800040008x

DeLaune, P.B., P.A. Moore, Jr., D.K. Carman, A.N. Sharpley, B.E. Haggard, and T.C. Daniel. 2004a. Development of a phosphorus index for pastures fertilized with poultry litter- factors affecting phosphorus runoff. J. Environ. Qual. 33:2183–2191. doi:10.2134/jeq2004.2183

DeLaune, P.B., P.A. Moore, Jr., D.K. Carman, A.N. Sharpley, B.E. Haggard, and T.C. Daniel. 2004b. Evaluation of the phosphorus source component in the phosphorus index for pastures. J. Environ. Qual. 33:2192–2200. doi:10.2134/jeq2004.2192

DeLaune, P.B., P.A. Moore, Jr., and J.L. Lemunyon. 2006. Effect of chemical and microbial amendments on phosphorus runoff from composted poultry litter. J. Environ. Qual. 35:1291–1296. doi:10.2134/jeq2005.0398

DeLaune, P.B., and P.A. Moore, Jr. 2016. Copper and zinc runoff from land application of composted poultry litter. J. Environ. Qual. 45:1565–1571. doi:10.2134/jeq2015.09.0499

Doye, D.G., J.G. Berry, P.R. Green, and P.E. Norris. 1992. Broiler production: Considerations for potential growers. Okla. Coop. Ext. Ser. Fact Sheet 202. Oklahoma State University, Stillwater, OK.

Edwards, D.R., and T.C. Daniel. 1992. Environmental impacts of on-farm poultry waste disposal-A review. Bioresource Technology 41: 9-33.

Edwards, D.R., and T.C. Daniel. 1993a. Effects of poultry litter application rate and rainfall intensity on quality of runoff from fescuegrass plots. J. Environ. Qual. 22:361–365. doi:10.2134/jeq1993.00472425002200020017x

Edwards, D.R., and T.C. Daniel. 1993b. Runoff quality impacts of swine manure applied to fescue plots. Trans. ASAE 36:81–86. doi:10.13031/2013.28317

Espinoza, L., N. Slaton, M. Mozaffari, and M. Daniels. 2007. The use of poultry litter in row crops. University of AR Fact Sheet FSA2147. University of Arkansas, Little Rock, AR. https://www.uaex.edu/publications/PDF/FSA-2147.pdf. (assessed Nov. 2017).

Eugene, B., P.A. Moore, Jr., H. Li, D. Miles, S. Trabue, R. Burns, and M. Buser. 2015. Effect of alum additions to poultry litter on in-house ammonia and greenhouse gas concentrations and emissions. J. Environ. Qual. 44:1530–1540. doi:10.2134/jeq2014.09.0404

FSA, Feedlot Services Australia. 2007. Feedlot services. Australia Pty Ltd. http://www.fsaconsulting.net/ (assessed Nov. 2017).

Gilley, J.E., and L.M. Risse. 2000. Runoff and soil loss as affected by the application of manure. Trans. ASABE 43:1583–1588. doi:10.13031/2013.3058

Goodwin, H.L., F.T. Jones, S.E. Watkins, and J.S. Hipp. 2003. New Arkansas laws regulate use and management of poultry litter and other nutrients. Arkansas Coop. Ext. Ser. FSA29. University of Arkansas, Little Rock, AR.

Gollehon, N., M. Caswell, M. Ribaudo, R. Kellogg, C. Lander, and D. Letson. 2001. Confined animal production and manure nutrients. Agricultural Information Bulletin No. 771. U.S. Department of Agriculture, Resources Economics Division, Economic Research and Service, Washington D.C.

Hao, X., and Z. He. 2019. Pelletizing animal manures for on- and off-farm use, In: H.M. Waldrip, P. H. Pagliari, and Z. He, editors Animal manure: Production, characteristics, environmental concerns and management. ASA Special Publication 67. ASA and SSSA, Madison, WI. p. 297–318.

He, Z., P.H. Pagliari, and H.M. Waldrip. 2016. Applied and environmental chemistry of animal manure: A review. Pedosphere 26:779–816. doi:10.1016/S1002-0160(15)60087-X

He, Z., and D.C. Olk. 2011. Manure amino compounds and their bioavailability. In: Z. He, editor, Environmental chemistry of animal manure. Nova Science Publishers, Inc., N.Y. p. 179–199.

He, Z., and Z.N. Senwo. H., Zou, I.A. Tazisong, and D.A. Martens, 2014. Amino compounds in poultry litter, litter-amended pasture soils and grass shoots. Pedosphere 24: 178-185.

Haggard, B.E., P.A. Vadas, D.R. Smith, P.B. DeLaune, and P.A. Moore, Jr. 2005. Effect of extraction ratios on soluble phosphorus content and its relation with runoff phosphorus concentrations. Biosystems Eng. 92:409–417. doi:10.1016/j.biosystemseng.2005.07.007

Hutchinson, G.L., and F.G. Viets, Jr. 1969. Nitrogen enrichment of surface water by absorption of ammonia volatilized from cattle feedlots. Science 166:514–515. doi:10.1126/science.166.3904.514

Jarvis, S.C., and B.F. Pain. 1990. Ammonia volatilization from agricultural land. The Fertiliser Society Proceedings. No. 298. p. 1-35. The Fertiliser Society, London.

Keshavarz, K., and R.E. Austic. 2004. The use of low-protein, low-phosphorus, amino acid- and phytase-supplemented diets on laying hen performance and nitrogen and phosphorus excretion. Poult. Sci. 83:75–83. doi:10.1093/ps/83.1.75

Kleinman, P.J.A., A.N. Sharpley, A.M. Wolf, D.B. Beegle, and P.A. Moore, Jr. 2002. Measuring water-extractable phosphorus in manure as an indicator of phosphorus in runoff. Soil Sci. Soc. Am. J. 66:2009–2015. doi:10.2136/sssaj2002.2009

Kleinman, P., D. Sullivan, A. Wolf, R. Brandt, Z. Dou, H. Elliot, J. Kovar, A. Leytem, R. Maguire, P. Moore, L. Saporito, A. Sharpley, A. Shober, T. Sims, J. Toth, G. Toor, H. Zhang, and T. Zhang. 2007. Selection of a water-extractable phosphorus test for manures and biosolids as an indicator of runoff loss potential. J. Environ. Qual. 36:1357–1367. doi:10.2134/jeq2006.0450

Kornegay, E.T. 1996. Nutritional, environmental and economic considerations for using phytase in pig and poultry diets. In: E.T. Kornegay, editor, Nutrient management of food animals to enhance and protect the environment. Lewis Publishers, Boca Raton, FL. p. 277–302.

Locatelli, A., K.L. Hiett, A.C. Caudill, and M.J. Rothrock. 2017. Do fecal and litter microbiomes vary within the major areas of a commercial poultry house, and does this affect sampling strategies for whole-house microbiomic studies? J. Appl. Poult. Res. 26:325–336. doi:10.3382/japr/pfw076

Leytem, A.B., P.W. Plumstead, R.O. Maguire, P. Kwanyuen, and J. Brake. 2007. What aspect of dietary modification in broilers controls litter water-soluble phosphorus: Dietary phosphorus, phytase, or calcium? J. Environ. Qual. 36:453–463. doi:10.2134/jeq2006.0334

Marshall, S.B., C.W. Wood, L.C. Braun, M.L. Cabrers, M.D. Mullen, and E.A. Guertal. 1998. Ammonia volatilization from tall fescue pastures fertilized with broiler litter. J. Environ. Qual. 27:1125–1129. doi:10.2134/jeq1998.00472425002700050018x

Meisinger, J.J., and W.E. Jokela. 2000. Ammonia volatilization from dairy and poultry manure. In: NRAES, editor, Managing nutrients and pathogens from animal agriculture (NRAES-130). Natural Resource, Agriculture, and Engineering Service, Cooperative Extension, Ithaca, NY. p. 334-354.

Modderman, C. 2019. Composting with or without organic residual additives. In: H.M. Waldrip, P.H. Pagliari, and Z. He, editors, Animal manure: Production, characteristics, environmental concerns and management. ASA Special Publication 67. Special, Madison, WI. p. 219–228.

Miles, D.M., S.L. Brannon, and B.D. Lott. 2004. Atmospheric ammonia is detrimental to the performance of modern commercial broilers. Poult. Sci. 83:1650–1654. doi:10.1093/ps/83.10.1650

Miles, D.M., P.A. Moore, Jr., D.R. Smith, D.W. Rice, H.L. Stilborn, D.R. Rowe, B.D. Lott, S.L. Branton, and J.D. Simmons. 2003. Total and water-soluble phosphorus in broiler litter over three flocks with alum litter treatment and dietary inclusion of high available phosphorus corn and phytase supplementation. Poult. Sci. 82:1544–1549. doi:10.1093/ps/82.10.1544

Miles, D.M., W.W. Miller, S.L. Brannon, W.R. Maslin, and B.D. Lott. 2006. Ocular responses to ammonia in broiler chickens. Avian Dis. 50:45–49. doi:10.1637/7386-052405R.1

Moore, P.A., Jr. 2011. Improving the sustainability of animal agriculture by treating manure with alum. In: Z. He, editor, Environmental chemistry of animal manure. Nova Science Publishers, Inc., Hauppauge, NY. p. 349–381.

Moore, P.A., Jr., T.C. Daniel, D.R. Edwards, and D.M. Miller. 1995a. Effect of chemical amendments on ammonia volatilization from poultry litter. J. Environ. Qual. 24:293–300. doi:10.2134/jeq1995.00472425002400020012x

Moore, P.A., Jr., T.C. Daniel, D.R. Edwards, and D.M. Miller. 1996. Evaluation of chemical amendments to reduce ammonia volatilization from poultry litter. Poult. Sci. 75:315–320. doi:10.3382/ps.0750315

Moore, P.A., Jr., T.C. Daniel, A.N. Sharpley, and C.W. Wood. 1998. Poultry manure management In: R.J. Wright, W.D. Kemper, P.D. Millner, J.F. Power, and R.F. Korcak, editors, Agricultural uses of municipal, animal, and industrial byproducts. USDA Agricultural Research Service, Washington, D.C. p. 60–77.

Moore, P.A., Jr., T.C. Daniel, and D.R. Edwards. 1999. Reducing phosphorus runoff and improving poultry production with alum. Poult. Sci. 78:692–698. doi:10.1093/ps/78.5.692

Moore, P.A., Jr., T.C. Daniel, and D.R. Edwards. 2000. Reducing phosphorus runoff and inhibiting ammonia loss from poultry manure with aluminum sulfate. J. Environ. Qual. 29:37–49. doi:10.2134/jeq2000.00472425002900010006x

Moore, P.A., Jr., D.M. Miles, R. Burns, K. Berg, and I.H. Choi. 2011. Ammonia emission factors from broiler litter in barns, in storage, and after land application. J. Environ. Qual. 40:1395–1404. doi:10.2134/jeq2009.0383

Moore, P.A., Jr., T.C. Daniel, A.N. Sharpley, and C.W. Wood. 1995b. Poultry manure management: Environmentally sound options. J. Soil Water Conserv. 50:321–327.

Moss, D.P., B.J. Chambers, and T.J. van der Weerden. 1995. Measurement of ammonia emissions from land application of organic manures. Asp. Appl. Biol. 43:221–228.

Mullen, J., U. Bekchanov, B. Karali, D. Kissel, M. Risse, K. Rowles, and S. Collier. 2011. Assessing the market for poultry litter in Georgia: Are subsidies needed to protect water quality? J. Agric. Appl. Econ. 43:553–568. doi:10.1017/S1074070800000079

Nahm, K.H. 2002. Efficient feed nutrient utilisation to reduce pollutants in poultry and swine manure. Crit. Rev. Environ. Sci. Technol. 32:1–16. doi:10.1080/10643380290813435

NASS, National Agricultural Statistical Service. 2015. USDA Poultry Production Data. NASS, Washington, D.C. https://www.usda.gov/sites/default/files/documents/nass-poultry-stats-factsheet.pdf (assessed November 2017).

NASS, National Agricultural Statistical Service. 2016. Poultry- Production and value. NASS, Washington, D.C. http://usda.mannlib.cornell.edu/usda/current/PoulProdVa/PoulProdVa-04-28-2017.pdf (assessed Nov. 2017).

National Chicken Council. 2015. Per capita consumption of poultry and livestock, 1965 to estimated 2012, in pounds. National Chicken Council, Washington, D.C. http://www.nationalchickencouncil.org/about-the-industry/statistics/per-capita-consumption-of-poultry-and-livestock-1965-to-estimated-2012-in-pounds/.

National Chicken Council. 2018. Per capita consumption of poultry and livestock, 1965 to estimated 2018, in pounds. National Chicken Council, Washington, D.C. https://www.nationalchickencouncil.org/about-the-industry/statistics/per-capita-consumption-of-poultry-and-livestock-1965-to-estimated-2012-in-pounds/ (Accessed 21 Feb. 2019).

Nelson, T.S., T.R. Shieh, R.J. Wodzinski, and J.H. Ware. 1968. The availability of phytate phosphorus in soybean meal before and after treatment with mold phytase. Poult. Sci. 47:1842–1848. doi:10.3382/ps.0471842

Nelson, T.S., T.R. Shieh, R.J. Wodzinski, and J.H. Ware. 1971. Effect of supplemental phytase on the utilization of phytate phosphorus by chicks. J. Nutr. 101:1289–1293. doi:10.1093/jn/101.10.1289

NRC, National Research Council. 1994. Nutrient Requirements of Poultry: Ninth Revised ed. National Academies Press, Washington, D.C.

Pagliari, P.H., and C.A.M. Laboski. 2014. Effects of manure inorganic and enzymatically hydrolyzable phosphorus on soil test phosphorus. Soil Sci. Soc. Am. J. 78:1301–1309. doi:10.2136/sssaj2014.03.0104

Patterson, P.H., P.A. Moore, Jr., and R. Angel. 2005. Phosphorus and poultry nutrition. In: A.N. Sharpley, editor, Agriculture and phosphorus management: The Chesapeake Bay. CRC Press, Boca Raton, FL. p. 635–682.

Payne, J. 2012. Litter management strategies impact nutrient content. Poultry Practices. Okla. Coop. Ext. Ser., Stillwater, OK.

Perez, P.G., R. Zhang, X. Wang, J. Ye, and D. Huang. 2015. Characterization of the amino acid composition of soils under organic and conventional management after addition of different fertilizers. J. Soils Sediments 15:890–901. doi:10.1007/s11368-014-1049-3

Pote, D.H., T.C. Daniel, A.N. Sharpley, P.A. Moore, Jr., D.R. Edwards, and D.J. Nichols. 1996. Relating extractable soil phosphorus to phosphorus losses in runoff. Soil Sci. Soc. Am. J. 60:855–859. doi:10.2136/sssaj1996.03615995006000030025x

Pote, D.H., T.C. Daniel, D.J. Nichols, A.N. Sharpley, P.A. Moore, D.M. Miller, and D.R. Edwards. 1999a. Relationship between phosphorus levels in three ultisols and phosphorus concentrations in runoff. J. Environ. Qual. 28:170–175. doi:10.2134/jeq1999.00472425002800010020x

Pote, D.H., T.C. Daniel, D.J. Nichols, P.A. Moore, Jr., D.M. Miller, and D.R. Edwards. 1999b. Seasonal and soil-drying effects on runoff phosphorus relationships to soil phosphorus. Soil Sci. Soc. Am. J. 63:1006–1012. doi:10.2136/sssaj1999.6341006x

Pote, D.H., T.R. Way, K.R. Sistani, and P.A. Moore, Jr. 2009. Water-quality effects of a mechanized subsurface-banding technique for applying poultry litter to perennial grassland. J. Environ. Manage. 90:3534–3539. doi:10.1016/j.jenvman.2009.06.006

Pote, D.H., T.R. Way, P.J. Kleinman, P.A. Moore, Jr., J.J. Meisinger, K.R. Sistani, L.S. Saporito, A.L. Allen, and G.W. Feyereisen. 2011. Subsurface application of poultry litter in pasture and no-till soils. J. Environ. Qual. 40:402–411. doi:10.2134/jeq2010.0352

Raboy, V. 2002. Progress in breeding low phytate corps. J. Nutr. 132:503S–505S. doi:10.1093/jn/132.3.503S

Ro, K.S., P.A. Moore, Jr., A.A. Szogi, and P.D. Millner. 2017. Ammonia and nitrous oxide emissions from broiler houses with downtime windrowed litter. J. Environ. Qual. 46:498–504. doi:10.2134/jeq2016.09.0368

Rynk, R., M. van der Kamp, G.B. Willsson, M.E. Singley, T.L. Richard, J.J. Kolega, F.R. Gouin, L. Laliberty, Jr., D. Kay, D.W. Murphy, H.A.J. Hoitink, and W.F. Brinton. 1992. On-farm composting handbook. NRAES-54. Natural Resource, Agriculture, and Engineering Service, Cooperative Extension, Cornell University, Ithaca, NY.

Self-Davis, M.L. and P.A. Moore, Jr. 2000. Determining water-soluble phosphorus in animal manure. In: G.M. Pierzynski, editor, Methods of phosphorus analysis for soils, sediments, residuals and waters. Southern Coop. Ser. Bull. 396. North Carolina State Univ., Raleigh, NC. p. 74–76.

Sharpley, A.N. 1995. Dependence of runoff phosphorus on extractable soil phosphorus. J. Environ. Qual. 24:920–926. doi:10.2134/jeq1995.00472425002400050020x

Sharpley, A.N., S. Herron, C. West, and T.C. Daniel. 2007. Overcoming the challenges of phosphorus-based management in poultry farming. J. Environ. Qual. 62:375–389.

Sharpley, A.N., and B. Moyer. 2000. Forms of phosphorus in manures and composts and their dissolution during rainfall. J. Environ. Qual. 29:1462–1469. doi:10.2134/jeq2000.00472425002900050012x

Sharpley, A.N., S. Herron, C. West, and T.C. Daniel. 2009a. Outcomes of phosphorus-based nutrient management in the Eucha-Spavinaw watershed. In: A. Franzluebbers, editor, Farming with grass: Sustainable mixed agricultural landscapes in grassland environments. Soil and Water Conservation Society, Ankeny, Iowa. p. 192–204.

Sharpley, A.N., N. Slaton, T. Tabler, K. VanDevender, M. Daniels, F. Jones, and T. Daniel. 2009b. Nutrient analysis of poultry litter. Ark. Coop. Ext. Ser. Publ. number FSA9529. Arkansas Cooperative Extension, Little Rock, AR.

Schroder, J.L., H. Zhang, J.R. Richards, and Z. He. 2011. Sources and contents of heavy metals and other trace elements in animal manures. In: Z. He, editor, Environmental chemistry of animal manure. Nova Science Publishers, NY. p. 385–414.

Shreve, B.R., P.A. Moore, Jr., T.C. Daniel, D.R. Edwards, and D.M. Miller. 1995. Reduction of phosphorus in runoff from field-applied poultry litter using chemical amendments. J. Environ. Qual. 24:106–111. doi:10.2134/jeq1995.00472425002400010015x

Smith, D.R., P.A. Moore, Jr., D.M. Miles, B.E. Haggard, and T.C. Daniel. 2004. Decreasing phosphorus runoff from land-applied poultry litter with dietary modifications and alum addition. J. Environ. Qual. 33:2210–2216. doi:10.2134/jeq2004.2210

Sonzogni, W.C., S.C. Chapra, D.E. Armstrong, and T.J. Logan, 1982. Bioavailability of phosphorus inputs to lakes. J. Environ. Qual. 11: 555–562. doi:10.2134/jeq1982.00472425001100040001x

Sweeten, J.M. 1988. Composting manure sludge. p. 38-44. In: National Poultry Waste Management Symp., Ohio State Univ. Columbus, OH.

Tabler, G.T., Y. Liang, and K.W. VanDevender. 2009. Poultry litter production and associated challenges. Avian Advice 11:8–10.

Tasistro, A.S., D.E. Kissel, and P.B. Bush. 2004. Spatial variability of broiler litter composition in a chicken house. J. Appl. Poult. Res. 13:29–43. doi:10.1093/japr/13.1.29

Turner, B.L., and A.B. Leytem. 2004. Phosphorus compounds in sequential extracts of animal manures: Chemical speciation and novel fractionation procedure. Environ. Sci. Technol. 38:6101–6108. doi:10.1021/es0493042

Turner, B.L., M.J. Paphazy, P.M. Haygarth, and I.D. McKelvie. 2002. Inositol phosphates in the environment. Philos. Trans. R. Soc. Lond. B Biol. Sci. 357:449–469. doi:10.1098/rstb.2001.0837

USDA-Natural Resource Conservation Service (NRCS). 2006. Poultry litter manure transfer incentives through the Environmental Quality Incentives Program (EQIP). Oklahoma Information Sheet. NRCS Oklahoma, Stillwater, OK. https://view.officeapps.live.com/op/view. aspx?src=https%3A%2F%2Fwww.nrcs.usda.gov%2Fwps%2FPA_NRCSConsumption%2Fdownloa d%3Fcid%3DSTELPRDB1257149%26ext%3Ddoc

Van Breemen, N., P.A. Burrough, E.J. Velthorst, H.F. van Dobben, T. de Wit, T.B. Ridder, and H.F.R. Reijinders. 1982. Soil acidification from atmospheric ammonium sulphate in forest canopy throughfall. Nature 299:548–550. doi:10.1038/299548a0

VanDevender, K., J. Langston, and M. Daniels. 2000. Utilizing dry poultry litter– An overview. Arkansas Coop. Ext. Ser. FSA8000. University of Arkansas, Little Rock, AR.

Zheng, W., and M. Guo. M., and G. Czapar. 2019. Environmental fate and transport of veterinary antibiotics derived from animal manure. In: H.M. Waldrip, P.H. Pagliari, and Z. He., editors, Animal manure: Production, characteristics, environmental concerns and management. ASA and SSSA, Madison, WI, Spec. Publ. 67.

Zublena, J.P., J.C. Baker, and T.A. Carter. 1993. Poultry manure as a fertilizer source. Soil Facts fact sheet authored by North Carolina Coop. Ext. Serv., Raleigh. AG-439-5. https://content.ces. ncsu.edu/poultry-manure-as-a-fertilizer-source. (Accessed 21 Feb. 2019).

Nutrient Characteristics of Swine Manure and Wastewater

Melissa L. Wilson,* Suresh Niraula, and Erin L. Cortus

Abstract

Swine (*Sus scrofa domestica*) manure can be a valuable source of nutrients for crops, but overapplication can result in water and air quality issues as excess nutrients leave the plant root zone via leaching, run-off, or gaseous emissions (examples include ammonia volatilization or nitrate denitrification). Thus it is important to understand nutrient characteristics as well as the factors that influence them to optimize agricultural production systems. The American Society of Agricultural and Biological Engineers (ASABE) has published standardized values of manure properties for a variety of animals, including swine, although their latest update summarized data up through 2002. Changes in swine production and management practices since this time have resulted in a need for an updated literature review. The objectives of this chapter were to: i) summarize and compare as-excreted and as-removed swine manure characteristics in literature from North America since 2003 and ii) review current trends in swine production in North America and how these practices may impact the variability seen in manure characteristics. Besides a literature review, we also summarized nutrient data from two commercial manure analysis laboratories in the midwestern United States. Overall, manure property information was found for a variety of growth stages (phases) and manure storage types, but the most complete information was found for finishing swine manure. Data for all nutrients in swine manure were highly variable so it is difficult to make conclusions about site-specific practices. We highly recommend that producers regularly analyze their manure to more accurately understand nutrient content.

Introduction

In the United States, there are over 72 million swine (*Sus scrofa domestica*) in production at any given time of the year (USDA-NASS, 2018), which produce roughly 31.5 billion gallons of manure per year. When used at appropriate agronomic rates, manure is a valuable resource that can provide nutrients for crops and improve soil productivity. In fact, there are often more nutrients per unit on average in stored swine manure compared with dairy or beef manure. On the other hand, applying more manure than any particular crop can use in a given year can cause air and water quality issues. Consequently, it is important to understand the properties of swine manure to optimize its use and minimize potential risks associated with manure application.

Manure properties that influence crop and soil productivity include nutrients and organic matter. Major nutrients of interest include N and P, which occur in inorganic and organic forms in manure, as well as K, which is present in the inorganic form. Another nutrient of recent interest is sulfur (S). In prior decades, sufficient

Abbreviations: CL1, Agvise Laboratories; CL2, Stearns DHIA Laboratories; CP, crude protein; DDGS, distillers dried grains with solubles; TN, total nitrogen.

M.L. Wilson and S. Niraula, Dep. Soil, Water, and Climate, University of Minnesota, 439 Borlaug Hall, 1991 Upper Buford Circle, Saint Paul, MN 55108; E.L Cortus, Department of Bioproducts and Biosystems Engineering, University of Minnesota, 1390 Eckles Ave., St. Paul, MN 55108. *Corresponding author (mlw@umn.edu).

doi:10.2134/asaspecpub67.c6

Animal Manure: Production, Characteristics, Environmental Concerns and Management. ASA Special Publication 67. Heidi M. Waldrip, Paulo H. Pagliari, and Zhongqi He, editors.

Table 1. The life cycle phases and common production systems used for each phase of pig production in the North America.

Life Cycle Phase (Description) Days, Weight Range	Production System†					
	Farrow-to-Finish	Farrow-to-Nursery	Farrow-to-Wean	Nursery	Wean-to-Finish	Grow-Finish or Finishing
Gestation (Pregnant Sow) 114 d, Variable weight	X	X	X			
Farrowing (Lactating Sow and Piglets) 21 d, 2 to 15 lbs (piglets)	X	X	X			
Nursery (Piglets) Up to 56 d, 15 to 60 lb	X	X		X	X	
Growing-Finishing (Pigs) Up to 120 d, 60 to 280 lb (market)	X				X	X

† Adapted from NPB (2009, 2016).

S was delivered to croplands through atmospheric deposition (David et al., 2016). However, due to the Clean Air Act, less S is being emitted from industrial and transportation sources (USEPA, 2017), so, farmers are looking at manure as a source of this nutrient for their crops. The organic matter content of manure will influence the forms of nutrients that are present and thus impact availability for crop growth.

The American Society of Agricultural and Biological Engineers (ASABE) has published standardized values of manure properties for a variety of animals, including swine (2005). The focus was on two things: properties of manure "as-excreted" by livestock and properties of manure "as-removed" from storage or animal housing. The as-excreted values help with planning for storage of the manure while the as-removed values help with planning for how the manure will be used. The data for these standards was last compiled in 2002, however, and since this time, trends in swine production have changed. For example, the USDA's Economic Research Service (USDA-ERS) reported that between 1998 and 2009, swine production shifted to larger operations and regionally from the Southeast to the Heartland. Due to differences in climate of these regions, the types of manure storage systems also shifted from largely lagoon storage to pits or tanks (Key et al., 2011).

The majority of hogs in the North America are raised in confinement under controlled conditions, and operations are increasingly moving toward specializing in certain growth phases (Key et al., 2011; Brisson, 2014). This allows producers to maintain optimum growing conditions and maximize efficiency for each growing phase. The growth phases can impact manure properties.

The life cycle of a pig includes three to five typical stages (NPB, 2016). *Farrowing* is the time from when a sow delivers her litter to when the piglets are weaned. This period is typically three weeks, and during this time the sow nurses the piglets. *Nursery* is the phase that lasts approximately six to eight weeks after weaning, when the newly weaned piglets transition to a solid diet. *Growing-Finishing* or *Grow-Finish* describe the phase that lasts approximately 16 to 17 weeks after the nursery phase. Historically, growing was a period of faster weight gain, until approximately 54 kg (120 lb), and finishing was the period for more fat deposition (NPB, 2009). At the end of this growth phase, pigs are market weight, or approximately 127 kg (280 lb). Most pigs will leave the cycle after this stage, but in some systems, producers retain females for breeding stock. *Gilt Development* is the stage where female pigs (gilts) continue past market weight until they reach maturity and are bred. *Gestation* is the pregnancy stage that lasts 114 d. Most sows are bred using artificial insemination, collected from boars on a small sub-set of farms across the country. Table 1 describes the production systems used to raise pigs through these various phases. Each stage of production requires specific feed formulations to meet the energy, amino acid, mineral, and vitamin requirements based on the stage of production. Most diets are corn and soybean meal-based (Sotak-Peper et al., 2015; Schulz, 2016), and the growth of the ethanol industry since 2000 has spurred considerable use of distillers dried grains with solubles (DDGS) (Stein and Shurson, 2009).

Swine production systems may employ open manure management systems, like pastures, but more commonly use confinement or barn systems (NPB, 2009). Swine manure is often collected in solid form when bedding is used. However, in most of confinement systems in North America, pigs are housed with either fully or partially slatted floors through which manure (feces + urine) can fall for short (< 1 d) or long-term (> 6 mo) storage before transfer to outdoor storage or field application (USDA-NRCS, 2011). The transfer process may include flushing or scraping the manure from underfloor pits.

The majority of swine manure in North America is stored in one of four ways: in deep pits underneath the production barn or outdoors in an earthen pit, storage tank, or lagoons. Earthen pits and lagoons require sufficient volume to store or treat precipitation and runoff in addition to manure collected in the storage. Deep pits, earthen pits, and storage tanks are typically used for storage of raw manure until land application, while lagoons are used to treat manure anaerobically before land application. Anaerobic lagoons allow denitrification and evaporation to occur, reducing the nitrogen and moisture content. Since this process is temperature dependent, lagoons are used more often in the southeastern United States where anaerobic treatment can happen year round. In fact, 90% of manure is stored in lagoons in this region while only 20 to 30% was stored in this manner in other regions (Key et al., 2011). In the United States as a whole, approximately 60% of hogs were produced on systems using pits or tanks, and only 34% are on systems using lagoons, as of 2009 (Key et al., 2011). Each system will impact nutrient and other manure characteristics differently.

Due to changes in the swine production industry, typical manure properties have likely changed since the ASABE standards were developed. Properties may be influenced by a multitude of factors: animal age and genetics, diet, housing type, storage, climate, and environmental management. Animal age and genetics as well as diet have likely impacted the as-excreted values of manure while changes in housing type, storage, climate, and environmental management have likely influenced the as-removed values. The objectives of this chapter are to: i) summarize and compare as-excreted and as-removed swine manure characteristics in literature from the United States since 2003 and ii) review current trends in swine production in the United States and how these practices may impact the variability seen in manure characteristics.

Methods
Literature Data

A survey of peer-reviewed published literature was conducted to identify articles related to nutrient content of swine manure: i) as-excreted and ii) as-removed from storage and/or applied to croplands. The search was limited to studies conducted in North America from 2003 to 2018. As-excreted literature data were included in the analysis if as-is manure composition data were available or could be calculated, or if the distribution of nutrient output was separated between feces and urine. Additional relevant papers were found by searching through the reference lists of papers already selected for the analysis. After the final screening of articles, we identified a total of 35 papers that presented as-excreted data and 46 papers that were used in the analysis of as-removed manure characteristic data. In most as-excreted cases, papers reported manure characteristics for multiple diets, and in some as-removed cases, papers reported multi-year or multi-site nutrient analysis values in their study. The as-excreted and as-removed manure nutrient values from ASABE (2005) were included in this review for comparison to earlier data. In addition, data based on swine growth stages from the Midwest Plan Service (MWPS, 2004) and data compiled by researchers at Pennsylvania State University (PSU) (Kephart, 2009) were also reported for comparison. It should be noted that this literature review focused on liquid swine manure as 86% of manure is handled as liquid in the United States (Key et al., 2011).

Laboratory Data

Manure nutrient analyses reports (excluding farm identifiable information) were obtained from two commercial laboratories, Agvise Laboratories (CL1) and Stearns DHIA Laboratories (CL2) located in the United States Midwest region and compiled by the authors. The swine manure nutrient data ranged from 2012 to 2017 for CL1 and from 2013 to 2017 for CL2. The data from CL1 and CL2 were combined to obtain the summary statistics on "Laboratory" values of selected nutrients. The estimates from laboratories provided information on livestock species and manure type; however, it did not provide the information on manure storage, housing, and feed inputs in diet.

Data Analysis

For as-removed cases, liquid swine manure with moisture levels greater than 90% were used for the summary statistics. Data collected from literature and laboratories on swine manure were converted to be expressed in units of g L^{-1}. The density, when not provided, was assumed 1 kg L^{-1} for urine, feces, and manure. The conversion factors for total P and K to P_2O_5 and K_2O are 2.29 and 1.2, respectively (Rieck-Hinz et al., 2011). Summary statistics on data from literature and laboratories, where available, are reported in Tables 2 through 4. The extreme observations were removed as outliers. Laboratory data are assumed to fall under the "as-removed" category and are placed under "unspecified housing/storage" section of Table 4. In addition, the published data on swine manure nutrients from ASABE standards, MWPS, and PSU are presented in Table 6.

Results
Manure "As-excreted" from the Animal
Characteristics

As-excreted manure is the combined urine and feces mixture prior to any storage. The ASABE (2005) Standard provides baseline swine manure characteristic values based on an average animal mass for each stage of production. ASABE (2005) further

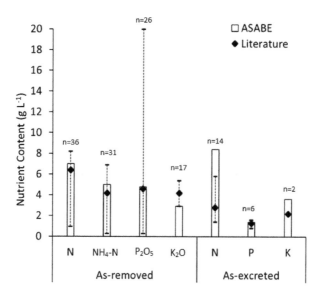

Fig. 1. Comparison of the American Society of Agricultural and Biological Engineers (ASABE) standard manure nutrient characteristics for finishing manure (as-removed and as-excreted swine and from pit storage) with data from a literature review of studies from 2003 to 2018 in North America. Dotted lines represent the maximum and minimum values and *n* is number of samples for literature review data.

promotes diet-based estimates of N and P excretion (grow-finish and sows only) and dry matter excretion (grow-finish pigs only) using farm-specific growth, feed, and carcass quality characteristics. Separately, urine and feces production and composition enable investigation into feedstuff digestibility and conversion. However, advanced manure treatment technologies like solid-liquid separation make use of the manure component characteristics. Table 2 summarizes manure production and concentration characteristics for as-excreted swine feces, urine, and manure reported from research studies in North America since 2003. In addition, Table 3 provides the average values as percentage of nutrient intake for swine feces, urine, and manure at various growth stages. Since the most complete data set was found for swine finishing manure, ASABE standards were compared directly with data from literature in Fig. 1.

Dry Matter Content

The rate of manure production increases with mass and feed intake. As-excreted swine manure is approximately 10% solids, with 80 to 90% of the solids organic or volatile solids (ASABE, 2005). Limited data suggests the feces are 36% solid (Ziemer et al., 2009; Jacobs et al., 2011; Spiehs et al., 2012). The range of total solids of combined liquid and feces was 2 to 14.5% in reviewed literature data, with an average of 5.9% for finishing pigs, which makes sense when approximately 75% of the total production is urine (Table 2).

Nitrogen

American Society of Agricultural and Biological Engineers (2005) provides standard estimates of N excretion per day, that when divided by average manure production rates result in total N concentrations ranging from 6.4 to 8.5 g L^{-1} (Table 2). The limited,

Table 2. The average daily production and concentration of selected nutrients in as-excreted swine manure from literature data (Note– averages represent a range of experimental diets).

Manure Components	Manure production	Total solids	Manure composition Nutrients N	P†	K†	S	Ca	References‡
	g animal⁻¹	%			g L⁻¹			
			Feces					
Nursery pig	58 (n = 5)§¶ 42–72#	–	–	18.1(n = 5)¶ 14.1–22.6	–	–	18.1(n = 5)¶ 16.8–20.5	4
Growing pig	213(n = 28) 50–533	38.2(n = 12) 34.4–42.0	–	12.2(n = 30) 5.2–31.4	–	–	–	1, 2, 5, 12, 14, 20
Finishing pig	790(n = 12) 509–917	34.6(n = 7) 30.6–39.1	11.1(n = 5) 9.5–13.1		–	–	6.3(n = 5) 1.2–26.3	–	7, 13, 19, 21
Gestating sow	375(n = 4)¶ 287–565	61.1(n = 4)¶ 60.8–61.4	–	–	–	–	–	15
			Urine					
Nursery pig	2650(n = 8)¶ 1750–4160	–	–	–	–	–	–	16
Growing pig	3099(n = 16) 831–5170	–	–	0.2(n = 6) 0.1–0.32	–	–	–	1, 5, 6, 12, 17
Finishing pig	5191(n = 8) 4770–5860	–	–	–	–	–	–	13,19
Gestating sow	5825(n = 4)¶ 3200–7400	–	–	–	–	–	–	15
			Manure					
Growing pig	4701(n = 8)¶ 3800–5630	–	–	–	–	–		5
Finishing pig	6304(n = 12) 3060–8958	5.9(n = 18) 2–14.5	2.8(n = 14) 1.4–5.8	1.3(n = 6) 0.8–1.6	2.2(n = 2)¶ 2.1–2.2	0.39(n = 10) 0.2–1.0	1.3(n = 4)	8, 9, 10, 11, 18, 19, 21
Gestating sow	6200(n = 4)¶ 3765–7687	–	–	–	–	–	–	15
			ASABE Manure Data					
Nursery pig	1333	10	8.5	1.4	3.3	–	–	
Finishing pig	4667	10	8.4	1.4	3.6			
Lactating sow	12000	10	7.1	2.1	2.1	–	–	3
Gestating sow	5000	10	6.4	1.8	4.4	–	–	
Boar	3800	10	7.3	2.5	4.6	–	–	

† To compare with excreted forms of P and K, convert oxide forms to elemental forms. Multiply P_2O_5 by 0.437 to get P and K_2O by 0.83 to get K.

‡ Reference list includes: [1]Adhikari et al. (2016); [2]Almeida and Stein (2010); [3]ASABE (2005); [4]Hanson et al. (2012); [5]Hill et al. (2009); [6]Ige et al. (2006); [7]Jacobs et al. (2011); [8]Kerr et al. (2006); [9]Kerr et al. (2018); [10]Li et al. (2017); [11]Li et al. (2015); [12]Nyachoti et al. (2006); [13]Pilcher et al. (2015); [14]Powers et al. (2006); [15]Renteria-Flores et al. (2008); [16]Rincker et al. (2005); [17]Shaw et al. (2006); [18]Shriver et al. (2003); [19]Spiehs et al. (2012); [20]Thacker et al. (2006); [21]Ziemer et al. (2009).

§ n represents number of samples.

¶ Only one reference contributed to average and range values.

Minimum and maximum values found in the study.

recent literature that provided feces, urine, and manure concentration data report the N content of total manure is less at 3 g L^{-1} for finishing pigs (Kerr et al., 2006, 2018; Ziemer et al., 2009; Li et al., 2015, 2017), which represents 45% of N intake on average (Table 3). Nitrogen in urine is predominantly urea that can rapidly transform to NH_4–N in the presence of the urease enzyme, which is contributed by fecal material (Ndegwa et al., 2008). Fecal N is mainly organic (in the feces). Ammoniacal-N was 16% of total N in the study by Kerr et al. (2006) for finishing pig manure.

Phosphorus

Standard P excretion estimates are between 1.4 and 2.5 g L^{-1} of manure, depending on stage of production (ASABE, 2005). Literature data (Li et al., 2015, 2017; Kerr et al., 2018) for finishing pigs agrees well with the standard estimates at approximately 1.3 g L^{-1} (Fig. 1). As a proportion of P intake, the majority of P intake in the feed is excreted in the feces (Rincker et al., 2005; Veum et al., 2009; Liu et al., 2012; Spiehs et

Table 3. The average daily production of selected nutrients in as-excreted swine manure from literature data as a percentage of nutrient intake (Note–averages represent a range of experimental diets).

Manure Component	Percentage of manure nutrient relative to feed intake					References†
	N	P	K	S	Ca	
			%—			
Feces						
Nursery pig	19.8(n = 9)‡§ 16.9–23.9¶	33.3(n = 14)§ 23.4–46.7		–	21.2(n = 14) 13.4–26.6	20,25
Growing pig	17.3(n = 45) 8.6–28.2	45.7(n = 46) 22.1–75.6	27.9(n = 4)§ 25.9–30.1	–	58.0(n = 23) 14.6–86.1	1, 4, 5, 7, 8, 9, 13, 14, 15, 16, 21, 24
Finishing pig	13.0(n = 27) 6.5–19.2	53.3(n = 6) 50.6–60.5		–	–	2, 3, 17, 22, 23, 26
Gestating sow	15.4(n = 4)§ 13.6–17.1	–		–	–	19
Urine						
Nursery pig	17.9(n = 9)§ 14.3–22.7	3.8(n = 14) 2.1–6.5		–	25.4(n = 14) 5.4–47.8	20,25
Growing pig	38.0(n = 40) 12.4–52.5	4.1(n = 46) 0.1–21.9	64.4(n = 4)§ 57.3–68.0	–	8.5(n = 15) 1.5–34.9	1, 4, 5, 7, 8, 13, 14, 15, 16, 21
Finishing pig	40.7(n = 27) 16.2–61.8	10.0(n = 6) 1.2–16.3		–	–	2, 3, 17, 22, 23, 26
Gestating sow	38.1(n = 4)§ 34.3–41.9	–		–	–	19
Manure						
Nursery pig	36.9(n = 9)§ 32.1–42.2	35.1(n = 19) 27.9–48.8		–	47.0(n = 14) 27–72.4	6, 19, 24
Growing pig	55.6(n = 33) 25.4–73.6	56.3(n = 29) 38.7–78.9	92.3(n = 4)§ 84.4–98.1	–	62.3(n = 13) 40.1–104.5	5, 7, 13, 14, 15, 16, 21
Finishing pig	45.1(n = 30) 21.7–69.8	56.9(n = 12) 29.9–70.2	50.2(n = 6) 44.1–57.2	44.2(n = 2)§ 44–44.3	2, 10, 11, 12, 17, 18, 22, 23, 26
Gestating sow	53.5(n = 4)§ 48.1–58.5	–		–	–	19

† Reference list includes: [1]Adhikari et al. (2016); [2]Atakora et al. (2011b); [3]Atakora et al. (2011a); [4]Emiola et al. (2009); [5]Guthrie et al. (2004); [6]Hanson et al. (2012); [7]Hill et al. (2009); [8]Htoo et al. (2007); [9]Jha and Leterme (2012); [10]Kerr et al. (2006); [11]Kerr et al. (2018); [12]Li et al. (2017); [13]Liu et al. (2012); [14]Nortey et al. (2007); [15]Nyachoti et al. (2006); [16]Pedersen et al. (2007); [17]Pilcher et al. (2015); [18]Pomar et al. (2014); [19]Renteria-Flores et al. (2008); [20]Rincker et al. (2005); [21]Shaw et al. (2006); [22]Shriver et al. (2003); [23]Spiehs et al. (2012); [24]Thacker et al. (2006); [25]Veum et al. (2007); [26]Widyaratne and Zijlstra (2007).
‡ *n* represents number of samples
§ Only one reference contributed to average and range values.
¶ Minimum and maximum values found in the study.

al., 2012; Adhikari et al., 2016; Kerr et al., 2018), and represents approximately 33% of P intake in the feed for nursery and 46% of the P intake in the feed for grow-finish pigs.

Potassium

Very little recent research has documented K concentrations in as-excreted manure, or partitioning between urine and feces. Hill et al. (2009) reported that 22% and 78% of K excreted were in the feces and urine, respectively. For finishing swine manure, the average concentration of total K was approximately 1.6-fold higher in the ASABE standard compared with the literature average (Fig. 1).

Other

Sulfur and calcium are reported in limited studies. Calcium is often reported in tandem with P, given its role in binding P. Sulfur is frequently studied relative to diets with DDGS, with the potential to increase S intake, excretion and odor production (Spiehs et al., 2012).

Influence of Diet

One of the largest influences on as-excreted manure is the feed. Historically, livestock producers fed their animals above typically recommended nutrient levels as a way to ensure dietary needs were being met. This means that higher levels of nutrients were being excreted in the manure than necessary. As feed costs and environmental concerns have risen, researchers and producers alike have searched for more efficient diets. Fertilizer prices also influence decisions, as manure nutrients can effectively replace commercial fertilizers. Generally speaking, manure characteristics can be manipulated by one or both of the following ways: altering the diet relative to the nutrient needs of the animal, and by improving digestibility. Other potential influences on manure characteristics are alternative feed sources and feed additives.

Accurately Meeting Nutritional Needs

Nutrient excretion is the difference between nutrient intake and nutrient retention. While 100% nutrient retention from the diet is unrealistic, Sutton and Lander (2003) estimated that formulating the diet closer to nutritional needs could reduce N and P excretions by 10 to 15%. Swine diets are formulated based on amino acid balance needs of the pig. While some feedstuffs like soybean meal provide the required distribution of amino acids, adding synthetic amino acids can more closely meet the pig requirements, reducing excretion. Synthetic amino acid additions are shown to reduce NH_4–N levels (Kerr et al., 2006; Ziemer et al., 2009; Liu et al., 2017a), as well as K levels (Li et al., 2015). Amino acids contribute to the crude protein (CP) makeup of a diet. Studies on the effect of dietary CP on total nitrogen (TN) is mixed, however. Shriver et al. (2003) and Atakora et al. (2011a) found that manure TN content (on a dry matter basis) was impacted by CP levels in the diet, while Kerr et al. (2006) and Li et al. (2015) did not find significant differences. Ziemer et al. (2009) found that increased CP in the diet only influenced manure TN levels when cellulose was also increased in the diet. As for P levels in manure, CP does not appear to be an influential factor (Sutton and Lander, 2003; Li et al., 2015).

The nutritional needs of swine varies through different life stages (NRC, 2012); thus, manure characteristics will change over time. To meet these nutritional changes, phase feeding, the practice of changing the diet formulation as the pig grows, is often employed. In a study to evaluate low-phytate soybean meal on pig growth, Powers et al. (2006) formulated diets to meet the nutritional needs of four different growth phases. Despite these formulations, there were still significant differences in P excretion across phases. In another study, Powers et al. (2007) observed phase effects on excreted N and NH_4–N; initially, levels rose through the first two grower phases, then decreased as pigs aged through the finishing phases. In general, phase feeding can reduce N and P by approximately 5% (Sutton and Lander, 2003). However, Knowlton et al. (2004) estimated that moving from one-phase feeding to three-phase feeding system, and better meeting the exact nutritional needs of the growing pigs, P excretion was reduced by 12.5%.

Improving Digestibility

Improving the digestibility of the diet is another method in improving feed efficiency. In plants, most P is held in the form phytate, which cannot be broken down by the swine digestive system. By adding the enzyme phytase to the diet to help break down phytate into a digestible form, P efficiency can be improved. It has been well documented that adding phytase can reduce P excreted by pigs (Hill et al., 2009; Almeida and Stein, 2010; Adhikari et al., 2016) by 34% (Brumm et al., 2002) and N excretion by 2 to 5% (Sutton and Lander, 2003). The addition of phytase also reduces the need for synthetic sources of P in the diet formulation, which has feed cost implications.

Besides adding the enzyme phytase, there are several other ways to improve feed efficiency. For instance, there are several crops, including corn and soybean, which have been genetically engineered to have low phytate content. Wienhold and Miller (2004) found that low-phytate corn decreased total P levels in manure. Hill et al. (2009) reported that both low-phytate corn and low-phytate soybean significantly reduced P excretion in pigs by 12% and 15.5%, respectively, compared with traditional corn and soybean.

Another way to improve digestibility is to change the physical characteristics of the diet. For example, pelletizing or finely grinding the diet can decrease dry matter and nutrients excreted by swine (Ferket et al., 2002). Luo (2016) reported that by decreasing the particle size of food fed to pigs from coarse to fine, the percent N excreted in fecal matter was decreased by 18%, but the percent S was not affected. The process of pelletizing food includes heat treatment, which helps reduce antinutritional factors that reduce the availability of nutrients (Ferket et al., 2002). Rojas et al. (2014) studied diets that were extruded, pelletized, or both and found that especially in higher fiber diets, N digestibility was increased.

Alternative Feed Sources

Due to rising feed costs, researchers and swine producers have looked to novel feed sources. One such example that has been evaluated is distillers dried grains with solubles (DDGS), a fermented product with low concentrations of phytate-bound P (Stein and Shurson, 2009), that is increasingly available as a byproduct of ethanol production. It has been found that adding DDGS to swine diets increases the amount of manure produced (Xu et al., 2006; Trabue and Kerr, 2014; Kerr et al., 2018). This means that while concentrations of nutrients can be reduced in manure when adding DDGS to the diet, total excreted nutrients can actually increase in some cases. For example, Spiehs et al. (2012) reported that P excreted by pigs was not affected by a diet including 20% DDGS, despite evidence that DDGS increased P digestibility. However, a review of the literature by Stein and Shurson (2009) found that in most cases, P concentrations in manure are reduced by adding DDGS to the diet as long as the total dietary P is adjusted to account for the higher digestibility of P in DDGS. As for N, studies are generally in agreement that increasing levels of DDGS increase the levels of N excreted by pigs (Jha and Leterme, 2012; Spiehs et al., 2012; Trabue and Kerr, 2014; Pilcher et al., 2015). The amount of S in diets containing DDGS tends to be higher than conventional diets, thus more S is excreted in manure as well (Trabue and Kerr, 2014). Kerr et al. (2018) also reported increased S concentrations in swine manure due to DDGS in the diet compared with conventional corn–soybean meal (0.98 vs. 0.62 g S L^{-1}, respectively).

Other examples of novel feed sources include fishmeal, lemna protein concentrate, and dried skim milk. Rojas et al. (2014) evaluated the replacement of soybean meal with either fishmeal or lemna protein concentrate (a product made from the plant commonly known as duckweed) in diets formulated to meet the energy requirements of the pigs. They found that fishmeal increased swine P excretion compared with soybean meal, while lemna protein concentration reduced P excretion. This is likely due to changes in P intake in the various diets, however, since the authors reported that percent P digestibility did not differ between treatments. Yen et al. (2004) reported no differences in manure N concentrations between pigs fed traditional diets versus a diet with 10% dried skim milk to replace some of the soybean meal. Algae (Urriola et al., 2018) and insects (Surendra et al., 2016) have also been studied as protein sources for swine and other livestock production in the United States, but the impact on manure has not been evaluated.

Feed Additives

There are a variety of feed additives to promote growth or improve the health of swine, but few studies have evaluated the impact of feed additives on manure characteristics. The ionosphore narasin, an anticoccidial drug, has been shown to improve the growth performance of pigs (Kerr et al., 2017), and it did not significantly impact manure characteristics, with the exception of an increase in carbon (Kerr et al., 2018). Other additives such as tylosine phosphate, an antimicrobial, and ractopamine HCl, a nutrient portioning agent, did not impact manure N excretion in swine (Pilcher et al., 2015).

Manure and Wastewater "As-removed" Rrom Storage for Land Application

Characteristics

As-removed manure is the combination of livestock feces and urine with wasted feed, bedding material, and wastewater in the animal confinement. Additionally, there are many transformations and losses that occur during handling and storage of manure that causes changes in characteristics from the original, as-excreted form. As an example, manure stored in lagoons from finishing growth stages may differ in composition and characteristics compared with the lagoons from nurseries. Moreover, the concentration of manure nutrients may differ with component (wastewater vs. sludge) due to vertical stratification of either physical or chemical parameters in swine lagoons.

The average concentration of selected nutrients in swine lagoons and pit storage based on growth stages and/or physical components from research studies in North America since 2003 is summarized in Table 4. Manure characteristics according to the ASABE (2005) Standards and Midwest Plan Service (MWPS, 2004) are shown in Table 5 and include information prior to 2002 and 2004, respectively. Summarized data from PSU (Kephart, 2009) are also included for each phase of production common

Table 4. The average concentration of selected nutrients in swine manure from data obtained from literature and commercial laboratories in the U.S. Midwest region.

Storage and Housing	Manure nutrient concentration								References‡
	Total N	NH$_4$–N	P$_2$O$_5$†	K$_2$O†	Ca	Na	Mg	S	
	—————————————————————g L^{-1}—————————————————————								
Lagoons (literature data)									
Wastewater/slurry									
Nursery §	0.48(n = 6)¶ 0.40–0.52#	0.19	0.17(n = 6) 0.16–0.18	0.73	0.15	0.11	0.04	0.02	1
Gestation/ farrow	0.42(n = 7) 0.32–0.59	0.35	0.15(n = 7) 0.13–0.18	0.60(n = 2) 0.47–0.72	0.10(n = 2) 0.05–0.15	0.21	0.04(n = 2) 0.03–0.05	0.05	1,2
Finisher	2.16(n = 21) 0.40–6.20	2.32(n = 10) 0.31–5.0	0.70(n = 14) 0.02–4.08	1.34(n = 7) 0.48–2.09	0.46(n = 7) 0.06–1.40	0.33(n = 7) 0.10–0.52	0.26(n = 7) 0.02–0.94	0.03	1,3,4,5,6
Unspecified growth	0.28(n = 2) 0.16–0.39	0.23(n = 2) 0.13–0.33	2.26(n = 5) 0.07–4.58	0.30(n = 2) 0.17–0.44	0.05(n = 2) 0.04–0.07	0.12(n = 2) 0.05–0.19	0.03	–	7,8,9
Sludge									
Unspecified growth§	3.2(n = 2) 3.0–3.4	0.7(n = 2) 0.54–0.86	4.9(n = 2) 3.1–6.6	1.4(n = 2) 1.2–1.6	4.2(n = 2) 1.3–7.1	0.27(n = 2) 0.25–0.28	–	–	10
Pits (literature data)									
Sow/gestation/ farrow	2.2(n = 11) 0.77–3.4	1.4(n = 11) 0.40–2.7	1.8(n = 6) 0.50–3.0	–	–	–	–	–	11–15
Finisher/ farrow-finish	6.4(n = 36) 0.94–8.2	4.2(n = 31) 0.30–6.9	4.6(n = 26) 0.27–20.0	4.2(n = 17) 2.9–5.4	–	–	–	–	13–25
Unspecified growth	4.1(n = 13) 0.63–7.7	2.6(n = 15) 0.20–5.2	3.0(n = 13) 0.13–6.0	2.3(n = 11) 0.5–4.3	–	–	–	–	14,20,26,27
Unspecified growth/housing									
Literature data	7.8(n = 36) 0.77–30.0	2.0(n = 41) 0.36–4.7	5.7(n = 45) 0.11–26.8	9.7(n = 21) 1.3–29.6	–	–	–	–	28–46
Laboratory data	5.2(n = 26729) 0–21.9	3.6(n = 13356) 2.4–10.6	2.6(n = 26714) 5.6–18.3	3.6(n = 26721) 0.05–28.2	0.5(n = 9451) 0.001–4.7	0.4(n = 9451) 0.001–3.1	0.4(n = 9450) 0.001–2.5	0.1(n = 13502) 0.0–2.2	

† To compare with excreted forms of P and K, convert oxide forms to elemental forms. Multiply P$_2$O$_5$ by 0.437 to get P and K$_2$O by 0.83 to get K.

‡ Reference list includes: [1]Westerman et al. (2010), [2]McLaughlin et al. (2012), [3]Buckley et al. (2011), [4]Ball Coelho et al. (2007), [5]Kranz et al. (2005), [6]Kumaragamage et al. (2016), [7]Adeli et al. (2008), [8]Hubbard et al. (2011), [9]Kumaragamage et al. (2016), [10]Israel and Smyth (2015), [11]Bullock et al. (2016), [12]Loria et al. (2007), [13]Park et al. (2006), [14]Parker et al. (2013), [15]Higgins et al. (2004), [16]Chantigny et al. (2007), [17]Chantigny et al. (2008), [18]Chantigny et al. (2009), [19]Chantigny et al. (2010), [20]Karlen et al. (2004), [21]Maurer et al. (2017), [22]Schuster and Bartelt-Hunt (2017), [23]Smith et al. (2007b), [24]Woli et al. (2013), [25]Nicolaisen et al. (2007), [26]Moody et al. (2009), [27]Singer et al. (2008), [28]Cavanagh et al. (2011), [29]Ball Coelho et al. (2012), [30]Jarecki et al. (2009), [31]Jayasundara et al. (2010), [32]Karimi et al. (2017), [33]Lamb (2014), 34Long et al. (2018), [35]Loria and Sawyer (2005), [36]McAndrews et al. (2006), [37]Nikièma et al. (2013), [38]Parkin et al. (2006), [39]Priyashantha et al. (2007), [40]Rochette et al. (2009), [41]Schlegel et al. (2017), [42]Smith et al. (2007a), [43]Smith et al. (2008), [44]Tabbara (2003), [45]Thomas et al. (2016), [46]Springer et al. (2005).

§ Only one reference contributed to average and range values.

¶ n represents number of samples

Minimum and maximum values found in the study

Table 5. The average pH and electrical conductivity (EC) of selected nutrients in swine manure from data obtained from literature and commercial laboratories in the U.S. Midwest region.

Storage and Housing	pH	EC	References‡
		dS m⁻¹	
Lagoons (literature data)			
Wastewater/slurry			
Nursery §	7.5	7.5	1
Gestation/farrow	7.8(n = 2) 7.5–8.1	6.2(n = 2) 5.0–7.5	1,2
Finisher	7.3(n = 2) 6.7–7.8	11.6(n = 9) 4.9–18.0	1,3,4
Unspecified growth §	7.9	–	5
Sludge			
Unspecified growth§	7.3(n = 2) 7.0–7.6	–	6
Pits (literature data)			
Sow/gestation/farrow	7.3(n = 11) 5.8–8.4	4.3(n = 2) 4.3–4.4	7–11
Finisher/farrow-finish	7.7(n = 16) 6.9–8.7	21.4(n = 3) 7.8–42.4	9–17
Unspecified growth§	7.6(n = 4) 7.2–8.0	7.7(n = 4) 6.4–9.0	10
Unspecified growth/housing			
Literature data	7.5(n = 26) 6.3–8.8	8.0(n = 14) 2.3–19.8	18–28
Laboratory data	7.9(n = 440) 3.9–8.9	12.6(n = 252) 0.4–38	–

‡ Reference list includes: [1]Westerman et al. (2010); [2]McLaughlin et al. (2012); [3]Buckley et al. (2011); [4]Kumaragamage et al. (2016); [5]Adeli et al. (2008); [6]Israel and Smyth (2015); [7]Bullock et al. (2016); [8]Loria et al. (2007); [9]Park et al. (2006); [10]Parker et al. (2013); [11]Higgins et al. (2004); [12]Chantigny et al. (2007); [13]Chantigny et al. (2008); [14]Chantigny et al. (2009); [15]Maurer et al. (2017); [16]Schuster and Bartelt-Hunt (2017); [17]Nicolaisen et al. (2007); [18]Cavanagh et al. (2011); [19]Jarecki et al. (2009); [20]Lamb (2014); [21]Loria and Sawyer (2005); [22]McAndrews et al. (2006); [23]Nikièma et al. (2013); [24]Parkin et al. (2006); [25]Rochette et al. (2009); [26]Smith et al. (2007a); [27]Smith et al. (2008); [28]Thomas et al. (2016).

§ Only one reference contributed to average and range values.

¶ n represents number of samples

Minimum and maximum values found in the study

to Pennsylvania and the northeast region of the United States. For comparison purposes, the most complete information in the literature and the ASABE standards was for swine finishing manure in pits (Fig. 1). The differences found in Table 5 and Fig. 1 re-emphasize the need for updating the ASABE standards and adding additional categories for as-removed manure based on storage type and animal phase.

Total N

As shown in Tables 4 and 5, total N concentration across all reviewed data sources ranged from 0.28 to 7.8 g L⁻¹. The lowest values tended to be in lagoon systems, particularly the wastewater or slurry component, though there were differences among sources of data. Lagoon wastewater and/or slurry ranged from 0.28 to 0.48 g L⁻¹ in the literature, which was slightly lower than the ASABE standard at 0.6 g L⁻¹. Few studies evaluated lagoon sludge, but Israel and Smyth (2015) reported an average of 3.2 g L^{-1} while the ASABE standard was 2.6 g L^{-1}, a difference of about 23%. Swine manure from pit storages tended to have higher total N. The literature data showed that across growth phases, pit slurry averaged 4.2 g L^{-1}; however, the literature values were as high as 8.2 g L^{-1} (Woli et al., 2013). The MWPS and PSU data, across growth stages, ranged from 2.2 to 4.4 g L^{-1} and 1.8 to 7.0 g L^{-1}, respectively (Table 5). When comparing the finishing manure in pits, the literature data showed 9.4% less total N than the ASABE reported values, though variability was high in the literature data (Fig. 1). The summarized laboratory data, which is assumed to include all growth phases and unknown storage, showed an average of 5.2 g L^{-1} of total N (Table 3), much higher than ASABE standards across storage units, except the finisher slurry.

Ammonium N

The total NH_4–N concentration across all literature data sources ranged from 0.19 to 4.2 g L^{-1} (Table 4) and was similar to the range of values reported in the ASABE standards (0.4 to 5.0 g L^{-1}) in Table 5. The literature data showed that in lagoons, the manure from finisher barns tended to have 3 to 11 times more NH_4–N than the other growth stages while in deep pits the finisher-barn manure had approximately 2.5 times more NH_4–N. When comparing lagoons to deep pits in general, the wastewater/slurry NH_4–N concentrations were lower than in deep pits. The MWPS and PSU values on NH_4–N concentrations in pits across growth stages ranged from 1.6 to 3.1 g L^{-1} and 1.0 to 4.7 g L^{-1} (Table 5) and were close to the range of literature reported values of 1.4 to 4.2 g L^{-1} (Table 4). Literature data showed 19% less total NH_4–N than the ASABE reported values for finishing pig manure in pit (Fig. 1). The laboratory value found for NH_4–N was 3.6 g L^{-1} (Table 4), well within ASABE range across storage types.

Total P as P_2O_5

The P_2O_5 concentration across all available data sources ranged from 0.15 to 5.7 g L^{-1} (Table 4). According to PSU and MWPS data, the P_2O_5 concentration in manure stored in pits ranged from 1.6 to 3.1 g L^{-1} and 1.4 to 5.3 g L^{-1} respectively, across growth stages. These values were well within the range of ASABE standard values (0.5 to 5.7 g L^{-1}) on storage (Table 6). The ASABE reported value of 0.5 g L^{-1} for lagoon surface water was within the lagoon wastewater/slurry concentrations in literature values (0.15–2.26 g L^{-1}). The lagoon sludge P_2O_5 concentration was about 16% higher in ASABE compared with literature reported values. The average P_2O_5 values from both the ASABE standards and the literature for the finishing pig manure in pits were similar (~4.7 g L^{-1}), and the variability (standard deviation) in the literature data was fairly high (Fig. 1). However, the measure of variability was not provided in the ASABE standard for swine manure. The laboratory value on P_2O_5 was 2.6 g L^{-1}, well within the data range found in this study.

Total K as K_2O

The K_2O concentration across all data sources ranged from 0.30 to 9.7 g L^{-1} (Table 4). The PSU and MWPS data, which is from manure stored in pits, ranged from 1.3 to 2.6 g L^{-1} and 1.3 to 4.8 g L^{-1}, respectively, across growth stages (Table 6). The ASABE standard across storage types, including pits and lagoons, ranged from 0.7 to 2.9 g L^{-1}. As with total N, NH_4–N, and P_2O_5, the concentration of K_2O was lower in finisher lagoons than in manure stored in pits. The literature data showed K_2O

Table 6. The average concentration of selected nutrients in swine manure from data obtained from the American Society of Agricultural and Biological Engineers (ASABE), the Midwest Plan Service (MWPS), and Pennsylvania State University (PSU).

Storage and Housing	Manure nutrient concentration							
	Total N	NH_4–N	P_2O_5	K_2O	Ca	Na	Mg	S
	—————————————g L^{-1}—————————————							
ASABE (2005)								
Finisher slurry, wet-dry feeders	7.0	5.0	4.8	2.9	2.5	0.4	–	0.4
Flush building	2.0	1.4	1.6	2.0	0.4	0.3	0.3	0.2
Agitated solids and water	1.0	0.5	1.4	0.7	0.8	0.2	0.3	0.2
Lagoon surface water	0.6	0.4	0.5	0.8	0.1	0.2	0.1	0.04
Lagoon sludge	2.6	0.7	5.7	0.8	0.4	0.2	0.1	0.1
MWPS (MWPS, 2004b)– Pit storage								
Farrowing	1.8	1.0	1.4	1.3	–	–	–	–
Nursery	3.0	1.7	2.3	2.6	–	–	–	–
Grow-finish (deep pit)	6.0	4.0	5.0	3.6	–	–	–	–
Grow-finish (wet/dry feeder)	7.0	4.7	5.3	4.8	–	–	–	–
Grow-finish (earthen pit)	3.8	2.9	2.6	2.4	–	–	–	–
Breeding-gestation	3.0	1.4	3.0	2.9	–	–	–	–
Farrow-finish	3.4	1.9	2.9	2.8	–	–	–	–
Farrow-feeder	2.5	1.3	2.2	2.3	–	–	–	–
PSU (Kephart, 2009)– Pit storage								
Farrow to wean	2.2	1.6	2.2	1.3	–	–	–	–
Swine wean-finish	4.4	3.1	2.8	2.5	–	–	–	–
Swine grow-finish	3.7	2.8	2.9	2.6	–	–	–	–
Swine nursery	2.3	1.7	1.0	1.7	–	–	–	–

was lower in lagoon wastewater or slurry than in lagoon sludge, while the ASABE standard found the concentrations to be similar in the two components. Interestingly, when comparing the data for finishing swine manure stored in pits, the K_2O concentration found in the literature data was approximately 45% higher than the ASABE standard (Fig. 1). The laboratory value on K_2O was at 3.6 g L^{-1}, well within the ranges found in other data sources, particularly for manure stored in pits.

Other Nutrients

The means of available data from the laboratory summaries and the literature on Ca, Na, Mg, and S showed a range of values (Table 4). Literature data based on one study (Westerman et al., 2010) showed higher Ca levels of 0.46 g L^{-1} from finisher compared with lagoons from nursery lagoons and gestation and farrow systems. In contrast, the highest ASABE reported values for Ca was at 2.5 g L^{-1} from finisher slurry (wet-dry feeders) (Table 6). In terms of other manure nutrients, literature data suggested slightly higher concentrations of Na (0.21 g L^{-1}) and S (0.05 g L^{-1}) at gestation and/or farrow lagoons compared with the literature data on nursery and finisher lagoons. Laboratory values from unspecified storage or housing for Ca, Na, and S were within the values reported in ASABE standards.

Other Manure Characteristics

In the literature, the pH levels of swine manure varied from 7.3 to 7.8, and there did not appear to be any trends in storage type or growth stages (Table 5). The commercial laboratories report showed swine manure pH in the range of 3.9 to 8.9, with an average of 7.9. Understanding manure pH in important as it is linked to potential gaseous nutrient losses.

The average total soluble salts measured as electrical conductivity (EC) in the literature ranged 4.3 to 21.4 dS m^{-1} (Table 5) and showed more variability in manure from pits compared to lagoons. The EC of swine manure from laboratory data varied from 0.4 to 38 dS m^{-1} suggesting a wide range of variability in salt content of manure. The potential of manure-induced soil salinization may increase when manure with higher EC is applied to soil. An EC measurement may be of importance when irrigating liquid manure or effluent onto standing crops so as to avoid potential leaf burn.

Influence of Housing, Environmental Management, and Storage

Once manure is excreted, there are a variety of chemical and physical changes that can occur, which explains the differences seen in as-excreted and as-removed manure reported in Tables 2, 4, and 5. For instance, one of the main loss pathways for nutrients, particularly N and S, is through gaseous NH_3, N_2O, and H_2S emissions. This is highly dependent on manure pH levels since low pH (less than 7) tends to favor H_2S emissions (Dai et al., 2015; Liu et al., 2017b) while high pH (greater than 7) favors NH_3 emissions (Rotz, 2004; Higgins et al., 2005; Trabue and Kerr, 2014). Phosphorus and K can be influenced as solids settle out of the liquid portion. All of these factors are greatly influenced by housing, the environmental conditions the pigs live in, and how manure is stored and treated.

Housing and Environmental Management

In most of confinement systems in North America, pigs are housed with either fully or partially slatted floors through which manure (feces + urine) can fall. Not all manure makes it through the slats, however, and urine puddles have been identified as a significant source of NH_3 loss in swine housing (Cortus et al., 2008). In a small-scale study, Kai et al. (2006) estimated that 50 to 60% of the NH_3 emissions came from the manure pit while 40 to 50% came from the manure on slats in a fully slatted floor. The cleanliness of housing also appears to impact the manure properties during storage. Li et al. (2017) evaluated dirty housing (sprayed with manure) versus clean (rinsed daily), and found that when stored in simulated deep pits, the manure from the dirty housing system had increased NH_3, N_2O, and H_2S gas emissions.

Manure is generally considered to be made up of feces and urine, but it also includes wasted food and water. Since pigs tend to root in their food, spillage can occur and it is estimated that N and P in manure can increase by 1.5% for every 1% of food spilled (Ferket et al., 2002). While wasted water may not impact nutrient content of the manure, it can cause dilution which ultimately influences how the manure is handled. Wasted water may increase depending on water delivery device, housing temperature, and pig behavior (Patience, 2012). Typically, nipple drinkers have more wastage than cup drinkers or wet and/or dry feeders (Matlock et al., 2014) and there are estimates that proper management and drinker selection could reduce water wastage by 30% (Muhlbauer et al., 2011).

Management of environmental factors within housing, like temperature and humidity, have an impact on manure properties. Temperature and humidity not only affect the food intake (and thus manure production) of pigs (NRC, 2012), but they can influence gaseous emissions and nutrient transformations of manure stored in pits as well. In a meta-analysis of swine production facilities in the United States, Liu et al. (2014) reported that NH_3 emissions from all types of swine facilities (including deep pits and recharge pits) increased with increasing temperature, but this was not true of H_2S emissions. Thorne et al. (2009) reported that H_2S emissions in barn were impacted by number of pigs housed, wind, and relative humidity.

Ventilation systems in swine housing can impact gaseous emissions and thus nutrient characteristics of manure. Besides impacting temperatures in the barn, ventilation also removes NH_3 from the production house, and NH_3 losses are increased because of "reduced resistance to NH_3 transfer into the air above the manure" (Ndegwa et al., 2008). Higher air velocities over the manure surface, resulting from increased ventilation, also increase the rate of convective mass transfer (Cortus et al., 2008). Jacobson et al. (2005) reported that there were seasonal changes in NH_3 and H_2S gas emissions in two gestation sow barns, but these were primarily due to changes in ventilation from summer to winter, since ventilation is inversely related to gas concentrations.

Handling

Handling and moving manure into storage can impact manure characteristics, particularly nutrients and dry matter content. For example, in flush systems, water is used to transport manure from temporary storage under the confinement building to outside storage. When manure is flushed, there is a burst of gaseous H_2S losses, but flushing more frequently (daily instead of every 7 or 14 d) reduced overall H_2S and NH_3 gas losses (Lim et al., 2004). In flush systems with pit recharge (the pit is partially filled with water for the manure to fall into), the rate of pit recharge (or how often manure is removed) can also impact H_2S emissions (Liu et al., 2014). For example, the longer the manure is in the pit before being flushed, the more H_2S is emitted.

Agitation is often used to mix manure prior to application so that nutrients are redistributed uniformly. Unfortunately, this process promotes the aerial loss of nutrients. In a deep pit system, Hoff et al. (2006) found that the average H_2S concentration in the barn increased by approximately 62 times the before-agitation levels, while NH_3 increased by 10 times. Since P levels are associated with solids that settle to the bottom of pits or tanks, the length of agitation may impact P in the manure as it is removed. Higgins et al. (2004) reported that with 10 min of agitation in a 151,500 L (40,000 gal) deep pit (replicated over several seasons), P levels still increased toward the bottom of the pit, as shown in subsamples taken from every load as the pit was unloaded.

Storage

Typically, deep pits will retain more nutrients, especially N, than lagoons and earthen pits. One reason is because deep pits are not impacted by dilution with flush water or rainfall and are less likely to have evaporative losses (Lorimor et al., 2008). Normally, nutrient losses occur by gaseous emissions or by incomplete draining of the pit during cleanout. In a review of the literature, Griffing et al. (2007) reported that NH_3 emissions in slurry pits were approximately 21% of excreted total Kjehldal N. Phosphorus is not lost as a gas, but can be left behind in solids that have settled prior to cleanout. Higgins et al. (2004) reported that total P was highly correlated to total solids in swine manure, and that despite agitating several finishing pits prior to pumping them out, total P and total solids both increased toward the bottom of the pit. Total N, however, was uniform through the system, suggesting it is not related to total solids.

If using outdoor manure storage, producers must move the manure out of the building. Manure removal is aided by gravity, scrapers, or flushing with either fresh water or recycled wastewater. The method selected impacts N and S levels, both in the barn and in the storage. In addition, outdoor storage such as tanks or earthen pits are typically uncovered and thus further impacted by environmental conditions. In humid climates, dilution from precipitation or runoff is likely to happen, while in dry climates, evaporation rates will be higher and the manure nutrients may become more concentrated (Lorimor et al., 2008). The USEPA (2005) proposed emission factors for swine systems with lagoon storage and deep pit storage of 2.7 and 3.4 kg head^{-1} yr^{-1}, respectively. Ammonia emission measurements from a sow barn in North Carolina with lagoon storage suggest the average daily mean ammonia emissions from the animal housing range from 6.46 to 8.18 g NH_3 sow^{-1} d^{-1} (Robarge et al., 2010); the emissions from the lagoon storage for the sow barn ranged from 5 to 20 g NH_3 sow^{-1} d^{-1} in the winter months and 20 to 80 g NH_3 sow^{-1} d^{-1} in the summer (Grant and Boehm, 2010). Stinn et al. (2014) studied a swine production system in Iowa that included farrowing barns with outdoor earthen pit storage. Of the total NH_3 emissions from the farrowing barns, 88% came from the barns themselves while another 12% occurred in the earthen pits.

Manure lagoons can be impacted by some of the same conditions as other forms of outdoor storage, but are often diluted purposefully to aid in the treatment process. The solid portion of manure, called sludge, is allowed to settle to the bottom while the liquid portion, often called the wastewater, is anaerobically treated by naturally present bacteria. Of the total amount of N that enters a lagoon, over 70% can be lost (Rotz, 2004) and compared with deep pits, lagoons emit more N_2O (Powers and Capelari, 2017). Seasonality also plays a large role in lagoons as studies have shown significant differences in N and P dynamics depending on the time of the year (Westerman et al., 2010; McLaughlin et al., 2012, 2016). For instance, NH_3 losses were highest in the summer and lowest in the winter in North Carolina (James et al., 2012). For P, the trend is more variable, but generally there is a decline from April to May, followed by an increase over the summer, and then a slow decline over the winter (Westerman et al., 2010).

To reduce the environmental impacts of swine production, researchers have evaluated alternative manure storage systems. Generally speaking, these systems attempt to separately collect urine and feces to reduce the chances of NH_3 loss and to assist in P recovery. Koger et al. (2014) evaluated a conveyor belt system to facilitate the collection of feces separate from urine underneath slatted floors. They found that NH_3 gas emissions were 1.0 kg pig^{-1} yr^{-1} in their system compared with the 2.3 to 3.7 kg NH_3 pig^{-1} yr^{-1} losses reported by others. Von Bernuth et al. (2005) also constructed an alternative collection system where the manure pit underneath the slats was shaped like a "V" at the bottom. This allowed the solids to be collected via scrapers while the liquids drained to the bottom to be collected separately. In this manner they were able to capture 70% of the N, 98% of the P, and 89% of the K fed to the pigs.

Conclusions

Managing swine manure as a resource is a balancing game. On one hand, producers want to maximize the amount of nutrients retained in the manure so that it can be used as a nutrient source for crops. On the other hand, they are simultaneously trying to minimize the amount of nutrients excreted from the pigs while optimizing growth. The difficulty in this balancing act is that each nutrient behaves differently and thus must be managed in diverse ways. Nitrogen and S are mainly lost from manure as a gas, thus managing it to reduce these losses should focus on diet and environmental conditions of the production environment. While P and K can be manipulated via the diet, there are fewer loss pathways than with N. Adequately agitating manure prior to land application can help reduce issues with P and K removal from storage. There is less known about potential loss pathways for other secondary and micronutrients, like Ca, Mg, and Na, but thoroughly agitating manure still remains a good recommendation.

Due to the myriad of management practices, it can be expected that wide ranges in nutrient characteristics of swine manure exist. This was confirmed by our review of data obtained from the literature, as well as two commercial laboratories in the midwestern United States. However, there are some limitations to this study. First, many authors reported collecting manure in studies, but did not report manure nutrient analyses. Second, when manure characteristics were reported, there was limited information regarding housing type, diet, storage, and management practices that could potentially affect the nutrient concentration values. Nevertheless, this review could be relevant from the perspective of manure nutrient management at times when no site-specific manure data are available. However, site-specific information is always preferable, so it is our recommendation that producers test the manure from their operations on a regular basis to get an accurate idea of nutrient contents.

References

Adeli, A., C.H. Bolster, D.E. Rowe, M.R. McLaughlin, and G.E. Brink. 2008. Effect of long-term swine effluent application on selected soil properties. Soil Sci. 173(3):223–235. doi:10.1097/ss.0b013e31816408ae

Adhikari, P.A., J.M. Heo, and C.M. Nyachoti. 2016. Standardized total tract digestibility of phosphorus in camelina (Camelina sativa) meal fed to growing pigs without or phytase supplementation. Anim. Feed Sci. Technol. 214:104–109. doi:10.1016/j.anifeedsci.2016.02.018

Almeida, F.N., and H.H. Stein. 2010. Performance and phosphorus balance of pigs fed diets formulated on the basis of values for standardized total tract digestibility of phosphorus. J. Anim. Sci. 88(9):2968–2977. doi:10.2527/jas.2009-2285

ASABE. 2005. ASAE D384.2 Manure production and characteristics. ASABE, St. Joseph, MI.

Atakora, J.K.A., S. Moehn, and R.O. Ball. 2011a. Enteric methane produced by finisher pigs is affected by dietary crude protein content of barley grain based, but not by corn based, diets. Anim. Feed Sci. Technol. 166–167:412–421. doi:10.1016/j.anifeedsci.2011.04.029

Atakora, J.K.A., S. Moehn, J.S. Sands, and R.O. Ball. 2011b. Effects of dietary crude protein and phytase–xylanase supplementation of wheat grain based diets on energy metabolism and enteric methane in growing finishing pigs. Anim. Feed Sci. Technol. 166–167:422–429. doi:10.1016/j.anifeedsci.2011.04.030

Ball Coelho, B., R. Murray, D. Lapen, E. Topp, and A. Bruin. 2012. Phosphorus and sediment loading to surface waters from liquid swine manure application under different drainage and tillage practices. Agric. Water Manage. 104:51–61. doi:10.1016/j.agwat.2011.10.020

Ball Coelho, B.R., R.C. Roy, E. Topp, and D.R. Lapen. 2007. Tile water quality following liquid swine manure application into standing corn. J. Environ. Qual. 36(2):580–587. doi:10.2134/jeq2006.0306 von Bernuth, R.D., J.D. Hill, E. Henderson, S. Godbout, D. Hamel, and F. Pouliot. 2005. Efficacy of a liquid/solid isolation system for swine manure. Trans. ASAE 48(4):1537–1546. doi:10.13031/2013.19185

Brisson, Y. 2014. The changing face of the Canadian hog industry. Statistics Canada, Ottawa, ON. p. 1–11. https://www150.statcan.gc.ca/n1/en/pub/96-325-x/2014001/article/14027-eng.pdf?st=8QxYs2Hn (Accessed 11 Mar. 2019).

Brumm, M.C., C.A. Shapiro, and W.L. Kranz. 2002. Managing swine dietary phosphorus to meet manure management goals. University of Nebraska-Lincoln, Lincoln, NE.

Buckley, K.E., M.C. Therrien, and R.M. Mohr. 2011. Agronomic performance of barley cultivars in response to varying rates of swine slurry. Can. J. Plant Sci. 91(1):69–79. doi:10.4141/cjps09032

Bullock, E.L., D.R. Edwards, and R.S. Gates. 2016. Effects of chemical amendments to swine manure on runoff quality. Trans. ASABE 59(6):1651–1660. doi:10.13031/trans.59.11636

Cavanagh, A., M.O. Gasser, and M. Labrecque. 2011. Pig slurry as fertilizer on willow plantation. Biomass Bioenergy 35(10):4165–4173. doi:10.1016/j.biombioe.2011.06.037

Chantigny, M.H., D.A. Angers, G. Bélanger, P. Rochette, N. Eriksen-Hamel, S. Bittman, K. Buckley, D. Massé, and M.-O. Gasser. 2008. Yield and nutrient export of grain corn fertilized with raw and treated liquid swine manure. Agron. J. 100(5):1303–1309. doi:10.2134/agronj2007.0361

Chantigny, M.H., D.A. Angers, P. Rochette, G. Bélanger, D. Massé, and D. Côté. 2007. Gaseous nitrogen emissions and forage nitrogen uptake on soils fertilized with raw and treated swine manure. J. Environ. Qual. 36(6):1864–1872. doi:10.2134/jeq2007.0083

Chantigny, M.H., A.J. Douglas, M. Ae, C. Beaupré, A.E. Philippe, R. Ae, and D.A. Angers. 2009. Ammonia volatilization following surface application of raw and treated liquid swine manure. Nutr. Cycling Agroecosyst. 85:275–286. doi:10.1007/s10705-009-9266-7

Chantigny, M.H., P. Rochette, D.A. Angers, S. Bittman, K. Buckley, D. Massé, G. Bélanger, N. Eriksen-Hamel, and M.-O. Gasser. 2010. Soil nitrous oxide emissions following band-incorporation of fertilizer nitrogen and swine manure. J. Environ. Qual. 39(5):1545–1553. doi:10.2134/jeq2009.0482

Cortus, E.L., S.P. Lemay, E.M. Barber, G.A. Hill, and S. Godbout. 2008. A dynamic model of ammonia emission from urine puddles. Biosystems Engineering 99(3):390–402. doi:10.1016/j.biosystemseng.2007.11.004

Dai, X.-R., C.K. Saha, J.-Q. Ni, A.J. Heber, V. Blanes-Vidal, and J.L. Dunn. 2015. Characteristics of pollutant gas releases from swine, dairy, beef, and layer manure, and municipal wastewater. Water Res. 76:110–119. doi:10.1016/j.watres.2015.02.050

David, M.B., L.E. Gentry, and C.A. Mitchell. 2016. Riverine response of sulfate to declining atmospheric sulfur deposition in agricultural watersheds. J. Environ. Qual. 45(4):1313–1319. doi:10.2134/jeq2015.12.0613

Emiola, A., O. Akinremi, B. Slominski, and C.M. Nyachoti. 2009. Nutrient utilization and manure P excretion in growing pigs fed corn-barley-soybean based diets supplemented with microbial phytase. Anim. Sci. J. 80(1):19–26. doi:10.1111/j.1740-0929.2008.00590.x

Ferket, P., E. Van Heugten, T.A. Van Kempen, and R. Angel. 2002. Nutritional strategies to reduce environmental emissions from nonruminants. J. Anim. Sci. 80:E168–E182. doi:10.2527/animalsci2002.80E-Suppl_2E168x

Grant, R., and M.T. Boehm. 2010. National air emissions monitoring study: Data from the southeastern US pork production facility NC4A. Final report to the Agricultural Air Research Council. U.S. Environmental Protection Agency, West Lafayette, IN.

Griffing, E.M., M. Overcash, and P. Westerman. 2007. A review of gaseous ammonia emissions from slurry pits in pig production systems. Biosystems Engineering 97(3):295–312. doi:10.1016/j.biosystemseng.2007.02.012

Guthrie, T.A., G.A. Apgar, K.E. Griswold, M.D. Lindemann, J.S. Radcliffe, and B.N. Jacobson. 2004. Nutritional value of a corn containing a glutamate dehydrogenase gene for growing pigs. J. Anim. Sci. 82(6):1693–1698. doi:10.2527/2004.8261693x

Hanson, A.R., G. Xu, M. Li, M.H. Whitney, and G.C. Shurson. 2012. Impact of dried distillers grains with solubles (DDGS) and diet formulation method on dry matter, calcium, and phosphorus retention and excretion in nursery pigs. Anim. Feed Sci. Technol. 172(3–4):187–193. doi:10.1016/j.anifeedsci.2011.12.027

Higgins, S.F., M.S. Coyne, S.A. Shearer, and J.D. Crutchfield. 2005. Determining nitrogen fractions in swine slurry. Bioresour. Technol. 96(9):1081–1088. doi:10.1016/j.biortech.2004.08.018

Higgins, S., S. Shearer, M. Coyne, and J. Fulton. 2004. Relationship of total nitrogen and total phosphorus concentration to solids content in animal waste slurries. Appl. Eng. Agric. 20(3):355–364. doi:10.13031/2013.16066

Hill, B.E., A.L. Sutton, and B.T. Richert. 2009. Effects of low-phytic acid corn, low-phytic acid soybean meal, and phytase on nutrient digestibility and excretion in growing pigs. J. Anim. Sci. 87(4):1518–1527. doi:10.2527/jas.2008-1219

Hoff, S.J., D.S. Bundy, M.A. Nelson, B.C. Zelle, L.D. Jacobson, A.J. Heber, J. Ni, Y. Zhang, J.A. Koziel, and D.B. Beasley. 2006. Emissions of ammonia, hydrogen sulfide, and odor before, during, and after slurry removal from a deep-pit swine finisher. J. Air Waste Manag. Assoc. 56(5):581–590. doi:10.1080/10473289.2006.10464472

Htoo, J.K., W.C. Sauer, J.L. Yañez, M. Cervantes, Y. Zhang, J.H. Helm, and R.T. Zijlstra. 2007. Effect of low-phytate barley or phytase supplementation to a barley-soybean meal diet on phosphorus retention and excretion by grower pigs. J. Anim. Sci. 85(11):2941–2948. doi:10.2527/jas.2006-816

Hubbard, R.K., W.F. Anderson, G.L. Newton, J.M. Ruter, and J.P. Wilson. 2011. Plant growth and elemental uptake by floating vegetation on a single-stage swine wastewater lagoon. Trans. ASABE 54(3):837–845. doi:10.13031/2013.37108

Israel, D.W., and T.J. Smyth. 2015. Crop utilization of nitrogen in swine lagoon sludge. Commun. Soil Sci. Plant Anal. 46(12):1525–1539. doi:10.1080/00103624.2015.1043459

Jacobs, B.M., J.F. Patience, W.A. Dozier, K.J. Stalder, and B.J. Kerr. 2011. Effects of drying methods on nitrogen and energy concentrations in pig feces and urine, and poultry excreta. J. Anim. Sci. 89(8):2624–2630. doi:10.2527/jas.2010-3768

Jacobson, L.D., B.P. Hetchler, and V.J. Johnson. 2005. Spatial, diurnal, and seasonal variations of temperature, ammonia and hydrogen sulfide concentrations in two tunnel ventilated sow gestation buildings in Minnesota. Pro Livestock Environment VII, Beijing, China. 18-20 May 2005. American Society of Agricultural and Biological Engineers, St. Joseph, MI. p. 198.

James, K.M., J. Blunden, I.C. Rumsey, and V.P. Aneja. 2012. Characterizing ammonia emissions from a commercial mechanically ventilated swine finishing facility and an anaerobic waste lagoon in North Carolina. Atmos. Pollut. Res. 3(3):279–288. doi:10.5094/APR.2012.031

Jarecki, M.K., T.B. Parkin, A.S.K. Chan, T.C. Kaspar, T.B. Moorman, J.W. Singer, B.J. Kerr, J.L. Hatfield, and R. Jones. 2009. Cover crop effects on nitrous oxide emission from a manure-treated Mollisol. Agric. Ecosyst. Environ. 134(1–2):29–35. doi:10.1016/j.agee.2009.05.008

Jayasundara, S., C. Wagner-Riddle, G. Parkin, J. Lauzon, and M.Z. Fan. 2010. Transformations and losses of swine manure 15 N as affected by application timing at two contrasting sites. Can. J. Soil Sci. 90:55–73 www.nrcresearchpress.com (Accessed 20 Dec. 2018). doi:10.4141/CJSS08085

Jha, R., and P. Leterme. 2012. Feed ingredients differing in fermentable fibre and indigestible protein content affect fermentation metabolites and faecal nitrogen excretion in growing pigs. Animal 6(4):603–611. doi:10.1017/S1751731111001844

Kai, P., B. Kaspers, and T. Van Kempen. 2006. Modeling sources of gaseous emissions in a pig house with recharge pit. Trans. ASABE 49(5):1479–1486.

Karimi, R., W. Akinremi, and D. Flaten. 2017. Cropping system and type of pig manure affect nitrate-nitrogen leaching in a sandy loam soil. J. Environ. Qual. 46(4):785–792. doi:10.2134/jeq2017.04.0158

Karlen, D.L., C.A. Cambardella, and R.S. Kanwar. 2004. Challenges of managing liquid swine manure. Appl. Eng. Agric. 20:693–699. doi:10.13031/2013.17460

Kephart, K. 2009. Manure production, analysis, and estimates of nutrient excretion in swine. Pennsylvania State University, University Park, PA.

Kerr, B.J., S.L. Trabue, and D.S. Andersen. 2017. Narasin effects on energy, nutrient, and fiber digestibility in corn-soybean meal or corn-soybean meal-dried distillers grains with solubles diets fed to 16-, 92-, and 141-kg pigs. J. Anim. Sci. 95(9):4030. doi:10.2527/jas2017.1732</jrn>

Kerr, B.J., S.L. Trabue, M.B. Van Weelden, D.S. Andersen, and L.M. Pepple. 2018. Impact of narasin on manure composition, microbial ecology, and gas emissions from finishing pigs fed either a corn-soybean meal or a corn-soybean meal-dried distillers grains with solubles diets. J. Anim. Sci. 96:1317–1329. doi:10.1093/jas/sky053

Kerr, B.J., C.J. Ziemer, S.L. Trabue, J.D. Crouse, and T.B. Parkin. 2006. Manure composition of swine as affected by dietary protein and cellulose concentrations 1. J. Anim. Sci. 84:1584–1592 https://pdfs.semanticscholar.org/8225/64d53f82522fd5923260bc287409cfed5a6c.pdf (Accessed 18 June 2018). doi:10.2527/2006.8461584x

Key, N., W.D. Mcbride, M. Ribaudo, and S. Sneeringer. 2011. Trends and developments in hog manure management: 1998-2009. USDA-Economic Research Service, Washington, D.C.

Knowlton, K., J. Radcliffe, C. Novak, and D. Emmerson. 2004. Animal management to reduce phosphorus losses to the environment. J. Anim. Sci. 82:E173–E195 (Accessed 21 June 2018). doi:10.2527/2004.8213_supplE173x

Koger, J.B., B.K. O'Brien, R.P. Burnette, P. Kai, M.H.J.G. Van Kempen, E. Van Heugten, and T.A.T.G. Van Kempen. 2014. Manure belts for harvesting urine and feces separately and improving air quality in swine facilities. Livest. Sci. 162(1):214–222. doi:10.1016/j.livsci.2014.01.013

Kranz, W.L., C.A. Shapiro, B.E. Anderson, M.C. Brumm, and M. Mamo. 2005. Effect of swine lagoon water application rate and alfalfa harvest frequency on dry matter production and N harvest. Appl. Eng. Agric. 21(2):203–210. doi:10.13031/2013.18154

Kumaragamage, D., O.O. Akinremi, and G.J. Racz. 2016. Comparison of nutrient and metal loadings with the application of swine manure slurries and their liquid separates to soils. J. Environ. Qual. 45(5):1769–1775. doi:10.2134/jeq2016.04.0130

Lamb, S.J. 2014. Nutrient runoff following swine manure application. Unviersity of Lincoln-Nebraska, Lincoln, NE. https://digitalcommons.unl.edu/cgi/viewcontent.cgi?referer=https://www.google.com/&httpsredir=1&article=1008&context=envengdiss (Accessed 8 Mar. 2019).

Li, M.M., K.M. Seelenbinder, M.A. Ponder, L. Deng, R.P. Rhoads, K.D. Pelzer, J.S. Radcliffe, C.V. Maxwell, J.A. Ogejo, R.R. White, and M.D. Hanigan. 2017. Effects of dirty housing and a Salmonella Typhimurium DT104 challenge on pig growth performance, diet utilization efficiency, and gas emissions from stored manure. J. Anim. Sci. 95(3):1264–1276. doi:10.2527/jas.2016.0863

Li, Q.F., N. Trottier, and W. Powers. 2015. Feeding reduced crude protein diets with crystalline amino acids supplementation reduce air gas emissions from housing. J. Anim. Sci. 93(2):721–730. doi:10.2527/jas.2014-7746

Lim, T., A. Heber, J. Ni, D. Kendall, B. Richert, D.C. Kendall, and A.J. Heber. 2004. Effects of manure removal strategies on odor and gas emissions from swine finishing. Trans. ASAE 47(6): 2041–2050. (Accessed 24 June 2018).

Liu, S., J.-Q. Ni, J.S. Radcliffe, and C.E. Vonderohe. 2017a. Mitigation of ammonia emissions from pig production using reduced dietary crude protein with amino acid supplementation. Bioresour. Technol. 233:200–208. doi:10.1016/j.biortech.2017.02.082

Liu, S., J.-Q. Ni, J.S. Radcliffe, and C. Vonderohe. 2017b. Hydrogen sulfide emissions from a swine building affected by dietary crude protein. J. Environ. Manage. 204:136–143. doi:10.1016/j.jenvman.2017.08.031

Liu, Z., W. Powers, J. Murphy, and R. Maghirang. 2014. Ammonia and hydrogen sulfide emissions from swine production facilities in North America: A meta-analysis. J. Anim. Sci. 92:1656–1665. doi:10.2527/jas.2013-7160

Liu, P., L.W.O. Souza, S.K. Baidoo, and G.C. Shurson. 2012. Impact of distillers dried grains with solubles particle size on nutrient digestibility, DE and ME content, and flowability in diets for growing pigs. J. Anim. Sci. 90(13):4925–4932. doi:10.2527/jas.2011-4604

Long, C.M., R.L. Muenich, M.M. Kalcic, and D. Scavia. 2018. Use of manure nutrients from concentrated animal feeding operations. J. Great Lakes Res. 44(2):245–252. doi:10.1016/j.jglr.2018.01.006

Loria, E.R., and J.E. Sawyer. 2005. Extractable soil phosphorus and inorganic nitrogen following application of raw and anaerobically digested swine manure. Agron. J. 97(3):879–885. doi:10.2134/agronj2004.0249

Loria, E.R., J.E. Sawyer, D.W. Barker, J.P. Lundvall, and J.C. Lorimor. 2007. Use of anaerobically digested swine manure as a nitrogen source in corn production. Agron. J. 99(4):1119–1129. doi:10.2134/agronj2006.0251

Lorimor, J., W. Powers, and A. Sutton. 2008. Manure characteristics–Manure management systems: Series 2. Midwest Plan Service, Ames, IA.

Luo, Z. 2016. Role of dietary factors on nutrient excretion and manure characteristics of growing pigs. University of Minnesota, Saint Paul, MN. https://conservancy.umn.edu/handle/11299/178901 (Accessed 23 June 2018).

Matlock, M., B. Greg Thoma, P. Eric Boles Mansoor Leh, H. Sandefur Rusty Bautista, and P. Rick Ulrich. 2014. A life cycle analysis of water use in U.S. pork production: Comprehensive Report. Pork Checkoff, Fayetteville, AR.

Maurer, D.L., J.A. Koziel, and K. Bruning. 2017. Field scale measurement of greenhouse gas emissions from land applied swine manure. Front. Environ. Sci. Eng. doi:10.1007/s11783-017-0915-9

McAndrews, G.M., M. Liebman, C.A. Cambardella, and T.L. Richard. 2006. Residual effects of composted and fresh solid swine (Sus scrofa L.) manure on soybean [Glycine max (L.) Merr.] growth and yield. Agron. J. 98(4):873–882. doi:10.2134/agronj2004.0078

McLaughlin, M.R., J.P. Brooks, and A. Adeli. 2012. Temporal flux and spatial dynamics of nutrients, fecal indicators, and zoonotic pathogens in anaerobic swine manure lagoon water. Water Res. 46(16):4949–4960. doi:10.1016/j.watres.2012.06.023

McLaughlin, M.R., J.P. Brooks, A. Adeli, and J.N. Jenkins. 2016. Improving estimates of N and P loads in irrigation water from swine manure lagoons. Irrig. Sci. 34(3):245–260. 10.1007/s00271-016-0495-7

Moody, L.B., R.T. Burns, and K.J. Stalder. 2009. Efffect of anaerobic digestion on manure characteristics for phosphorus precipitation from swine waste. Appl. Eng. Agric. 25(1):97–102

Muhlbauer, R.V., L.B. Moody, R.T. Burns, J. Harmon, and K. Stalder. 2011. Water consumption and conservation techniques currently available for swine production National Pork Board, Des Moines, IA. p. 165–196.

MWPS. 2004. Manure characteristics. Midwest Plan Service, Ames, IA.

Ndegwa, P.M., A.N. Hristov, J. Arogo, and R.E. Sheffield. 2008. A review of ammonia emission mitigation techniques for concentrated animal feeding operations. Biosystems Engineering 100(4):453–469. doi:10.1016/j.biosystemseng.2008.05.010

Nicolaisen, J.E., J.E. Gilley, B. Eghball, and D.B. Marx. 2007. Crop residue effects on runoff nutrient concentrations following manure application. Trans. ASABE. 10.13031/2013.23158

Nikièma, P., K.E. Buckley, J.M. Enns, H. Qiang, and O.O. Akinremi. 2013. Effects of liquid hog manure on soil available nitrogen status, nitrogen leaching losses and wheat yield on a sandy loam soil of western Canada. Can. J. Soil Sci. 93(5):573–584. doi:10.4141/cjss2012-070

Nortey, T.N., J.F. Patience, P.H. Simmins, N.L. Trottier, and R.T. Zijlstra. 2007. Effects of individual or combined xylanase and phytase supplementation on energy, amino acid, and phosphorus digestibility and growth performance of grower pigs fed wheat-based diets containing wheat millrun. J. Anim. Sci. 85(6):1432–1443. doi:10.2527/jas.2006-613

NPB. 2009. Quick facts: The pork industry at a glance. National Pork Board, Ames, IA.

NPB. 2016. Life cycle of a market pig. Natl. Pork Board- Pork Checkoff, Ames, IA. https://www.pork.org/facts/pig-farming/life-cycle-of-a-market-pig/ (accessed 19 Dec. 2018).

NRC. 2012. Nutrient requirements of swine. National Academies Press, Washington, D.C.

Nyachoti, C.M., S.D. Arntfield, W. Guenter, S. Cenkowski, and F.O. Opapeju. 2006. Effect of micronized pea and enzyme supplementation on nutrient utilization and manure output in growing pigs. J. Anim. Sci. 84(8):2150–2156. doi:10.2527/jas.2004-467

Park, K.-H., A.G. Thompson, M. Marinier, K. Clark, and C. Wagner-Riddle. 2006. Greenhouse gas emissions from stored liquid swine manure in a cold climate. Atmos. Environ. 40(4):618–627. doi:10.1016/j.atmosenv.2005.09.075

Parker, D.B., J. Gilley, B. Woodbury, K.-H. Kim, G. Galvin, S.L. Bartelt-Hunt, X. Li, and D.D. Snow. 2013. Odorous VOC emission following land application of swine manure slurry. Atmos. Environ. 66:91–100. doi:10.1016/j.atmosenv.2012.01.001

Parkin, T.B., T.C. Kaspar, and J.W. Singer. 2006. Cover crop effects on the fate of N following soil application of swine manure. Plant Soil 289(1–2):141–152. doi:10.1007/s11104-006-9114-3

Patience, J.F. 2012. The importance of water in pork production. Anim. Front. 2(2): 28–35. https://academic.oup.com/af/article/2/2/28/4638608 (Accessed 25 June 2018).

Pedersen, C., M.G. Boersma, and H.H. Stein. 2007. Digestibility of energy and phosphorus in ten samples of distillers dried grains with solubles fed to growing pigs. J. Anim. Sci. 85(5):1168–1176. doi:10.2527/jas.2006-252

Pilcher, C.M., R. Arentson, and J.F. Patience. 2015. The interaction of fiber, supplied by distillers dried grains with solubles, with an antimicrobial and a nutrient partitioning agent on nitrogen balance, water utilization, and energy digestibility in finishing pigs. J. Anim. Sci. 93(3):1124. doi:10.2527/jas.2013-7309

Pomar, C., J. Pomar, F. Dubeau, E. Joannopoulos, and J.-P. Dussault. 2014. The impact of daily multiphase feeding on animal performance, body composition, nitrogen and phosphorus excretions, and feed costs in growing–finishing pigs. animal 8(05):704–713. doi: 10.1017/S1751731114000408.

Powers, W., and M. Capelari. 2017. Production, management, and the environment symposium: Measurement and mitigation of reactive nitrogen species from swine and poultry production. J. Anim. Sci. 95(5):2236–2240. doi:10.2527/jas.2016.1187

Powers, W.J., E.R. Fritz, W. Fehr, and R. Angel. 2006. Total and water-soluble phosphorus excretion from swine fed low-phytate soybeans. J. Anim. Sci. 84(7):1907–1915. doi:10.2527/jas.2005-656

Powers, W.J., S.B. Zamzow, and B.J. Kerr. 2007. Reduced crude protein effects on aerial emissions from swine. Appl. Eng. Agric. 23(4): 539–546. https://pubag.nal.usda.gov/pubag/downloadPDF.xhtml?id=17623&content=PDF (accessed 23 June 2018).

Priyashantha, K.R.S., C.P. Maule, and J.A. Elliott. 2007. Influence of slope position and hog manure injection on fall soil P and N distribution in an undulating landscape. Trans. ASABE 50(1):45–52. doi:10.13031/2013.22410

Renteria-Flores, J.A., L.J. Johnston, G.C. Shurson, and D.D. Gallaher. 2008. Effect of soluble and insoluble fiber on energy digestibility, nitrogen retention, and fiber digestibility of diets fed to gestating sows. J. Anim. Sci. 86(10):2568–2575. doi:10.2527/jas.2007-0375

Rieck-Hinz, A., T.G. Miller, and J.E. Sawyer. 2011. How to interpret your manure analysis. USDA-NRCS, Ames, IA.

Rincker, M.J., G.M. Hill, J.E. Link, A.M. Meyer, and J.E. Rowntree. 2005. Effects of dietary zinc and iron supplementation on mineral excretion, body composition, and mineral status of nursery pigs. J. Anim. Sci. 83(12):2762–2774. doi:10.2527/2005.83122762x

Robarge, W., K. Wang, B. Bogan, K. Wang, J. Kang, and A. Heber. 2010. National air emissions monitoring study: Emissions data from two sow gestation barns and one farrowing room in North Carolina-Site NC4B. Final report. U.S. Environmental Protection Agency, West Lafayette, IN.

Rochette, P., D.A. Angers, M.H. Chantigny, J.D. MacDonald, M.-O. Gasser, and N. Bertrand. 2009. Reducing ammonia volatilization in a no-till soil by incorporating urea and pig slurry in shallow bands. Nutr. Cycl. Agroecosyst. 84(1):71–80. doi:10.1007/s10705-008-9227-6

Rojas, O.J., Y. Liu, and H.H. Stein. 2014. Concentration of metabolizable energy and digestibility of energy, phosphorus, and amino acids in lemna protein concentrate fed to growing pigs. J. Anim. Sci. 92:5222–5229. doi:10.2527/jas.2014-8146

Rotz, C.A. 2004. Management to reduce nitrogen losses in animal production management to reduce nitrogen losses in animal production. J. Anim. Sci. 82:119–137. doi:10.2527/2004.8213_supplE119x

Schlegel, A.J., Y. Assefa, H.D. Bond, L.A. Haag, and L.R. Stone. 2017. Changes in soil nutrients after 10 years of cattle manure and swine effluent application. Soil Tillage Res. 172:48–58. doi:10.1016/j.still.2017.05.004

Schulz, K. 2016. Corn, soybean meal still king of the swine diets. Natl. Hog Farmer. FarmProgress, Irving, TX. https://www.nationalhogfarmer.com/nutrition/cheap-corn-soybeans-play-well-hogs-needs (Accessed 19 Dec. 2018).

Schuster, N.R., and S.L. Bartelt-Hunt. 2017. Runoff water quality characteristics following swine slurry application under broadcast and injected conditions. Trans. ASABE 60(1):53–66. doi:10.13031/trans.11370

Shaw, M.I., A.D. Beaulieu, and J.F. Patience. 2006. Effect of diet composition on water consumption in growing pigs. J. Anim. Sci. 84(11):3123–3132. doi:10.2527/jas.2005-690

Shriver, J.A., S.D. Carter, A.L. Sutton, and Richert. 2003. Effects of adding fiber sources to reduced-crude protein, amino. J. Anim. Sci 81(2):492–502. doi:10.2527/2003.812492x

Singer, J.W., C.A. Cambardella, and T.B. Moorman. 2008. Enhancing nutrient cycling by coupling cover crops with manure injection. Agron. J. 100(6):1735–1739. doi:10.2134/agronj2008.0013x

Smith, E., R. Gordon, C. Bourque, and A. Campbell. 2007a. Comparison of three simple field methods for ammonia volatilization from manure. Can. J. Soil Sci. 87:469–477 www.nrcresearchpress.com (Accessed 20 Dec. 2018). doi:10.4141/CJSS06038

Smith, E., R. Gordon, C. Bourque, and A. Campbell. 2008. Management strategies to simultaneously reduce ammonia, nitrous oxide and odour emissions from surface-applied swine manure. Can. J. Soil Sci. 88:571–584 www.nrcresearchpress.com (Accessed 20 Dec. 2018). doi:10.4141/CJSS07089

Smith, D.R., P.R. Owens, A.B. Leytem, and E.A. Warnemuende. 2007b. Nutrient losses from manure and fertilizer applications as impacted by time to first runoff event. Environ. Pollut. 147(1):131–137. doi:10.1016/j.envpol.2006.08.021

Sotak-Peper, K.M., J.C. Gonzalez-Vega, and H.H. Stein. 2015. Concentrations of digestible, metabolizable, and net energy in soybean meal produced in different areas of the United States and fed to pigs. J. Anim. Sci. 93(12):5694. doi:10.2527/jas.2015-9281

Spiehs, M.J., M.H. Whitney, G.C. Shurson, R.E. Nicolai, J.A. Renteria Flores, and D.B. Parker. 2012. Odor and gas emissions and nutrient excretion from pigs fed diets containing dried distillers grains with solubles. Appl. Eng. Agric. 28(3):431–437. doi:10.13031/2013.41492

Springer, T.L., C.M. Taliaferro, and J.A. Hattey. 2005. Nitrogen source and rate effects on the production of buffalograss forage grown with irrigation. Crop Sci. 45:668–672. doi:10.2135/cropsci2005.0668

Stein, H.H., and G.C. Shurson. 2009. Board-invited review: The use and application of distillers dried grains with solubles in swine diets. J. Anim. Sci. 87(4):1292–1303. doi:10.2527/jas.2008-1290

Stinn, J.P., H. Xin, T.A. Shepherd, H. Li, and R.T. Burns. 2014. Ammonia and greenhouse gas emissions from a modern U.S. swine breeding-gestation-farrowing system. Atmos. Environ. 98:620–628. doi:10.1016/j.atmosenv.2014.09.037

Surendra, K.C., R. Olivier, J.K. Tomberlin, R. Jha, and S.K. Khanal. 2016. Bioconversion of organic wastes into biodiesel and animal feed via insect farming. Renew. Energy 98:197–202. doi:10.1016/j.renene.2016.03.022

Sutton, A., and C. Lander. 2003. Effects of diet and feeding management on nutrient content. Washington, D.C.

Tabbara, H. 2003. Phosphorus loss to runoff water twenty-four hours after application of liquid swine manure or fertilizer. J. Environ. Qual. 32(3):1044–1052. doi:10.2134/jeq2003.1044

Thacker, P.A., B.G. Rossnagel, and V. Raboy. 2006. The effects of phytase supplementation on nutrient digestibility, plasma parameters, performance and carcass traits of pigs fed diets based on low-phytate barley without inorganic phosphorus. Can. J. Anim. Sci. 86: 245–254. doi: https://doi.org/10.4141/A05-077. doi:10.4141/A05-077

Thomas, B.W., J.K. Whalen, M. Sharifi, M. Chantigny, and B.J. Zebarth. 2016. Labile organic matter fractions as early-season nitrogen supply indicators in manure-amended soils. J. Plant Nutr. Soil Sci. 179(1):94–103. doi:10.1002/jpln.201400532

Thorne, P.S., A. Ansley, and S.S. Perry. 2009. Concentrations of bioaerosols, odors and hydrogen sulfide inside and downwind from two types of swine livestock operations. J. Occup. Environ. Hyg. 6(4):211–220. doi:10.1080/15459620902729184

Trabue, S., and B. Kerr. 2014. Emissions of greenhouse gases, ammonia, and hydrogen sulfide from pigs fed standard diets and diets supplemented with dried distillers grains with solubles. J. Environ. Qual. 43(4):1176. doi:10.2134/jeq2013.05.0207

Urriola, P.E., J.A. Mielke, Q. Mao, Y.-T. Hung, J.F. Kurtz, L.J. Johnston, G.C. Shurson, C. Chen, and M. Saqui-Salces. 2018. Evaluation of a partially de-oiled microalgae product in nursery pig diets. Transl. Anim. Sci. 2(2):169–183. doi:10.1093/tas/txy013

USDA- National Agricultural Statistics Service. 2018. Quarterly hogs and pigs. USDA-NASS, Washington, D.C.

USDA-NRCS. 2011. Agricultural waste management systems. In: L. Owens, editor, Agricultural waste management field handbook. USDA-Natural Resources Conservation Service, Fort Worth, TX.

USEPA. 2005. National Emission Inventory- Ammonia emissions from animal agricultural operations. Revised draft report. U.S. Environmental Protection Agency. Washington, D.C.

USEPA. 2017. Sulfur dioxide trends. U.S. Environmental Protection Agency, Washington, D.C. https://www.epa.gov/air-trends/sulfur-dioxide-trends (accessed 20 June 2018).

Veum, T.L., D.R. Ledoux, and V. Raboy. 2007. Low-phytate barley cultivars improve the utilization of phosphorus, calcium, nitrogen, energy, and dry matter in diets fed to young swine. J. Anim. Sci. 85(4):961–971. doi:10.2527/jas.2006-453

Veum, T.L., D.R. Ledoux, M.C. Shannon, and V. Raboy. 2009. Effect of graded levels of iron, zinc, and copper supplementation in diets with low-phytate or normal barley on growth performance, bone characteristics, hematocrit volume, and zinc and copper balance of young swine. J. Anim. Sci. 87(8): 2625–2634. (Accessed 27 June 2018).

Westerman, P.W., J. Arogo Ogejo, and G.L. Grabow. 2010. Swine anaerobic lagoon nutrient concentration variation with season, lagoon level, and rainfall. Appl. Eng. Agric. 26(1): 147. doi:10.13031/2013.29472

Widyaratne, G.P., and R.T. Zijlstra. 2007. Nutritional value of wheat and corn distiller's dried grain with solubles: Digestibility and digestible contents of energy, amino acids and phosphorus, nutrient excretion and growth performance of grower-finisher pigs. Can. Vet. J. 87(1):103-114.

Wienhold, B.J., and P.S. Miller. 2004. Phosphorus fractionation in manure from swine fed traditional and low-phytate corn diets. J. Environ. Qual. 33:389–393. http://digitalcommons. unl.edu/animalscifacpub (Accessed 23 June 2018). doi:10.2134/jeq2004.3890

Woli, K.P., S. Rakshit, J.P. Lundvall, J.E. Sawyer, and D.W. Barker. 2013. On-farm evaluation of liquid swine manure as a nitrogen source for corn production. Agron. J. 105(1):248–262. doi:10.2134/agronj2012.0292

Xu, G., G. He, K. Baidoo, and G. Shurson. 2006. Effect of feeding diets containing corn distillers dried grains with solubles (DDGS), with or without phytase, on nutrient digestibility and excretion in nursery pigs. J. Anim. Sci 84(2): 122. doi:10.2527/jas.2008-1403 (Accessed 20 June 2018).

Yen, J.T., J.E. Wells, and D.N. Miller. 2004. Dried skim milk as a replacement for soybean meal in growing-finishing diets: Effects on growth performance, apparent total-tract nitrogen digestibility, urinary and fecal nitrogen excretion, and carcass traits in pigs. J. Anim. Sci. 82(11):3338–3345. doi:10.2527/2004.82113338x

Ziemer, C.J., B.J. Kerr, S.L. Trabue, H. Stein, D.A. Stahl, and S.K. Davidson. 2009. Dietary protein and cellulose effects on chemical and microbial characteristics of swine feces and stored manure. J. Environ. Qual. 38(5):2138. doi:10.2134/jeq2008.0039

Anaerobic Digestion of Dairy and Swine Waste

J. H. Harrison* and P. M. Ndegwa

Abstract

Anaerobic digestion (AD) of animal manure is an effective method for conversion of undigested plant carbon to methane for use as a transport fuel or production of electricity. The resulting liquid and solid components after AD are valuable as a source of crop nutrients and soil amendments, as well as a bedding material. Treatment of manure with AD reduces pathogens and greenhouse gases emissions.

Benefits and Species of Animals

Anaerobic digesters (AD) provide many benefits including: odor management, reduced emission of greenhouse gases, pathogen control, and production of biogas (methane) for use as a fuel. The economic benefit of converting fibrous materials (manure, food waste, byproduct feeds) into methane is directly linked to the use as fuel to: generate electricity, for transport vehicles, and provide energy for home heating and cooking. In addition, the liquid after anaerobic digestion can be used as a nutrient source for crop production, while solids after anaerobic digestion can be used either as bedding for animals or sold as a soil amendment. One environmental and societal benefit of AD is the diversion of wastes away from landfills (Gould, 2012)

While ADs are most commonly associated with dairy farming, AD are also utilized in the management of manure and waste water at swine and poultry operations (USEPA, 2017).

Types of Anaerobic Digestion Designs

Anaerobic digesters can be classified as passive systems, low-rate systems, or high-rate systems (Hamilton, 2012). A passive system is when biogas recovery is achieved with an existing treatment system, such as would be accomplished with a covered manure lagoon (Fig. 1). A low-rate system consists primarily of manure as the main source of methane-forming organisms. Designs include complete mix and plug flow (Fig. 2A and 2B). A high-rate system exists when methane-forming microorganisms are trapped in the digester and are represented by solids recycling, fixed film, suspended media, and sequencing batch digesters (Fig. 3).

Cautions and Hazards

Since biogas is most often used as a fuel to generate electricity, the adoption of AD has been slow in areas with low electricity rates. Low electrical power rates

J.H. Harrison, Department of Animal Sciences, Washington State University, Puyallup, WA; P.M. Ndegwa, Department of Biological Systems Engineering, Washington State University, Pullman, WA. *Corresponding author (jhharrison@wsu.edu)

doi:10.2134/asaspecpub67.c13

Animal Manure: Production, Characteristics, Environmental Concerns and Management. ASA Special Publication 67. Heidi M. Waldrip, Paulo H. Pagliari, and Zhongqi He, editors.

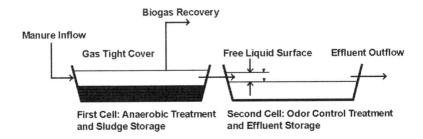

Fig. 1. Schematic of covered lagoon digestion system (Source: Hamilton, 2012).

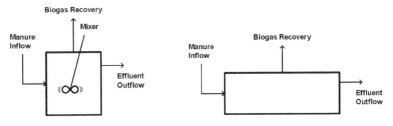

Fig. 2. Schematic of low rate systems, complete mix (A) and plug flow (B) digester (Source: Hamilton, 2012).

can make the profitability a challenge, and acceptance of additional feedstocks and their associated tipping fees can be an important component of the business model. As with any type of farming operation, running an AD has risks associated with it such as explosion, asphyxiation, and exposure to hydrogen sulfide.

Greenhouse Gases and Anaerobic Digestion

Gases, in the atmosphere, that absorb and emit heat (infrared radiation) are called greenhouse gases (GHG) because they function akin to the roof and walls of a greenhouse. As the sun's energy (i.e., solar radiation) travels to earth, some of it is reflected at both the earth's surface and atmosphere, while most is absorbed by the earth's surface, thus warming it. The warmed earth's surface then emits infrared radiation back into the atmosphere. Some of the infrared radiation escapes back to space, but the greenhouse gases, in the atmosphere, absorb some and re-emit it in all directions, which warms the earth's surface and the lower atmosphere. The dominant GHG in the earth's atmosphere are carbon dioxide (CO_2), methane (CH_4), nitrous oxide (N_2O), hydrofluorocarbons (HFCs), water vapor (H_2O), and ozone (O_3) (USEPA, 2017). The presence of GHG in the atmosphere thus keeps the earth's surface warm. The amount of heat trapped in the earth (and thus the earth's temperature) depends on the concentrations of the GHG in the atmosphere and recent increases in the global temperature has been associated with increase in release of these gases into the atmosphere. Industrialization is responsible for majority of this increase in GHG in the atmosphere, especially during the last 15 decades. In particular, agriculture contributes about 9% of GHG emissions, mainly from livestock, soils, and rice production (USEPA, 2017). There are two major potential GHG from

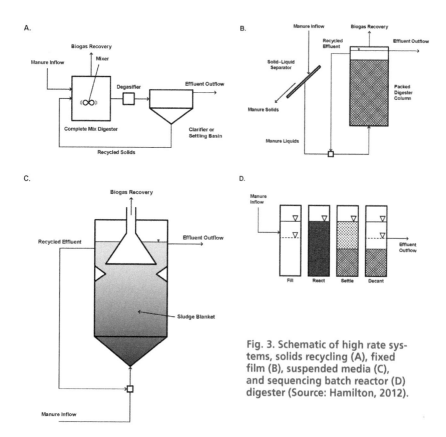

Fig. 3. Schematic of high rate systems, solids recycling (A), fixed film (B), suspended media (C), and sequencing batch reactor (D) digester (Source: Hamilton, 2012).

dairy and swine facilities, namely CH_4 and N_2O (Montes et al., 2013). Both CH_4 and N_2O are extremely potent GHGs with respective global warming potentials (GWP) of 25 and 298 times that of CO_2 (Montes et al., 2013; Solomon et al., 2007).

The manner in which dairy and swine manures are managed largely dictates CH_4 and N_2O emissions; hence, their respective contributions to GHG in the atmosphere. Different manure treatments and storage methods determine how much of these GHG are produced. Manure management accounts for about 15% of the total GHG emissions from the Agricultural sector in the United States (USEPA, 2017). The two main approaches to managing GHG emissions from livestock manure are: (i) controlling manure decomposition so as to reduce N_2O and CH_4 emissions, and (ii) capturing CH_4 from manure decomposition and utilizing it as renewable energy. While handling manure as a solid rather than a liquid, for example, would likely reduce CH_4 emissions; this practice may increase N_2O emissions. On the other hand, holding manure in anaerobic environments invariably produces CH_4, which if captured can be used as an energy substitute for fossil fuels.

Anaerobic digestion is the microbial decomposition of organic materials in the absence of oxygen, producing mainly CH_4 (55–65%) and CO_2 (30–35%) gases, commonly known as biogas (Montes et al., 2013; Gerber et al., 2013). Anaerobic digestion of livestock manure coupled with complete utilization of biogas produced is thus effective in reducing CH_4 emission because this diminishes readily available C for microbial processes to produce CH_4, either during subsequent storage or following land application of the digester effluent (Montes et al., 2013; Massé et al., 2007; Safley and Westerman, 1994; Massé et al., 1996). Furthermore, when CH_4 is collected and used as a renewable energy source, it substitutes for nonrenewable fossil fuels, further mitigating emissions of GHG, NOx, hydrocarbons, and particulate matter (Börjesson and Berglund, 2006). It is reasonable to consider the amount of CH_4 produced and utilized from managed or controlled AD as the equivalent amount of reduced CH_4 emissions from excreted manure. However, the amount of CH_4 produced and collected, in practice, does not directly translate into an equal amount of reduced CH_4 emissions because the undigested manure would not yield a corresponding same amount of CH_4 gas in uncontrolled environmental conditions. The two major impacts of anaerobic digestion in GHG emissions mitigations are from: (i) replaced fossil fuel consumption, and (ii) reduced emissions due to reduced fertilizer use and production (Kaparaju & Rintala, 2011). When anaerobic digestion of livestock manure is performed correctly, and the biogas is captured and fully utilized, CH_4 emissions from the digester effluent becomes a minor issue, whereas N_2O emissions usually dominates, especially following field application of the digester effluent (Montes et al., 2013; Wulf et al., 2002).

During the anaerobic digestion process, however, organic N-containing compounds in manure, such as proteins, amino acids, and urea, are reduced to total ammoniacal nitrogen (TAN); which consists of ammonium (NH_4^+) and/or ammonia (NH_3) (Bernet et al., 2000). Anaerobic digestion of manure, therefore, increases TAN, which may promote N_2O emissions. For N_2O emissions to occur, manure must first be handled aerobically where either NH_4^+-N or organic-N is converted to nitrate (NO_3^-) and nitrite (NO_2^-) during nitrification and then handled anaerobically where the NO_3^- and NO_2^- are reduced to dinitrogen gas (N_2), with intermediate productions of N_2O and nitric oxide (NO) through denitrification (USEPA, 2010). Most of the N_2O resulting from manure is produced in manure-amended soils through microbial nitrification under aerobic conditions and denitrification under anaerobic conditions, with denitrification generally producing the larger quantity of N_2O (VanderZaag et al., 2011; USEPA, 2010). Emissions of N_2O also may occur from surfaces of dry-lots, composting piles, biofilters, and permeable manure storage covers because the conditions, at these individual units, are similar to those in the manure-amended soils.

Although anaerobic digestion increases TAN, it also stabilizes the organic C in the feedstock (i.e., reduces the fraction of easily degradable C in manures); this reduces the energy source, which is necessary for the growth of N_2O-forming microorganisms, effectively reducing the potential of N_2O emissions when effluent is applied to soil (Petersen, 1999; Bertora et al., 2008). Research indicates that as much as 70% reductions in N_2O emissions is achievable from application of anaerobic digestion effluent compared with raw manure applications (Montes et al., 2013). Research also indicates that mineralization of organic-N, during anaerobic digestion, not only increases TAN but also increases manure pH, which promotes NH_3 volatilization (Petersen and Sommer, 2011). This implies that NH_3 volatilization may be higher in digested manure (Petersen and Sommer, 2011) thus reducing the amount of TAN available for nitrification–denitrification on application of effluent to the soil.

Several researchers have studied N_2O emissions from undigested versus digested livestock slurries applied to soil. Wulf et al. (2002) found the effect of anaerobic digestion on emissions of N_2O was small relative to the effect of application technique on both arable land and grassland. Petersen (1999) spring-applied a mixture of liquid manures (i.e., cattle and pig) and anaerobically digested manure to barley on loamy sand soil. The highest N_2O emissions in both years occurred with untreated manure. Based on their results, they proposed that anaerobic digestion could reduce direct N_2O emissions from liquid manure applied to soils by 20 to 40%. These N_2O emissions reductions are consistent with results of Clemens and Huschka (2001) who conducted a laboratory study using undigested and digested cattle slurry. During the first eight days after application, N_2O emissions were dominated by denitrification, a result attributed to the level of easily available organic substances in the digestate measured as biological oxygen demand (BOD). Nitrous oxide emissions correlated positively with BOD as well as soil moisture. Overall, N_2O emissions were reduced by anaerobic digestion at low (35%) and medium (54%) water filled pore space (WFPS), but there was no difference at a high WFPS (71%). Chantigny et al. (2007) research, applying digested liquid swine manure to a forage crop over 3 yr also corroborate these other studies. Relative to undigested manure, N_2O emissions were 54–69% less on loam soil and 17–71% less on

Table 1. Summary of digester project profiles (Source: USEPA, 2017).

Farm Name	Location	# Animals	Co-digestion	System Design	kW	Third party operation	Co-Products
AA Dairy	New York	600		Horizontal Plug Flow	130		
Baldwin Dairy	Wisconsin	1050		Modified mixed plug flow	200		
Barstow's Longview Farm	Massachusetts	250	Food waste (22,000 tons/yr)	Complete Mix	800	yes	
Faber Dairy	Oregon	350		Complete Mix	100		
Big Sky West	Idaho	4700	Organic wastes	Mixed plug flow	1500	yes	Magic Dirt
Blue Spruce Farm	Vermont	2100	Whey	Two stage mixed plug flow	680		
Butler Swine Farms	North Carolina	7890	Food waste	Covered lagoon	180		
Castelanelli Bros Dairy	California	3200		Covered lagoon	300		
Clover Hill Dairy	Wisconsin	2000		Two stage mixed plug flow	480		
Cottonwood Dairy	California	5000	Cheese plant waste water	Covered Lagoon	700		
Crave Bros Dairy	Wisconsin	1900	Cheese plant waste water	Complete Mix	633		Potting Mix
Dane County Community Digester	Wisconsin	~ 550	Food waste, restaurant grease and glycerin	Complete Mix	2200		
Danny Kluthe Swine Farm	Nebraska	8000		Complete Mix	80		CNG
Emerling Dairy Farms	New York	1100		Horizontal Plug Flow	230		
Fair Oaks Farms	Indiana	9000		Two stage mixed plug flow	700		CNG
		3000		Vertical Plug Flow	1060		
Five Star Dairy farm	Wisconsin	850	High fat food wastes	Complete Mix	775		
Foster Bros Farm	Vermont	490		Horizontal Plug Flow	85		

Table 2. Forms of nitrogen in fresh manure and anaerobic digestion effluent (Kirchmann and Witter, 1992; numbers in parenthesis demonstrate the percent distribution of total N).

Type of material	Total N (mg g^{-1} dry matter)	NH$_4$–N (mg g^{-1} dry matter)
Cattle		
Fresh manure	23.3	0.22 (0.9%)
Anaerobic effluent	41.5	21.17 (51%)
Swine		
Fresh manure	30.8	2.55 (8.3%)
Anaerobic effluent	42.5	21.64 (50.9%)
Poultry		
Fresh manure	41.0	4.19 (8.2%)
Anaerobic effluent	67.3	50.85 (75.6%)

sandy loam soil. Overall, there is consistent evidence that anaerobic digestion has the potential to reduce N$_2$O emissions by up to 70% on moderately well drained soils. The possible reduction of GHG together with the economic benefits of extracting renewable energy from anaerobic digestion, makes AD a viable GHG emission mitigation technology on dairy and swine operations.

Case Studies of Anaerobic Digesters
On farm Case Studies

Anaerobic digesters are common across the United States (see Fig. 4, EPA AgStar) and most often used on dairy and swine operations (see Table 1, Digester Summary, Project Profiles, and EPA AgStar). There are approximately 244 livestock ADs and the possibility of 8000 additional ADs could produce 257 billion ft^{-3} per year of biogas, or enough fuel for 2 million passenger cars (Elger, 2017). Although most ADs are owned and operated by farm management, a model that is gaining in popularity is third party ownership–operation. The advantages of third party ownership–operation is that the owner of the animal operation can continue to focus on traditional interests of animals and crops, leaving the management of manure to another entity. Additional benefits of third party ownership–operation are the capturing of products for eco-markets such as renewable gas to vehicle, and fiber products for bulk and retail sale. Third party ownership and management also provides a more focused model to divert food waste away from landfills and centralize the collection of biogas for processing to CNG. Another model of AD management, is the formation of a centralized AD, or community AD, for processing of manure or manure and additional feedstocks. The advantage of a community system is that smaller farms can participate without the large capital outlay required if installed on their own farm. An issue to be aware of in regard to community systems is biosecurity; since manure that is comingled will be returned to cooperating farms for use as a fertilizer. The comingled manure that is returned may contain pathogens that are of a biosecurity concern, such as *Mycobacterium avium* subsp. *paratuberculosis*.

The AD of feedstocks results in biogas that can be used to fuel an engine that will run a generator for production of electricity, or the biogas can be cleaned and compressed to provide CNG for vehicle and truck use, thus displacing the need for gasoline or diesel fuels. Additional co-products from AD of manure include fiber as a replacement for peat moss, fiber for formation of degradable pots for horticultural plants, and animal bedding. The processing of manure with an AD offers the opportunity to capture nutrient streams for use as crop fertilizers. After liquid–solids separation, the liquid stream is most often used for forage production, while nutrient rich streams such and struvite (Brown, 2016) (ammonium-magnesium-phosphate) and dissolved air flotation solids (DAF) can be exported off farm for use as nutrient sources or soil amendments.

Small-scale AD in Developing Regions

While developed countries have the concentrated animal density and infrastructure to promote the adoption of highly engineered AD systems, developing countries must rely on less sophisticated models (Larson, 2017; McCord et al., 2019). Common feedstocks for use in ADs in Uganda, Rwanda, and Bolivia include manure, food waste, slaughterhouse waste, and human waste (Vogeli et al., 2014). The designs usually consist of underground domes constructed of brick and cement, above ground poly-tanks, and bag digesters (Vogeli et al., 2014). Many of these digesters are simple pass-flow reactors that do not have mixing components, do not require elaborate continuous monitoring, and are adaptable to any tropical climate (Lansing et al., 2008). A common use of the biogas is for cooking, which provides for improved indoor air quality since it replaces wood (McCord et al., 2017). The replacement of wood with biogas translates into reduced deforestation, which is in general, a benefit for the environment. An additional technology that has been adopted is the use of absorption coolers for refrigeration that are run on biogas (Larson, 2017).

Changes in Manure Composition as a Result of AD

The nitrogen, (N), phosphorus (P), and potassium (K) are not lost or reduced due to the AD process, but are transformed (see Fig. 1) from organic forms to inorganic forms

Fig. 4. Location of anaerobic digesters across the United States. (Source: U.S. EPA).

while the carbon is converted to biogas (Table x). As a result, the levels of ammonium N and inorganic P increase as a percent of total N and total P when compared to raw manure. The increase in ammonium content will vary due to pre-digester management of the manure. Ammonia nitrogen has been shown to increase 37% and ortho-P increased 26% after anaerobic digestion (Aldrich, 2005). Similar average changes in inorganic P content were observed by Pagliari and Laboski (2013).

Nitrogen that enters a digester from dairy manure is either in the ammonium or organic form. Much of the organic nitrogen is converted via nitrogen mineralization during the digestion process to ammonium and free ammonia, raising the overall level of ammoniacal nitrogen in the effluent (Field et al., 1984). Although a small amount of ammonia gas will be lost to biogas, the total nitrogen leaving the digester is generally considered equal to that added to the digester (Topper et al., 2006).

Nutrient content of the AD input will vary depending on the species of the contributing manure and if there is any addition of other organic feedstocks (co-digestion). Kirchmann and Witter (1992; Table 2) evaluated fresh and anaerobically digested manure from three different species for nutrient concentration. They concluded that anaerobic digestion of manure resulted in higher ammoniacal N concentrations (50–75% of total N) in the digested material. Anaerobic digestion facilitates nitrogen mineralization, while carbon is converted to biogas. Additionally, carbon is partially removed from the digested material, reducing the C to N ratio (Kirchmann and Witter, 1992; Möller et al., 2008).

Nutrient speciation data collected from previous AD studies suggest that a high percentage of the P can be found in the inorganic form in the AD effluent (Wrigley et al., 1992; Bowers et al., 2007; Marti et al., 2008; Moody et al., 2009). Bowers et al. (2007) demonstrated total phosphorus (TP) content ranging from 238 to 323 ppm, from which ortho-phosphorus (OP) contributed 106 to 231 ppm of the TP in the post digestion effluent of a co-digestion dairy manure AD. Similarly, Moody et al. (2009) and colleagues demonstrated a 26% increase of inorganic P (PO_4^{3-}) in digested swine slurry compared to the raw swine slurry (1591 and 1256.2 mg L^{-1} of PO_4^{3-}, respectively). In addition, when evaluating the nutrient transformation of five different types of ADs in New York, the percent change in OP after digestion varied from 7–27% depending on the type of AD (Fig. 5; Gooch et al., 2006). The case study data from Cornell (Fig. 2) also demonstrates the percent change of total Kjeldahl nitrogen (TKN), ammonia nitrogen (NH_3–N), organic nitrogen (ON), and TP (Gooch et al., 2006). The positive percent change indicates a greater concentration of the nutrient in the post digested effluent compared to the influent vs. a negative percent change which indicates the nutrient is more concentrated in the influent before digested compared to after digestion. These data represent several farms with digesters, some with different digester models, and therefore variation would be expected. Also, one should not expect loss of N or P during digestion, and variability due to sampling and analyses could cause errors in mass balance calculations.

The percent changes was calculated as influent nutrient value minus effluent nutrient value; therefore even though there is a negative percent change of NH_3–N and OP, there is a greater amount in the digested effluent compared to the influent.

Value of AD Manure for Crop Production

Although there is a growing body of research on anaerobic digestion and crop nutrient availability, this technology has not been extensively studied. Most work has been focused on short term nutrient recovery (1–3 yr), and not the long term impacts (5+ years) of fertilizing with AD manures (Arthurson, 2009). In the review by Nkoa (2014), he suggests that AD effluents may pose environmental concerns due to higher NH_3 emission potential, and higher concentrations of copper (Cu) and zinc (Zn). Pagliari and Laboski (2013) found that when using AD manure, changes in soil test P were mainly a function of soil clay content and inorganic P + total enzymatically hydrolyzable P in manure.

Anaerobically digested manure (AD) has been shown to have the same positive effects on yield and crop production when applied at equal rates of plant-available N as synthetic fertilizers or raw manures in corn and forage production systems (Morris and Lathwell, 2004; Loria et al., 2007; Saunders et al., 2012), while soil quality and fertility indicators are improved relative to synthetic fertilizers (de Boer, 2008; Arthurson, 2009).

Since anaerobically digested dairy manure could provide more plant available nitrogen than untreated manure (Kirchmann and Witter, 1992; Michel et al., 2010), the potential exists for increasing agricultural efficiencies (Morris and Lathwell, 2004; Möller and Stinner, 2010). Increased concentrations of ammoniacal-N in AD manure would increase the potential loss of N via free ammonia (NH_3–N) volatilization in the field, so best management practices would be required to take advantage of the higher ammoniacal-N content.

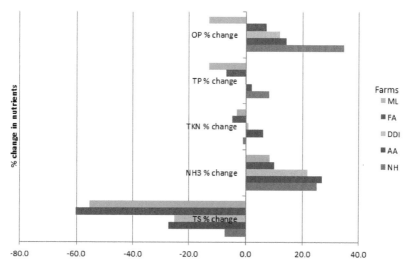

Nutrient Transformation Changes During Anaerobic Digestion of Manure

Fig. 5. Nutrient transformation (% change) of five different farms with digesters from a case study in NY state (Source: Gooch et al., 2006).

Application of AD dairy manure to corn has been shown to produce similar total plant N uptake and equivalent or greater yields than inorganic fertilizer. Acidic soils have been shown to exhibit distinct lower yields compared to alkaline soils when receiving AD dairy manure, which demonstrated an advantage in the early stages for corn growth over synthetic fertilizers (Morris and Lathwell, 2004). Application of digested manure that is rich in ammonium to acidic soils (pH < 7.0) restricts ammonia loss resulting in greater plant uptake of nitrogen (Nelson, 1982).

Manure serves as a useful, low-cost source of nutrients for crop production (Sommerfeldt et al., 1988; Jokela, 1992; Ferguson et al., 2005; Nyiraneza and Snapp, 2007). Anaerobically digested manure provides sufficient nutrients to support biomass and crop yields equivalent to synthetic fertilizers and raw manures (Bittman et al., 1999; Loria et al., 2007; Saunders et al., 2012). Some studies (Rubæk et al., 1996; Chantigny et al., 2007; de Boer, 2008) have found increased yield and nitrogen availability with application of anaerobically digested material as compared to non-digested material, possibly due to increased nitrogen content and reduced carbon content, which can result in nitrogen immobilization by microbes. In addition, manure applications to soils have enhanced soil quality and fertility compared to soils receiving synthetic fertilizers (de Boer, 2008; Arthurson, 2009). A crop would typically recover < 50% of applied fertilizer nitrogen (Stevens et al., 2005). Up to 46% of applied manure nitrogen may be left over in the soil at the end of the growing season, increasing the potential for loss, after multiple applications during a season (Muñoz et al., 2003). Over-application of manure nitrogen in excess of crop uptake can result in nitrate leaching (Angle et al., 1993). Some studies have indicated that manure nitrogen poses an equal or slightly less risk to leaching than synthetic fertilizers (Jokela, 1992; Trindade et al., 2009; Saunders et al., 2012). Others have determined manure increases nitrate leaching (Jemison and Fox, 1994). During winter months when plants are dormant, nitrate leaching can be the main source of N loss (Bakhsh et al., 2007). The shift in organic to inorganic nutrients during the AD process should be considered when developing a farm nutrient management plan.

Conclusion

Anaerobic digesters (AD) provide many benefits including: odor management, reduced emission of greenhouse gases, pathogen control, and production of biogas (methane) for use as a fuel. In addition, the liquid after anaerobic digestion can be used as a nutrient source for crop production, while solids after anaerobic digestion can be used as either bedding for animals or sold as a soil amendment. Due to the reduction in pathogens after AD, treated manure should provide an advantage for integration of livestock and food production systems with minimal risk of food contamination.

References

Aldrich, B.S. 2005. Anaerobic digestion of dairy manure: Implications for nutrient management planning. North East Branch of the American Society of Agronomy Annual Meeting, July 11-13, 2005, Storrs, CT. Cornell University, Ithaca, NY.

Angle, J.S., C.M. Gross, R.L. Hill, and M.S. McIntosh. 1993. Soil nitrate concentrations under corn as affected by tillage, manure, and fertilizer applications. J. Environ. Qual. 22:141–147. doi:10.2134/jeq1993.00472425002200010018x

Arthurson, V. 2009. Closing the global energy and nutrient cycles through application of biogas residue to agricultural land- potential benefits and drawbacks. Energies 2:226–242. doi:10.3390/en20200226

Bakhsh, A., R.S. Kanwar, C. Pederson, and T.B. Bailey. 2007. N-source effects on temporal distribution of NO3-N leaching losses to subsurface drainage water. Water Air Soil Pollut. 181:35–50. doi:10.1007/s11270-006-9274-z

Bernet, N., N. Delgenes, J.C. Akunna, J.P. Delgenes, and R. Moletta. 2000. Combined anaerobic–aerobic SBR for the treatment of piggery wastewater. Water Res. 34:611–619. doi:10.1016/S0043-1354(99)00170-0

Bertora, C., F. Alluvione, L. Zavattaro, J.W. van Groenigen, G. Velthof, and C. Grignani. 2008. Pig slurry treatment modifies slurry composition, N2O, and CO2 emissions after soil incorporation. Soil Biol. Biochem. 40:1999–2006. doi:10.1016/j.soilbio.2008.03.021

Bittman, S., C.G. Kowalenko, D.E. Hunt, and O. Schmidt. 1999. Surface-banded and broadcast dairy manure effects on tall fescue yield and nitrogen uptake. Agron. J. 91:826–833. doi:10.2134/agronj1999.915826x

Börjesson, P., and M. Berglund. 2006. Environmental systems analysis of biogas systems—Part I: Fuel-cycle emissions. Biomass Bioenergy 30:469–485. doi:10.1016/j.biombioe.2005.11.014

Bowers, K.E., T.X. Zhang, and J.H. Harrison. 2007. Phosphorus removal by struvite crystallization in various livestock wastewaters. American Society of Agricultural and Biological Engineers International Air and Waste Symposium, Broomfield, Colorado, 16-19 Sept. InTech, London.

Brown, K.M., J.H. Harrison, K. Bowers, R.G. Stevens, A.I. Bary, and K. Harrison. 2016. Agronomic response of crops fertilized with struvite derived from dairy manure. Water Air Soil Pollut (2016) 227:388 doi:10.1007/s11270-016-3093-7

Chantigny, M.H., D.A. Angers, P. Rochette, G. Belanger, D. Massé, and D. Cote. 2007. Gaseous nitrogen emissions and forage nitrogen uptake on soils fertilized with raw and treated swine manure. J. Environ. Qual. 36:1864–1872. doi:10.2134/jeq2007.0083

Clemens, J., and A. Huschka. 2001. The effect of biological oxygen demand of cattle slurry and soil moisture on nitrous oxide emissions. Nutr. Cycl. Agroecosyst. 59(2):193–198. doi:10.1023/A:1017562603343

de Boer, H.C. 2008. Co-digestion of animal slurry can increase short-term nitrogen recovery by crops. J. Environ. Qual. 37:1968–1973. doi:10.2134/jeq2007.0594

Elger, N. 2017. On-farm Anaerobic Digestion in the U.S. U.S. Environmental Protection Agency, Agstar, Mankato, MN. https://www.epa.gov/sites/production/files/2017-03/documents/agstar-midwest-manure-management-summit-2017.pdf (Accessed 24 May 2017)

Ferguson, R.B., J.A. Nienaber, R.A. Eigenberg, and B.L. Woodbury. 2005. Long-term effects of sustained beef feedlot manure application on soil nutrients, corn silage yield, and nutrient uptake. J. Environ. Qual. 34:1672–1681. doi:10.2134/jeq2004.0363

Field, J.A., J.S. Caldwell, S. Jeyanayagam, R.B. Reneau, Jr., W. Kroontje, E.R. Collins, Jr. 1984. Fertilizer recovery from anaerobic digesters. Trans. ASAE. 27 (6): 1871-1876. doi:10.13031/2013.33060

Gerber, P., B. Henderson, and H. Makkar, editors. 2013. Mitigation of greenhouse gas emissions in livestock production– a review of technical options for non-CO2 emissiosn. FAO, Rome, Italy.

Gooch, C.A., S.F. Inglis, and P.E. Wright. 2006. Biogas distributed generation systems evaluation and technology transfer–Interim report. NYSERDA Project No. 6597. New York State Energy Research and Development Authority, Albany, NY.

Gould, C. 2012. Introduction to anaerobic digestion. Bioenergy Training, University of Wisconsin Extension, Madison, WI.

Hamilton, D. 2012. Types of anaerobic digesters. Farm Energy Extension, Washington, D.C. https://articles.extension.org/pages/30307/types-of-anaerobic-digesters#Passive_Systems (Accessed 9 May 2019).

Jemison, J.M., and R.H. Fox. 1994. Nitrate leaching from nitrogen- fertilized and manured corn measured with zero-tension pan lysimeters. J. Environ. Qual. 23:337–343. doi:10.2134/jeq1994.00472425002300020018x

Jokela, W.E. 1992. Nitrogen-fertilizer and dairy manure effects on corn yield and soil nitrate. Soil Sci. Soc. Am. J. 56:148–154. doi:10.2136/sssaj1992.03615995005600010023x

Kaparaju, P., and J. Rintala. 2011. Mitigation of greenhouse gas emissions by adopting anaerobic digestion technology on dairy, sow, and pig farms in Finland. Renew. Energy 36:31–41. doi:10.1016/j.renene.2010.05.016

Kirchmann, H., and E. Witter. 1992. Composition of fresh, aerobic and anaerobic farm animal dungs. Bioresour. Technol. 40:137–142. doi:10.1016/0960-8524(92)90199-8

Lansing, S., R.B. Botero, and J.F. Martin. 2008. Waste treatment and biogas quality in small-scale agricultural digesters. Bioresour. Technol. 99:5881–5890. doi:10.1016/j.biortech.2007.09.090

Larson, R. 2017. Anaerobic digestion systems webinar. Livestock and Poultry Environmental Learning Center, University of Nebraska, Lincoln, NE. https://learn.extension.org/events/3142 (Accessed 9 May 2019).

Loria, E.R., J.E. Sawyer, D.W. Barker, J.P. Lundvall, and J.C. Lorimor. 2007. Use of anaerobically digested swine manure as a nitrogen source in corn production. Agron. J. 99:1119–1129. doi:10.2134/agronj2006.0251

Massé, D.I., F. Croteu, and L. Massé. 2007. The fate of crop nutrients during digestion of swine manure in psychrophilic anaerobic sequencing batch reactors. Bioresour. Technol. 98:2819–2823. doi:10.1016/j.biortech.2006.07.040

Massé, D. I., N.K. Patni, R.I. Droste, K.J. Kennedy. 1996. Operation strategies for psychrophilic anaerobic digestion of swine manure slurry in sequencing batch reactors. Can. Journal of Civil Engin. 23: 1285–1294. doi:10.1139/l96-937

Marti, N., A. Bouzas, A. Seco, and J. Ferrer. 2008. Struvite precipitation assessment in anaerobic digestion processes. Chem. Eng. J. 141:67–74. doi:10.1016/j.cej.2007.10.023

Michel, J., A. Weiske, and K. Müller. 2010. The effect of biogas digestion on the environmental impact and energy balances in organic cropping systems using the life-cycle assessment methodology. Renewable Agriculture and Food Systems 25:204–218. doi:10.1017/S1742170510000062

McCord, A.I., S.A. Stefanos, V. Tumwesige, D. Losto, A.H. Meding, A. Adong, J.J. Schauer, and R.A. Larson. 2017. The impact of biogas and fuelwood use on institutional kitchen air quality in Kampula, Uganda. Indoor Air 27:1067–1081. doi:10.1111/ina.12390

McCord, A.I., S.A. Stefanos, V. Tumwesige, D. Lsoto, M. Kawala, J. Mutebi, I. Nansubuga, and R.A. Larson. 2019. Anaerobic digestion in Uganda: Risks and opportunities for integration of waste management and agricultural systems. Int. J. Env. Sci. Technology. In Press.

Möller, K., W. Stinner, A. Deuker, and G. Leithold. 2008. Effects of different manuring systems with and without biogas digestion on nitrogen cycle and crop yield in mixed organic dairy farming systems. Nutr. Cycl. Agroecosyst. 82:209–232. doi:10.1007/s10705-008-9196-9

Möller, K., and W. Stinner. 2010. Effects of organic wastes digestion for biogas production on mineral nutrient availability of biogas effluents. Nutr. Cycling Agroecosyst. 87:395–413. doi:10.1007/s10705-010-9346-8

Montes, F., R. Meinen, C. Dell, A. Rotz, A.N. Hristov, J. Oh, G. Waghorn, P.J. Gerbrer, B. Henderson, H.P.S. Makkar, and J. Dijkstra. 2013. Mitigation of methane and nitrous oxide emissions from animal operations: II. A review of manure management mitigation options. J. Anim. Sci. 91:5070–5094. doi:10.2527/jas.2013-6584

Moody, L., R. Burns, and K.J. Stalder. 2009. Effect of anaerobic digestion on manure characteristics for phosphorus precipitation from swine waste. Appl. Eng. Agric. 25:97–102. doi:10.13031/2013.25430

Morris, D.R., and D.J. Lathwell. 2004. Anaerobically digested dairy manure as fertilizer for maize in acid and alkaline soils. Commun. Soil Sci. Plant Anal. 35:1757–1771. doi:10.1081/CSS-120038567

Muñoz, G.R., J.M. Powell, and K.A. Kelling. 2003. Nitrogen budget and soil N dynamics after multiple applications of unlabeled or (15) Nitrogen-enriched dairy manure. Soil Sci. Soc. Am. J. 67:817–825. doi:10.2136/sssaj2003.0817

Nelson, D.W. 1982. Gaseous losses of nitrogen other than through denitrification. In: F.J. Stevenson, J.M. Bremner, R.D. Hauck, and D.R. Keeney, editors. Nitrogen in agricultural soils. Agronomy monograph 22. American Society of Agronomy, Madison, WI. p. 327–363.

Nkoa, R. 2014. Agricultural benefits and environmental risks of soil fertilization with anaerobic digestates: A review. Agron. Sustain. Dev. 34:473–492. doi:10.1007/s13593-013-0196-z

Nyiraneza, J., and S. Snapp. 2007. Integrated management nitrogen and efficiency of inorganic and organic in potato systems. Soil Sci. Soc. Am. J. 71:1508–1515. doi:10.2136/sssaj2006.0261

Pagliari, P.H., and C.A.M. Laboski. 2013. Dairy manure treatment effects on manure phosphorus fractionation and changes in soil test phosphorus. Biol. Fertil. Soils 49:987–999. doi:10.1007/s00374-013-0798-2

Petersen, S.O., and S.G. Sommer. 2011. Ammonia and nitrous oxide interactions: Roles of manure organic matter management. Anim. Feed Sci. Technol. 166–167:503–513. doi:10.1016/j.anifeedsci.2011.04.077

Petersen, S.O. 1999. Nitrous oxide emissions from manure and inorganic fertilizers applied to spring barley. J. Environ. Qual. 28:1610–1618. doi:10.2134/jeq1999.00472425002800050027x

Rubæk, G.H., K. Henriksen, J. Petersen, B. Rasmussen, and S.G. Sommer. 1996. Effects of application technique and anaerobic digestion on gaseous nitrogen loss from animal slurry applied to ryegrass (Lolium perenne). J. Agric. Sci. 126:481–492. doi:10.1017/S0021859600075572

Safley, L.M., Jr., and P.W. Westerman. 1994. Low-temperature digestion of dairy and swine manure. Bioresour. Technol. 47:165–171. doi:10.1016/0960-8524(94)90116-3

Saunders, O., A.-M. Fortuna, J.H. Harrison, E. Whitefield, C. Cogger, A.C. Kennedy, and A.I. Bary. 2012. Comparison of raw dairy manure slurry and anaerobically digested slurry as N sources for grass forage production. Int. J. Agron. Article ID 101074. doi:10.1155/2012/101074

Solomon, S., D. Qin, M. Manning, K. Averyt, and M. Marquis, editors. 2007. Climate change 2007-the physical science basis: Working group I contribution to the fourth assessment report of the IPCC. Vol. 4. Cambridge University Press, Cambridge, U.K.

Sommerfeldt, T.G., C. Chang, and T. Entz. 1988. Long-term annual manure applications increase soil organic-matter and nitrogen, and decrease carbon to nitrogen ratio. Soil Sci. Soc. Am. J. 52:1668–1672. doi:10.2136/sssaj1988.03615995005200060030x

Stevens, W.B., R.G. Hoeft, and R.L. Mulvaney. 2005. Fate of nitrogen-15 in a long-term nitrogen rate study: II. Nitrogen uptake efficiency. Agron. J. 97:1046–1053. doi:10.2134/agronj2003.0313

Topper, P.A., R.E. Graves, and T. Richard. 2006. The fate of nutrients and pathogens during anaerobic digestion of dairy manure. Pennsylvania State, Department of Agricultural and Biological Engineering, University Park, PA.

Trindade, H., J. Coutinho, S. Jarvis, and N. Moreira. 2009. Effects of different rates and timing of application of nitrogen as slurry and mineral fertilizer on yield of herbage and nitrate-leaching potential of a maize/Italian ryegrass cropping system in north-west Portugal. Grass Forage Sci. 64:2–11. doi:10.1111/j.1365-2494.2008.00664.x

USEPA. 2010. Inventory of U.S. greenhouse gas emissions and sinks: 1990–2008. USEPA, Washington, D.C. www.epa.gov/climatechange/emissions/usinventoryreport.html

USEPA. 2017. Inventory of U.S. greenhouse gas emissions and sinks. U.S. Environmenal Protecion Agency, Washington, D.C. https://www.epa.gov/ghgemissions/inventory-us-greenhouse-gas-emissions-and-sinks (Accessed 20 June 2018).

VanderZaag, A.C., S. Jayasundara, and C. Wagner-Riddle. 2011. Strategies to mitigate nitrous oxide emissions from land applied manure. Anim. Feed Sci. Technol. 166–167:464–479. doi:10.1016/j.anifeedsci.2011.04.034

Vogeli, Y, C R Lohri, A Gallardo, S Diener, and C Zurbugg. 2014. Anaerobic digestion of biowaste in developing countries. Practical information and case studies. Eawag– Swiss Federal Institute of Aquatic Science and Technology, Department of Water and Sanitation in Developing Countries (Sandec), Dubendorf, Switzerland.

Wrigley, T., K. Webb, and H. Venkitachalm. 1992. A laboratory study of struvite precipitation after anaerobic digestion of piggery wastes. Biores. Techn. 41:117–121. St Joseph, MI.

Wulf, S., M. Mateing, and J. Clemens. 2002. Application technique and slurry co-fermentation effects on ammonia, nitrous oxide, and methane emissions after spreading. J. Environ. Qual. 31:1795–1801. doi:10.2134/jeq2002.1795

Farming Characteristics and Manure Management of Small Ruminant and Cervid Livestock

A.B. Norris and W.B. Smith*

Abstract

Small ruminants (e.g., sheep and goats) represent a small, but rapidly growing, sector of the United States livestock industry. Likewise, the farming of cervid species (e.g., deer) represents a significant financial contributor to agricultural receipts. Thus, management of wastes from both production systems is of interest to the greater population. Unlike other livestock species, where diet is intensively managed (thereby creating a consistent byproduct through excreta), the diet of sheep, goats, and deer is highly variable, leading to inconsistencies in manure composition and use. Inclusion of anti-nutritional factors, commonly found in browse that would exist in these diets, also lends itself to considerations in use of small ruminant manure. This chapter will seek to address the intricacies that exist in characterization of excreta from sheep, goats, and deer.

Small ruminant species are of great importance within production systems globally, as they are an essential source of milk, meat, and fiber, especially within developing countries. However, sheep and goat products are not limited to areas historically recognized for product utilization as demand within developed countries, such as the United States; it continues to rise as a result of increased immigration and its influence on local culture. Expansion of small ruminant production is largely driven by animal adaptability to differing ecotypes and reduced infrastructure required for smallholder production. High levels of diversity enable survival and production within many environment types, with small ruminant production being present within ecotypes varying from desert shrub to tropical savannahs and forests. The associated production systems differ greatly as well, with systems diverging from the historical animal-based pastoral systems to crop-based and confined feeding systems. The type of management system and primary use of small ruminants is largely dependent on location, as climate and socioeconomical factors greatly influence animal production for use as subsistence or commercial purposes. Small ruminant production is rich with diversity, as some areas use minimal to no grazing within intensive dairy operations, whereas other regions use pastoral systems based on nomadism and transhumance. It is this innate diversity that makes small ruminant production significant on a global scale.

Digestive Anatomy and Physiology

Any holistic discussion of manure from a given class of animals must first begin with the anatomy and physiology of the animal before progressing into an

Author list and affiliations: A. B. Norris, Department of Soil and Crop Sciences, Texas A&M University, College Station; and W. B. Smith, Department of Animal Science and Veterinary Technology, Tarleton State University, Stephenville, TX. *Corresponding author (wbsmith@tarleton.edu)

doi:10.2134/asaspecpub67.c7

Animal Manure: Production, Characteristics, Environmental Concerns and Management. ASA Special Publication 67. Heidi M. Waldrip, Paulo H. Pagliari, and Zhongqi He, editors.

examination of nutrition. Only then may one understand the system under which manure is produced and can begin to imagine avenues for use and treatment. Thus, to understand this chapter, dedicated to the small and exotic ruminant species, one must first understand the nature of a ruminant animal. This is best summarized by Dehority (2002, p. 145), who stated that in order for the "herbivores to utilize forages and plants as their sole energy source is dependent on microorganisms living at various sites within their gastrointestinal tract." As such, we will begin our discussion of these species with a discussion of the anatomical and microbiological niches through which ruminants carry out their place in the ecosystem.

A ruminant animal is unique in that is has a four-chambered stomach capable of holding a large amount of material and populated with a microbiome specially equipped to break down the cell walls of plants. These compartments are the reticulum, the rumen, the omasum, and the abomasum. The abomasum is the gastric, or "true," stomach of the animal and functions similarly to that of the remaining members of Kingdom Animalia. The foregut, principally the reticulorumen, is believed to have been an evolutionary derivation of the esophagus and cardiac regions of the stomach and serves as the nongladular, microbial vat (Dehority, 2002).

Hofmann (1989) elucidated the primary difference among ruminant animals. According to this paper, "grass and roughage eaters", or grazers, developed in a coevolutionary fashion with range plants with a high proportion of plant cell wall carbohydrates (Hofmann, 1989). Implied within this definition is that these animals are better adapted to consuming low-quality diets; Hofmann (1989) went on to say that these animals exhibited long feeding periods followed by low periods of rest and rumination. Included in this group are cattle (*Bos taurus*), water buffalo (*Bubalus bubalis*), and banteng (*Bos javanicus*; Hofmann, 1989), though there is disagreement as to the inclusion of sheep (*Ovis aries*). At the opposite end of the spectrum of ruminant animals, Hofmann (1989) describes an animal that selects plants with a high proportion of cell soluble contents based on primarily olfactory cues, known as "concentrate selectors," or browsers; in his description, there are no domesticated species within this category (though several game species of major interest fit squarely within this definition). Hofmann (1989) finally describes a group of animals intermediate to each of these definitions, termed the intermediate feeders. These include the red deer (*Cervus elaphus*) and goat (*Capra hircus*).

Ruminant size can have a great impact on the function of the digestive tract. Some of the effect of size can be simply linked to the capacity of the rumen (35 to 100 L in cattle versus 3 to 15 L in sheep; Dehority, 2002). In addition to rumen capacity, size and function (grazer versus browser) can influence the overall anatomy of the animal. Proportional weight of the parotid salivary gland increases linearly with body size, though browsers nearly always have larger glands than grazers, presumably to accommodate a diet higher in cell solubles (Robbins et al., 1995). Intake, particle size, specific gravity, and concentration of solids are all known to impact digestion and passage rates, but these rates can differ between sheep and cattle under the same conditions (Dehority, 2002). While cattle tend to have ruminal turnover times of 1.3 to 3.7 d (31 to 89 h), similar diets in sheep have turnover times in the range of 0.8 to 2.2 d (19 to 53 h; Dehority, 2002). Similarly, while both large and small ruminants rely on particle size and specific gravity for ingesta to pass through the rumen into the omasum, the critical particle size in cattle (2 to 5 mm) has been shown to be quite larger than that in sheep (1 to 2 mm; Dehority, 2002). Robbins et al. (1995) demonstrated that fiber digestive capability increased with increases in body size among ruminant species.

Small Ruminant Livestock
Sheep

As of 2016, there are 551 million meat-producing sheep worldwide, with an additional 249 million milk-producing sheep (FAOSTAT, 2016). Distributions of meat- and milk-producing sheep worldwide are presented in Fig. 1 and 2, respectively. For comparison, the United States ranked 46th in meat production and had no reported data for milk production (FAOSTAT, 2016)

Management

As evidenced by the primary regions of production, systems utilized are greatly contrasting among and within regions. This is largely a result of regional environmental and socioeconomic factors that drive production of specific products (meat, milk, and wool). The primary management systems utilized are extensive production or traditional pastoralism for wool and meat production and intensive dairy and finishing operations (FAO, 1995).

Within extensive production, meat, wool, or both are commonly produced with milk and other products being highly utilized within more traditional pastoral systems. In extensive systems, animals spend substantial, if not all, time on pasture and rely on grazing of forages for nutrients with only minor supplementation during times of environmental stress. These systems are more prevalent within semiarid open rangelands where daily interaction is not commonly feasible: semiarid grasslands of the Americas, hilly, semiarid regions of the Middle East,

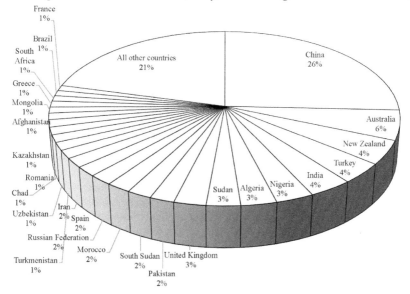

Fig. 1. Distribution of production (number of animals) of meat-producing sheep worldwide. Data were adapted from FAOSTAT (2016).

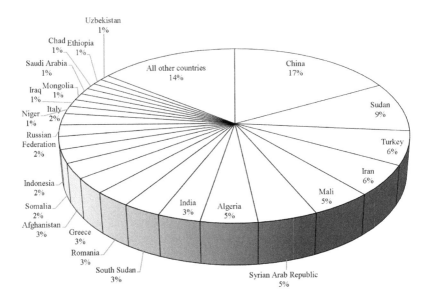

Fig. 2. Distribution of production (number of animals) of milk-producing sheep worldwide. Data were adapted from FAOSTAT (2016).

and arid deserts or North Africa. The degree of extensiveness varies based on environmental conditions and production goals, with some systems still having high management inputs in the form of supplementation, winter and lambing housing, transhumance, and animal processing. Reduced extensiveness can be beneficial to production from a welfare stance as animals become accustomed to human contact, reducing stress during handling procedures that can be detrimental to production. Breeding and management for animals able to survive adverse conditions, lamb without assistance, and reduced shepherding, ('easy-care' sheep) is a method that can be utilized to reduce inputs. However, development of these sheep has welfare issues. Major issues that can arise within extensive systems are predation, parasites, and disease as infrequent shepherding reduces animal observation and may require greater inputs within herd health programs and use of guard animals.

As with extensive systems, intensive systems have variable intensities dependent on production goals and locale. Within highly intensified dairy and finishing operations, animals may be kept exclusively within housing or confinement and fed harvested feeds, however, in many cases pasture-based systems are still greatly utilized, either wholly or partly. Often times, intensive operations more readily utilize innovation and technology to improve individual animal production. In dairy operations, this is commonly in the form of milking machines, improved breeds, estrus synchronization, and multiple mating or insemination periods. Within meat production systems, intensification is often similar but may also include strict culling for reproductive and growth traits, induction of twin and triplet lambing, early weaning, out of season lambing, twice a year lambing, and stringent parasite and disease control. Intensive systems cost more to operate and therefore are highly dependent on commodity pricing to enable implementation of specific management regimes.

Manure Composition Relative to Diet

As discussed within the management section, large variability exists in the type of management utilized within sheep production based on forage base and production goals. The diversity within feedstuffs directly impacts the composition of animal excreta as ruminal kinetics greatly determines the extent of feedstuff degradation and subsequent absorption within the lower-gastrointestinal tract. As fibrous constituents within feedstuffs increase, reduced rate of passage is often noted and, as a result, digestibility is often increased, but total nutrient consumption and absorption per day is decreased (Table 1). In alfalfa hay, NDFD ranged from 33–57%, whereas pangola hay ranged from 55–73%. Unfortunately, it is not possible to make comparisons on the total NDF excretion, but it is evident that as cell wall constituents increase within the diet, NDF excretion per unit of NDF intake will likely be reduced; however, total NDF and concentration within manure is likely highly variable.

When looking at nitrogen digestibility highly fermentable forages commonly have greater nitrogen digestibility due to increased proportion of nitrogen being soluble, whereas in lower value forages nitrogen digestion tends to decrease due to association with the cellulo-lignin matrix. Thus, as fermentability decreases nitrogen excretion would increase due to reduced digestibility. When comparing pangola grass of contrasting maturities we can see that nitrogen digestibility is 36% greater in the immature versus the mature hay (Watson and Norton, 1982). However, due to increased organic matter intake, total nitrogen excreted was also greater in the immature forage, but the proportion of fecal nitrogen and urinary nitrogen is much lower (~ 1:1 vs. ~2:1). This is indicative of increased digestibility that enabled absorption of nitrogen. When looking at highly fermentable legumes such as alfalfa, lablab, sirato, and stylo, it is evident that most of the nitrogen is readily digested as the proportion of fecal to urinary nitrogen ranges from roughly 1:3–1:6. Although increased digestibility of nitrogen can be beneficial to the animal, increasing excretion of nitrogen via the urine can be greatly detrimental to the environment due to rapid conversion of urea to volatile nitrogenous compounds such as ammonia and nitrous oxide.

Additionally, secondary compounds readily found within feedstuffs consumed by sheep can also alter ruminal kinetics and digestibility, impacting manure composition. The primary compounds of interest are condensed tannins as they have demonstrated the capacity to be both beneficial and detrimental to digestion. These compounds are of great interest due to their capacity to ruminally protect nitrogen, reducing ruminal ammonia production and increasing nitrogen use efficiency by increasing amino acid absorption within the small intestine. Therefore, in diets containing condensed tannins a shift in the route of nitrogen excretion from the urine to the feces is often exhibited but is highly dependent on condensed tannin type and affinity for the constituents of consumed feedstuffs. However, condensed tannins will often result in decreased fiber digestibility leading to reduced energy attained from the diet and greater levels of fiber within feces. It is evident that the composition of manure is highly variable and largely dependent on the type of diet consumed and the presence of plant secondary compounds that may impact digestion.

Pagliari and Laboski (2012) undertook measures to characterize manure phosphorus contents across livestock species. In collecting these data, a reference table was generated whereby a general idea of the physical and chemical properties of

Table 1. Feedstuff digestibility and manure excretion from sheep.

Feedstuff	Feeding level†	Intake	DMD‡	OMD	NDFD	ADFD	GED	CPD	Fecal N	Urinary N	Tannins	Citation
		g kg BW⁰·⁷⁵			g kg⁻¹					g d⁻¹		
Cassia fistula L. plus supplement	A	43.8	500	530	450	.	.	520	.	.	Yes	Salem et al., 2006
Chamaecrista rotundifolia (Pers.) Greene	A	12.1	550	580	.	.	.	740	1.1	10.2	Yes	Mupangwa et al., 2000
Chorisia speciosa (A. St.-Hil. et al.) Ravenna plus supplement	A	52.9	520	550	510	.	.	520	.	.	Yes	Salem et al., 2006
Cynodon dactylon (L.) Pers./ *Mucuna pruriens* (L.) DC.	A	71.2	640	.	.	.	630	590	6.4	3.9	No	Loyra-Tzab et al., 2011
Cynodon dactylon (L.) Pers. plus supplement	A	42.4	670	690	520	.	.	710	.	.	No	Maia et al., 2012
Glycine max (L.) Merr./ *Sorghum bicolor* (L.) Moench ssp. bicolo	G	85.4	640	660	570	.	660	610	8.4	7.8	No	Lima et al., 2011
Hay (*Digitaria eriantha* Steud.)	A	48.8	550	550	550	.	.	400	.	.	.	Tomkins et al., 1991
Hay (*Digitaria eriantha* Steud.; immature)	A	47.8§	.	610	730	630	.	600	7.4	7.0	No	Watson and Norton, 1982
Hay (*Digitaria eriantha* Steud.; mature)	A	31.9§	.	530	640	530	.	240	4.3	2.0	No	Watson and Norton, 1982
Hay (*Medicago sativa* L.)	M	.	560	590	330	370	560	730	.	.	No	Isac et al., 1994
Hay (*Medicago sativa* L.)	M	46.0	640	670	530	490	650	830	.	.	No	Alcaide et al., 2000
Hay (*Medicago sativa* L.) plus supplement	M	46.0	680	710	550	490	660	780	.	.	No	Alcaide et al., 2000
Hay (*Medicago sativa* L.) plus supplement	M	40.0	730	770	700	780	740	760	.	.	No	Alcaide et al., 2000
Hay (*Medicago sativa* L.) plus supplement	A	41.5	730	760	570	.	760	800	5.6	18.2	No	López and Fernandez, 2014
Hay (*Medicago sativa* L.) plus supplement	A	42.4	700	730	600	.	730	750	6.3	17.5	No	López and Fernandez, 2014
Eucalyptus camaldulensis Dehnh. plus supplement	A	37.8	510	510	430	.	.	490	.	.	Yes	Salem et al., 2006
Hyparrhenia Andersson ex Fourn.	A	35.0	540	590	.	.	570	440	4.5	3.6	No	Gihad, 1976
Lablab Adans.	A	48.2	640	650	.	.	.	710	4.8	23.1	Yes	Mupangwa et al., 2000
Macroptilium atropurpureum (Moc. & Sessé ex DC.) Urb.	A	52.6	580	590	.	.	.	810	4.8	29.1	Yes	Mupangwa et al., 2000
Schinus molle L. plus supplement	A	41.0	510	550	490	.	.	500	.	.	Yes	Salem et al., 2006
Silage (*Lolium perenne* L. ssp. perenne)	M	28.7	690	710	710	.	.	.	4.7	10.5	No	Deaville et al., 2010
Silage (*Glycine max* (L.) Merr./*Sorghum bicolor* (L.) Moench ssp. bicolor)	G	80.5	710	740	620	.	730	720	5.2	8.5	No	Lima et al., 2011
Silage (*Zea mays* L.)	A	74.3	640	660	520	520	660	530	3.0	1.2	No	Dos Santos et al., 2011
Sorghum bicolor (L.) Moench ssp. drummondii (Nees ex Steud.) de Wet & Harlan plus supplement	A	40.4	670	740	700	610	.	740	2.1	4.9	No	Hentz et al., 2012
Straw (*Avena sativa* L.) plus supplement	M	.	680	710	630	590	680	580	4.8	4.0	Yes	Dentinho et al., 2014
Straw (*Vicia* L.)	M	.	550	570	510	530	540	460	.	.	No	Isac et al., 1994
Stylosanthes guianensis (Aubl.) Sw.	A	50.9	580	620	.	.	.	850	4.2	23.9	Yes	Mupangwa et al., 2000

† A, ad libitum; G, growth; M, maintenance.

‡ DMD, dry matter digestibility; OMD, organic matter digestibility; NDFD, neutral detergent fiber digestibility; ADFD, acid detergent fiber digestibility; GED, gross energy disappearance; CPD, crude protein digestibility.

§ Organic matter basis

manure could be ascertained. In this assessment, authors obtained manure samples from two difference sheep operations. Dry matter content was approximately 386 g kg⁻¹, total carbon was 364 g kg⁻¹, and total nitrogen was 24 g kg⁻¹ (of which, 21% was attributed to NH_4–N), and pH was 8.5 (Pagliari and Laboski, 2012).

Goats

As of 2016, 459 million meat-producing goats worldwide, with an additional 202 million milk-producing goats (FAOSTAT, 2016). Distributions of meat- and milk-producing goats worldwide are presented in Fig. 3 and 4, respectively. By comparison, the United States ranked 62nd in milk production and had no reported data in meat production (FAOSTAT, 2016).

Management

Goat production systems are very similar to sheep in many respects. Primary differences between sheep and goats are location, dietary adaptation, and average number of offspring. The majority of goat production tends to occur within the tropics and dry zones of developing countries. This is largely due to their innate ability to tolerate heat stress and lower quality diets, most especially those with increased plant secondary compounds.

The use of goats for meat production can be split into two primary categories: subsistence and commercial production. Within developing countries, the former is common, although the trade or sale of animals for goods can comprise a substantial portion. Typically, small amounts of animal products, milk, meat, skin, and hair, are utilized by producers, while the rest are sold or traded within local markets. In these production scenarios, animals are commonly utilized for both milk and meat, with use of goats within small farms being common. Husbandry practices are typically small family herds comprised of indigenous breeds that feed on poor pastures and rangelands, as well as graze crop residues following harvest. Although dependent on region, horizontal

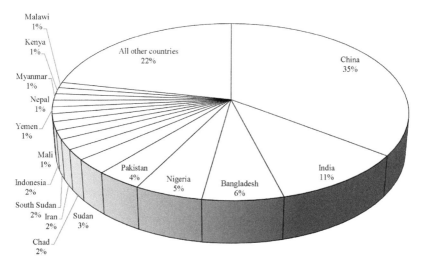

Fig. 3. Distribution of production (number of animals) of meat-producing goats worldwide. Data were adapted from FAOSTAT (2016).

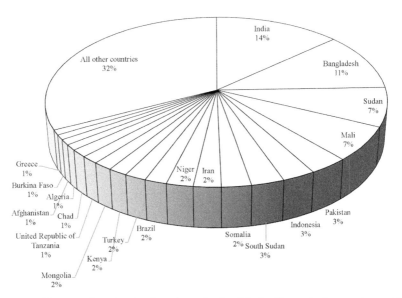

Fig. 4. Distribution of production (number of animals) of milk-producing goats world-wide. Data were adapted from FAOSTAT (2016).

transhumance will occur with animals grazing highlands during the spring and low-lands during the fall. In other regions, vertical transhumance, daily movement of animals, is commonplace as herds are taken to grazing areas during the day and returned to homesteads at night. Nomadism remains largely rooted in goat production as shepherds move animals along common grazing routes. Within nomadic production, animals typically winter on rangeland with minimal supplementation, whereas stationary and transhumant flocks are wintered on the producer's farm. These systems are commonly considered extensive due to decreased capacity for inputs.

Within commercial settings, animals can still be produced in an extensive manner; however, the use of supplementation and health programs are more read-ily available. As with sheep, most commercial production is aimed at producing specific products (milk, cheese, meat, and hair) for sale, with animal productiv-ity and efficiency being a primary goal. However, inputs are typically lower with smallholder producers. Within intensive systems, modern technology such as genetic selection, defined breeding seasons, disease management, and supplemen-tation for production are utilized. Regions of Europe, such as Spain, France, Italy, and Greece, are responsible for approximately 20% of global milk production, but only own roughly 2.5% of the worlds herd. In these regions, milk holds substantial economic value, either for direct consumption or for production of cheese. These systems vary, with some regions being more extensive than others, but are largely based on utilization of modern technology to maintain efficient production.

Manure Composition Relative to Diet

As with sheep, goat manure composition is largely based on the diet consumed. Although ruminal kinetics between these species vary, ruminal retention of feed will largely determine the total amount of dietary constituents excreted within feces. When high-quality forages are provided, digestibility does not alter greatly from those exhibited within sheep; however, when diets are comprised of lower-quality forages, digestibility tends to be greater for goats (Table 2). Looking at pangola hay of contrasting maturities, NDFD of the immature forage was relatively similar for sheep and goat, 73 vs. 75% respectively; however, when fed the mature forage, NDFD differed rather substantially, 64% vs. 71% respectively (Watson and Norton, 1982). This may not truly be a function of digestive function, but rather a result of increased selection exhibited within goats. As discussed within the management section, globally, most goats are exposed to low-quality forages and browse. Consumption of low-quality diets by small ruminants results in a trade-off between digestibility and intake, as diet quality decreases, ruminal retention tends to increase and subsequently reduces total daily intake. This can be highly detrimental to animal health and production as daily nutrient requirements may not be able to be met if passage rate is too greatly reduced. Therefore, the capacity for these animals to select diets of

Table 2. Feedstuff digestibility and manure excretion from goats.

Feedstuff	Feeding level†	Intake	DMD‡	OMD	NDFD	ADFD	GED	CPD	Fecal N	Urinary N	Tannins	Citation
		g kg BW$^{-0.75}$			g kg^{-1}				g d^{-1}			
Cajanus cajan (L.) Millsp.	A	.	700	720	530	.	.	530	.	.	.	Brown et al., 1988
Cassia fistula L. plus supplement	A	41.9	550	510	480	.	.	570	.	.	Yes	Salem et al., 2006
Chorisia speciosa A. St.-Hil. plus supplement	A	52.7	610	570	520	.	.	540	.	.	Yes	Salem et al., 2006
Dactylis glomerata L. plus supplement	A	75.1	730	.	700	740	.	750	.	.	No	Luginbuhl et al., 2000
Eucalyptus camaldulensis Dehnh. plus supplement	A	36.6	520	560	450	.	.	510	.	.	Yes	Salem et al., 2006
Hay (Digitaria eriantha Steud.; immature)	A	.	.	630	750	670	.	640	5.9	6.3	No	Watson and Norton, 1982
Hay (Digitaria eriantha Steud.; mature)	A	.	.	580	710	630	.	300	3.5	1.7	No	Watson and Norton, 1982
Hay (Medicago sativa L.)	M	52.0	630	670	520	470	640	820	.	.	No	Alcaide et al., 2000
Hay (Medicago sativa L.)	M	.	550	570	300	320	540	700	.	.	No	Isac et al., 1994
Hay (Medicago sativa L.) plus supplement	M	52.0	670	710	540	480	650	780	.	.	No	Alcaide et al., 2000
Hay (Medicago sativa L.) plus supplement	M	45.0	730	760	680	720	740	750	.	.	No	Alcaide et al., 2000
Hay (Medicago sativa L.) plus supplement	A	121.3	680	690	370	.	690	690	.	.	No	López and Fernandez, 2013
Hay (Medicago sativa L.) plus supplement	A	125.0	640	650	50	.	660	710	.	.	No	López and Fernandez, 2013
Hay (Sorghum bicolor (L.) Moench ssp. bicolor)	A	36.0	.	.	.	.	.	.	.	.	No	Huston et al., 1988
Hyparrhenia Andersson ex Fourn.	A	40.5	540	590	.	.	570	430	2.7	3.3	No	Gihad, 1976
Pennisetum purpureum Schumach.	A	.	600	650	650	.	.	710	.	.	.	Brown et al., 1988
Schinus molle L.	A	41.6	530	570	500	.	.	540	.	.	Yes	Salem et al., 2006
Straw (Triticum aestivum L.)	A	42.6	440	.	.	.	.	120	4.9	2.7	No	El-Meccawi et al., 1994
Straw (Vicia L.)	M	.	580	590	530	530	550	500	.	.	No	Isac et al., 1994

† A, ad libitum; G, growth; M, maintenance.

‡ DMD, dry matter digestibility; OMD, organic matter digestibility; NDFD, neutral detergent fiber digestibility; ADFD, acid detergent fiber digestibility; GED, gross energy disappearance; CPD, crude protein digestibility.

higher nutritional value is of great importance, most especially when low-quality forages and secondary plant compounds are abundant within the diet.

The provision of higher quality diets, as with intensive dairy production, tends to follow closely with that of sheep. Digestibility of the historically high-quality forage, alfalfa, is rather low as NDFD ranged from 30 to 68%. This variation could be due to numerous variables, including processing of feed and nutritive composition; however, the gross energy digested was 54 to 74%, indicating that nonfibrous carbohydrates were relatively high, and thus the passage rate did not allow adequate fiber degradation. Therefore, higher-quality diets will promote faster rate of passage and subsequently enable greater daily intake, resulting in greater fibrous constituents within the manure as a percent of fiber intake. These forages will also have increased digestion and metabolism of nitrogen, resulting in greater excretion within the urine as compared with feces. It is evident when looking at pangola forages of differing maturities as the proportion of fecal to urinary nitrogen excretion is ~ 1:1 in mature forage and 1:3 for immature forage.

In their assessment of physical and chemical manure properties, Pagliari and Laboski (2012) obtained manure samples from two difference goat operations. Dry matter content was approximately 312 g kg^{-1}, total carbon was 420 g kg^{-1}, total nitrogen was 34 g kg^{-1} (of which, 38% was attributed to NH_4–N), and pH was 7.7 (Pagliari and Laboski, 2012).

Cervid Livestock
Extent of Management Within the United States

The Agricultural and Food Policy Center at Texas A&M University has sought to characterize the extent of cervid farming within the United States (Anderson et al., 2007). At the time of that publication, there were in excess of 7800 deer farms in the United States, one thousand of which were in the state of Texas. The species included in the assessment were white-tailed deer (*Odocoileus virginianus*), elk (*Cervus canadensis*), fallow deer (*Dama dama*), and red deer (*Cervus elaphus*). For breeding operations, whitetail operations averaged 21 males, 25 females, and 23 fawns; elk farms included 18 males, 24 females, and 16 fawns; fallow deer operations averaged 43 males, 94 females, and 50 fawns; and red deer farms averaged 46 males, 57 females, and 35 fawns (Anderson et al., 2007). As a result of these populations, the cervid livestock industry generated an estimated $893.5 million in direct expenditures (Anderson et al., 2007).

Red Deer/Elk

Management

These species are an important part of ecosystems as they typically serve as native large-ruminant grazers within ecosystems. However, interest in the utilization of these animals for production within the United States is on the rise due to increased consumer demand for healthier meat alternatives, as well as byproduct utilization. Within Australia and New Zealand, the farming of these animals began in the 1970s with the New Zealand industry valuing approximately $257 million in 2000. These animals perform well on marginal land not fit for other livestock production and are able to fit into grazing operations; the number of deer farms will likely increase in the near future. Production of these animals requires specialty infrastructure, including taller fences and working facilities. Within the United States, a major concern is chronic wasting disease; therefore, areas with high incidence of the disease requires increased management to ensure transmission is minimized. Additionally, in most locations production of these animals will require specialty permits.

Manure Composition Relative to Diet

Literature concerning these species is rather sparse; however, being primarily grazers, they tend to select diets with increased grasses and forbs. Therefore, intake of fibrous constituents will be higher in some regards; but, as discussed with sheep and goat, diet composition is the major driver of manure composition. It is clear that poor-quality forages are not digested as efficiently, with NDFD being relatively low in most diets except for those of high quality, such as alfalfa, ryegrass, and wheat (Table 3). Forages tend to be digested to a greater extent during the spring

Table 3. Feedstuff digestibility and manure excretion from red deer and elk.

Feedstuff	Feeding level†	Season	Intake	DMD‡	OMD	NDFD	ADFD	GED	CPD	Tannins	Citation
			g kg BW$^{-0.75}$			g kg^{-1}					
Acer L.	A	.	40.7	410	.	300	270	420	.	Yes	Mould and Robbins, 1982
Agrostis L./ *Festuca* L.	A	Spring	70.0§	.	460	390	.	.	390	No	Milne et al., 1978
Agrostis L./ *Festuca* L.	A	Summer	46.3§	.	430	380	.	.	380	No	Milne et al., 1978
Agrostis L./ *Festuca* L.	A	Winter	38.3§	.	410	360	.	.	200	No	Milne et al., 1978
Bromus inermis Leyss.	A	.	67.9	480	.	500	480	450	.	No	Mould and Robbins, 1982
Calluna Salisb.	A	Spring	54.9§	.	300	300	.	.	270	No	Milne et al., 1978
Calluna Salisb.	A	Summer	29.5§	.	460	300	.	.	180	No	Milne et al., 1978
Calluna Salisb.	A	Winter	34.1§	.	470	300	.	.	120	No	Milne et al., 1978
Chamerion angustifolium (L.) Holub	A	.	63.3	460	.	330	350	430	.	Yes	Mould and Robbins, 1982
Dactylis glomerata L.	A	.	50.1	500	.	540	520	430	.	Yes	Mould and Robbins, 1982
Festuca idahoensis Elmer	A	.	46.2	290	.	410	390	290	.	No	Mould and Robbins, 1982
Hay (*Digitaria eriantha* Steud.)	A	.	48.9	520	530	570	.	.	370	No	Tomkins et al., 1991
Medicago sativa L.	A	Summer	62.5	570	600	450	.	.	.	No	Dominique et al., 1991
Medicago sativa L.	A	Winter	46.7	550	590	400	.	.	.	No	Dominique et al., 1991
Medicago sativa L.	A	.	58.8	610	.	510	500	580	.	No	Mould and Robbins, 1982
Pellets (*Lolium perenne* ssp. perenne)	A	Summer	82.3§	.	570	570	.	.	590	No	Milne et al., 1978
Pellets (*Lolium perenne* L. ssp. perenne)	A	Winter	57.3§	.	630	570	.	.	610	No	Milne et al., 1978
Phalaris L.	A	.	40.8	500	.	450	420	510	.	No	Mould and Robbins, 1982
Phleum pratense L.	A	.	64.8	550	.	550	510	520	.	No	Mould and Robbins, 1982
Straw (*Poa* L.)	A	.	35.1	360	.	480	480	400	.	No	Mould and Robbins, 1982
Triticum aestivum L.	A	.	84.9	740	.	820	800	720	.	No	Mould and Robbins, 1982
Triticum aestivum L.	A	.	49.3	850	.	870	840	850	.	No	Mould and Robbins, 1982

† A, ad libitum.

‡ DMD, dry matter digestibility; OMD, organic matter digestibility; NDFD, neutral detergent fiber digestibility; ADFD, acid detergent fiber digestibility; GED, gross energy disappearance; CPD, crude protein digestibility.

§ Organic matter basis

Table 4. Feedstuff digestibility and manure excretion from white-tailed deer.

Feedstuff	Feeding level†	Season	Intake	DMD‡	OMD	NDFD	ADFD	GED	CPD	Tannins	Citation
			g kg BW$^{-0.75}$		g kg^{-1}						
Bromus inermis Leyss.	.	.	.	540	.	580	550	530	.	No	Robbins et al., 1975
Medicago sativa L.	.	.	.	550	.	370	350	540	.	No	Robbins et al., 1975
Pellets	A	.	54.0	760	.	820	.	780	950	No	Barnes et al., 1991
Pellets	.	.	.	600	.	380	260	580	.	No	Robbins et al., 1975
Pellets	.	.	.	550	.	.	.	600	660	No	Wheaton and Brown, 1983
Pellets (*Medicago sativa* L.)	A	.	67.7	470	.	370	330	460	570	No	Galbraith et al., 1998
Secale cereale L.	.	.	.	710	.	730	660	670	770	No	Ullrey et al., 1987
Senegalia berlandieri Britton & Rose	A	Autumn	32.8	380	.	70	.	360	430	Yes	Barnes et al., 1991
Senegalia berlandieri Britton & Rose	A	Spring	30.1	480	.	230	.	460	560	Yes	Barnes et al., 1991
Senegalia berlandieri Britton & Rose	A	Summer	34.4	410	.	80	.	390	510	Yes	Barnes et al., 1991
Senegalia berlandieri Britton & Rose	A	Summer	46.1	350	.	0	.	460	560	Yes	Barnes et al., 1991
Senegalia berlandieri Britton & Rose	A	Winter	50.6	420	.	180	.	390	470	Yes	Barnes et al., 1991

† A = ad libitum.

‡ DMD, dry matter digestibility; OMD, organic matter digestibility; NDFD, neutral detergent fiber digestibility; ADFD, acid detergent fiber digestibility; GED, gross energy disappearance; CPD, crude protein digestibility.

and summer. More specific knowledge concerning these animals is required if utilization of these animals on a commercial scale is to continue.

White-tailed Deer

Management

In contrast to red deer, white-tailed deer production is not based on harvest of animals for distribution. Instead, white-tailed deer production revolves around the sport of harvesting the animal and is split into two categories: the breeding of animals for hunting ranches and the operation of hunting ranches. Although some commercial meat production is present within the United States, the vast majority of venison is harvested for individual use and not purchased, although the meat is a low-fat alternative to other common meat sources such as beef and sheep. As with red deer, these animals perform very well on areas that are considered marginal or low value to other livestock or production practices. As white-tailed deer are primarily browsers, diets will consist largely of browse, forbs, and mast, with some grasses of greater digestibility. However, infrastructure must be in place for controlled breeding and management of animals. As these animals can often be detrimental to the environment in areas of high density, utilization of game-fences is often utilized to preserve lands and enable proper herd management. Management of these animals often entails maintaining populations below carrying capacity to improve availability of nutrients and decrease winter death loss that is often exhibited in northern regions of North America. Supplementation of high energy feed and protein is often performed during times of low feed availability and stress; however, this can artificially inflate carrying capacity and lead to major death losses if caution is not taken.

Manure Composition Relative to Diet

Manure composition will vary greatly dependent on diet and season. When tannins are present in the diet, NDFD is greatly reduced, ranging from 0 to 23% (Table 4). In contrast, high-quality diets such as alfalfa, commercial feeds, rye, and brome are digested rather well. Unfortunately, literature on white-tailed deer is lacking, so estimation of excreta composition cannot be determined accurately estimated based on feedstuffs.

Manure Nutrients

With the exception of Pagliari and Laboski (2012), seldom has a single publication sought to bring together tables of composition of manure from various ruminant species, especially with inclusion of the small ruminants. Such a compilation is presented in Table 5. Much like the data presented in the preceding sections, manure composition is highly variable and likely related to dietary composition.

Table 5. Manure composition from various ruminant species.

Form	DM†	N	P	K	C/N	pH	Citation
		—————g kg^{-1}—————					
			Sheep				
fresh	.	16	4.2	6.9	.	.	Abdelrazzag, 2002
fresh	.	17.6	13.8	.	28.7	6.9	Jalali and Ranjbar, 2009
fresh	676	13	8	11	.	.	Martí-Herrero et al., 2015
fresh	213	.	8.0	.	.	8.3	McDowell and Stewart, 2005
fresh	386	24	8.9	24	15.2	8.5	Pagliari and Laboski, 2012
fresh	.	1.64	0.0014	0.0018	.	8.0	Pavlou et al., 2007
2 wk vermicompost	356	22.6	29.9	6.8	9.3	8.9	Albanell et al., 1988
6 wk vermicompost	388	23.5	30.8	7.2	8.8	8.4	Albanell et al., 1988
12 wk vermicompost	420	21.2	44.6	10.3	8.9	8.2	Albanell et al., 1988
			Goats				
fresh (male)	946	28.1	4.2	9.3	.	.	Osuhor et al., 2002
fresh (female)	946	27.8	4.3	9.3	.	.	Osuhor et al., 2002
fresh	.	49	41	19	.	.	Awodun et al., 2007
fresh	470.8	11.7	11.9	23.6	36.1	8.5	Cho et al., 2017
fresh	516.6	18.2	10.8	28.9	25.1	9.2	Cho et al., 2017
fresh	839.1	25.5	20.6	13.1	16.9	8.0	Cho et al., 2017
fresh	827.0	23.1	26.8	12.1	19.0	7.4	Cho et al., 2017
fresh	451.2	16.0	13.5	9.9	27.8	7.9	Cho et al., 2017
fresh	391.2	6.7	8.2	6.4	65.1	8.0	Cho et al., 2017
fresh	312	34	11.7	33	12.4	7.7	Pagliari and Laboski, 2012
			Deer				
fresh	407	.	7.4	.	.	8.2	McDowell and Stewart, 2005
fresh	707.6	18.5	5.7	.	52.2	.	Wang et al., 2018

† DM, dry matter; N, nitrogen; P, phosphorus; K, potassium.

However, manure from small ruminant and cervid species remains a viable option for inclusion in agronomic protocols.

References

Abdelrazzag, A. 2002. Effect of chicken manure, sheep manure and inorganic fertilizer on yield and nutrients uptake by onion. Pak. J. Biol. Sci. 5:266–268. doi:10.3923/pjbs.2002.266.268

Albanell, E., J. Plaixats, and T. Cabrero. 1988. Chemical changes during vermicomposting (Eisenia fetida) of sheep manure mixed with cotton industrial wastes. Biol. Fertil. Soils 6:266–269. doi:10.1007/BF00260823

Alcaide, E.M., A.I.M. García, and J.F. Aguilera. 2000. A comparative study of nutrient digestibility, kinetics of degradation and passage and rumen fermentation pattern in goats and sheep offered good quality diets. Livest. Prod. Sci. 64:215–223. doi:10.1016/S0301-6226(99)00149-9

Anderson, D.P., B.J. Frosch, and J.L. Outlaw. 2007. Economic impact of the United States cervid farming industry. APFC Res. Rep. 07-4. Department of Agricultural Economics, Texas A&M University, College Station, TX. https://www.afpc.tamu.edu/research/publications/480/rr-2007-04.pdf (Accessed 12 Nov. 2018).

Awodun, M.A., L.I. Omonijo, and S.O. Ojeniyi. 2007. Effect of goat dung and NPK fertilizer on soil and leaf nutrient content, growth and yield of pepper. Int. J. Soil Sci. 2:142–147. doi:10.3923/ijss.2007.142.147

Barnes, T.G., L.H. Blankenship, L.W. Varner, and J.F. Gallagher. 1991. Digestibility of guajillo for white-tailed deer. J. Range Manage. 44:606–610. doi:10.2307/4003045

Brown, D., M. Salim, E. Chavalimu, and H. Fitzhugh. 1988. Intake, selection, apparent digestibility and chemical composition of Pennisetum purpureum and Cajanus cajan foliage as utilized by lactating goats. Small Rumin. Res. 1:59–65. doi:10.1016/0921-4488(88)90044-2

Cho, W.-M., B. Ravindran, J.K. Kim, K.-H. Jeong, D.J. Lee, and D.-Y. Choi. 2017. Nutrient status and phytotoxicity analysis of goat manure discharged from farms in South Korea. Environ. Technol. 38:1191–1199. doi:10.1080/09593330.2016.1239657

Deaville, E.R., D.I. Givens, and I. Mueller-Harvey. 2010. Chestnut and mimosa tannin silages: Effects in sheep differ for apparent digestibility, nitrogen utilisation and losses. Anim. Feed Sci. Technol. 157:129–138. doi:10.1016/j.anifeedsci.2010.02.007

Dehority, B.A. 2002. Gastrointestinal tracts of herbivores, particularly the ruminant: Anatomy, physiology and microbial digestion of plants. J. Appl. Anim. Res. 21:145–160.

Dos Santos, R.D., L.G.R. Pereira, A.L.A. Neves, G.G.L. de Araújo, A.S.L. de Aragão, and M.L. Chizzotti. 2011. Intake and total apparent digestibility in lambs fed six maize varieties in the Brazilian semiarid. Rev. Bras. Zootec. 40:2922–2928. doi:10.1590/S1516-35982011001200040

El-Meccawi, S., M. Kam, A. Brosh, and A. A. Degen. 2009. Energy intake, heat production and energy and nitrogen balances of sheep and goats fed wheat straw as a sole diet. Livest. Sci. 125:88–91. doi:10.1016/j.livsci.2009.02.018

FAO. 1995. A Classification of livestock production systems. FAO, Rome, Italy. http://www.fao.org/DOCREP/V8180T/v8180T0y.htm (Accessed 30 Sept. 2018).

FAOSTAT. 2016. Livestock primary. FAO, Rome, Italy. http://www.fao.org/faostat/en/#data/QL. (Accessed 30 Sept. 2018).

Francoise Domingue, B. M., D. W. Dellow, P. R. Wilson, and T. N. Barry. 1991. Comparative digestion in deer, goats, and sheep. New Zealand J. Agric. Res. 34:45-53. doi:10.1080/00288233.1991.10417792.

Galbraith, J.K., G.W. Mathison, R.J. Hudson, T.A. McAllister, and K.-J. Cheng. 1998. Intake, digestibility, methane and heat production in bison, wapiti and white-tailed deer. Can. J. Anim. Sci. 78:681–691. doi:10.4141/A97-089

Gihad, E.A. 1976. Intake, digestibility and nitrogen utilization of tropical natural grass hay by goats and sheep. J. Anim. Sci. 43:879–883. doi:10.2527/jas1976.434879x

Hentz, F., G.V. Kozloski, T. Orlandi, S.C. Ávila, P.S. Castagnino, C.M. Stefanello, and G.F. Estivallet Pacheco. 2012. Intake and digestion by wethers fed a tropical grass-based diet supplemented with increasing levels of canola meal. Livest. Sci. 147:89–95. doi:10.1016/j.livsci.2012.04.007

Hofmann, R.R. 1989. Evolutionary steps of ecophysiological adaptation and diversification of ruminants: A comparative view of their digestive system. Oecologia 78:443–457. doi:10.1007/BF00378733

Huston, J.E., B.S. Engdahl, and K.W. Bales. 1988. Intake and digestibility in sheep and goats fed three forages with different levels of supplemental protein. Small Rumin. Res. 1:81–92. doi:10.1016/0921-4488(88)90047-8

Isac, M.D., M.A. García, J.F. Aguilera, and E. Molina Alcaide. 1994. A comparative study of nutrient digestibility, kinetics of digestion and passage and rumen fermentation pattern in goats and sheep offered medium quality forages at the maintenance level of feeding. Arch. Anim. Nutr. 46:37–50. doi:10.1080/17450399409381756

Jalali, M., and F. Ranjbar. 2009. Rates of decomposition and phosphorus release from organic residues related to residue composition. J. Plant Nutr. Soil Sci. 172:353–359. doi:10.1002/jpln.200800032

Lima, R., R.F. Díaz, A. Castro, and V. Fievez. 2011. Digestibility, methane production and nitrogen balance in sheep fed ensiled or fresh mixtures of sorghum-soybean forage. Livest. Sci. 141:36–46. doi:10.1016/j.livsci.2011.04.014

López, M.C., and C. Fernández. 2013. Energy partitioning and substrate oxidation by Murciano-Granadina goats during mid lactation fed soy hulls and corn gluten feed blend as a replacement for corn grain. J. Dairy Sci. 96:4542–4552. doi:10.3168/jds.2012-6473

López, M.C., and C. Fernández. 2014. Energy partitioning and substrate oxidation by Guirra ewes fed soy hulls and corn gluten feed blend as a replacement for barley grain. Anim. Feed Sci. Technol. 189:11–18. doi:10.1016/j.anifeedsci.2013.12.005

Loyra-Tzab, E., L.A. Sarmiento-Franco, C.A. Sandoval-Castro, and R.H. Santos-Ricalde. 2011. Nutrient digestibility and metabolizable energy content of Mucuna pruriens beans fed to growing Pelibuey lambs. Anim. Feed Sci. Technol. 169:140–145. doi:10.1016/j.anifeedsci.2011.06.003

Luginbuhl, J., M. Poore, and A. Conrad. 2000. Effect of level of whole cottonseed on intake, digestibility, and performance of growing male goats fed hay-based diets. J. Anim. Sci. 78:1677–1683. doi:10.2527/2000.7861677x

Maia, M.D.O., I. Susin, E.M. Ferreira, C.P. Nolli, R.S. Gentil, A.V. Pires, and G.B. Mourão. 2012. Intake, nutrient apparent digestibility and ruminal constituents of sheep fed diets with canola, sunflower or castor oils. Rev. Bras. Zootec. 41:2350–2356. doi:10.1590/S1516-35982012001100008

Martí-Herrero, J., R. Alvarez, R. Cespedes, M.R. Rojas, V. Conde, L. Aliaga, M. Balboa, and S. Danov. 2015. Cow, sheep, and llama manure at psychrophilic anaerobic co-digestion with low cost tubular digesters in cold climate and high altitude. Bioresour. Technol. 181:238–246. doi:10.1016/j.biortech.2015.01.063

McDowell, R.W., and I. Stewart. 2005. Phosphorus in fresh and dry dung of grazing dairy cattle, deer, and sheep: Sequential fraction and phosphorus-31 nuclear magnetic resonance analyses. J. Environ. Qual. 34:598–607. doi:10.2134/jeq2005.0598

Milne, J.A., J.C. MacRae, A.M. Spence, and S. Wilson. 1978. A comparison of the voluntary intake and digestion of a range of forages at different times of the year by the sheep and the red deer. Br. J. Nutr. 40:347–357. doi:10.1079/BJN19780131

Mould, E.D., and C.T. Robbins. 1982. Digestive capabilities in elk compared to white-tailed deer. J. Wildl. Manage. 46:22–29. doi:10.2307/3808404

Mupangwa, J.F., N.T. Ngongoni, J.H. Topps, T. Acamovic, H. Hamudikuwanda, and L.R. Ndlovu. 2000. Dry matter intake, apparent digestibility and excretion of purine derivatives in sheep fed tropical legume hay. Small Rumin. Res. 36:261–268. doi:10.1016/S0921-4488(99)00125-X

Osuhor, C.U., J.P. Alawa, and G.N. Akpa. 2002. Research note: Manure production by goats grazing native pasture in Nigeria. Trop. Grassl. 36:123–125.

Pagliari, P.H., and C.A.M. Laboski. 2012. Investigation of the inorganic and organic phosphorus forms in animal manure. J. Environ. Qual. 41:901–910. doi:10.2134/jeq2011.0451

Pavlou, G.C., C.D. Ehaliotis, and V.A. Kavvadias. 2007. Effect of organic and inorganic fertilizers applied during successive crop seasons on growth and nitrate accumulation in lettuce. Sci. Hortic. (Amsterdam) 111:319–325. doi:10.1016/j.scienta.2006.11.003

Robbins, C.T., D.E. Spalinger, and W. van Hoven. 1995. Adaptation of ruminants to browse and grass diets: Are anatomical-based browser-grazer interpretations valid? Oecologia 103:208–213. doi:10.1007/BF00329082

Robbins, C. T., P. J. Van Soest, W. W. Mautz, and A. N. Moen. 1975. Feed analyses and digestion with reference to white-tailed deer. J. Wildl. Manage. 39:67-79.

Salem, A.Z.M., M.Z.M. Salem, M.M. El-Adawy, and P.H. Robinson. 2006. Nutritive evaluations of some browse tree foliages during the dry season: Secondary compounds, feed intake and in vivo digestibility in sheep and goats. Anim. Feed Sci. Technol. 127:251–267. doi:10.1016/j.anifeedsci.2005.09.005

Tomkins, N.W., N.P. Mcmeniman, and R.C.W. Daniel. 1991. Voluntary feed-intake and digestibility by red deer (Cervus elaphus) and sheep (Ovis ovis) of Pangola grass (Digitaria decumbens) with or without a supplement of leucaena (Leucaena leucocephala). Small Rumin. Res. 5:337–345. doi:10.1016/0921-4488(91)90071-W

Ullrey, D.E., J.T. Nellist, J.P. Duvendeck, P.A. Whetter, and L.D. Fay. 1987. Digestibility of vegetative rye for white-tailed deer. J. Wildl. Manage. 51:51–53. doi:10.2307/3801628

Wang, H., J. Xu, L. Sheng, and W. Liu. 2018. Effect of addition of biogas slurry for anaerobic fermentation of deer manure on biogas production. Energy 165:411–418. doi:10.1016/j.energy.2018.09.196

Watson, C., and B. W. Norton. 1982. The utilization of Pangola grass hay by sheep and angora goats. Proc. Aust. Soc. Anim. Prod. 14:467–470.

Wheaton, C., and R.D. Brown. 1983. Feed intake and digestive efficiency of south Texas white-tailed deer. J. Wildl. Manage. 47:442–450. doi:10.2307/3808517

Generation and Management of Manure from Horses and Other Equids

Michael L. Westendorf,* Carey A. Williams, Stephanie Murphy, Laura Kenny, and Masoud Hashemi

Abstract

Horse manure presents challenges in management and disposal that are different than manure from other species. Horse farms often have small numbers of animals and limited acreages and producers may be less familiar with manure management and disposal best management practices (BMPs) or with conservation assistance opportunities available. The average adult horse will produce about 10 tons of manure per year, or about 9000 L in volume (12 cubic yards). This manure contains nitrogen (N) and phosphorus (P), which can be an environmental concern, and microorganisms, which may influence human and animal health. In addition, horse manure is a source of greenhouse gas emissions. The most common manure management technology used for managing horse manure is composting, although there are opportunities for other methods including anaerobic digestion. The nutrients in horse manure can be affected by diet and feeding management. Often the levels of nutrients in a standard diet are above requirements, creating opportunities for management programs that can reduce nutrients in the waste through diet modifications. Properly grazed pastures, particularly when pastures are rotationally grazed, recycle nutrients faster than ungrazed pastures, presenting an opportunity for reducing waste excretion and runoff. Best management practices such as manure storages, fencing, properly located feeding, watering, and sheltering areas and nutrient management programs can result in improvements in soil, water, and environmental quality. There are a variety of methods available to equine producers and their advisors for implementing manure management and conservation programs on farm.

According to the 2012 USDA Agricultural Census (USDA-NASS, 2012), there are 3,910,000 horses, ponies, and mules dwelling on 603,000 farms in the United States (this number does not include all farms, only those surveyed by the USDA National Agricultural Statistics Service). These numbers have either increased or remained the same since the 1997 census. This is an average of less than 7 animals per farm. There is no good estimate of the amount of acreage dedicated to equine production and use in the United States.

Horses are predominantly used as pleasure and sport animals and not for meat, dairy, or fiber products, like most other livestock. Equestrians who breed, raise, train, and otherwise keep horses are often unfamiliar with typical farming operations. Large horse farms may have acres of pasture and hay fields with operators trained in agronomy, but smaller facilities may simply purchase feed (hay, grain, supplements) for their horses and/or have manure hauled away for disposal. This provides a disconnection from the land and its best management practices, whereas the waste products of horses require understanding of proper use and disposal to protect health and environmental quality.

M.L. Westendorf,* Carey A. Williams, Department of Animal Science, Rutgers, The State University of New Jersey, New Brunswick, NJ; S. Murphy, Soil Testing Laboratory, Rutgers, The State University of New Jersey; L. Kenny, Cooperative Extension Services, Pennsylvania State University; M. Hashemi, Stockbridge School of Agriculture, University of Massachusetts, Amherst, MA. *Corresponding Author (michael.westendorf@rutgers.edu)

doi:10.2134/asaspecpub67.c8

Animal Manure: Production, Characteristics, Environmental Concerns and Management. ASA Special Publication 67. Heidi M. Waldrip, Paulo H. Pagliari, and Zhongqi He, editors. © 2019. ASA and SSSA, 5585 Guilford Rd., Madison, WI 53711, USA.

Equine operations can have multiple environmental effects that go beyond a specific farm. For instance, soil erosion is a major contributor to nonpoint source pollutants originating from equine pastures, stables, and trails. Soil displaced by horse hoof traffic associated with poor pasture grass management or trail maintenance often leads to problematic sedimentation offsite with the downstream transport of nitrogen (N), phosphorus (P), and bacteria. Another consideration is the proper disposal of manure produced on a farm. Since manure may contain parasites, many equine producers are hesitant to apply stall waste to their pastures. Most stall waste also includes a high percentage of bedding material such as wood shavings. This added material can significantly increase the volume of waste being spread and can impact the growth and development of pasture grasses.

A survey of 230 equine facility owners or managers found that many horse owners fail to get information about best management practices for small farms; less than half of respondents used conservation techniques on their farm other than pasture rotation or dry lots, and a large majority (> 73%) had never been in contact with someone from a conservation agency (Marriott et al., 2012). There is an intense need to educate owners and managers of horse facilities about manure management for sustaining environmental quality as well as animal health. In another survey focused on horse farm operators in Maryland, respondents reported high knowledge of, but limited implementation of best management practices (BMPs) (Fiorellino et al., 2013). The respondents reported highest knowledge of appropriate stocking density, rotational grazing, and appropriate manure management, but among the least utilized BMPs were storage of manure on impervious surface, use of sacrifice lot within a pasture, rotational grazing, and managing roof runoff. In general, respondents reported lesser knowledge about soil fertility, nutrient management plans, erosion control and conservation plans, and other pasture quality topics. Another survey of 69 small horse facilities (63 with 10 or fewer horses; Wickens et al., 2017) also found that 63.2% owners were unfamiliar with horse manure BMPs. Challenges to implementing BMPs include not only knowledge but also costs (equipment, hauling) and lack of equipment (front end loader, spreader).

This paper provides an overview of current manure management and disposal methods on horse farms, as well as concerns about soil, water, and environmental quality, and health risks related to parasites and other microorganisms. Current and new manure treatment technologies, feeding management to limit nutrient losses, and pasture and grazing management systems to limit waste losses and promote environmental quality will be covered. Hopefully, horse farmers and their advisors will consider the many opportunities for conservation management on farms.

Manure Generation

An adult horse will produce up to 25 kg of manure daily or about 10 tons per year containing 4.5 kg of N, 1.7 kg of P_2O_5, and 2.4 kg of K_2O per ton. Volume of manure is 25 L per day or 9125 L per year (12 cubic yards) (Table 1 ASAE, 2005; Westendorf and Krogmann 2013). A horse kept in a stall will require about 5 to 10 kg of bedding per day that must be replaced on a regular basis. Dirty bedding may increase the volume of manure by two or three times, depending on the type of bedding used and management practices. Manure plus bedding will be removed daily and replaced with unsoiled bedding. Soiled bedding will have a volume of 60 to 80 L per day, (2–3 cubic feet per day; Westendorf and Krogmann, 2013; Wheeler and Zajaczkowski, 2009).

A variety of bedding sources are used for horses (Westendorf et al., 2010). These are reviewed in the Horse Facilities Handbook (MWPS, 2005) and by Westendorf and Krogmann (2013). (See Table 2 and 3.) Wood products are the bedding of choice for many equine producers. Ward et al. (2001) studied the use of bedding and determined bedding effectiveness. These researchers found that processed newspaper could be an effective bedding source. The advantage of straw for availability and ease of disposal was hurt by the fact it tended to be less absorbent in a stall. Pratt et al. (1999) found ammonia in stalls to be a health concern. In some regions of the country, straw bedded horse manure is in demand on mushroom farms (Westendorf and Krogmann, 2013). It is composted and makes an excellent medium for growing mushrooms. The chief disadvantage of wood products is the slow breakdown of wood in the manure in a compost pile, or in the soil. Manure bedded heavily with wood shavings or chips may actually have a negative impact on the soil. Because it will take so long to break down, it may affect other processes in the soil.

One concern with using hardwood shavings for bedding is the possibility of laminitis or founder. Standing on black walnut (*Juglans nigra* L.) shavings can cause laminitis or founder; because of this, all hardwood shavings are often avoided (Martinson et al., 2007). Other types of hardwoods are probably acceptable, but care should be taken that black walnut is not mixed in (Westendorf and Krogmann, 2013).

When managed correctly, the nutrients present in horse manure can be beneficial to growing crops. However, poor management when storing or land-applying manure can lead to significant environmental harm. Nitrogen and phosphorus may enter surface water, enriching it and causing eutrophication, a process which can lead to dangerously low dissolved oxygen levels and subsequent fish kills. Eutrophication will be discussed in more detail in the Soils and Water section of this chapter. It has been a major problem in the Chesapeake Bay, and it is the primary reason we want to minimize nutrient excretion from animals, nutrient buildup in soils, and nutrient runoff from fields.

Since paddocks are especially susceptible to nutrient runoff losses to nearby waterways, Parvage et al. (2017) studied several organic materials (straw, peat, wood chips) for possible use in paddocks to reduce these losses. Wood chips were most effective in reducing P and carbon (C) losses from outdoor paddock soils but did not reduce N loss.

Manure Storage and Treatment

The use or implementation of appropriate manure management practices can help to further reduce the risks that animal waste will contaminate water or air. Manure on

Table 1. Manure Production and Characteristics. (ASAE, 2003). ‡

	Total solids	Volatile solids	BOD†	N	P	K	Ca	Mg	Total	Volume	H₂O
					—kg d⁻¹—					d⁻¹	%
Sedentary-500 kg	3.8	3.0	0.48	0.089	0.013	0.027	0.023	0.009	25	25	85
Horse- Intense exercise- 500 kg	3.9	3.1	0.49	0.15	0.033	0.095	0.069	0.018	26	26	85

† Biochemical Oxygen Demand

‡ N, Nitrogen; P, Phosphorus; K, Potassium; Ca, Calcium; Mg, Magnesium.

Table 2. Density of bedding materials (Horse Facilities Handbook, MWPS, 2005).

a. Loose bedding	Density kg m⁻³ (lb ft⁻³)
Straw	40.05 (2.5)
Wood shavings	144.2 (9)
Sawdust	192.2 (12)
Non-legume hay	16.02– 20.82 (1.0–1.3)
b. Baled bedding	
Straw	80.09 (5)
Wood shavings	320.37 (20)
Peat moss	28.83– 40.05 (1.8–2.5)
c. Chopped bedding	
Straw	112.13 (7)
Newspaper	224.26 (14)

Table 3. Absorption of bedding materials (Horse Facilities Handbook, MWPS, 2005).

Material	Water Absorption- kg water absorbed per kg bedding lb water absorbed per lb bedding
Pine shavings	2.0
Pine sawdust	2.5
Pine chips	2.0
Hardwood chips	1.5
Oat straw	2.5
Wheat straw	2.2
Shredded newspaper	1.6

horse farms is generally stored solid. The Horse Facilities Handbook (MWPS, 2005) describes manure storages and the many options for suitable manure storage. They all have a few characteristics in common, including not being stored in low areas prone to flooding or on steep ground. Manure storage structures should be sized for the amount of manure and bedding being produced during a certain time frame. The storage should be convenient for depositing and removing manure; three walls and a solid base allow for use of a tractor and bucket loader when pushing or removing the pile. It should be designed to prevent nutrient-laden fluid from infiltrating or running off using either an impermeable base or a cover to keep it dry. Any leachate should drain into an appropriately sized filter strip (Kelly and Westendorf, 2014). In some cases, the easiest method of manure storage is a dumpster or bin that is regularly hauled away.

Soil amendments containing manure must be managed to reduce environmental effects. Maltais-Landry et al. (2018a) compared emissions losses from chicken, turkey, and horse manure. Manures were incubated at different temperatures and water holding capacities. Horse manure incubated at 20 °C and 60% water holding capacity had a 13- to 130-fold increase in methane (CH_4) and a 4- to 70-fold increase in nitrous oxide (N_2O) emissions. Methane emissions reached maximum at 120% water holding capacity. Horse manure incubated at 20 °C and 60% water holding capacity maximized N availability and minimized greenhouse gas emissions. Maltais-Landry et al. (2018b) conducted a six-month decomposition experiment using chicken (broiler), turkey, and horse manure. Horse manure had increases in P, potassium (K), and N_2O emissions, but not carbon dioxide or CH_4

emissions when exposed to increased rainfall. They conclude that manure storages should be designed to provide maximum rainfall protection.

Biochar, a byproduct of biomass that has undergone pyrolysis (Mackie et al., 2015) was added to horse manure and municipal biosolids and applied to tall fescue plots (Williams and Edwards, 2017). Runoff samples were tested for nutrients and fecal coliforms. Under simulated rainfall conditions, when applied with horse manure biochar decreased concentrations of ammonia-nitrogen (NH_3–N), total suspended solids, and fecal coliforms. This study used soil hydraulic conditions (USDA-NRCS, 1986) as a co-variate and found that biochar effectiveness may be affected by a soils hydrologic characteristics.

Some form of composting is often conducted on horse farms. Composting is a managed process; microorganisms, including bacteria, actinomycetes, and fungi will break down organic materials at elevated temperatures (Krogmann et al., 2006). Composting has the advantage of reducing the volume of material, breaking down indigestible material into a plant-usable form, reducing moisture levels, eliminating many odors, and resulting in a product that is easier to manage and may potentially be marketable (NRAES, 1999). The final product can be spread on pastures according to a plan accounting for nutrient uptake by crops. The chief disadvantage environmentally is that the breakdown of organic matter during the composting process will result in the loss of N as ammonia. This will result in atmospheric ammonia loss and in a reduced ratio of N to P for plant uptake whenever the compost pile is turned (NRC, 2007). In addition, the organic N that remains after the composting process is more recalcitrant, contributing further to reduced amount and rate of N mineralization from the composted manure (Eghball et al., 2002). Knowledge of the composting process as well as equipment, space, and attentive management is required to produce high-quality compost.

Turning the composting material achieves proper aeration and ensures that all parts of the manure pile reach elevated temperatures for a certain time period (NRAES, 1999; Krogmann et al., 2006). A horse manure pile should be adjusted for moisture and oxygen availability (NRAES, 1999). Moisture content in a compost pile should be 55 to 60% and oxygen availability can be maintained through ideal pile structure (porosity) and proper and regular turning of the compost pile. Maintaining a proper ratio of carbon/nitrogen N (C/N) is critical. A proper C to N ratio is 20:1 to 30:1. Table 4 compares the C to N ratios of manure and bedding sources.

When greater amounts of woody products are used (Krogmann et al., 2006), they should be supplemented with higher N cosubstrates. Horse stalls bedded with wood shavings or chips may have a very high C to N ratio and will result in a poor compost, especially when bedded heavily. Adding N in the form of dairy manure,

Table 4. Compost Carbon/Nitrogen (C/N) Ratios for Various Feedstocks (NRAES, 1999).

Feedstock	C/N
Horse manure (unbedded)	19:1
Dairy manure	20:1
Straw	40:1- 100:1
Wood shavings	500:1– 600:1
Sawdust	400:1
Grass clippings	17:1

grass clippings, food waste, or even an N fertilizer may be useful when dealing with higher C to N ratios (Krogmann et al., 2006) to support microbial decomposition.

One study looking at various wood and straw bedding products demonstrated that incorporating a simple aerobic composting system on horse farms may greatly reduce the overall volume of manure and yield a material that is beneficial for land application in pasture-based systems (Komar et al., 2012). The pile temperatures in the study indicate that composting did occur after a 100-d process. However, only the temperatures observed in the straw materials (long stem and pelleted) remained elevated long enough to reduce the persistence of pathogens, parasites and weed seeds. The chemical and physical characteristics of the composted material indicate that straw-based materials may be better suited for composting and subsequent land application over wood-based products (Komar et al., 2012).

Keskinen et al. (2017) compared compostability of three different bedding types: peat, wood shavings, and pelleted straw. Pelleted straw had superior composting characteristics than peat or wood shavings, comparable to the results of Komar et al. (2012). Total dry matter declined in pelleted straw by 50% compared to 30% in wood shavings and 20% in peat. Pelleted straw also reached active composting temperatures of 60 °C, while only wood shavings reached temperatures above 40 °C. Leaching losses of P were greatest in peat manure, both fresh and composted. Active aeration was required in all treatments. These authors recommended that all composts and storages be covered to reduce leaching losses.

Anaerobic digestion for biogas production has been researched for managing horse manure. Böske et al. (2015) determined the biochemical CH_4 potentials in different horse manure mixtures (horse manure/bedding material 2:1) in thermophilic conditions. Horse manure plus wheat straw had the highest CH_4 yield, 245.7 kg^{-1} of volatile solids compared with 114.5, 142.3, and 71.6 kg^{-1} for flax, hemp, and wood chips, respectively. When compared with a previous study (Böske et al., 2014) this thermophilic system resulted in a 59.8% increase in CH_4 yield and a 58.1% increase in CH_4 production rate when compared with a mesophilic system. Horse manure bedding materials often have a high content of ligno-cellulosic materials that can limit or slow breakdown. Zhang et al. (2017) developed a three-stage table-top digester that allowed for high solids digestion and a methanogenesis chamber under wet anaerobic conditions. Food waste and horse manure were codigested. The table-top model increased wet anaerobic digestion, resulting in an increase in CH_4 output of 11 to 23%. Sequencing analysis found increases in hydrolyzing bacteria, acidogenic bacteria, and methanogenic bacteria in the three chambers of the digester, respectively. Methanogens were increased by 0.8 to 1.28 times, compared with controls. This table-top digester was successful in linking high-solids anaerobic digestion with wet anaerobic digestion.

Bedding sources high in woody-type ligno-cellulosic materials may limit or slow breakdown processes. The composting studies described above (Komar et al., 2012; Keskinen et al., 2017) and the one anaerobic digestion study (Böske et al., 2015) found that straw materials, both long and pelleted straw break down or are digested more quickly than other sources. Wood shavings (Westendorf et al., 2010) is often the bedding of choice on horse farms, improved disposal options for different types of bedding and similar materials need to be studied.

Biomass from horse manure has significant potential as an energy source. Svanberg et al. (2018) suggest that the supply chains for linking horse manure with energy production are insufficient. These authors believe that current logistics make energy production from horse manure a challenge. Limitations of converting horse manure into energy include the geographical size and spread of stables, the quality of the product, transportation, and the development of appropriate end users.

Eriksson et al. (2016) completed a Life-Cycle Assessment (LCA) for horse manure disposal treatments. The technologies compared were unmanaged composting, managed composting, large scale incineration, drying and small-scale combustion, and liquid anaerobic digestion with thermal pre-treatment. (Unmanaged composting includes no aeration or mixing). Results were variable, no clear conclusions were drawn except the climate impacts of anaerobic digestion was preferable to all other treatments. The analysis was completed using computational modeling software and calculated emissions to air and water using an environmental assessment methodology. Other results were variable or affected by insufficient data. Hadin et al. (2017) completed an analysis based on a field survey and the use of modeling software. Anaerobic digestion resulted in reduced environmental impact when compared with unmanaged composting. According to Svanberg et al. (2018) transportation problems and geographical concerns limit the transfer of manure to end-users. However, Hadin et al. (2017) suggests that increases in transportation costs may be balanced by the reduced environmental impact of anaerobic digestion.

Anaerobic digestion results in lower emissions (especially climate impact emissions) than other management technologies (composting, incineration, combustion). Assuming horse manure must be transported to centralized areas (Svanberg et al., 2018) for processing; transportation limitations may limit the technology depending on the levels of environmental benefits (Hadin et al., 2017). However, it is noted that the dry nature of horse manure enables it to be more easily "exported" to be utilized in manufacturing of organic fertilizer, and in the Chesapeake Bay watershed, horse manure is often marketable to mushroom growers (Kleinman et al., 2012).

Diet Management to Reduce Nutrient Waste

Overfeeding of diets can lead to increased nutrient excretion in manure (NRC, 2007). Research with dairy cattle has shown that P levels can be reduced in the diet without a decline in animal performance (Valk et al., 2000; Wu et al., 2001; Knowlton and Herbein, 2002). Dietary modifications in swine and poultry (Maguire et al., 2006), and/or the use of low phytate feedstuffs (Powers et al., 2006), or phytase enzymes (Angel et al., 2005) can reduce the level of N and P excretion in the waste. Phytate is an indigestible form of P in feed than cannot be utilized by monogastrics such as swine or poultry, but it can be broken down in the rumen of cattle and sheep due to microbial action (NRC, 2001). Phytase is an enzyme that can be added to the diet to break down the indigestible phytate found in cereal grain; this releases the P for use by swine or poultry (Angel et al., 2005). Phytase has been used in equine diets with minor effects on P excretion (Hainze et al., 2004; van Doorn et al., 2004).

Not as much research has been completed with horses and animal feeding and nutrient excretion into the environment, although the Horse NRC (2007) indicates that diet composition influences the amount and composition of waste. Topliff and Potter (2002) said that diet influenced the excretion of N and P in performance horses. Rations are often formulated to provide a safety factor that will cover dietary shortfalls; this may result in excess nutrients excreted into the environment.

Westendorf et al. (2013) surveyed horse farmers in New Jersey and found that 60% of those questioned either balanced equine diets on their own or had no feeding plan at all; most of those who did get advice used a veterinarian or a feed store.

One third of the respondents indicated reducing P in the diet to reduce excretion, but it was not clear what measures were taken. Over 75% of respondents rotationally grazed their pastures; this is a recommended management practice that will reduce water runoff, soil erosion, and improve soil health and quality when done correctly. Nearly 90% of those who rotationally grazed also made use of a sacrifice lot or a heavy use area to concentrate animal use for shelter, water, and feeding.

Another project focused on a feed management program for horses (Westendorf and Williams, 2015) that is similar to the USDA-NRCS 592 Feed Management Program (Harrison et al., 2012). Twenty-one farms cooperated in the program and most had little knowledge of environmentally friendly feeding practices and even after a year in the project were still overfeeding. One of the challenges with horses is that in a typical boarding stable, there could be numerous owners and all may have different feeding strategies. The authors concluded that a program for horses should focus on monitoring animal condition, pasture management, and feed and forage testing.

Harper et al. (2009) found that horse farmers were feeding P and N at 185% and 159% of requirements, respectively. Westendorf and Williams (2015) showed that overfeeding a P supplement (NaH_2PO_4), formulated to provide 4.5 times daily P requirement, increased fecal P from 3.6 g kg^{-1} in the low P group to 8.1 g kg^{-1} in the high P group. Since the P supplement used in this study was a highly soluble form of P (NaH_2PO_4), samples were analyzed for Water Extractable Phosphorus (WEP), WEP in horse manure increased from 2.1 g kg^{-1} in the low P group to 6.8 g kg^{-1} in the high P group. Increased WEP levels may be linked with P runoff when manure is land-applied (Kleinman et al., 2002). This test has been used in other species (cattle, dairy, swine, poultry; Wolf et al., 2005; Kleinman et al., 2005), but never before in horse manure. Westendorf and Williams (2015) confirmed that when horses are overfed P, there will be an increase in fecal P, including the level of WEP and therefore an increased P runoff risk.

Excess N from manure can also cause health and environmental issues. Ammonia levels have been found to be elevated in the air surrounding horse stalls as well as found to be higher in manure when increased crude protein levels are fed to horses (Williams et al., 2011). In this study, the treatment group of horses was supplemented with 700 g d^{-1} of soybean meal top dressed on 500 g of sweet feed per day (TRT; 1042 g protein d^{-1} DM total), while the control group received the sweet feed meals without the soybean meal (CON; 703 g protein d^{-1} total). Both groups were also fed 8 kg d^{-1} of a grass hay mix (562 g protein d^{-1} DM), water and salt ad libitum. Results showed in manure, horses fed the TRT diet excreted more N and NH_3 than horses fed the CON diet (Williams et al., 2011).

Bott et al. (2016) lists several strategies for reducing the environmental impacts of N from horses, including reducing N intake and improving the management of outputs. Precision feeding or feeding to meet but not exceed NRC (2007) requirements may reduce N impact (Bott et al., 2016). One of the challenges with horses is the numerous feeding strategies on farms (Westendorf et al., 2010, 2013). Every owner may insist on feeding their horses differently, and many owners cannot recognize an overweight horse (Morrison et al., 2017). Overfeeding horses may be common throughout the industry (Harper et al., 2009, Williams and Burk, 2010). Other livestock industries have more precision feeding data–this is lacking in equine feeding and management (Harrison et al., 2012). Even if this data existed, it may not be utilized. A better approach for equine might be to focus on pasture management and management of dry lots and surface areas to reduce mud accumulation and water runoff in feeding, shelter, and watering areas. On most horse farms, maintaining animal condition and avoiding overconditioning and obesity may be more manageable than to try to apply precision feeding. Barnyard and farm Best Management Practices (BMPs) should be implemented to manage stormwater, reduce surface erosion and runoff, and manage watersheds to prevent manure and nutrient contamination of surface waters around equine farms; this approach may be the most useful nutrient management strategy on equine farms.

In addition to the dietary effects on waste excretion, overall feed management can also influence nutrient management (Westendorf and Williams, 2007). Design of feeders, availability of water, and quality of feed and forage can all influence nutrients excreted in the waste. Feeding in elevated square bale feeders or utilizing a round bale feeder is one of the best strategies to reduce feed loss (Martinson et al., 2012; Grev et al., 2014). Animals fed on the ground will require more feed, and feed wasted on the ground could add up to 57% of the hay fed (Martinson et al., 2012), which would probably end up in the manure pile contributing to additional waste needing to be removed from the farm.

Pasture Management

Grazed pastures, particularly rotational grazing of pastures, recycle nutrients faster than ungrazed pastures. Nutrients on pasture land enter through animal waste, and waste feed or fertilizer; they leave through removal of forage, leaching, runoff, and erosion, or animal product and waste removal. Taking away the animal component removes about half of the inputs needed to recycle the nutrients. Dietary N, P, and (K) are required for basic maintenance of horses; however, not all of what is consumed is used by the animal, therefore the dietary concentrations of these nutrients will impact the nutrient cycling. Digestibility of N, P, and K in horses is approximately 80, 25, and 75%, respectively (NRC 2007). What does not get digested will end up excreted back into the soil.

Two major environmental concerns due to poor grazing management are soil erosion and contamination of ground and surface water by unmanaged manure. These are sources of nonpoint-source pollution, contaminants that enter surface water through runoff rather than direct discharge. Not only is it a problem when animals defecate directly into streams, but contaminated stream water may channel into ponds or lakes. Also, the loss of vegetation from animal trampling and forage stand loss related to poor pasture management can lead to groundwater contamination (Miner et al., 2000). This loss of vegetation destroys the natural filtering effect that a good grass cover would provide (Butler et al., 2007). The long-term goal of pasture management should be to maintain a dense grass and legume cover throughout a pasture. This will utilize more manure nutrients for growth, buffer and filter nutrient flow, reduce losses into water, and provide better nutrition for horses. Miner et al. (2000), Atwill et al. (2002) and Tate et al. (2004) all showed that vegetative cover was effective in filtering pathogens. Butler et al. (2007) demonstrated overall 85% reductions of N export from vegetated pasture amended with beef feces and urine compared with bare, compacted soil, even with as little as 45% vegetative cover. Good vegetative growth on pastures is essential for effective horse production benefiting both horses and the environment.

Stocking rate, grazing duration, rotational grazing, distribution of manure on pastures in relation to water and shelters, and type of forage grasses and legumes will also affect the interface between horses, manure, and the environment on pasture. More intensive grazing creates an increased rate of nutrient cycling due to the added animal inputs on the land. Even though no horse-related studies have been performed on this topic, studies in cattle have found that the plant-available N levels double when cattle were rotationally grazed with five grazing periods per season instead of three (Baron et al., 2002).

When rotationally grazing horses, each grazing unit should be mowed and dragged immediately after the animals are removed (Gilley et al., 2002). Horses are selective grazers and will leave some areas uneaten, and mowing ensures that all plants are the same height and return to the same stage of growth when the recovery period begins. Dragging or harrowing pastures helps to spread out manure piles, making the manure dry faster and distributing manure nutrients more evenly around the field. When spreading out manure piles, the risk of internal parasite larvae ingestion increases, so horses must be on an effective deworming protocol (see below). The rest period for each rotational unit also provides an ideal time for applying soil amendments such as lime and fertilizer or herbicides to control weeds (Henning et al., 2000; Undersander et al., 2002). Butler et al. (2007) suggest using riparian or otherwise sensitive areas for grazing during dry periods of the year, allowing adsorption and immobilization of nutrients before rainfall increases risks of N export; when there is rain immediately after manure application, loss is more likely.

As mentioned previously, the most effective way to reduce erosion and runoff is to maintain a high vegetative cover on pastures (Gilley et al., 2002). While the cover threshold to minimize erosion is 70% (Costin, 1980; Sanjari et al., 2009), plant attributes like type and habit can influence runoff. Decomposing litter from the vegetation becomes organic matter in the soil, which helps to decrease runoff potential by increasing water holding capacity and infiltration rates.

Below are some Best Management Practices (BMPs) which can be used when pasturing horses to protect water quality.

- *Removing manure.* The most effective way to ensure that nutrients from manure do not enter surface water is to remove it frequently. This is most important in unvegetated stress lots where soil is often compacted and less permeable, soil surface is not protected from erosive forces, and no plant uptake of nutrients is occurring.

- *Constructing heavy use pads or stress (or sacrifice) lots.* Heavy use pads are designed to improve drainage in animal concentration areas where intense trampling compacts soils and prohibits plant growth. Excavating the topsoil and constructing a base layer of large stone and a top layer of footing allows water to infiltrate through the structure and collect at the bottom rather than ponding at the surface. The water can either infiltrate from the base or be diverted with piping. Heavy use pads have been shown to improve drainage and control mud in animal congregation areas (Westendorf, 2008).

- *Using vegetated buffers.* A strip of vegetation, generally grass, is effective at filtering nutrients and sediments from runoff as the plants slow the flow of water, allowing sediment to settle, and take up nutrients. Farms should maintain vegetated buffers or filter strips around all unvegetated areas such as stress lots, barnyards, driveways, manure storages, and riding arenas (Welsch, 1991).

- *Keep clean water clean.* Stormwater that falls on a roof has not been contaminated by manure or sediment and is therefore "clean." This water should be prevented from coming into contact with manure or loose sediment often present around barns and sheds. This can be accomplished using French drains or gutters that divert clean stormwater to a safe place to infiltrate or even collect it to use for irrigation or arena watering.

- *Construct berms and waterways to divert water flow.* Another way to reroute clean water is using diversions such as berms (raised mounds of earth to stop water flow) or waterways (channels to encourage water flow). These can be used together to prevent stormwater on a steep slope from entering surface water or a contaminated area such as a stress lot.

- *Fence off open water and wet areas.* While horses do not tend to wallow in surface water, defecating directly into it, it is ideal to limit access to open waters. Crossing and drinking from streams erodes streambanks, dislodging sediment to flow downstream. In cases where a stream bisects a pasture or property, the ideal solution is to fence off most of it and install properly engineered stream crossings or culverts to ensure access to the entire farm.

More equine specific studies need to be performed looking at how grazing systems and equine diets affect nutrient cycling and how horse farm owners can utilize this to best manage their farm for optimal nutrient utilization.

Internal Parasite Concerns

Manure on pasture creates equine health risks in addition to environmental risks. The major gastrointestinal parasite of concern for adult horses is the small strongyle or cyathostomin, and its life cycle depends on eggs in manure developing into infective larvae and being consumed by horses on pasture. While horses generally avoid grazing near manure piles, runoff can transport larvae downhill, and the often-recommended practice of harrowing pastures to spread manure nutrients also spreads out the larvae. The various free-living stages of small strongyles can be quite resistant to frost, desiccation, and heat, so environmental conditions alone are not sufficient to clean a pasture of parasites. Other options for reducing infectivity of pastures include removing manure frequently, avoiding rotation immediately after deworming, and switching pastures to hay or other livestock grazing for at least a year (Reinemeyer and Nielson, 2013).

Parasite populations on pasture can be used to lessen the effect of parasite resistance to anthelmintics. The old industry standard practice of rotating anthelmintic classes every eight weeks has led to small strongyle and ascarid (generally a concern in foals) populations that are resistant to these drugs. The new AAEP deworming recommendations involve less frequent treatment of horses that naturally shed few eggs so that a portion of the parasite population on pasture is not exposed to anthelmintics (refugia) and dilutes the resistant population. Horse owners are encouraged to perform fecal egg count reduction tests before and after deworming to evaluate the level of resistance to different anthelmintic classes on their farms (AAEP, 2013).

Bacteria and Microbes

According to the 2003 ASAE Manure Production and Characteristics document, a 455-kg horse (1000 lb) excretes 420 million fecal coliform colonies and 260 billion fecal streptococci colonies. According to the Equine NRC (2007), these may be used as indicators when determining water quality risks from horses. Bilotta et al. (2007) indicate manure-borne bacterial pathogens such as *Eschericia coli* and *Salmonella* spp., viruses such as *Rotavirus* spp., and protozoan parasites *Cryptosporidium* and *Giardia* spp. as threats to human health. Horses are not likely to be sources of the more toxic form of *E. coli* (0157:H7) or of *Listeria* (Equine NRC, 2007). Kistemann et al. (2002) found increases in bacterial colony counts in runoff water, especially fecal streptococci and *Clostridium perfringens* after heavy rainfall in agricultural communities. Miner et al. (2000) found that 95% of fecal streptococcus and coliforms in feedlot runoff were removed through a vegetative filter. They also said that indicator bacteria such as fecal coliform or fecal streptococcus seldom travel more than six feet in a well-established pasture or grazing area.

The USDA-NAHM (2001) sampled 972 farms with 8417 horses for the incidence of Salmonella shedding. Nationwide, 1.8% of operations had at least one horse that shed salmonella, but only 0.8% of horses overall shed the organism in their feces. On farms with greater than 20 horses, 9.4% had at least one horse culture positive, while only 2.9% of farms with 6 to 19 horses, and less than 0.1% of farms with less than six horses had at least one fecal culture positive horse. The study found 14 different salmonella serotypes, some not associated with illness in horses. The Merck Veterinary Manual (2016) estimated the level of shedding of *Salmonella eteritica* to be < 2% nationally (the proportion of shedding in hospitalized horses is approximately 8%).

There is some concern about protozoal diseases such as cryptosporidiosis and giardiasis in horses. One study found 0.33% of horses were carrying *Cryptosporidiosum parvum*, and the same study found 0.66% of horses carrying *Giardia* (Forde et al., 1997). In a study of backcountry recreation horses, which might be expected to have access to *Giardia* or *Cryptosporidia*, Johnson et al. (1997) found that none of the 91 horses had *Cryptosporidium* or *Giardia* oocysts. Ostoja et al. (2014) described how pack animals (horses, mules, burros) can have detrimental environmental effects. Pack stock management should "minimize bare ground, maximize plant cover, maintain species composition of native plants, minimize trampling, especially on wet soils and stream banks, and minimize direct urination and defecation by pack stock into water." The authors state there is limited information about pack stock behavior and suggest research identifying specific environmental variables.

Another project looked at the use of buffer strips and showed that when slopes were less than 20% with good ground cover (they used tall fescue grass) and a buffer strip length of three meters or greater, this was sufficient to remove 99.9% of *Cryptosporidium parvum* oocysts (Atwill et al., 2002). Tate et al. (2004) also found that the use of vegetative buffers can reduce the spread of waterborne *Cryptosporidium parvum* on extensively grazed grasslands.

The use of appropriate BMPs can reduce the risk of waterborne contamination of pathogens. Inamdar et al. (2002) indicated that although the introduction of BMPs (manure storages, stream fencing, water troughs, and nutrient management programs) resulted in improvements in water quality (as measured by fecal coliforms or streptococci), BMPs alone may not ensure water quality. On the other hand, Miner et al. (2000) recommends a buffer strip of ≥ 50 feet between manure application areas and waterways for adequate protection; they also indicate that fecal coliforms and streptococci seldom travel more than six feet in well-managed pastures.

Soil and Water Quality

Land applications of horse manure or horse manure composts must be based on prior soil and water quality assessments (Edwards et al., 1999; Pernes-Debuyser and Tessier, 2004; Flores et al., 2005; Ferreras et al., 2006). With regard to land applications of horse manure, soil quality is associated with increases in soil organic matter, water infiltration rate, and water retention. Pernes-Debuyser and Tessier (2004), found that horse manure resulted in more plant-available water, less runoff, and reduced losses of nutrients. Soil organic C has many benefits to chemical, physical, and biological processes of soil (Blanco-Canqui et al., 2013), including positive effects on water-stable aggregation, macroporosity, bulk density, buffering capacity, microbial diversity, and others. Manure provides C as well as valuable nutrients; before manufactured fertilizers became common, manure was considered a most valuable resource to farmers to maintain soil fertility (McNeill and Winiwarter, 2006; Finck, 2006). In general, nearly all K in manure is immediately available to plants and microbes; inorganic forms of P in horse manure are immediately available while organic P must be mineralized to become bioavailable (Pagliari and Laboski, 2014). Many factors complicate nitrogen cycling, making it more difficult to estimate availability (Eghball et al., 2002). Horse manure should be tilled into the soil when spread or dragged when applying to pastures to increase adsorption of nutrients.

However, waste from equine or other livestock operations is a substantial source of nutrients and microbes to surface waters contaminated by runoff. Animal waste, excreted and reaching surface waters, can lead to eutrophication from enhanced inputs of P and N. In terms of environmental consequences of improper waste management, eutrophication is the most common impairment of surface waters in the United States (USEPA, 2015). This can lead to excessive biological growth and result in consequences such as over-production of aquatic vegetation leading to toxic algal blooms, anoxia, fish kills, loss of biodiversity, and other damages (Carpenter et al., 1998). Overproduction of aquatic vegetation including algae not only causes unsightly scums and poor light penetration but consumes excessive amounts of dissolved oxygen, especially during decomposition. Low oxygen levels in the water may impact some aquatic organisms. So proper manure management to minimize risk of these impacts is in the best interest of human, animal, and environmental health.

Waste from livestock can result in elevated nutrient concentrations in stormwater. Small equine operations, as well as large, represent a potential risk to receiving waters. Pagliari and Laboski (2012) demonstrated that about 45% of the total P in horse manure is water-soluble, and most (91%) of water-soluble P in horse manure is in the inorganic form. This emphasizes the hazard of eutrophication from runoff transported from horse paddocks and pasture into waterways. Airaksinen et al. (2007) measured nutrient concentrations in surface run-off water from paddocks that had been cleaned after seven months of use. Measurements, taken in the spring, were 3.4 to 18.8 mg L^{-1} for total P, 3.0 to 15.0 mg L^{-1} for phosphate and 18.3 to 140.0 mg L^{-1} for total N. The cleaned paddocks had significantly lower concentrations of nutrients in runoff compared with runoff from uncleaned paddocks, especially considering the highest concentrations coming from the feeding areas of paddocks that were not cleaned.

Although the dissolved inorganic form of P is dominant in manure compared with dissolved organic P (Pagliari and Laboski, 2012), dissolved organic P can significantly contribute to P mobility and leaching in certain heavily manured soils as P adsorption capacity has been approached or exceeded (Eghball et al., 1996); soil mineralogy, pH, and physical properties (texture, pore-size distribution) would be factors affecting adsorption and mobility of phosphorus forms as well as the total P load (Lentz and Lehrsch, 2018). Total organic P levels are relatively low in horse manure compared to most other non-ruminant livestock manures, likely due to phytase production by microflora in the horse gut, similar to ruminant digestion (Pagliari and Laboski, 2012). Dissolved organic P can be transported to subsoil and even to groundwater where coarse soils and/or high-water tables exist (Eghball et al., 1996). This dissolved organic P would also be susceptible to runoff losses to water bodies (Daniel et al., 1993). Measurement of WEP may be a tool to help manage P runoff on horse farms (Wolf et al., 2005; Westendorf and Williams, 2015). Once mineralized in the soil environment, inorganic P has strong tendencies to react or be adsorbed ("fixed") to sesquioxide, silicate, and carbonate minerals and become much less mobile. Yet elevated P concentrations (soil test levels) in the soil still result in a greater potential for P pollution to aquatic ecosystems where soil erosion is possible (Fluck et al., 1992). The dominant phase for P in runoff from tilled land is adsorbed to eroded soil particles, and grassed land, being much less susceptible to erosion, has greater percentage loss as dissolved P (Sharpley et al., 1994). These soil particles may be detached and then suspended in runoff during storm events and carried to lakes, streams and rivers, where the P diffuses in response to concentration gradients. In most natural, oligotrophic fresh water systems, P is the limiting nutrient for growth and the primary cause of eutrophication when added.

Nitrogen is found in runoff in dissolved and particulate phases. The particulate phase will typically contain N mostly in an organic form, such as in soil organic matter or partially decomposed manure. Humified organic matter is often adsorbed onto soil particles of relatively high density, whereas partially undigested, particulate organic matter of manure is low density and more readily carried in runoff sheet or rill flow. The immediate inorganic N product of organic matter breakdown, ammonium (NH_4^+), is soluble but subject to adsorption in soil by cation exchange capacity (due to net negative charges of most clay minerals), and so may be transported with eroded soil particles. In soil, biological nitrification of ammonium produces (ultimately) nitrate (NO_3^-), which is also water soluble and considered more mobile because of its anionic properties, not being adsorbed to cation exchange sites on soil minerals. Therefore, the ammonium species of N is often found bound to clays and organic sediment while nitrate is usually found in the dissolved phase (Davis et al., 2001). Therefore, N travels in surface water but can also infiltrate and travel through the soil to groundwater shallow. Nitrogen is the limiting nutrient for primary growth in temperate estuaries and coastal ecosystems, (Howarth, 1988) and often the cause of eutrophication in those situations. Coastal regions with sandy coastal plain soils, shallow groundwater and high density of horses (NJ, FL) may be most at risk for eutrophication of estuaries, bays, and shorelines by N derived from horse manure (or otherwise).

Many resources have been developed to assist horse owners and stable operators with effective, sustainable horse manure management. Besides the list of BMPs provided above, the Council of Bay Area Resource Conservation Districts (CA) provides a series of fact sheets relating specifically to equine manure management (www.marincounty.org/depts/pw/divisions/creeks-bay-and-flood/mcstoppp/community-resources/horse-owners), and on the East Coast, additional examples are Maryland's Horse Outreach Workgroup (https://mda.maryland.gov/resource_conservation/Pages/horse_pasture_manure_info.aspx), Rutgers Equine

Science Center resources (https://esc.rutgers.edu/resources/library/), and eXtension Horse Quest (https://articles.extension.org/horses).

References

AAEP. 2013. Parasite control guidelines. Parasite Control Subcommittee, AAEP Infectious Disease Committee. American Association of Equine Practitioners, Lexington, KY. https://aaep.org/sites/default/files/Guidelines/AAEPParasiteControlGuidelines_0.pdf (Accessed 27 Feb. 2019).

Airaksinen, S., M.-L. Heiskanen, and H. Heinonen-Tanski. 2007. Contamination of surface run-off water and soil in two horse paddocks. Bioresour. Technol. 98(9):1762–1766. doi:10.1016/j.biortech.2006.07.032

Angel, R.C., W.J. Powers, T.J. Applegate, N.T.M. Tamim, and M.C. Christman. 2005. Influence of phytase on water-soluble phosphorus in poultry and swine manure. J. Environ. Qual. 34:563–571. doi:10.2134/jeq2005.0563

ASAE. 2003. Manure production and charactersitics. American Society of Agricultural Engineers. ASAE D384. http://www.manuremanagement.cornell.edu/Pages/General_Docs/Other/ASAE_Manure_Production_Characteristics_Standard.pdf (Accessed 27 Feb. 2019).

ASAE. 2005. Manure production and characteristics. American Society of Agricultural Engineers. ASAE D384. http://www.agronext.iastate.edu/immag/pubs/manure-prod-char-d384-2.pdf (Accessed 27 Feb. 2019).

Atwill, E.R., L. Hou, B.M. Karle, T. Harter, K.W. Tate, and R.A. Dahlgren. 2002. Transport of Cryptosporidium parvum oocysts through vegetated buffer strips and estimated filtration efficiency. Appl. Environ. Microbiol. 68(11):5517–5527. doi:10.1128/AEM.68.11.5517-5527.2002

Baron, V.S., E. Mapfumo, A.C. Dick, M.A. Naeth, E.K. Okine, and D.S. Chanasyk. 2002. Grazing intensity impacts on pasture carbon and nitrogen flow. J. Range Manage. 55:535–541. doi:10.2307/4003996

Bilotta, G.S., R.E. Brazier, and P.M. Haygarth. 2007. The impacts of grazing animals on the quality of soils, vegetation, and surface waters in intensively managed grasslands. Adv. Agron. 94:237–280. doi:10.1016/S0065-2113(06)94006-1

Blanco-Canqui, H., C.A. Shapiro, C.S. Wortmann, R.A. Drijber, M. Mamo, T.M. Shaver, and R.B. Ferguson. 2013. Soil organic carbon: The value to soil properties. J. Soil Water Conserv. 68(5):129A–134A. doi:10.2489/jswc.68.5.129A

Böske, J., B. Wirth, F. Garlipp, J. Mumme, and H. Van den Weghe. 2014. Anaerobic digestion of horse dung mixed with different bedding materials in an upflow solid-state (UASS) reactor at mesophilic conditions. Bioresour. Technol. 158:111–118. doi:10.1016/j.biortech.2014.02.034

Böske, J., B. Wirth, F. Garlipp, J. Mumme, and H. Van den Weghe. 2015. Upflow anaerobic solid-state (UASS) digestion of horse manure: Thermophilic vs. mesophilic performance. Bioresour. Technol. 175:8–16. doi:10.1016/j.biortech.2014.10.041

Bott, R.C., A. Woodward, E.A. Greene, N.L. Trottier, C.A. Williams, M.L. Westendorf, A.M. Swinker, S.L. Mastellar, and K.M. Martinson. 2016. Environmental implications of nitrogen output on horse operations: A review. J. Equine Vet. Sci. 45:98–106 http://www.j-evs.com/article/S0737-0806(15)00549-3/pdf. doi:10.1016/j.jevs.2015.08.019

Butler, D.M., N.N. Ranells, D.H. Franklin, M.H. Poore, and J.T. Green, Jr. 2007. Ground cover impacts on nitrogen export from manured riparian pasture. J. Environ. Qual. 36:155–162. doi:10.2134/jeq2006.0082

Carpenter, S.R., N.F. Caraco, D.L. Correll, R.W. Howarth, A.N. Sharpley, and V.H. Smith. 1998. Nonpoint pollution of surface waters with phosphorus and nitrogen. Ecological Applications 8(3):559–568. doi:10.1890/1051-0761(1998)008[0559:NPOSWW]2.0.CO;2

Costin, A.B. 1980. Runoff and soil and nutrient losses from an improved pasture at Ginninderra, Southern Tablelands, New South Wales. Aust. J. Agric. Res. 31:533–546. doi:10.1071/AR9800533

Daniel, T.C., D.R. Edwards, and A.N. Sharpley. 1993. Effect of extractable soil surface phosphorus on runoff water quality. Trans. ASAE 36:1079–1085. doi:10.13031/2013.28437

Davis A.P., M. Shokouhian, H. Sharma, and C. Minami. 2001. Laboratory study of biological retention for urban stormwater management. Water Environ. Res. 73(1):5-14.

Edwards, D.R., P.A. Moore, S.R. Workman, and E.L. Bushee. 1999. Runoff of metals from alum-treated horse manure and municipal sludge. J. Am. Water Resour. Assoc. 35:155–165. doi:10.1111/j.1752-1688.1999.tb05460.x

Eghball, B., G.D. Binford, and D.D. Baltensperger. 1996. Phosphorus movement and adsorption in a soil receiving long-term manure and fertilizer application. J. Environ. Qual. 25:1339–1343. doi:10.2134/jeq1996.00472425002500060024x

Eghball, B., B.J. Wienhold, J.E. Gilley, and R.A. Eigenberg. 2002. Mineralization of manure nutrients. J. Soil Water Conserv. 57:470–473.

Eriksson, O., Å. Hadin, J. Hennessy, and D. Jonsson. 2016. Life cycle assessment of horse manure treatment. Energies 9:1011. doi:10.3390/en9121011

Ferreras, L., E. Gomez, S. Toresani, I. Firpo, and R. Rotondo. 2006. Effect of organic amendments on some physical, chemical and biological properties in a horticultural soil. Bioresour. Technol. 97:635–640. doi:10.1016/j.biortech.2005.03.018

Finck, A. 2006. Soil nutrient management for plant growth. In: B.P. Warkentin, editor, Footprints in the soil: People and ideas in soil history. Elsevier. Amsterdam, The Netherlands. p. 427–453.

Fiorellino, N.M., K.M. Wilson, and A.O. Burk. 2013. Characterizing the use of environmentally friendly pasture management practices by horse farm operators in Maryland. J. Soil Water Conserv. 68:34–40. doi:10.2489/jswc.68.1.34

Flores, P., I. Castellar, and J. Navarro. 2005. Nitrate leaching in pepper cultivation with organic manure and supplementary additions of mineral fertilizer. Commun. Soil Sci. Plant Anal. 36:2889–2898. doi:10.1080/00103620500306072

Fluck, R.C., C. Fonyo, and E. Flaig. 1992. Land-use-based phosphorus balances for Lake Okeechobee, Florida, drainage basins. Appl. Eng. Agric. 8:813–820. doi:10.13031/2013.26118

Forde, K.N., A.M. Swinker, J.L. Traub-Dargatz, and J.M. Cheney. 1997. The prevalence of Cryptosporidiu, giardia in the trail horse population utilizing public lands. Journal of Equine Veterinary Science 18(1):38–40.

Gilley, J.E., L.M. Risse, and B. Eghball. 2002. Managing runoff following manure application. J. Soil Water Conserv. 57:530–533.

Grev, A.M., E.C. Glunk, M.R. Hathaway, W.F. Lazarus, and K.L. Martinson. 2014. The effect of small square-bale feeder design on hay waste and economics during outdoor feeding of adult horses. J. Equine Vet. Sci. 34:1269–1273. doi:10.1016/j.jevs.2014.09.004

Hadin, Å., K. Hillman, and O. Eriksson. 2017. Prospects for increased energy recovery from horse manure - A case study of management practices, environmental impact and costs. Energies 10:1935. doi:10.3390/en10121935

Hainze, M.T.M., R.B. Muntifering, C.W. Wood, C.A. McCall and B.H. Wood. 2004. Fecal phosphorus excretion from horses fed typical diets with and without added phytase. Anim. Feed Sci. Technol. 117:265.

Harper, M., A. Swinker, W. Staniar, and A. Welker. 2009. Ration evaluation of Chesapeake Bay watershed horse farms from a nutrient management perspective. J. Equine Vet. Sci. 29(5):401–402. doi:10.1016/j.jevs.2009.04.101

Harrison, J., R. White, V. Ishler, G. Erickson, A. Sutton, T. Applegate, B. Richert, T. Nennich, R. Koelsch, R. Burns, D. Meyer, R. Massey, and G. Carpenter. 2012. Case study: Implementation of feed management as part of whole-farm nutrient management. Prof. Anim. Sci. 28:364–369. doi:10.15232/S1080-7446(15)30369-7

Henning, J., G. Lacefield, M. Rasnake, J. Burris, R. Johns, and L. Turner. 2000. Rotational grazing. ID-143. University of Kentucky, Cooperative Extension Service, Lexington, KY.

Howarth, R. 1988. Nutrient limitation of net primary production in marine ecosystems. Annual Review of Ecology and Systematics, 19:89-110. http://www.jstor.org/stable/2097149

Inamdar, S.P., S. Mostaghimi, M.N. Cook, K.M. Brannan, and P.W. McClellen. 2002. A long-term, watershed-scale, evaluation of the impacts of animal waste BMPs on indicator bacteria concentrations. J. Am. Water Resour. Assoc. 38(3):819–833. doi:10.1111/j.1752-1688.2002.tb00999.x

Johnson E., E.R. Atwill, M.E. Filkins, and J. Kalush. 1997. The prevalence of shedding of Cryptosporidium and Giardia spp. based on a single fecal sample collection from each of 91 horses used for backcountry recreation. J. Vet. Diagn. Invest. 9(1):56–60.

Kelly, F., and M. Westendorf. 2014. Storing manure on small horse and livestock farms. Fact Sheet # FS1192. Rutgers Cooperative Extension, New Brunswick, NJ. http://njaes.rutgers.edu/pubs/fs1192 (Accessed 27 Feb. 2019).

Keskinen, R., M. Saastamoinen, J. Nikama, S. Särkijärvi, M. Myllymäki, T. Salo, and J. Uusi-Kämppä. 2017. Recycling nutrients from horse manure: Effects of bedding type and its compostability. Agric. and Food Sci. 26:68–79.

Kistemann, T., Classen, T., Koch, C., Dangendorf, F., Fischeder, R., Gebel, J., et al. (2002). Microbial load of drinking water reservoir tributaries during extreme rainfall and runoff. Appl. Environ. Microbiol. 68, 2188–2197. doi:10.1128/AEM.68.5.2188-2197.2002

Kleinman, P.J.A., K.S. Blunk, R. Bryant, L. Saporito, D. Beegle, K. Czymmek, Q. Ketterings, T. Sims, J. Shortle, J. McGrath, F. Coale, M. Dubin, D. Dostie, R. Maguire, R. Meinen, A. Allen, K. O'Neill, L. Garber, M. Davis, B. Clark, K. Sellner, and M. Smith. 2012. Managing manure for sustainable livestock production in the Chesapeake Bay watershed. J. Soil Water Conserv. 67:54A–61A. doi:10.2489/jswc.67.2.54A

Kleinman. P.J.A., A.N. Sharpley, A.M. Wolf, D.B. Beegle, and P.A. Moore, Jr. 2002. Measuring water-extractable phosphorus in manure as an indicator of phosphorus in runoff. Soil Sci. Soc. Am. J. 66:2009–2015. https://www.researchgate.net/publication/43270577_Measuring_Water_Extractable_Phosphorus_in_Manure_as_an_Indicator_of_Phosphorus_in_Runoff (Accessed 27 Feb. 2019).

Kleinman, P.J.A., A.M. Wolf, A.N. Sharpley, D.B. Beegle, and L.S. Saporito. 2005. Survey of water-extractable phosphorus in livestock manures. Soil Sci. Soc. Am. J. 69:701–708. doi:10.2136/sssaj2004.0099

Knowlton, K.F., and J.H. Herbein. 2002. Phosphorus partitioning during early lactation in dairy cows fed diets varying in phosphorus content. J. Dairy Sci. 85:1227–1236. doi:10.3168/jds.S0022-0302(02)74186-6

Komar, S., R. Miskewitz, M. Westendorf, and C.A. Williams. 2012. Effects of bedding type on compost quality of equine stall waste: Implications for small horse farms. J. Anim. Sci. 90:1069–1075. doi:10.2527/jas.2010-3805

Krogmann, U., M.L. Westendorf, and B.F. Rogers. 2006. Best management practices for horse manure composting on small farms. Rutgers Cooperative Extension. New Jersey Agricultural Experiment Station. Bulletin Series. E307. Rutgers University, New Brunswick, NJ. http://njaes.rutgers.edu/pubs/publication.asp?pid=E307 (Accessed 27 Feb. 2019).

Lentz, R.D., and G.A. Lehrsch. 2018. Mineral fertilizer and manure effects on leached inorganic nitrogen, nitrate isotopic composition, phosphorus, and dissolved organic carbon under furrow irrigation. J. Environ. Qual. 47:287–296. doi:10.2134/jeq2017.09.0384

Mackie, K.A., S. Marhan, F. Ditterich, H.P. Schmidt, and E. Kandeler. 2015. The effects of biochar and compost amendments on copper immobilization and soil microorganisms in a temperate vineyard. Agric. Ecosyst. Environ. 201:58–69. doi:10.1016/j.agee.2014.12.001

Maguire, R.O., P.W. Plumstead, and J. Brake. 2006. Impact of diet, moisture, location, and storage on soluble phosphorus in broiler breeder manure. J. Environ. Qual. 35(3):858–865. doi:10.2134/jeq2005.0435

Maltais-Landry, G., N. Bertoni, W. Valley, N. Grant, Z. Nesic, and S.M. Smukler. 2018a. Greater impacts of incubation temperature and moisture on carbon and nitrogen cycling in poultry relative to horse manure-based soil amendments. J. Environ. Qual. 47:914–921. doi:10.2134/jeq2017.11.0420

Maltais-Landry, G., K. Neufeld, D. Poon, N. Grant, Z. Nesic, and S. Smukler. 2018b. Protection from wintertime rainfall reduces nutrient losses and greenhouse gas emissions during the decomposition of poultry and horse manure-based amendments. J. Air Waste Manag. Assoc. 68:377–388. doi:10.1080/10962247.2017.1409294

Marriott, J.M., A. Shober, P. Monaghan, and C. Wiese. 2012. Equine owner knowledge and implementation of conservation practices. J. Extension 50:5RIB4. www.joe.org/joe/2012october/rb4.php?pdf=1 (Accessed 5 Sept. 2018).

Martinson, K., L. Hovda, and M. Murphy. 2007. Plants poisonous or harmful to horses in the North Central United States. University of Minnesota Extension, Minnesota Racing Commission, and College of Veterinary Medicine, Saint Paul, MN. http://pss.uvm.edu/pdpforage/Materials/AnimalDisorders/PlantsPoisonousHorses_Un_Minn.pdf (Accessed 27 Feb. 2019).

Martinson, K., J. Wilson, K. Cleary, W. Lazarus, W. Thomas, and M. Hathaway. 2012. Round-bale feeder design affects hay waste and economics during horse feeding. J. Anim. Sci. 90:1047–1055.

McNeill, J.R., and V. Winiwarter. 2006. Soils, soil knowledge and environmental history: An introduction. In: J.R. McNeill and V. Winiwarter, editors, Soils and societies: Perspectives from environmental history. The White Horse Press, Isle of Harris, U.K. p. 1–6.

Miner, J.R., F.J. Humenik, and M.R. Overcash. 2000. Managing livestock wastes to preserve environmental quality. Iowa State University, Ames, IA. p. 246–248.

Morrison, P.K., P.A. Harris, C.A. Maltin, D. Grove-White, C.F. Barfoot, and C.McG. Argo. 2017. Perceptions of obesity and management practices in a UK population of leisure-horse owners and managers. Journal of Equine Veterinary Science 53:19–29 doi:10.1016/j.jevs.2017.01.006

MWPS. 2005. Horse facilities handbook. MidWest Plan Service. Iowa State University. Ames, IA.

NRAES. 1999. Field guide to on-farm composting. NRAES-114. Natural Resource, Agriculture, and Engineering Service, Ithaca, NY.

NRC. 2001. Nutrient requirements of dairy cattle. 7th rev. ed. National Academy Press. National Research Council, Washington, D.C.

NRC. 2007. Nutrient requirements of horses. 7th rev. ed. National Academy Press. National Research Council, Washington, D.C.

Ostoja, S.M., M.L. Brooks, P.E. Moore, E.L. Berlow, R. Blank, J. Roche, J. Chase, and S. Haultain. 2014. Potential environmental effects of pack stock on meadow ecosystems of the Sierra Nevada, USA. Rangeland J. 36:411–427. doi:10.1071/RJ14050

Pagliari, P.H., and C.A.M. Laboski. 2014. Effects of manure inorganic and enzymatically hydrolyzable phosphorus on soil test phosphorus. Soil Sci. Soc. Am. J. 78(4):1301–1309. doi:10.2136/sssaj2014.03.0104

Pagliari, P.H., and C.A.M. Laboski. 2012. Investigation of the inorganic and organic phosphorus forms in animal manure. J. Environ. Qual. 41:901–910. doi:10.2134/jeq2011.0451

Parvage, M.M., B. Ulen, and H. Kirchmann. 2017. Can organic materials reduce excess nutrient leaching from manure-rich paddock soils? J. Environ. Qual. 46:105–112. doi:10.2134/jeq2016.06.0223

Pernes-Debuyser, A., and D. Tessier. 2004. Soil physical properties affected by long-term fertilization. Eur. J. Soil Sci. 55:505–512. doi:10.1111/j.1365-2389.2004.00614.x

Powers, W.J., E.R. Fritz, W. Fehr, and R. Angel. 2006. Total and water-soluble phosphorus excretion from swine fed low-phytate soybeans. J. Anim. Sci. 84(7):1907–1915. doi:10.2527/jas.2005-656

Pratt, S.E., L.M. Lawrence, T. Barnes, D. Powell, and K. Warren. 1999. Measurement of ammonia concentrations in horse stalls. In: Proc.16th Equine Nutr. Physiol. Symp., Raleigh, NC. p. 334.

Reinemeyer, C.R., and M.K. Nielson. 2013. Handbook of equine parasite control. Wiley-Blackwell, Ames, IA.

Sanjari, G., B. Yu, H. Ghadiri, C.A.A. Ciesiolka, and C.W. Rose. 2009. Effects of time-controlled grazing on runoff and sediment loss. Aust. J. Agric. Res. 47:796–808.

Sharpley, A.N., S.C. Chapra, R. Wedepohl, J.T. Sims, T.C. Daniel, and K.R. Reddy. 1994. Managing agricultural phosphorus for protection of surface waters: Issues and options. J. Environ. Qual. 23:437–451. doi:10.2134/jeq1994.00472425002300030006x

Svanberg, M., C. Finnsgård, J. Flodén, and J. Lundgren. 2018. Analyzing animal waste-to-energy supply chains: The case of horse manure. Renew. Energy 129:830–837. doi:10.1016/j.renene.2017.04.002

Tate, K.W., M.D. Pereira, and E.R. Atwill. 2004. Efficacy of vegetated buffer strips for retaining Cryptosporidium parvum. J. Environ. Qual. 33(6):2243–2251. doi:10.2134/jeq2004.2243

Topliff, D. R. and G. D. Potter. 2002. Comparison of dry matter, nitrogen, and phosphorus excretion from feedlot steers and horses in race/performance training. Written testimony submitted to the United States Environmental Protection Agency, Washington, D.C.

Undersander, D., B. Albert, D. Cosgrove, D. Johnson, and P. Peterson. 2002. Pastures for profit: A guide to rotational grazing. Cooperative Extension Publishing, University of Wisconsin-Extension, Madison, WI.

USDA-NAHM. 2001. Salmonella and the U.S. Horse Population (2001). NAHMS Equine '98. USDA National Animal Health Monitoring Service, Washington, D.C.

USDA-NASS. 2012. Census of agriculture–United States data. USDA National Agricultural Statistics Service, Washington, D.C. https://agcensus.usda.gov/Publications/2012/Full_Report/Volume_1,_Chapter_1_US/st99_1_028_031.pdf (Accessed 27 Feb. 2019).

USDA-NRCS. 1986. Urban hydrology for small watersheds. TR-55. USDA Natural Resources Conservation Service, Washington, DC.

USEPA. 2015. Preventing eutrophication: Scientific support for dual nutrient criteria. EPA– 820–S– 15–001. United States Environmental Protection Agency, Washington, D.C. https://www.epa.gov/sites/production/files/documents/nandpfactsheet.pdf (Accessed 27 Feb. 2019).

Valk, H., J.A. Metcalf, and P.J.A. Withers. 2000. Prospects for minimizing phosphorus excretion in ruminants by dietary manipulation. J. Environ. Qual. 29:28–36. doi:10.2134/jeq2000.00472425002900010005x

van Doorn, D.A., H. Everts, H. Wouterse, and A.C. Beynen. 2004. The apparent digestibility of phytate phosphorus and the influence of supplemental phytase in horses. J. Anim. Sci. 82:1756–1763. doi:10.2527/2004.8261756x

Ward, P.L., J.E. Wohlt, and S.E. Katz. 2001. Chemical, physical, and environmental properties of pelleted newspaper compared to wheat straw and wood shavings as bedding for horses. J. Anim. Sci. 79:1359–1369. doi:10.2527/2001.7961359x

Welsch, D.J. 1991. Riparian forest buffers: Function and design for protection and enhancement of water resources. U.S. Dep. of Agriculture Forest Service, Radnor, PA. Publication # NA-PR-07-91. http://www.na.fs.fed.us/spfo/pubs/n_resource/riparianforests/index.htm (Accessed 26 Feb. 2019).

Westendorf, M.L., T. Joshua, S.J. Komar, C.A. Williams, and R. Govindasamy. 2010. Case study: Manure management practices on New Jersey equine farms. Prof. Anim. Sci. 26:123–129. doi:10.15232/S1080-7446(15)30565-9

Westendorf, M.L. 2008. What is a sacrifice or exercise lot for horses? eXtension and the Livestock Poultry Environmental Learning Center, Kansas City, MO. http://www.extension.org/faq/27505 (Accessed 27 Feb. 2019).

Westendorf, M.L., V. Puduri, C.A. Williams, T. Joshua, and R. Govindasamy. 2013. Dietary and manure management practices on equine farms in two New Jersey watersheds. J. Equine Vet. Sci. 33:601–606 http://www.sciencedirect.com/science/article/pii/S0737080612007848. doi:10.1016/j.jevs.2012.09.007

Westendorf, M.L., and C.A. Williams. 2015. Effects of excess dietary phosphorus on fecal phosphorus excretion and water extractable phosphorus in horses. J. Equine Vet. Sci. 35:495–498. doi:10.1016/j.jevs.2015.01.020

Westendorf, M. L. and C. A. Williams. 2007. Nutrient management on livestock farms: Tips for feeding. FS1064. Rutgers Cooperative Extension. New Jersey Agricultural Experiment Station. Rutgers University, New Brunswick, NJ.

Westendorf, M. L. and U. Krogmann. 2013. Horse manure management: Bedding use. FS537. Rutgers Cooperative Extension. New Jersey Agricultural Experiment Station. Rutgers University, New Brunswick, NJ.

Wheeler, E. and J. S. Zajaczkowski. 2009. Horse stable manure management. FS537. Penn State Cooperative Extension. Pennsylvania State Agricultural Experiment Station. Rutgers University, New Brunswick, NJ.

Wickens, C., C. LaRiche, J. Hinton, M. Lusk, J. Wallace, and V. Harwood. 2017. Manure management Practices and Educational Needs of Florida Small Scale Equine Operations. (Poster abstract). 2017 International Meeting of American Society of Agronomy, Crop Science Society of America, Soil Science Society of America, Madison, WI.

Williams, C.A., and A.O. Burk. 2010. Nutrient intake during an elite level three-day event competition is correlated to inflammatory markers and antioxidant status. Equine Vet. J. 42(Suppl. 38):116–122.

Williams, C.A., C. Urban, and M.L. Westendorf. 2011. Dietary protein affects nitrogen and ammonia excretion in horses. J. Equine Vet. Sci. 31:305–306. doi:10.1016/j.jevs.2011.03.135

Williams, R.E., and D.R. Edwards. 2017. Effects of biochar treatment of municipal biosolids and horse manure on quality of runoff from fescue plots. Trans. ASABE 60:409–417. doi:10.13031/trans.11891

Wolf, A.M., P.J.A. Kleinman, A.N. Sharpley, and D.B. Beegle. 2005. Development of a water-extractable test for manure: An interlaboratory study. Soil Sci. Soc. Am. J. 69:695–700. doi:10.2136/sssaj2004.0096

Wu, Z., L.D. Satter, A.J. Blohowiak, R.H. Stauffacher, and J.H. Wilson. 2001. Milk production, estimated phosphorus excretion, and bone characteristics of dairy cows fed different amounts of phosphorus for two or three years. J. Dairy Sci. 84:1738–1748. doi:10.3168/jds.S0022-0302(01)74609-7

Zhang, J., K.-C. Loh, J. Lee, C.-H. Wang, Y. Dai, and Y.W. Tong. 2017. Three-stage anaerobic co-digestion of food waste and horse manure. Sci. Rep. 7:1269. doi:10.1038/s41598-017-01408-w

Organic Animal Farming and Comparative Studies of Conventional and Organic Manures

Zhongqi He*

Abstract

Organic agriculture emerged as a reaction to the industrialization of agriculture and its associated environmental and social problems. Organic animal (livestock and poultry) farming production is an important part of organic agriculture. There is plenty of research on comparisons of conventional and organic animal production systems on different aspects of sustainability. However, there are limited research publications on the comparison of the manure characteristics and properties between conventional and organic animal farming systems. To better support the organic farming industry and to promote organic manure research, this chapter reviewed the general organic animal farming practices and comparative studies of conventional and organic manures. The limited manure studies have evaluated the difference between conventional and organic manures in three aspects: i) microbiological hazards and antibiotic resistance, ii) greenhouse gas emissions and iii) chemical composition and characterization. Organic manure may show differences from conventional manure, but the observations from these studies are not always consistent. A comparative case study of dairy manure found that difference in spectroscopic features between the two types of manure could be used as the traceable markers in certification and authentification of organic farming management practices. More research on organic manure chemistry would also be helpful in addressing whether conventional manure meets the criteria of organic fertilizers for organic crop farms.

Organic agriculture, sometimes called biological or ecological agriculture, emerged as a reaction to the industrialization of agriculture and its associated environmental and social problems (Röös et al., 2018). In other words, the purposes of organic agriculture are not only for sustainable food security and food safety, but also include caring for and protecting the environment (landscapes, climate, habitats, biodiversity, air and water). While it began at low levels, the global trend of consumer demand for organic products has been extremely positive with an increase of more than three fold since the turn of the century (Reganold and Wachter, 2016). Organic animal (livestock and poultry) farming production is an important part of organic agriculture. It is estimated that global demand for animal source food will be over 50% higher in 2030 compared with 2000 because of growth of the world population, increased incomes, and urbanization, mostly in developing regions (van Wagenberg et al., 2017). Organic animal production features cultural, biological, and mechanical methods to ensure environmentally safe and chemical residue-free foods, along with high animal welfare standards. In contrast, conventional animal farming is associated with technologies for increased productivity, such as high yielding breeds, modern feeding techniques and veterinary health products, and synthetic pesticides (van Wagenberg et al., 2017). There is considerable

USDA-ARS Southern Regional Research Center, 1100 Robert E. Lee Blvd. New Orleans, LA 70124. *Corresponding author (Zhongqi.he@usda.gov)

doi:10.2134/asaspecpub67.c9

Key words: Conventional manure, Organic agriculture, Organic manure, Sustainability, Traceable marker

Animal Manure: Production, Characteristics, Environmental Concerns and Management. ASA Special Publication 67. Heidi M. Waldrip, Paulo H. Pagliari, and Zhongqi He, editors.

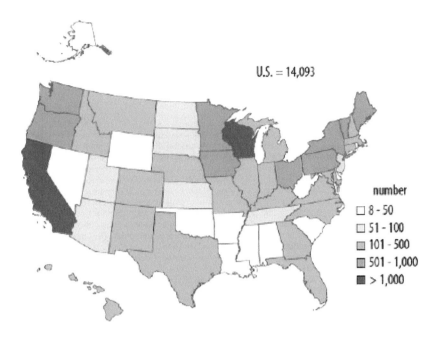

Fig. 1. Distribution of organic farms (certified and exempt) in the United States per USDA 2014 Organic Survey (USDA, 2015).

research on comparison of conventional and organic animal production systems available as reviewed by van Wagenberg et al. (2017), Smith-Spangler et al. (2012), De Vries et al. (2015), and Hovi et al. (2003). However, there are limited research publications on the comparison of the animal manure characteristics and properties between conventional and organic animal farming systems. It is not uncommon that the term "organic manure" is used to refer to manure samples collected from conventional animal farms (Abbasi et al., 2007; Antonious, 2018; Olesen et al., 2007; Yadav et al., 2013). Furthermore, the practice of application of animal manure into cropland is typical termed as an "organic amendment" so that the relevant crop products are called "organic" (Lynch et al., 2012; Sullivan et al., 2014; Wander et al., 2007; Wild et al., 2011). On the other hand, with the advance in manure and nutrient chemistry (He, 2011, 2012; He and Zhang, 2014; He et al., 2016), there is awareness of characteristic differences between the manure samples from conventional and organic livestock production farms; the terms "conventional manure" and "organic manure" are used to distinguish the two types of animal manure from each other (He and Waldrip, 2015; He et al., 2009; Oelofse et al., 2013). Therefore, for promoting organic manure research, this chapter reviewed the general organic animal farming practices and limited comparative studies of conventional and organic manures.

Organic Animal Farming Practices

Organic Animal Farming

The U.S. Department of Agriculture (USDA) (USDA, 2011, p. 1) defines "organic" as "a labeling term that indicates that the food or other agricultural product has been produced through approved methods. These methods integrate cultural, biological, and mechanical practices that foster cycling of resources, promote ecological balance, and conserve biodiversity. Synthetic fertilizers, sewage sludge, irradiation, and genetic engineering may not be used". For example, the organic dairy farming should include the requirements in management practices as: (i) zero grazing is not permitted, and the maximum livestock density is limited to two livestock units per hectare for dairy cows; (ii) at least half of the total housing floor must be solid and not of slatted or grid construction; (iii) dry litter must be given in the rest area; (iv) livestock must be fed on organically produced feedstuffs and 60% of the ruminant diet must come from forage; and (v) the use of synthetic amino acids and growth promoters is forbidden, as is the use of veterinary drugs in the absence of illness (Novak and Fiorelli, 2010).When appropriate medications and antimicrobial treatments must be applied to restore sick animals to health (USDA, 2011), these animals will lose their organic status (Sato et al., 2005b).

By the end of 2014, 43.7 million hectares are under organic agricultural management worldwide (Willer and Lernoud, 2016). In 2014, the United States had 14,093 organic farms with 3.7 million acres, producing $5.5 billion in organic products (Fig. 1) (USDA, 2015). Of these 14,093 farms, 12,634 were certified organic farms and 1459 were exempt from certification because they grossed less than $5,000 annually from organic sales. The top organic sector in 2014 sales was livestock and poultry products, primarily milk and eggs, followed by vegetables grown in the open and fruits, tree nuts, and berries (Table 1). Of the $5.5 billion in 2014 organic sales, $3.3 billion, or 60%, came from the sale of crops, 28% came from livestock and poultry product sale (primarily milk and eggs), and 12% came directly from sales of livestock and poultry. While the global data of annual amounts of animal manure are available (He et al., 2016), no such global data are available specifically for organic manure. Figure 2 shows the case of Denmark's organic manure production per the data reported by Oelsfse et al. (2013). Organic manure accounts only about 0.6% to 9.7% in the four types of manure with an average of 4.7%.

Table 1. Organic sales ($ millions) by sector and top commodities in the United States per United States 2014 organic survey (USDA, 2015).

Sector	Sale
Livestock and poultry products	1,504
Milk	1082
Eggs	420
Vegetables in the open	1,250
Lettuce	264
Spinach	117
Broccoli	79
Carrots	69
Sweet potatoes	68
Fruits, tree nuts, and berries	1,032
Apples	250
Grapes	195
Strawberries	89
Blueberries	61
Almonds	32
Field crops	718
Corn for grain	155
Hay	139
Wheat	102
Soybean	72
Rice	35
Livestock and poultry	660
Broiler chickens	372
Other cattle	131
Milk cows	69
Turkeys	50
Beef cows	16
Mushrooms	109
Vegetables under protection	76
Others	106
US total	5,455

Table 2. Selected comparative studies of microbiological hazards and antibiotic resistance of conventional manure and organic manure.

Reference	Study region	Manure type, farm and sample	Objective	Main observation
Alali et al. (2010)	North Carolina, USA	Broiler fecal droppings C: 4–40 O: 3–30	Prevalence of Salmonella	Prevalence of fecal Salmonella lower in organic birds than in conventionally raised birds
Berge et al. (2010)	California, Oregon, and Washington, USA	Cattle feces, C:11–607 O:7–345	Multiple antimicrobial resistance	No differences in resistance profiles of isolates
Bonde and Sørensen (2012)	Denmark	Pig feces C:11–449 O:11–534	Presence of Salmonella	The serological test result was a significant predictor of Salmonella shedding at slaughter in individual pigs from conventional systems, but not in pigs from organic systems
Cho et al. (2006a)	Minnesota, USA	Dairy feces and piled manure C:45–1750 O:13–458	Prevalence of shiga toxin-encoding bacteria	Their prevalence greater in organic farms compared to conventional farms
Cho et al. (2006b)	Minnesota, USA	Dairy feces C: 18–271 O: 8–166	Occurrence of E. coli O157 s	Organic farming may have an impact on fecal shedding of *E. coli* O157 which was isolated more frequently from organic farms than from conventional farms..
Franz et al. (2007)	The Netherlands	Dairy manure C:9–9 O:16–16	Prevalence of shiga toxin-producing E. coli genes	Relatively more conventional farms positive for shiga toxin-producing genes
Gerzova et al. (2015)	Sweden, Denmark, France and Italy	Pig feces C:N/A-233 O: N/A-235	Antibiotic resistance genes, microbiota composition	Geographical location's impact > farm's organic or conventional status.
Halbert et al. (2006a)	Michigan, Minnesota, New York, Wisconsin, USA	Dairy feces C:N/A-912 O: N/A-304	Use of antimicrobials	Conventional: more tetracycline resistant isolates ($p < 0.01$)
Halbert et al. (2006b)	Michigan, Minnesota, New York, Wisconsin, USA	Dairy feces C: N/A-570 O: N/A-460	Reduced tetracycline susceptibility of Campylobacter	Conventional: more tetracycline resistant isolates ($p = 0.007$)
Johnston (2002)	Minnesota, USA	Cow feces C:5–30 O:5–30	Antibiotic resistance of bacteria	No difference in minimum inhibitory concentration
Kassem et al. (2017)	Ohio, USA	Poultry feces C: 3–180 O: 3–180	Prevalence of antimicrobial-resistant Campylobacter	Organic farming can potentially impact the emergence of Campylobacter
Kuhnert et al. (2005)	Switzerland	Dairy feces C: 60–485 O: 60–481	Prevalence of Shiga toxigenic Escherichia coli	O157:H7 present in more organic farms than conventional farms, but no statistically significant differences
McKinney et al. (2010)	Western USA	Poultry, pig and cattle manure lagoons C: 6–63 O: 2–87	Use of antimicrobials, antibiotic resistance genes	A marginal benefit of organic and small dairy operations also was observed compared to conventional and large dairies, respectively
Nulsen et al. (2008)	New Zealand	Swine feces C: 3-N/A O: 1-N/A	Antibiotic resistance	A higher level of antimicrobial resistance in the E. coli and Enterococcus spp. cultured from the feces of pigs compared with an organic farm which used no antibiotics
Österberg et al. (2016)	Denmark, France, Italy and Sweden	Pig feces C: 112–299 O: 111–299	Antibiotic Resistance	Resistance in intestinal E. coli was less common in organic than in conventional pigs
Ray et al. (2006)	Michigan, Minnesota, New York, and Wisconsin	Dairy feces C: 97–97 O:32–32	Antimicrobial susceptibility of Salmonella	Salmonella isolates with conventional farming more resistance to streptomycin and sulfamethoxazole
Sapkota et al. (2014)	Mid-Atlantic USA	Poultry litter C: 5–10 O: 5–10	Prevalence of antibiotic-resistant Salmonella	Lower levels of Salmonella with organic farms
Sato et al. (2005b)	Wisconsin, USA	Dairy feces C: 30–60 O:30–60	Parasite burden	Higher parasite burden on organic farms
Sato et al. (2005a)	Wisconsin, USA	Dairy feces C: 30–60 O:30–60	Antimicrobial susceptibility of Escherichia coli	E. coli isolates from organic dairy herds have significantly higher prevalences of resistance to 7 antimicrobials
Van Overbeke et al. (2006)	Belgium	Poultry feces C: 11–30 O: 9–30	Antibodies against poultry diseases	No differences in prevalence of Salmonella. Organic flocks are more frequently infected with Campylobacter than conventional flocks
Walk et al. (2007)	Wisconsin, USA	Dairy feces C: 30–300 O: 30–300	Antibiotic susceptible E. coli populations	Dairy farming practices have a proportionately large, negative effect on the prevalence of multidrug resistant strains.

Comparison of Characteristics of Animal Manure of Conventional and Organic Farming Practices
Microbiological Hazards and Antibiotic Resistance

In conventional animal farming, especially with confined animal feeding operations, veterinary pharmaceuticals are frequently used to maintain healthy and productive livestock and poultry (Song and Guo, 2014). More than 70% of the consumed veterinary pharmaceuticals are antibiotics-chemicals that can inhibit the growth of other microorganisms even at extremely low concentrations. Organic animal production systems are featured with veterinary pharmaceuticals-free farming practices. Thus, most publications (21 at least) on comparison of characteristics of conventional and organic manures are with evaluation of the impact of organic farming practices on microbiological hazards and antibiotic resistance of animal health and farm environments (Table 2). Most of them collected fecal samples, but a few studies also collected manure and litter samples. Campylobacter and Salmonella are pathogenic microorganisms transmissible to humans. The prevalence of the two representative microorganisms were comparatively studied most often. By fecal examination for the presence of worm eggs, Van Overbeke et al. (2006) found no significant differences in prevalence of Salmonella between organic and conventional broilers at slaughter. In contrast, Campylobacter infections at slaughter were significantly higher in organic flocks. On the other hand, Alali et al. (2010) reported that Salmonella prevalences in poultry fecal samples were 5.6% (10/180) and 38.8% (93/240) from organic and conventional farms, respectively. In other words, the prevalence of fecal Salmonella was lower in certified-organic birds than in conventionally raised birds. Bonde and Sørensen (2012) examined the presence of Salmonella in organic, conventional outdoor and indoor finishing pig herds in a Danish survey with participation of 34 herds. They found no significant differences between farming systems with regards to Salmonella shedding on-farm and at slaughter and no difference between systems in seroprevalence. However, their serological test result was a significant predictor of Salmonella shedding at slaughter in individual pigs from conventional systems, but not in pigs from organic systems.

E. coli O157 is frequently studied as the serotype of high-toxin producing *Escherichia coli* (STEC), an important group of zoonotic human pathogens. The STEC strains are generally carried asymptomatically by cattle and shed in their feces, serving as a means of maintenance and spread of these pathogens among cattle herds (Franz et al., 2007). Cho et al. (2006a, 2006b) collected fecal samples from dairy cattle and their farm environment and from manure piles at Minnesota county fairs from 2001 to 2002 for detection and isolation of shiga toxin-encoding bacteria (STB), especially STEC. Their results suggested that the STB prevalence was greater in organic farms compared with conventional farms, especially at the individual sample level. Organic farming seemed to have an impact on fecal shedding of *E. coli* O157, which was isolated more frequently from organic farms than from conventional farms. Franz et al. (2007) collected manure samples from 16 organic and 9 low-input conventional (LIC) Dutch dairy farms to determine the natural prevalence of the *E. coli* O157-specific rfbE gene and STEC virulence genes stx1 (coding for Shiga toxin 1), stx2 (coding for Shiga toxin 2), and eaeA (coding for intimin) in manure. The prevalence of rfbE was higher at organic farms (61%) than at conventional farms (36%). However, relatively more conventional farms were positive for

Table 3. Comparative studies of emissions of conventional manure and organic manures.

Reference	Study region	Manure type, farm and sample	Objective	Main observation
Sneath et al. (2006)	France, England, the Netherlands	Dairy manure C: 3–3 O: 3–3	Greenhouse gas emissions	Slightly lower methane emission rate with organic farms
Thomassen et al. (2008)	The Netherlands	Dairy manure C: 21-whole O: 11-whole	Life cycle assessment of milk production	Higher ammonia, methane, and nitrous oxide emissions occur on farm per kilogram organic milk than for conventional milk
Weiske et al. (2006)	Five European regions	Dairy manure C:-7-whole O: 7-whole	Greenhouse gas emissions	Mitigation potential of conventional farms > organic farms

STEC virulence genes eaeA, stx1, and stx2, which can potentially form a highly virulent combination. Kuhnert et al. (2005) studied the prevalence of STEC, and specifically O157:H7, in Swiss dairy cattle by collecting feces from approximately 500 cows from 60 organic farms and 60 integrated conventional farms. Serotypes of high-toxin producing *E. coli* were detected in all farms; O157:H7 were present in 25% of organic farms and 17% of conventional farms. However, in general, there were no significant differences between the two farm types concerning prevalence or risk for carrying STEC or O157:H7.

In conventional farms, veterinary pharmaceuticals, mainly antibiotics, are frequently administered to animals and their residues are excreted to manure. There are over 150 antibiotics in use today, of which more than 90% are natural products of bacteria and fungi, as well as semisynthetic modifications of natural products (Song and Guo, 2014). Because no antibiotics are applied in normal organic animal farming, it is interesting to compare the differences in antibiotic resistance between the farming practices. Some studies evaluated one representative antibiotic, such as Penicillin G (Johnston, 2002). Other studies evaluated more antibiotics (Table 2). For example, Sato et al. (2005a) tested 17 antimicrobials by means of a microbroth dilution test. They compared *E. coli* in cattle fecal samples for susceptibility of *E. coli* isolated from fecal samples to these antimicrobials between organic and conventional dairy farms. *Escherichia coli* solates from organic dairy herds had significantly lower prevalence of resistance to 7 antimicrobials; however, prevalence of resistance was not significantly different for the 10 other antimicrobials. Walk et al. (2007) investigated the influence of antibiotic selection on genetic composition of *E. coli* populations in fecal samples from 30 conventional and 30 organic dairy farms. Their results suggested that organic farming practices not only change the frequency of resistant strains but also impact the overall population genetic composition of the resident *E. coli* flora. Specifically, phylogroup B1 strains with low multidrug resistance were significantly associated with organic farms, implying organic farming had a proportionately large, negative effect on the prevalence of multidrug-resistant strains. McKinney et al. (2010) determined the prevalence and examined the behavior of tetracycline and sulfonamide resistance genes in a broad cross-section of livestock lagoons within the same semiarid western watershed of the United States. In all lagoons, sulfonamide resistance genes were generally more recalcitrant than tetracycline resistance genes, and there was a marginal benefit of organic operations compared with conventional dairy farming. With observations of other studies on poultry and pigs (Berge et al., 2010; Gerzova et al., 2015; Kassem et al., 2017), a general trend is that microbial organisms more often showed higher multidrug resistance in conventional systems while organic farming demonstrated equal or lower likelihood of antibiotic resistance.

Greenhouse Gas Emissions

There were only three publications found, and all were on dairy farms (Table 3). Emission of methane occurs in two ways: during enteric fermentation of a cow and from manure management. Thomassen et al. (2008) compared the integral assessment of the environmental impact of conventional and organic milk production systems based on 10 conventional and 11 organic farms in the Netherlands. For the organic system, they assumed an emission during fermentation of 128 kg CH_4 cow^{-1}yr^{-1} whereas for the conventional system 113 kg CH_4 cow^{-1} yr^{-1}. Emission from manure management was 0.0018 kg CH_4 kg^{-1} of manure yr^{-1} for liquid manure and 0.00037 kg CH_4 kg^{-1} of manure yr^{-1} for solid manure production in animal houses. On the farm level, animals and manure contributed 68% and managed soils 24% in the conventional system; in the organic system animals and manure contributed 76% and managed soils 16%. For volatilization of ammonia from manure, manure in raising stable, outside storage, and during pasture grazing contributed 52% and during application of fertilizer 41% in the conventional system whereas manure in stable, storage, and during pasture contributed 62% and during application of fertilizer 30% in the organic system.

Sneath et al. (2006) compared greenhouse emissions from manure storage on organic and conventional dairy farms in France, England, and the Netherlands. They monitored the emission rates of methane and nitrous oxide from manures in covered and uncovered slurry stores and farmyard manure heaps. Methane emission rate from the uncovered slurry stores on the conventional farm and the organic farm ranged from 14.4 to 49.6 and from 12.4 to 42.3 g C m^{-3} d^{-1} with the mean emission rates of 35 and 26 g C m^{-3} d^{-1}, respectively, On the other hand, nitrous oxide emission was detected close to zero on both farms. In the meantime, methane emission from the indoor organic farmyard manure in summer was 17.1 g C m^{-3} d^{-1} and the nitrous oxide emission was 411 mg N m^{-3} d^{-1}. Weiske et al. (2006) analyzed mitigation potentials of greenhouse gas emissions in conventional and organic dairy farming with representative dairy model farms in five European regions. They calculated the potential of emission reduction per kg milk to compare organic and conventional production systems and to investigate region and system specific differences. The simulated greenhouse gas emissions ranged from 1.3 to 1.7 kg CO_2–eqivalent kg^{-1} milk for conventional, and from 1.2 to 2.0 kg CO_2–eqivalent kg^{-1} milk for organic dairy farms. On average for all European dairy regions, the emissions from organic production systems were approximately 1.6 kg CO_2–eqivalent kg^{-1} milk and thus 10% higher compared to the conventional model farms (1.4 kg CO_2–eqivalent kg^{-1} milk). However, the authors cautioned that it is not a general conclusion, since organic farms showed both the highest and lowest emissions. Frequent removal of manure from animal housing into outside covered storage reduced farm emissions by up to 7.1%. Anaerobic digestion of manure for biogas production could be a very efficient and cost-effective option to reduce greenhouse gas emissions. With all potential measures and strategies, it is estimated a mitigation potential of 50 Mt CO_2–eqivalent Yr^{-1} for conventional and 3.2 Mt CO_2–eqivalent Yr^{-1} for organic specialized dairy farms, scaled up to the level of the whole European regions.

Table 4. Selected comparative studies of chemical composition of conventional manure and organic manures.

Reference	Study region	Manure type, and farm and sample	Objective	Main observation
Gustafson et al. (2003)	Northern Sweden	Dairy manure and urine C: 1–9 (m), 4 (u) O: 1–8 (m), 4 (u)	Concentrations of K, P, and Zn	No clear differences between the two farming systems
Gustafson et al. (2007)	Northern Sweden	Dairy manure and urine C: 1–24 (m), 20 (u) O: 1–24 (m), 20 (u)	Concentrations of Ca, Cu, K, Mg, Mn, N, P, S and Zn	Less differences in element concentrations between the two systems than forage
He et al. (2009)	Maine, USA	Dairy manure C: 1–1 O: 1–1	P forms	Organic manure contained more Ca and Mg species of P
He and Ohno (2012)	Maine, USA	Dairy manure C: 7–7 O:15–15	Organic matter characterization	Organic manure contained more hydrophobic aliphatic groups, less soluble C and N compounds
He and Wang (2012)	Maine, USA	Dairy manure C: 4–4 O: 5–5	Pyroslysis-gas chromatography characterization of water extracts	Hormone/antibiotic-related peaks may be used for monitoring the usage of chemical drugs
He et al. (2015)	Maine, USA	Dairy manure C: 4- 4 O: 15–4	Traceable markers of organic matter	A forage-related ingredient could be a traceable marker of organic farming
Waldrip et al. (2012)	Maine, USA	Dairy manure C: 1–1 O: 13–13	Effects on soil phosphatases and P availability	In a manner similar to conventional manure
Wang et al. (2011)	Maine, USA	Dairy manure C: 4–4 O: 5–5	Pyroslysis-gas chromatography characterization of whole manure	No obvious difference in chromatogram profiles

Chemical Composition

Per several farm-scale nutrient budget surveys, Watson et al. (2002) reported that mean N content was 5 kg N Mg^{-1} on a fresh weight basis with a range of 2 to 10 N Mg^{-1} for cattle farm yard manure and 2.5 kg N m^{-3} with a range of 1.1 to 4.1 kg N m^{-3} for cattle slurry collected in organic farming systems. These values were about 15% lower than manure from conventional systems. Steineck et al. (2000) surveyed 80 different farms all over Sweden. Based on 40 samples of solid manure and urine from organic and conventional dairy farms, they found that the DM concentrations of Zn in organic manure and of K in organic urine were significantly lower than the corresponding values from conventional farms. Table 4 listed eight publications of two case studies of chemical composition of organic manure conducted in Northern Sweden and Maine. Gustafson et al. (2003) presented one-year data of flows and balances of K, P, and Zn at a barn level for an organic and a conventional farming system on the coastal plain of Northern Sweden. The concentrations of K, P, and Zn in feeds, bedding material, milk, manure, and urine were all analyzed. The concentrations of K, P and Zn were 23.8, 6.2 and 0.143g kg^{-1} of dry matter in organic solid manure, and 22.1, 6.3 and 0.154g kg^{-1} of dry matter for conventional solid manure. The concentrations of K, P and Zn in urine were 243, 1.4, and 0.096 g L^{-1} of organic dairy, and 252, 1.5, and 0.082 g L^{-1} of conventional dairy. While minor differences in the data were observed, they were not statistically significant ($p > 0.05$). On a follow-up study, Gustafson et al. (2007) calculated barn balance of Ca, Cu, K, Mg, Mn, N, P, S and Zn in the two farming systems based on three-year data. The average concentrations of these elements in manure and urine were generally in agreement with those in the literature (Safley et al., 1985; Schroder et al., 2011; Steineck et al., 2000) and without significant difference between the two systems. Whereas Cu, Mn, P, and Zn were excreted in feces but in very small amounts in urine, Gustafson et al. (2007) found higher coefficients of variation for Ca, Cu, Mn, P, and Zn in urine probably due to contamination by manure. In summary, the element content of manure and urine calculated as [inputs – milk] would have underestimated the amount of Cu, Mn, and Zn, and overestimated the amount of K and N in manure. The differences between the organic and conventional systems related more to differences in forage to protein-rich feed ratio and home-grown to purchased ratio, and less to differences in element concentrations of the feed ingredients.

The Maine case study was focused on molecule-level comparison of chemical composition of organic and conventional manures (Table 4). Using solution and solid state ^{31}P NMR analysis of one organic and one conventional dairy manure sample, He et al. (2009) reported that organic dairy manure contained about 10% more inorganic phosphate than conventional dairy manure. Whereas organic dairy manure did contain slightly more phytate P, it contained 30 to 50% less monoester P than conventional dairy manure. Conventional dairy manure contained relatively higher contents of soluble inorganic P species and stable metal phytate species. In contrast, organic dairy manure contained more Ca and Mg species of P. Waldrip et al. (2012) investigated the effects of organic manure on three phosphatase activities, available soil P, and plant growth in a greenhouse study with sorghum-sudangrass (*Sorghum bicolor* subsp. *drummondii*). They found no difference in biomass production by sorghum-sudangrass plants fertilized with equivalent N rates from 13 organic manures, one conventional manure, or inorganic N fertilizer. The available soil P level and phosphatase activities were also similar each other between the two types of manure, but different from that soil receiving N fertilizer. Therefore, Waldrip et al. (2012) concluded that best management practices and application rates typical for conventional manure are likely to be also appropriate for manures from under organic management.

He and Ohno (2012) comparatively characterized 15 organic and seven conventional dairy manure samples by Fourier transform infrared (FT-IR) and fluorescence spectroscopies. They found no clearly distinct value ranges in whole and water-extractable organic matter between the two types of dairy manure with respect to C and N contents and FT-IR and fluorescence spectral features. However, based on the average values, the whole organic manure contained less soluble C and N compounds but more hydrophobic aliphatic groups on a dry weight basis. The soluble organic matter in organic samples contained more stable humic- and lignin-related components and less amino/protein N-related components. He and Ohno (2012) attributed these differences to more forage feedstuffs in organic dairy farming management and more protein additives in conventional dairy feedstuffs. Thus, He et al. (2015) further characterized these manure samples by ultraviolet-visible and solid state ^{13}C NMR spectroscopy. Solid state ^{13}C NMR analysis provided more information on manure composition than ultraviolet-visible spectra. Their data demonstrated that the ^{13}C NMR characteristics of triple peaks around 30 ppm were more evident in organic samples than in their conventional dairy manure samples

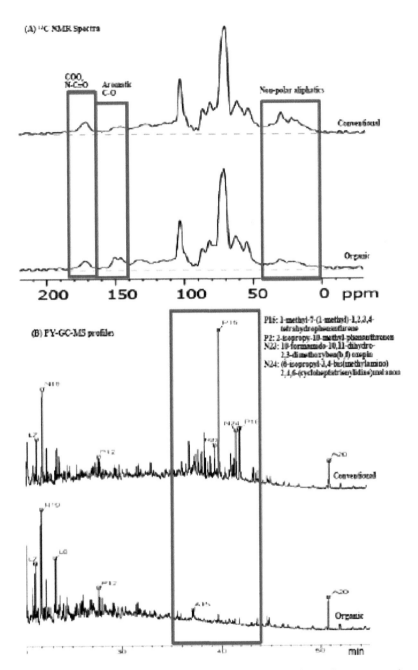

Fig. 2. Solid State ^{13}C NMR spectra (A) of whole conventional and organic manures and total ion current profiles of PY-GC–MS (B) of their water extracts. The red rectangles with relevant chemical compositions are highlighting the differences in spectral features between the two types of manure. Compiled per He and Wang (2012) and He et al. (2016, 2015).

(Fig. 2). They assigned that this ^{13}C NMR peak feature to the waxes and cuticular polysaccharides in manure, residues from forage stuffs. The plant cuticular matter is in high concentrations in forages and less digestible than polysaccharides and proteins that are prevalent in grains and other feedstuffs commonly used in conventional diet. Thus, examination of the relative intensity of cuticular matter–relevant triple peaks would serve as a scientific evidence for appropriate feeding in organic dairy farming practices. In addition, the ^{13}C NMR peaks around 140 to 184 ppm associated to aromatic syringyl units (hardwood), guaiacyl unit (softwood) and/or peptides could be used to identify proteinaceous dietary ingredients and/or the bedding materials (sawdust) (He et al., 2015) (Fig. 2).

Analytic pyrolysis gas chromatography–mass spectroscopy (PY-GC–MS) was also used to examine the whole and water extracts of the two types of dairy manure (He and Wang, 2012; Wang et al., 2011). While there are no obvious differences in the total ion chromatogram profiles between the manures (Wang et al., 2011), differences were observed in the water extracts' profile around 40 min of elution time (He and Wang, 2012) (Fig. 2). There are more peaks observed in the region in the conventional samples than in the organic sample. The two highest peaks were assigned to three-ring compounds: 1-methyl-7-(1-methyl)-1,2,3,4-tetrahydrophenanthrene (P15) and 2-isopropy-10-methyl-phenanthrene (P16) which could be derived from steroid hormones, antibiotic-related metabolites, or derivatives produced during animal or microbial metabolism or the analytical pyrolysis process itself. Thus, the absence of the peak clusters in the pyrolysis–gas chromatograms of organic manures could be indicative that pharmaceuticals were not in use in organic dairy farming.

Implication of Organic Manure Studies on Organic Farming Management and Future Studies

Organic Practices and Certification

Organic farmers are increasingly turning to certified organic farming systems as a way to provide verification of production methods to capture high-value markets and premium prices and boost farm income. As the certification standards continue to evolve with changing technologies and socioecological conditions, some requirements are based on scientific evidence, whereas others are driven by ideology (Reganold and Wachter, 2016). The characteristic ^{13}C NMR peaks and PY-GC–MS peaks in the Maine case study (Fig. 2) may be used as traceable marker for monitoring the usage of forage and chemical drugs for cows in organic farming. Indeed, based on the limited (five) organic manure samples examined, it was found that one sample (OD7) was more similar to conventional samples than to other organic manures in both solid ^{13}C NMR spectra and PY-GC–MS total ion current profiles (He and Wang, 2012; He et al., 2015). Thus, He and Waldrip (2015) further calculated the statistical significance between organic and conventional manures with or without OD7 (Table 5). The *P* values of the significance levels in the difference between organic and conventional manures improved to < 0.05 in three forage and protein feed regions. The *P* values of the significance levels of aromatic region (140–164 ppm) related bedding materials improved from 0.101 to 0.0603. This result strongly suggested OD7 is an abnormal organic manure sample. He et al. (2015) proposed three possible explanations on the difference

Table 5. Relative and average abundance (%) of C functional groups in organic (OD) and conventional (CD) dairy manures obtained by integrations of solid state ^{13}C NMR peak areas. Reprinted from He and Waldrip (2015).

Chemical shift (ppm) range	0–46	46–60	60–94	94–108	108–140	140–164	164–184	184–220	0–108	108–164
Assignment	Nonpolar aliphatics	NCH & OCH$_3$	O-Alkyl C	O-C-O anomerics	Aromatic C-C & C-H	Aromatic C-O	COO & N-C=O	ketone, quinone, aldehyde	Total aliphatic	Total aromatic
OD1	14.1	8.7	46.5	11.2	10.2	3.9	4.4	1.0	69.3	14.1
OD2	15.2	9.2	45.3	10.9	9.6	4.4	5.0	0.3	69.7	14.0
OD7	9.1	9.2	51.7	11.2	11.0	5.3	2.2	0.4	70.0	16.3
OD13	16.4	8.7	45.7	11.3	8.3	4.3	5.0	0.4	70.8	12.6
Average	13.7	9.0	47.3	11.2	9.8	4.5	4.2	0.5	70.0	14.3
Average (-OD7) †	15.2	8.9	45.8	11.1	9.4	4.2	4.8	0.6	69.9	13.6
SD	3.2	0.3	3.0	0.2	1.1	0.6	1.3	0.3	0.6	1.5
SD (-OD7) †	1.2	0.3	0.6	0.2	1.0	0.3	0.3	0.4	0.8	0.8
CD1	9.2	9.7	49.3	11.4	10.6	6.0	3.3	0.5	68.2	16.6
CD2	10.1	9.8	51.5	11.7	9.0	4.3	3.2	0.4	71.4	13.3
CD3	10.0	10.3	47.1	10.3	10.6	6.4	4.1	1.2	67.4	17.0
CD4	10.6	9.8	49.0	10.5	9.7	5.4	4.4	0.7	69.4	15.1
Average	10.0	9.9	49.2	11.0	10.0	5.5	3.8	0.7	69.1	15.5
SD	0.6	0.3	1.8	0.7	0.8	0.9	0.6	0.4	1.7	1.7
P-value (P > F, OD vs. CD)	0.060	0.003	0.311	0.636	0.781	0.101	0.602	0.492	0.394	0.313
P-value [OD (-OD7) vs. CD] †	< 0.001	0.005	0.028	0.718	0.842	0.063	0.042	0.652	0.481	0.131

† Data calculated without OD7's.

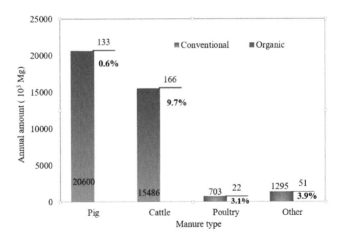

Fig. 3. Production of animal manure, from conventional and organic farms, in Denmark. Other type includes manure from fur animal manure, horses, sheep, deer, and goats. Values with each column show the amount of that type of manure. Percentages show the corresponding proportions of organic manure. Compiled per Oelsfse et al. (2013).

in traceable marker components between OD7 and other three organic samples: i) OD7 was from a conventional sample, but was mislabeled as "organic"; 2) OD7 was from an organic farm that contained cows under medical treatments permitting use of drugs for short periods, perhaps with less forage feedstuff, to restore an animal to health; and 3) OD7 was from an organic OD farm which had not observed USDA-approved organic farming practices. The first cause (mislabeled sample) could be attributed to a simple, correctable experimental error. For the second case (short-term use of nonorganic medicinals), the organic dairy status should be temporarily suspended for this farm using USDA organic farming guidelines. The third scenario was related to organic farming management misconduct and should be investigated and monitored to ensure compliance to the organic standards. Thus, this case study demonstrated that the differences in the PY-GC–MS profiles and solid state ^{13}C NMR spectra should help prevent misconduct and could provide a scientific basis for the evaluation and authentication of organic dairy farming programs.

While there is only one case study available, the further studies should be focused on surveying the organic animal manure production and utilization in regional and national scales. In other words, organic and conventional manure samples should be collected in regional and national scales and comparatively characterize them accordingly with wet chemistry and advanced instrumental analysis. It is also recommended to conduct guided feedstuffs–manure studies to confirm the link of animal characteristics to the diet composition and/or specific management practices.

Animal Manure Utilization in Organic Crop Agriculture

Soil fertility in organic farms is largely managed through efficient on-farm cycling of nutrients and with cropping systems that include legumes in rotation (Lynch et al., 2012). It is reported that 67% of all organic farms are applied green or animal manures, which are not necessarily from organic farms (USDA, 2011). For example, on the Prairies and eastern Canada where organic farms typically do not have livestock and primarily engaged in crop production, the sources of organic livestock manure are limited. Commercially-produced organic amendments (e. g., composts and pelletized dehydrated poultry manure) are options for supplemental N for organic crop producers (Alam et al., 2018). Thus, because the number one production practice of organic farms is the use of outsourced green or animal manures, organic fertilizer purchases rank fifth in production expenses after i) feed, ii) labor, iii) supplies and repairs, and iv) cash, rent, and lease (Verlinden et al., 2017).

Not only in North America, but also globally, organic crop production is often dependent on animal manure imported from conventional animal farms (Knudsen et al., 2011; Oelofse et al., 2011). The data in Fig. 3 clearly indicated that the supply of organic manure is in shortfall. On the other hand, nonorganic manure input from conventional farms has been questioned for concerns of the sustainability and environmental impact featured organic agriculture (He et al., 2016; Oelofse et al., 2011). Under the current USDA national organic program, manure fertilizer from conventional animal farms can be used in organic crop production (USDA, 2011, 2018). Under the European Union regulation (Regulation, 2007), it is permitted to the use of 170 kg N ha^{-1} from animal manure. However, farmers must provide documentation for the need to use manure from a conventional source. As an extreme case, the organic sector in Denmark has recently decided to gradually phase out, and ultimately ban, the use of conventional manures and straws in organic crop agriculture by 2021 (Knudsen et al., 2014; Oelofse et al., 2013). With the controversial approaches, more scientific information is needed on the differences in characteristics between conventional and organic manures so that relevant regulatory agencies, advocates, and growers would make evidence-based decisions in considering the suitability of different manure types for use in organic systems. While trace metals and pharmaceuticals residues are concerned, limited amount of additional treated conventional manure may be used in organic crop agriculture due to limited available organic manure (Alam et al., 2018). For example, composting (Qian et al., 2018; Selvam and Wong, 2017; Wallace et al., 2018) may be applied to degrade antibiotics in conventional manure to meets the criteria of organic fertilizers for organic crop farms.

Conclusions

There are numerous studies on characterization of animal manure from conventional animal farms (conventional manure). However, there are only a few studies on animal manure from organic animal farms (organic manure) conducted to support the organic farming industry. The limited studies have comparatively evaluated the differences between conventional and organic manures in three aspects: i) microbiological hazards and antibiotic resistance, ii) greenhouse gas emissions and iii) chemical composition and characterization. The observations from these studies are not always consistent, suggesting more research is needed for better understanding of organic manure and impact by organic farming practices. Future research should be focused on producing management tools whereby traceable marker components in manure are used to certify or authenticate the relevant animal farms that are managed according to guidelines and requirements for specific organic farming practices (e.g., natural-type, high-forage, veterinary pharmaceutical-free systems). As the practice of application of conventional manure into cropland is traditionally termed as an "organic amendment", future work should also address whether conventional manure with certain additives (e.g., pharmaceuticals), with treatment or not, meets the criteria of organic fertilizers for organic crop farms. Such studies would be helpful in management of conventional manure used in organic agriculture, and may also increase the value of organic manure. In addition, comparative evaluation of the long-term impact of conventional and organic manures on soil salinity, trace elements, and heavy metals are needed as their higher concentrations may harm crop yield and may have detrimental effects on soil biology such as earthworms (Nazarizadeh et al., 2017).

References

Abbasi, M.K., M. Hina, A. Khalique, and S.R. Khan. 2007. Mineralization of three organic manures used as nitrogen source in a soil incubated under laboratory conditions. Commun. Soil Sci. Plant Anal. 38:1691–1711. doi:10.1080/00103620701435464

Alali, W.Q., S. Thakur, R.D. Berghaus, M.P. Martin, and W.A. Gebreyes. 2010. Prevalence and distribution of Salmonella in organic and conventional broiler poultry farms. Foodborne Pathog. Dis. 7:1363–1371. doi:10.1089/fpd.2010.0566

Alam, M.Z., D. Lynch, G. Tremblay, R. Gillis-Madden, and A. Vanasse. 2018. Optimizing combining green manures and pelletized manure for organic spring wheat production. Can. J. Soil Sci. 98: 638–649. doi:10.1139/cjss-2018-0049.

Antonious, G.F. 2018. Biochar and animal manure impact on soil, crop yield and quality. Agricul. Waste Residues. IntechOpen. p. 45–67. doi:10.5772/intechopen.77008

Berge, A.C., D.D. Hancock, W.M. Sischo, and T.E. Besser. 2010. Geographic, farm, and animal factors associated with multiple antimicrobial resistance in fecal *Escherichia coli* isolates from cattle in the western United States. J. Am. Vet. Med. Assoc. 236:1338–1344. doi:10.2460/javma.236.12.1338

Bonde, M., and J.T. Sørensen. 2012. Faecal Salmonella shedding in fattening pigs in relation to the presence of Salmonella antibodies in three pig production systems. Livest. Sci. 150:236–239. doi:10.1016/j.livsci.2012.09.002

Cho, S., F. Diez-Gonzalez, C.P. Fossler, S.J. Wells, C.W. Hedberg, J.B. Kaneene, P.L. Ruegg, L.D. Warnick, and J.B. Bender. 2006a. Prevalence of shiga toxin-encoding bacteria and shiga toxin-producing Escherichia coli isolates from dairy farms and county fairs. Vet. Microbiol. 118:289–298. doi:10.1016/j.vetmic.2006.07.021

Cho, S., J.B. Bender, F. Diez-Gonzalez, C.P. Fossler, C.W. Hedberg, J.B. Kaneene, P.L. Ruegg, L.D. Warnick, and S.J. Wells. 2006b. Prevalence and characterization of Escherichia coli O157 isolates from Minnesota dairy farms and county fairs. J. Food Prot. 69:252–259. doi:10.4315/0362-028X-69.2.252

De Vries, M., C.E. Van Middelaar, and I.J.M. De Boer. 2015. Comparing environmental impacts of beef production systems: A review of life cycle assessments. Livest. Sci. 178:279–288. doi:10.1016/j.livsci.2015.06.020

Franz, E., M.A. Klerks, O.J. De Vos, A.J. Termorshuizen, and A.H.C. van Bruggen. 2007. Prevalence of Shiga toxin-producing *Escherichia coli* stx(1), stx(2), eaeA, and rfbE genes and survival of E-coli O157: H7 in manure from organic and low-input conventional dairy farms. Appl. Environ. Microbiol. 73:2180–2190. doi:10.1128/AEM.01950-06

Gerzova, L., V. Babak, K. Sedlar, M. Faldynova, P. Videnska, D. Cejkova, A.N. Jensen, M. Denis, A. Kerouanton, and A. Ricci. 2015. Characterization of antibiotic resistance gene abundance and microbiota composition in feces of organic and conventional pigs from four EU countries. PLoS One 10:e0132892. doi:10.1371/journal.pone.0132892

Gustafson, G.M., E. Salomon, and S. Jonsson. 2007. Barn balance calculations of Ca, Cu, K, Mg, Mn, N, P, S and Zn in a conventional and organic dairy farm in Sweden. Agric. Ecosyst. Environ. 119:160–170. doi:10.1016/j.agee.2006.07.003

Gustafson, G.M., E. Salomon, S. Jonsson, and S. Steineck. 2003. Fluxes of K, P, and Zn in a conventional and an organic dairy farming system through feed, animals, manure, and urine- a case study at Ojebyn, Sweden. Eur. J. Agron. 20:89–99. doi:10.1016/S1161-0301(03)00077-7

Halbert, L.W., J.B. Kaneene, P.L. Ruegg, L.D. Warnick, S.J. Wells, L.S. Mansfield, C.P. Fossler, A.M. Campbell, and A.M. Geiger-Zwald. 2006a. Evaluation of antimicrobial susceptibility patterns in Campylobacter spp isolated from dairy cattle and farms managed organically and conventionally in the midwestern and northeastern United States. J. Am. Vet. Med. Assoc. 228:1074–1081. doi:10.2460/javma.228.7.1074

Halbert, L.W., J.B. Kaneene, J. Linz, L.S. Mansfield, D. Wilson, P.L. Ruegg, L.D. Warnick, S.J. Wells, C.P. Fossler, and A.M. Campbell. 2006b. Genetic mechanisms contributing to reduced tetracycline susceptibility of Campylobacter isolated from organic and conventional dairy farms in the midwestern and northeastern United States. J. Food Prot. 69:482–488. doi:10.4315/0362-028X-69.3.482

He, Z., editor. 2011. Environmental chemistry of animal manure. Nova Science Publishers, NY. p. 1–459.

He, Z., editor. 2012. Applied research of animal manure: Challenges and opportunities beyond the adverse environmental concerns. Nova Science Publishers, New York. p. 1–325.

He, Z., and J.J. Wang. 2012. Characterization of plant nutrients and traceable marker components in dairy manure for organic dairy farming management evaluation. In: Z. He, editor, Applied research of animal manure: Challenges and opportunities beyond the adverse environmental concerns. Nova Science Publishers, New York. p. 3–19.

He, Z., and T. Ohno. 2012. Fourier transform infrared and fluorescence spectral features of organic matter in conventional and organic dairy manure. J. Environ. Qual. 41:911–919. doi:10.2134/jeq2011.0226

He, Z., and H. Zhang, editors. 2014. Applied manure and nutrient chemistry for sustainable agriculture and environment. Springer, Amsterdam, the Netherlands. p. 1–379. doi:10.1007/978-94-017-8807-6

He, Z., and H.W. Waldrip. 2015. Composition of whole and water-extractable organic matter of cattle manure affected by management practices. In: Z. He and F. Wu, editors, Labile organic matter-chemical compositions, function, and significance in soil and the environment. SSSA Spec. Publ. 62. Soil Science Society of America, Madison, WI. p. 41–60. doi:10.2136/sssaspecpub62.2014.0034

He, Z., P.H. Pagliari, and H.M. Waldrip. 2016. Applied and environmental chemistry of animal manure: A review. Pedosphere 26:779–816. doi:10.1016/S1002-0160(15)60087-X

He, Z., C.W. Honeycutt, T.S. Griffin, B.J. Cade-Menun, P.J. Pellechia, and Z. Dou. 2009. Phosphorus forms in conventional and organic dairy manure identified by solution and solid state P-31 NMR spectroscopy. J. Environ. Qual. 38:1909–1918. doi:10.2134/jeq2008.0445

He, Z., M. Zhang, X. Cao, Y. Li, J. Mao, and H.M. Waldrip. 2015. Potential traceable markers of organic matter in organic and conventional dairy manure using ultraviolet-visible and solid-state 13C nuclear magnetic resonance spectroscopy. Organic Agriculture 5:113–122. doi:10.1007/s13165-014-0092-0

Hovi, M., A. Sundrum, and S.M. Thamsborg. 2003. Animal health and welfare in organic livestock production in Europe: Current state and future challenges. Livest. Prod. Sci. 80:41–53. doi:10.1016/S0301-6226(02)00320-2

Johnston, J.R. 2002. A comparison of antibiotic resistance in bacteria isolated from conventionally versus organically raised livestock. Bios (Florence, AL, U. S.) 73:47–51.

Kassem, I.I., O. Kehinde, A. Kumar, and G. Rajashekara. 2017. Antimicrobial-resistant Campylobacter in organically and conventionally raised layer chickens. Foodborne Pathog. Dis. 14:29–34. doi:10.1089/fpd.2016.2161

Knudsen, M.T., G.F. de Almeida, V. Langer, L.S. de Abreu, and N. Halberg. 2011. Environmental assessment of organic juice imported to Denmark: A case study on oranges (Citrus sinensis) from Brazil. Organic Agriculture 1:167. doi:10.1007/s13165-011-0014-3

Knudsen, M.T., A. Meyer-Aurich, J.E. Olesen, N. Chirinda, and J.E. Hermansen. 2014. Carbon footprints of crops from organic and conventional arable crop rotations– using a life cycle assessment approach. J. Clean. Prod. 64:609–618. doi:10.1016/j.jclepro.2013.07.009

Kuhnert, P., C.R. Dubosson, M. Roesch, E. Homfeld, M.G. Doherr, and J.W. Blum. 2005. Prevalence and risk-factor analysis of Shiga toxigenic Escherichia coli in faecal samples of organically and conventionally farmed dairy cattle. Vet. Microbiol. 109:37–45. doi:10.1016/j.vetmic.2005.02.015

Lynch, D.H., M. Sharifi, A. Hammermeister, and D. Burton. 2012. Nitrogen management in organic potato production. In: Z.L. He, R. Larkin, and W. Honeycutt, editors, Sustainable potato production: Global case studies. Springer, Amsterdam, The Netherlands. p. 209–231. doi:10.1007/978-94-007-4104-1_12

McKinney, C.W., K.A. Loftin, M.T. Meyer, J.G. Davis, and A. Pruden. 2010. Tet and sul antibiotic resistance genes in livestock lagoons of various operation type, configuration, and antibiotic occurrence. Environ. Sci. Technol. 44:6102–6109. doi:10.1021/es9038165

Nazarizadeh, M., F. Raiesi, and H.R. Motaghian. 2017. Response of earthworm Eisenia fetida to the stresses induced by salinity and lead pollution in a soil amended with cow manure. Journal of Water and Soil 31:1355–1370.

Novak, S.M., and J.L. Fiorelli. 2010. Greenhouse gases and ammonia emissions from organic mixed crop-dairy systems: A critical review of mitigation options. Agron. Sustain. Dev. 30:215–236. doi:10.1051/agro/2009031

Nulsen, M., M. Mor, and D. Lawton. 2008. Antibiotic resistance among indicator bacteria isolated from healthy pigs in New Zealand. N. Z. Vet. J. 56:29–35. doi:10.1080/00480169.2008.36801

Oelofse, M., L.S. Jensen, and J. Magid. 2013. The implications of phasing out conventional nutrient supply in organic agriculture: Denmark as a case. Organic Agriculture 3:41–55.

Oelofse, M., H. Høgh-Jensen, L.S. Abreu, G.F. Almeida, A. El-Araby, Q. Yu-Hui, T. Sultan, and A. de Neergaard. 2011. Organic farm conventionalization and farmer practices in China, Brazil and Egypt. Agron. Sustainable Dev. 31:689–698. doi:10.1007/s13593-011-0043-z

Olesen, J.E., E.M. Hansen, M. Askegaard, and I.A. Rasmussen. 2007. The value of catch crops and organic manures for spring barley in organic arable farming. Field Crops Res. 100:168–178. doi:10.1016/j.fcr.2006.07.001

Österberg, J., A. Wingstrand, A.N. Jensen, A. Kerouanton, V. Cibin, L. Barco, M. Denis, S. Aabo, and B. Bengtsson. 2016. Antibiotic resistance in Escherichia coli from pigs in organic and conventional farming in four European countries. PLoS One 11:e0157049. doi:10.1371/journal.pone.0157049

Qian, X., J. Gu, W. Sun, X.-J. Wang, J.-Q. Su, and R. Stedfeld. 2018. Diversity, abundance, and persistence of antibiotic resistance genes in various types of animal manure following industrial composting. J. Hazard. Mater. 344:716–722. doi:10.1016/j.jhazmat.2017.11.020

Ray, K.A., L.D. Warnick, R.M. Mitchell, J.B. Kaneene, P.L. Ruegg, S.J. Wells, C.P. Fossler, L.W. Halbert, and K. May. 2006. Antimicrobial susceptibility of Salmonella from organic and conventional dairy farms. J. Dairy Sci. 89:2038–2050. doi:10.3168/jds.S0022-0302(06)72271-8

Reganold, J.P., and J.M. Wachter. 2016. Organic agriculture in the twenty-first century. Nat. Plants (London, U.K.) 2:15221. doi:10.1038/NPLANTS.2015.221

Regulation, C. 2007. No 834/2007 of 28 June 2007 on organic production and labelling of organic products and repealing Regulation (EEC) No 2092/91. Off. J. Eur. Union L 189:1–23.

Röös, E., A. Mie, M. Wivstad, E. Salomon, B. Johansson, S. Gunnarsson, A. Wallenbeck, R. Hoffmann, U. Nilsson, and C. Sundberg. 2018. Risks and opportunities of increasing yields in organic farming. A review. Agron. Sustain. Dev. 38:14. doi:10.1007/s13593-018-0489-3

Safley, L.M., P.W. Westerman, and J.C. Barker. 1985. Fresh dairy manure characteristics and barnlot nutrient losses. In: V.W. Ruttan, editor, Agricultural waste utilization and management: Proc. Int. Symp. Agric. Wastes. 5th. Chicago, IL. ASAE, St. Joseph. MI. p. 191–199.

Sapkota, A.R., E.L. Kinney, A. George, R.M. Hulet, R. Cruz-Cano, K.J. Schwab, G. Zhang, and S.W. Joseph. 2014. Lower prevalence of antibiotic-resistant Salmonella on large-scale U.S. conventional poultry farms that transitioned to organic practices. Sci. Total Environ. 476-477:387–392. doi:10.1016/j.scitotenv.2013.12.005

Sato, K., P.C. Bartlett, and M.A. Saeed. 2005a. Antimicrobial susceptibility of Escherichia coli isolates from dairy farms using organic versus conventional production methods. J. Am. Vet. Med. Assoc. 226:589–594. doi:10.2460/javma.2005.226.589

Sato, K., P.C. Bartlett, R.J. Erskine, and J.B. Kaneene. 2005b. A comparison of production and management between Wisconsin organic and conventional dairy herds. Livest. Prod. Sci. 93:105–115. doi:10.1016/j.livprodsci.2004.09.007

Schroder, J.L., H. Zhang, J.R. Richards, and Z. He. 2011. Sources and contents of heavy metals and other trace elements in animal manures. In: Z. He, editor, Environmental chemistry of animal manure. Nova Science Publishers, NY. p. 385–414.

Selvam, A., and J.W.C. Wong. 2017. Degradation of antibiotics in livestock manure during composting. In: J.W.C. Wong, R.D. Tyagi, and A. Pandey, editors, Current developments in biotechnology and bioengineering. Elsevier, Amsterdam, The Netherlands. p. 267–292. doi:10.1016/B978-0-444-63664-5.00012-5

Smith-Spangler, C., M.L. Brandeau, G.E. Hunter, J.C. Bavinger, M. Pearson, P.J. Eschbach, V. Sundaram, H. Liu, P. Schirmer, C. Stave, I. Olkin, and D.M. Bravata. 2012. Are organic foods safer or healthier than conventional alternatives? A systematic review. Ann. Intern. Med. 157:348–366. doi:10.7326/0003-4819-157-5-201209040-00007

Sneath, R.W., F. Beline, M.A. Hilhorst, and P. Peu. 2006. Monitoring GHG from manure stores on organic and conventional dairy farms. Agric. Ecosyst. Environ. 112:122–128. doi:10.1016/j.agee.2005.08.020

Song, W., and M. Guo. 2014. Residual veterinary pharmaceuticals in animal manures and their environmental behaviors in soils. In: Z. He and H. Zhang, editors, Applied manure and nutrient chemistry for sustainable agriculture and environment. Springer, Amsterdam, the Netherlands. p. 23–52. doi:10.1007/978-94-017-8807-6_2

Steineck, S., G. Gustafson, A. Andersson, M. Tersmeden, and J. Bergstrom. 2000. Plant nutrients and trace elements in livestock wastes in Sweden. Report 5111. Swedish Environmental Protection Agency, Stockholm, Sweden.

Sullivan, D.M., D.R. Bryla, and R.C. Costello. 2014. Chemcial characteristics of custom compost for high blueberry. In: Z. He and H. Zhang, editors, Applied manure and nutrient chemistry for sustainable agriculture and environment. Springer, Amsterdam, the Netherlands. p. 293–311.

Thomassen, M.A., K.J. van Calker, M.C. Smits, G.L. Iepema, and I.J. de Boer. 2008. Life cycle assessment of conventional and organic milk production in the Netherlands. Agric. Syst. 96:95–107. doi:10.1016/j.agsy.2007.06.001

USDA. 2011. Organic production and handling standards. United States Department of Agriculture, Washington, D.C. http://www.ams.usda.gov/AMSv1.0/getfile?dDocName=STELDEV3004445 (Accessed 19 Feb. 2019).

USDA. 2015. Organic farming- Results from the 2014 Organic Survey. ACH12-29. United States Department of Agriculture, Washington, D.C. www.agcensus.usda.gov (Accessed 19 Feb. 2019).

USDA. 2018. National organic program. United States Department of Agriculture, Washington, D.C. https://www.ams.usda.gov/about-ams/programs-offices/national-organic-program (Accessed 19 Feb. 2019).

Van Overbeke, I., L. Duchateau, L. De Zutter, G. Albers, and R. Ducatelle. 2006. A comparison survey of organic and conventional broiler chickens for infectious agents affecting health and food safety. Avian Dis. 50:196–200. doi:10.1637/7448-093005R.1

van Wagenberg, C., Y. De Haas, H. Hogeveen, M. van Krimpen, M. Meuwissen, C. van Middelaar, and T. Rodenburg. 2017. Animal board invited review: Comparing conventional and organic livestock production systems on different aspects of sustainability. Animal 11:1839–1851. doi:10.1017/S175173111700115X

Verlinden, S., L. McDonald, J. Kotcon, and S. Childs. 2017. Long-term effect of manure application in a certified organic production system on soil physical and chemical parameters and vegetable yields. HortTechnology 27:171–176. doi:10.21273/HORTTECH03348-16

Waldrip, H.M., Z. He, and T.S. Griffin. 2012. Effects of organic dairy manure on soil phosphatase activity, available soil phosphorus, and growth of sorghum-sudangrass. Soil Sci. 177:629–637. doi:10.1097/SS.0b013e31827c4b78

Walk, S.T., J.M. Mladonicky, J.A. Middleton, A.J. Heidt, J.R. Cunningham, P. Bartlett, K. Sato, and T.S. Whittam. 2007. Influence of antibiotic selection on genetic composition of *Escherichia coli* populations from conventional and organic dairy farms. Appl. Environ. Microbiol. 73:5982–5989. doi:10.1128/AEM.00709-07

Wallace, J.S., E. Garner, A. Pruden, and D.S. Aga. 2018. Occurrence and transformation of veterinary antibiotics and antibiotic resistance genes in dairy manure treated by advanced anaerobic digestion and conventional treatment methods. Environ. Pollut. 236:764–772. doi:10.1016/j.envpol.2018.02.024

Wander, M.M., W. Yun, W.A. Goldstein, S. Aref, and S.A. Khan. 2007. Organic N and particulate organic matter fractions in organic and conventional farming systems with a history of manure application. Plant Soil 291:311–321. doi:10.1007/s11104-007-9198-4

Wang, J.J., S.K. Dodla, and Z. He. 2011. Application of analytical pyrolysis-mass spectrometry in characterization of animal manure. In: Z. He, editor, Environmental chemistry of animal manure. Nova Science Publishers, NY. p. 3–24.

Watson, C., H. Bengtsson, M. Ebbesvik, A.K. Løes, A. Myrbeck, E. Salomon, J. Schroder, and E. Stockdale. 2002. A review of farm-scale nutrient budgets for organic farms as a tool for management of soil fertility. Soil Use Manage. 18:264–273. doi:10.1079/SUM2002127

Weiske, A., A. Vabitsch, J.E. Olesen, K. Schelde, J. Michel, R. Friedrich, and M. Kaltschmitt. 2006. Mitigation of greenhouse gas emissions in European conventional and organic dairy farming. Agric. Ecosyst. Environ. 112:221–232. doi:10.1016/j.agee.2005.08.023

Wild, P.L., C. van Kessel, J. Lundberg, and B.A. Linquist. 2011. Nitrogen availability from poultry litter and pelletized organic amendments for organic rice production. Agron. J. 103:1284–1291. doi:10.2134/agronj2011.0005

Willer, H., and J. Lernoud. 2016. The world of organic agriculture. Statistics and emerging trends 2016. Research Institute of Organic Agriculture FiBL and IFOAM Organics International, Frick, Switzerland.

Yadav, A., R. Gupta, and V.K. Garg. 2013. Organic manure production from cow dung and biogas plant slurry by vermicomposting under field conditions. International Journal of Recycling of Organic Waste in Agriculture 2:21. doi:10.1186/2251-7715-2-21

Fate and Transport of Estrogens and Estrogen Conjugates in Manure-Amended Soils

Xuelian Bai*

Abstract

Animal feeding operations release enormous amounts of animal manure globally, and the storage, disposal, and recycling of animal manure are of significant concern. Land application of animal manure can introduce considerable amounts of estrogenic compounds to the environment that are endocrine-disrupting chemicals. Estrogens can cause adverse effects on aquatic wildlife at part-per-trillion levels, and estrogen conjugates can behave as precursors to free estrogens in the environment. Herein, the behaviors of estrogens and their conjugates associated with extensive manure application need to be fully investigated. This chapter reviews the excretion and types of estrogens and estrogen conjugates in animal manure, the ecological risks, the persistence and mobility in manure-amended soils and the adjacent water systems, and the fate and transport in soil–water–plant ecosystems. This chapter summarizes state-of-the-art knowledge on the occurrence, fate, and transport of estrogens and estrogen conjugates, which provides critical information to better understand the environmental risks and behaviors of manure-borne estrogenic compounds.

Estrogens and Estrogen Conjugates in Animal Manure

In the United States, animal feeding operations produce approximately 453 million Mg of manure annually (Kellogg et al., 2000). In 2012, there were 63% fewer dairies in the United States than in 1997 as a result of the conglomeration of the dairy industry into concentrated animal feeding operations (CAFOs) (Pollard & Morra, 2017). Concentrated animal feeding operations have an average daily manure production of over 27,000 kg for a 500-head dairy, and therefore manure storage and disposal are serious management and environmental concerns. Land application is widely used as an economical way of disposal of animal manure and recycling nutrients. Concentrated animal feeding operations are considered a major source of steroidal hormones, accounting for 90% of the total estrogen load to the environment of the United States (Maier et al., 2000). Lange et al. (2002) reported that the total estrogens released by farm animals in the European Union and the United States were 33 and 49 metric tons per year, respectively. In the United States, cattle (*Bos taurus*), swine (*Sus scrofa domesticus*), and poultry (*Gallus domesticus*) contributed 45, 0.8, and 2.7 Mg of estrogens per year, respectively (Lange et al., 2002). More recently, He et al. (2016) reported the manure production in 2013 of cattle, pigs, chickens, and sheep and goats to be 1166.4, 91.35, 164.33, and 5.17 $\times 10^6$ Mg, respectively, in the United States. These naturally occurring

Division of Hydrologic Sciences, Desert Research Institute, Las Vegas, NV. *Corresponding author (xuelian.bai@dri.edu)

doi:10.2134/asaspecpub67.c14

Animal Manure: Production, Characteristics, Environmental Concerns and Management. ASA Special Publication 67. Heidi M. Waldrip, Paulo H. Pagliari, and Zhongqi He, editors.
© 2019.ASA and SSSA, 5585 Guilford Rd., Madison, WI 53711, USA.

Table 1. Structure and properties of natural steroidal hormones†.

Compound	Structure	Molecular Formula	Molecular Weight	log K$_{ow}$	pKa	Sw
			g mol^{-1}			mg L^{-1}
17β-Estradiol (E2)		C$_{18}$H$_{24}$O$_2$	272.4	4.01	10.7	3.6
17α-Estradiol (E2α)		C$_{18}$H$_{24}$O$_2$	272.4	4.0	–	3.9
Estrone (E1)		C$_{18}$H$_{22}$O$_2$	270.4	3.13	10.8	30
Estriol (E3)		C$_{18}$H$_{24}$O$_3$	288.38	2.45	10.5	27.3
Testosterone		C$_{19}$H$_{28}$O$_2$	288.4	3.32	-0.88, 19.1	23.4

† Values obtained from PubChem.

estrogenic hormones can pose a potential risk to aquatic wildlife and humans after land application of manure (Burkholder et al., 2007; Thorne 2007).

All species, sexes, and classes of animals release estrogenic hormones in urine and feces, but different estrogens are associated with different livestock species (Hanselman et al., 2003). Cattle excrete 17α-estradiol (E2α), 17β-estradiol (E2), and estrone (E1) as free and conjugated metabolites (Ivie et al., 1986). Swine and poultry mainly release E2, E1, and estriol (E3), as well as their conjugates in excreta (Moore et al., 1982). Different animal species release estrogens through different routes. Cattle excrete estrogens mostly in feces (58%), whereas swine (96%) and poultry (69%) excrete estrogens mostly in urine (Palme et al., 1996). Additionally, urinary estrogens are mainly in conjugated forms, whereas fecal estrogens are excreted as unconjugated free steroids (Palme et al., 1996). The daily total estrogen (i.e., free and conjugated) excretions of a cow were 145.23 to 179.27 µg, mainly through feces (92%), whereas a pig excreted 42.56 to 219.25 µg d^{-1} of estrogens mainly through urine (98% to 99%) (Zhang et al., 2014a). Estrogen conjugates contributed 14.6% to 48.8% to the total estrogen excretions in cattle feces and more than 98% in swine urine (Zhang et al., 2014a). A chicken excreted 0.66 to 12.78 mg d^{-1} of total estrogens through feces, among which 34.2% to 100% was contributed by conjugated estrogens (Zhang et al., 2014a). Another study reported the presence of E2α, E2, 17α-dihydroequilin, and E1 in biosolids and manure at concentrations ranging from 6 to 462 ng g^{-1} dry solids (Andaluri et al., 2012).

Animal manure is generally collected and temporarily stored in tanks, piles, or lagoons, where estrogens are either present in the aqueous phase or sorbed onto the solid phase of the storage systems. It is reported that E2 concentrations in dairy cattle, swine, and poultry manure ranged from below detectable limits (BDL) to 239 ± 30 mg kg^{-1}, BDL to 1215 ± 275 mg kg^{-1}, and 33 ± 13 to 904 mg kg^{-1}, respectively (Hanselman et al., 2003). The total free estrogen levels (E1, E2, E3, and E2α) were measured at 1000 to 21000 ng L^{-1} in swine lagoons, 1800 to 4000 ng L^{-1} in poultry lagoons, 370 to 550 ng L^{-1} in dairy cattle lagoons, and 22 to 24 ng L^{-1} in beef cattle lagoons, respectively (Hutchins et al., 2007). For estrogen conjugates, the highest levels measured in various lagoons were sulfated forms at concentrations of 2 to 91 ng L^{-1} for estrone-3-sulfate (E1–3S), 8 to 44 ng L^{-1} for 17β-estradiol-3-sulfate (E2–3S), 141-182 ng L^{-1} for 17α-estradiol-3-sulfate (E2α-3S), and 72 to 84 ng L^{-1} for 17β-estradiol-17-sulfate (E2–17S) (Hutchins et al., 2007). Additionally, swine, poultry, and cattle excrete approximately 96%, 69%, and 42% of estrogens as conjugates, respectively (Hanselman et al., 2003). In dairy cattle waste lagoons, 57% of the total estrogens were detected as conjugates, whereas in poultry lagoons, nearly all (95%) estrogens were conjugates (Hutchins et al., 2007).

Physicochemical Properties

Natural steroidal estrogens (E1, E2, and E3) have common steroid structures that are composed of four rings: a phenol, two cyclohexanes, and a cyclopentane. Steroidal estrogens have relatively low aqueous solubility and they are nonvolatile and hydrophobic (Table 1). Steroidal estrogens are released by humans and animals primarily as sulfate or glucuronide conjugates, which allows them to be easily excreted in urine or bile because of the increased water solubility (Johnson & Sumpter, 2001). Conjugated estrogens have a sulfate and/or glucuronide moiety attached at the C-3 and/or C-17 position of the parent compound. Glucuronide conjugation of steroidal estrogens in humans and animals is catalyzed by uridine 5′-diphospho-glucuronosyltransferase enzymes (Kiang et al., 2005), and sulfate conjugation is catalyzed by sulfotransferases using 3′-phosphoadenosine-5′-phosphosulfate as the sulfur donor (Gomes et al., 2009a; Shrestha et al., 2011). Unlike free estrogens, conjugates are not biologically active (Desbrow et al., 1998) because they do not bind to estrogen receptors (Hobkirk, 1985). Estrogen sulfate conjugates are more persistent and are detected more frequently than glucuronides in municipal sewage systems (Isobe et al., 2003; Schlüsener & Bester 2005, Gomes et al., 2005; D'Ascenzo et al., 2003).

Ecological Risks

Endocrine disrupting chemicals (EDCs) can have adverse effects on aquatic wild-life at trace concentrations and are becoming an increasing concern for water quality and environmental science studies. Naturally-occurring estrogenic hor-mones are the most potent EDCs because they can have adverse effects on the reproduction systems of aquatic wildlife at part-per-trillion levels (Jobling et al., 1998, Panter et al., 1998, Irwin et al., 2001). Among all steroidal hormones, E2 and E1 are of primary concern because they exert their toxic effects at lower concen-trations compared with many other EDCs. The lowest observable adverse effect level (LOAEL) of E2 on aquatic organisms is 10 ng L^{-1} (Routledge et al., 1998).

The adverse effects on aquatic wildlife exposed to estrogens have been stud-ied since the mid-1990s. Vitellogenin is a protein normally produced only in female fish, but male fish can have very high plasma vitellogenin concentrations after exposure to estrogenic chemicals in sewage effluents (Purdom et al., 1994). An in vivo study of estrogenic responses in male rainbow trout (*Oncorhynchus mykiss*) and common roach (*Rutilus rutilus*) was conducted by Routledge et al. (1998). After being exposed to E2 or E1 for 21 d, the elevated vitellogenin concentrations in male fish indicated that environmentally relevant concentrations of such estro-gens could induce vitellogenin production (Routledge et al., 1998). Panter et al. (2000) investigated the effects of intermittent exposure to estrogenic substances by exposing male fathead minnows (*Pimephales promelas*) to E2 at 30, 60, and 120 ng L^{-1} continuously or 120 ng L^{-1} intermittently for 21 and 42 d, respectively. They found that plasma vitellogenin levels from intermittent exposure were equal to those in response to continuous exposure at the same concentration. Irwin et al. (2001) measured E2 levels ranging from 0.05 to 1.8 ng L^{-1} in farm ponds near live-stock pastures and found that vitellogenin in male painted turtles (*Chrysemys picta*) could not be induced when exposed to 9.45 ng L^{-1} E2 for 28 d. A short-term analysis (11 d) of the effects of E2 and E1 on soil microbial activity and bacterial community structure was performed by Zhang et al. (2014b). They found that E1 or E2 alone markedly promoted soil dehydrogenase activity and stimulated the growth of partial bacteria strains, and the estrogen load might serve as a nutrient substance for soil microorganisms (Zhang et al., 2014b). Furthermore, Orozco-Hernández et al. (2018) reported that E2 induced cytogenotoxicity in the blood cells of common carp (*Cyprinus carpio*) at 1 ng, 1 µg, and 1 mg per liter.

So far, only a few studies have examined the relationships between manure-borne estrogens from animals and their adverse effects on aquatic wildlife. Verderame et al. (2016) studied the effects of manure application in organic farm-ing on lizards (*Podarcis sicula*). Lizards from the two organic farms with and without manure application displayed hepatic biosynthetic alterations typical of an estrogenic contamination. The hepatocytes contained both vitellogenin and estrogen receptor α transcripts and proteins, which were detected by in situ hybridization and immunocytochemistry. These findings suggest that exoge-nous estrogens, arising from the use of manure, could affect wild animals and lead to the bioaccumulation of estrogens in food chain, with possible risks to human consumers (Verderame et al., 2016). More investigations on the effects of manure-borne estrogens on aquatic and terrestrial wildlife are needed. Additionally, the ecological and human health risks of conjugated estrogens are still unclear, although many reports stated that there are no risks associated with conjugated estrogens because they are not biologically active.

Occurrence, Mobility, and Persistence in Manure-Amended Soils

Numerous studies have reported the persistence and transport of estrogens and their conjugates in soils, surface runoff, and leachate in agricultural fields that have been fertilized with manure. Selected monitoring and measurement studies are summarized in Table 2. Nichols et al. (1997) measured E2 at 133 and 102 mg kg^{-1} in normal and aluminum-treated poultry litter, and reported the maximum concentration of E2 in surface runoff at 1280 ng L^{-1} after litter application. Another study (Finlay-Moore et al., 2000) found that the background E2 concentrations in surface runoff from ungrazed pasture were 50 to 150 ng L^{-1}, and after poultry litter was applied, E2 concentrations increased to 20 to 2530 ng L^{-1} in the surface runoff and E2 levels in soils rose from 55 to 675 ng kg^{-1}. In addition, E2 was detected at a frequency of 37% and concentrations up to 1910 ng L^{-1} throughout a 2 m soil profile in an agricultural field where liquid swine manure was applied (Schuh et al., 2011). More than 88% of testosterone was found to be held by the applied manure and/or soil matrix even under the rainfall intensity of a 100-yr return frequency (Qi and Zhang, 2016). A one-year study (Olsen et al., 2007) examined the transport of E1 and E2 from manure to tile drainage systems at two field sites on structured, loamy soil. The estrogens leached from the root zone to tile drainage water in concentrations exceeding the LOAEL for as long as three months after application, with the maximum recorded concentration of E1 and E2 at 68.1 and 2.5 ng L^{-1}, respectively (Olsen et al., 2007). Additionally, Zhang et al. (2015) found that steroids were detected in the soils with detection frequencies from 3.13% to 100% and concentrations ranging from BDL to 109.7 mg kg^{-1}. Lafrance & Caron (2013) measured up to 58 ng L^{-1} of E1 shortly after manure application in a small agricultural watershed. Gall et al. (2015) monitored E2α, E2, E1, and E3 in a tile drain and receiving ditch for one year on a working farm in north central Indiana and found that repeated animal manure applications led to a frequent detection of hormones (> 50% in tile drain; > 90% in the ditch). All of these findings suggest that the application of manure to agricultural soils poses a potential estrogen contamination risk in the environment.

Different manure application methods and rates may affect estrogen load and detection in agricultural lands. Sosienski (2017) studied the spatial and temporal distribution of 11 hormones in an agricultural field with manure applied and in an adjacent stream, and detected hormones mainly in the top 0 to 5 cm of the soil. Subsurface application of poultry litter showed promise for reducing hormone transport in surface runoff. Hormones also showed little vertical and lateral movement in the soil, whereas E1 persisted at detectable levels during the study in all treatments (Sosienski, 2017). Mina et al. (2016) compared two methods of dairy cattle manure application (surface broadcast and shallow disk injection) and observed that estrogen concentrations in surface runoff from the broadcast plots were several orders of magnitude higher (> 5000 ng L^{-1}) than from the shallow disk injection plots (< 10 ng L^{-1}). Biswas et al. (2013) found that 96% less 17α-ethinylestradiol (EE2) mass transport from disk-tilled plots compared with no-till plots. The greatest loss of EE2 was 156 and 6 mg ha^{-1} from no-till and disked plots, respectively, demonstrating that a single-pass disk tillage treatment can limit the overland transport of steroid hormones from crop production areas.

In field-scale lysimeters treated with cattle manure, hormones were detected in only 5% of the leachate samples, with the greatest detected being progesterone (20 ng L^{-1}) (van Donk et al., 2013). Steroid hormones or metabolites were detected in 10% of the soil samples, where the majority of detections (74%) was within the top 1.2 m of the soil and the steroid detected most frequently was E2 (4%), with a maximum concentration of 4.3 ng g^{-1} (van Donk et al., 2013). Pinheiro et al. (2013) determined the presence of three hormones and six veterinary antibiotics in the surface runoff and drainage flow from volumetric lysimeters to which pig slurry was applied, and concentrations of the hormones and antibiotics found were between 4.6 and 1350.8 ng L^{-1}. In soil monoliths treated with pig slurry, estrogens were transported to a 1 m depth in loamy and sandy soil (Lægdsmand et al., 2009).

When animal manure is applied to agricultural lands, steroidal hormones can enter freshwater through surface runoff and groundwater through leaching. Despite steroidal estrogens possessing relatively low water solubility and high sorption affinity to soils and sediments, these contaminants are ubiquitously found in surface and subsurface water systems (Shore et al., 1993, Nichols et al., 1997, Peterson et al., 2000, Kolpin et al., 2002, Kolodziej et al., 2004). According to a nation-wide reconnaissance (Kolpin et al., 2002) of organic contaminants in 139 streams of 30 states in the United States, the median concentration of E2 and E1 was 160 and 27 ng L^{-1}, respectively; and the frequency of detection was 10.6% and 7.1% for E2 and E1, respectively. Matthiessen et al. (2006) determined estrogenic activity in streams running through livestock farms, where E1 and E2 were almost ubiquitous across all sites with E2 equivalents ranging from 0.04 to 3.6 ng L^{-1}. More recently, E1, E2α, E2, and E3 were found less than 1 ng L^{-1} in three headwater streams within a CAFO in upstate New York (Zhao et al., 2010). The low concentrations were likely because of degradation during the long residence time (~8 mo) of manure storage, during which 99.8% of the excreted estrogens were degraded (Zhao et al., 2010). Tremblay et al. (2018) demonstrated that estrogenic activity and steroid estrogens were prevalent in the waterways within the studied dairy watersheds, where E1 was the predominant steroid because it is a degradation byproduct of the main dairy cattle estrogen, E2α. For subsurface waters, E2 was detected from 6 to 66 ng L^{-1} in five groundwater springs in northwest Arkansas that were affected by nearby CAFOs (Peterson et al., 2000). In a more recent urban watershed monitoring study in Denver, Colorado, Bai et al. (2018) reported that the frequency of detection of nat-ural estrogens ranged from 0.7% to 11.4% with median concentrations of 112 to 612 ng L^{-1} in 311 surface water samples collected monthly in 2014 and 2015.

For the occurrence and transport of conjugated estrogens in manure-amended soils, Dutta et al. (2012) found that when poultry manure was applied to an agricultural field, no glucuronide conjugates were detected in surface runoff; only sulfate conjugates were found. In addition, E2–17S was at higher levels (i.e., up to107 ng L^{-1}) than the other sulfate conjugates (i.e., E2–3S, E2α-3S, and E1–3S), and runoff concentrations of E2–17S (i.e., 0.3-3.9 ng L^{-1}) were higher than free E2 (i.e., 0.5-1.9 ng L^{-1}) (Dutta et al., 2010).

Fate and Transport Processes

Sorption

Free estrogens are nonvolatile and relatively hydrophobic compounds, and they are sorbed rapidly on soils and sediments (Casey et al., 2003, Lee et al., 2003). A series of laboratory-based experiments were conducted by Lai et al. (2000) to determine the partitioning of natural and synthetic estrogens between water and sediments. The synthetic estrogens, EE2 and mestranol, with higher K_{ow} values, were found to have greater sorption coefficients and were more rapidly removed from the aqueous phase compared with the natural estrogens, E2, E1, and E3. The authors also reported that sorption of estrogens is correlated with soil organic carbon content (OC), particle size distribution, and salinity (Lai et al., 2000). Colucci et al. (2001) reported that after 3-d incubation in loam, sandy loam, and silt loam soil, the nonextractable E2 and E1 was 90.7%, 70.3%, and 56.0%, respectively, indicating a rapid removal from the aqueous phase. A further study (Colucci & Topp 2002) demonstrated that E2 at part-per-trillion levels was expected to rapidly dissipate in agricultural soils through soil binding and the formation of nonextractable residues.

Freundlich sorption coefficients of E2 were reported to range from 86 to 6670 L Kg^{-1}, as determined by batch equilibrium studies using four types of soils (Casey et al., 2003). The sorption affinity of E2 was highly correlated with silt content and soil OC and also associated with surface area and/or cation exchange capacity (Casey et

Table 2. Selected monitoring studies of estrogens and estrogen conjugates in manure-applied soils, surface water, and groundwater.

Study	Environment	Compound	Concentration
Nichols et al. (1997)	Poultry litter applied land	E2	133 ng g^{-1} in poultry litter; 1280 ng L^{-1} in runoff
Finlay-Moore et al. (2000)	Poultry litter applied land	E2	675 ng kg^{-1} in soil; 2530 ng L^{-1} in runoff
Dutta et al. (2012)	Agricultural watershed receiving poultry litter	Free and conjugated estrogens	Up to 57.5 ng L^{-1} of E1, 12.0 ng L^{-1} of E2α, and 19.2 ng L^{-1} of E3; up to 107 ng L^{-1} of E2-17S, 26.8 ng L^{-1} of E2-3S, and 29.2 ng L^{-1} E1-3S
Schuh et al. (2011)	Swine manure applied land	E2	1910 ng L^{-1} in soil
Lafrance & Caron (2013)	Hog/Dairy manure applied land	E1	Up to 58 ng L^{-1} in runoff
Pinheiro et al. (2013)	Pig slurry applied land	Hormones and antibiotics	4.6 to 1350.8 ng L^{-1} in runoff and drainage water
van Donk et al. (2013)	Cattle manure applied land	E2	Up to 4.3 ng g^{-1}
Zhang et al. (2015)	Vegetable cultivation area	E1, E2a, E2, E3, testosterone, androstendione and progesterone	Up to 109.7 ng g^{-1} in soils; up to 2.38 ng L^{-1} in groundwater; up to 14 ng L^{-1} in drainage water
Mina et al. (2016)	Dairy manure applied land	E1, E2a, E2 and E3	> 5000 ng L^{-1} in broadcast plots; < 10 ng L^{-1} from shallow disk injection plots
Olsen et al. (2007)	Pig slurry applied land	E1 and E2	E1 at 68.1 ng L^{-1} and E2 at 2.5 ng L^{-1} in tile drainage water
Kolpin et al. (2002)	Streams	E1 and E2	Median concentration of E1 at 27 ng L^{-1}; E2 at 160 ng L^{-1};
Matthiessen et al. (2006)	Streams	E2 equivalent	0.04 to 3.6 ng L^{-1}
Zhao et al. (2010)	Streams	E1, E2 a, E2 and E3	Less than 1 ng L^{-1}
Peterson et al. (2000)	Groundwater springs	E2	6 to 66 ng L^{-1}
Bai et al. (2018)	Streams	E1, E2, and EE2	Median concentrations from 112 to 612 ng L^{-1}

al., 2003). Another study reported that Freundlich sorption coefficients of E2 were 3.56 and 83.2 L Kg^{-1} in two soils (Lee et al., 2003). Moreover, equilibrium sorption of E2 and E1 in soils was achieved between 5 and 24 h following linear sorption isotherms, and the OC normalized sorption coefficient (log K_{oc}) was 2.49 for E2 and 2.99 for E1 (Casey et al., 2005). Sorption affinity of estrogen conjugates to soils is reported to be less than free estrogens. Bai et al. (2015) found that log K_{oc} of E2–17S was 2.20 and 2.45 in sandy soils with 1.29% and 0.26% OC, respectively. Additionally, E1–3S was reported to have log K_{oc} values of 1.73 to 2.08 in agricultural soils collected from New Zealand (Scherr et al., 2008). The formation of soil-bound residues may significantly reduce the risks of estrogens and estrogen conjugates to surface and subsurface waters near agricultural lands treated with manure.

Degradation

Mineralization and degradation of estrogens have been widely studied under various conditions. Degradation of free estrogens is reported to occur quickly in soils with half-lives less than 1 d (Colucci et al., 2001). Mineralization of estrogens and testosterone was investigated using biosolids by Layton et al. (2000), where 70% to 80% of E2 was mineralized to CO_2 within 24 h. Jacobsen et al. (2005) found that E2 was quickly converted to E1 within a few days in manured and unmanured soils, and the negligible mineralization rates of E2 in sterile soils indicated that the process was dependent on soil microorganisms. The degradation half-lives of E2 in aerobic soils and sediments ranged from 0.8 to 9.7 d, and the primary product was found to be E1 (Lee et al., 2003). Also, E1 and E2 had half-lives of 5 to 25 d in grassland soils amended with cattle and sheep manure, and the degradation rates in manure-amended soils were greater compared with unamended soils, demonstrating that animal manure could effectively remove estrogens in soils (Lucas and Jones 2006). Although most literature reports estrogen degradation to be a biotic process, abiotic transformation of E2 can also occur (Colucci et al., 2001). Sheng et al. (2009) demonstrated that E1 could be produced from E2 via abiotic oxidation mediated by naturally occurring MnO_2. More recently, the two isomers E2α and E2 were found to degrade at the same rates in soils with half-lives between 4 and 12 h, and E1 was the predominant metabolite produced via microbial processes (Mashtare et al., 2013). Goeppert et al. (2014) found that E2 was transformed to E1 and E1–3S in nonautoclaved soil and to E1 in autoclaved soil. The formation of E1–3S was biologically driven, and the transformation of E2 to E1 did not require biological interactions. In a further study, Goeppert et al. (2015) reported that the transport of E2 and E1 is subject to strong retardation and degradation in soil columns, and the transport of E1–3S is less retarded and only affected to a minor degree by degradation, which was triggered by arylsulfotransferase.

Deconjugation of Estrogen Conjugates

After being released to the environment, estrogen conjugates are not considered a risk unless they deconjugate to yield the active parent estrogen (Ingerslev & Halling-Sørensen 2003). Deconjugation is a common enzymatic hydrolysis process in the environment, which is governed by the bacterial enzymes β-glucuronidase and sulfatase for glucuronide and sulfate conjugates, respectively (Khanal et al., 2006). Estrogen conjugates can act as precursors to free estrogens and increase the total estrogen load in the environment via deconjugation. A key research question about estrogen conjugates is the uncertainty of whether and to what extent deconjugation occurs in the environment. Most existing studies are limited to field-monitoring in sewage waste influent and effluent, manure storage systems, and surface runoff to indirectly assess the behaviors of estrogen conjugates. Limited information is available on a thorough investigation of deconjugation using controlled experiments. Deconjugation potentials of estrogen conjugates have been studied in municipal sewage systems (D'Ascenzo et al., 2003; Gomes et al., 2009b; Kumar et al., 2012). However, estrogen conjugates may behave differently in agricultural soils because of different microbial populations and water contents.

Laboratory microcosm studies were conducted to determine aerobic degradation of E1–3S in three pasture soils at three temperatures (Scherr et al., 2008). The results showed that E1–3S was degraded rapidly without a lag phase in all soils and formed E1 as a primary metabolite. A further study (Scherr et al., 2008) reported that degradation of E2–3S followed first-order kinetics with half-lives ranging from 0.424 to 7.69 h. Two primary metabolites, E1–3S and E2, and one secondary metabolite, E1, were formed during the incubation; and furthermore, soil arylsulphatase activity played a major role in the transformation of E2–3S (Scherr et al., 2008). Additionally, Gomes et al. (2009b) found that 74% to 94% of sulfate conjugates still persisted in activated sludge after 8 h, suggesting that glucuronide conjugates dissipated more quickly than sulfate conjugates. Deconjugation also depends on conjugation positioning: D-ring glucuronides are more resistant than A-ring glucuronides (Gomes et al., 2009b). A further study reported that the glucuronide conjugate, 17β-estradiol-3-glucuronide (E2–3G), was quickly transformed to free E2 and E1 in soil-water slurries, which may contribute to free estrogens significantly in the environment (Shrestha et al., 2012).

Several microcosm studies have been conducted to provide a thorough understanding on the deconjugation mechanisms of estrogen conjugates. Zheng et al. (2013) demonstrated that the half-lives of E2α-3S at temperatures from 15 to 45 °C varied from 1.70 to 415 d and 22.5 to 724 d under aerobic and anaerobic conditions, respectively, indicating that this contaminant may accumulate in anaerobic or anoxic environments. Zheng et al. (2013) also identified three degradation products of E2α-3S, with E1–3S and E2α as the primary metabolites and E1 as the secondary metabolite. For aerobic degradation, oxidation at C-17 was a major mechanism; however, deconjugation of the E2α-3S thioester bond at C-3 was a predominant pathway under anaerobic conditions. Furthermore, Bai et al. (2013, 2015) studied the transformation pathways of E2–17S in agricultural soils of different OC. They found that hydroxylation was the primary metabolism pathway of E2–17S forming mono- and di-hydroxy E2–17S. Deconjugation occurred on the soil surface to a much lower extent compared with hydroxylation, forming E2 as a product and E2 was subsequently oxidized to E1 (Fig. 1). Additionally, E2–17S dissipation was significantly affected by soil OC, with half-lives of 4.9 to 26 h in high OC soil (i.e., 1.29%) and 64 to 173 h in low OC soil (i.e., 0.26%). Therefore, E2–17S was found to be recalcitrant to releasing free estrogens in agricultural soils and may not be a significant concern in terms of endocrine disrupting potentials. More recently, Ma & Yates (2017) reported that the aerobic degradation of E2–3G and E2–3S followed first-order kinetics in agricultural soils and the degradation rates were inversely related to their initial concentrations. The major degradation pathway of E2–3G and E2–3S was oxidation, yielding the primary metabolites estrone-3-glucuronide (E1–3G) and E1–3S, respectively; the second primary metabolite E2; and the

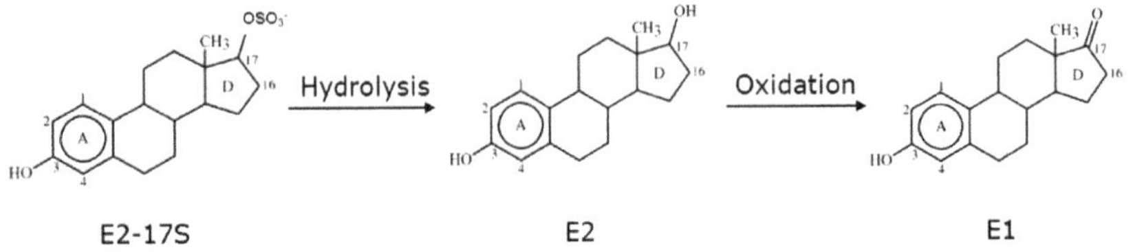

Fig. 1. Transformation pathways of E2–17S in soil–water systems.

secondary metabolite E1. Also, ring B unsaturated estrogens and their sulfate conjugates were tentatively proposed as minor metabolites. A following study (Ma & Yates 2018) determined degradation of E2–3G in river water at an environmentally relevant level (i.e., 25 ng L^{-1}) and observed first-order kinetics with the formation of E2 and E1. In contrast, E2–3S was slowly converted to E1–3S. Degradation of the two conjugates was much faster in sediments than in river water, which was likely because of the relatively high population densities of microorganisms in sediment. Deconjugation of the thioester bond at the C-3 position and oxidation at the C-17 position were the predominant mechanisms for E2–3G and E2–3S, respectively (Ma & Yates, 2018). In conclusion, the up-to-date studies have shown that estrogen conjugates on the C-3 position are readily converted to free estrogens and could be responsible for the frequent detections of free estrogens in surface and subsurface water. However, the C-17 conjugate E2–17S may not be an important source of free estrogens in soils after manure application.

Modeling Fate and Transport Processes

Mathematical models have been used to simulate the behaviors of steroid hormones and their conjugates. Das et al. (2004) applied forward modeling (i.e., predictive model) to simulate sorption and degradation of steroid hormones in soil columns. This two-region modeling approach consisted of advective-dispersive transport with nonequilibrium, two-site sorption, and first-order transformation mechanisms. Their results suggested that first-order kinetics was sufficient to modeling hormone degradation, but not accurate. Casey et al. (2003) used two convective-dispersive transport models, with and without transformation, and two-site kinetic Freundlich sorption to fit the breakthrough curves of E2. The results provided a good description of the experimental data, but the solutions were nonunique and the parameter estimates had low confidence. Another study by Casey et al. (2004) applied a one-site fully kinetic convective-dispersive model with sorbed phase transformation and Freundlich sorption to simulate the fate of testosterone in soils, which resulted in a satisfactory fit and reasonable parameter estimates.

These previous studies considered a two-phase system (i.e., aqueous vs. reversibly sorbed phase) for sorption and degradation of hormones. However, later studies reported that a significant fraction of the applied steroid hormones could be irreversibly bound to soil (Colucci et al., 2001; Fan et al., 2007). Without considering irreversible sorption, the models may give rise to inaccurate descriptions. Additionally, the previous models provided parameter estimates of relatively low confidence. To improve the modeling techniques for steroid hormones, Fan et al. (2008) developed a one-site, kinetic sorption and first-order transformation model to simulate the distribution of E2 in the aqueous, reversibly sorbed, and irreversibly sorbed phases. The model was solved inversely using a global optimization method, the stochastic ranking evolutionary strategy (SRES) (Runarsson and Yao, 2000), instead of the traditional local optimization parameter estimation method, and the one-site model resulted in satisfactory fits and unique solutions (Fan et al., 2008). Bai et al. (2014) further developed a conceptual model to simultaneously simulate the coupled sorption and transformation of E2–17S in various soil–water systems. The simulated processes included multiple transformation pathways (i.e., hydroxylation, hydrolysis, and oxidation) and mass transfer between the aqueous, reversibly sorbed, and irreversibly sorbed phases of four soil conditions for E2–17S and its four identified metabolites (Fig. 2). The model was inversely solved using finite difference to estimate process parameters. A global optimization method was successfully applied for the inverse analysis along with variable model restrictions to estimate 36 parameters with satisfactory confidence intervals.

A state-space mixing-cell model (Steiner et al., 2010) was developed to describe the transport of E1 and E2 by three transport processes in parallel. The inverse modeling of the leaching data did not support the hypothesis that antecedent concentrations of estrogens could be responsible for the observed breakthrough curves, but it did confirm that estrogens transported mainly via preferential and/or macropore flow and also via an enhanced movement by colloids. Most recently, Zhao and Lung (2018) adopted two simple models, the wash-off model and the empirical model, to quantify the pathways of E1, E2, and E2α from agricultural lands to the receiving water following rainfall events. The two models were calibrated and validated using the data obtained from three artificial rainfall

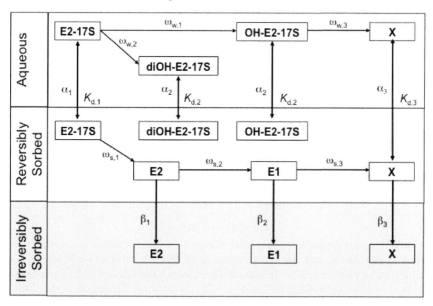

Fig. 2. Conceptual model to simulate fate of E2-17S in soil-water systems. Parameters include linear sorption coefficient Kd (L g⁻¹); first-order transformation rate coefficients in the aqueous and reversibly sorbed phase ww and ws (h⁻¹); mass transfer rate constant between the aqueous and reversibly sorbed phase a (h⁻¹); and mass transfer rate constant between the reversibly and irreversibly sorbed phase b (h⁻¹) (Bai et al. 2014).

events from the literature. Both models can closely reproduce the mass loads of estrogens during rainfall events and the wash-off model shows a better overall performance than the empirical model in this study.

Plant Uptake

Plants are likely exposed to environmentally persistent estrogens, which may result in food safety and human health concerns. Card et al. (2012) evaluated the sorption of natural and synthetic estrogens to dried root tissue of maize seedlings, where the sorption isotherms were nonlinear at aqueous concentrations below 0.1 μM and linear above that limit. After 22 d of growing maize seedlings, zeranol and zearalanone concentrations decreased by more than 96% in hydroponic solutions, and E2 and E1 were undetectable. All four estrogens were detected in root tissues at concentrations up to 0.19 mmol g^{-1}, and only E2 and zeranol were detected in shoots (Card et al., 2012). Additionally, testosterone was found in tomatoes (4.2 ng g^{-1}) and potatoes (5 ng g^{-1}) by Sabourin et al. (2012), and E1 (0.4–8.4 ng g^{-1}) and testosterone (14–115 ng g^{-1}) were both found present in lettuce plants (*Lactuca sativa* L.) by Shargil et al. (2015). Bircher et al. (2015) also documented that E2, EE2, zeranol, and trenbolone acetate were rapidly removed from 2 mg L^{-1} hydroponic solutions by more than 97% after 10 d of exposure to full poplar plants or live excised poplars (cut-stem, no leaves). Removals by sorption to dead poplar roots that had been autoclaved were significantly less, 71% to 84%. Major transformation products (i.e., E1 and E3 for E2) were detected in the root tissues of all three poplar treatments. These findings show that poplars may be effective in controlling the movement of hormonally active compounds from agricultural fields and preventing runoff to streams. Zheng et al. (2016) studied the uptake, translocation, accumulation, and depuration of three steroid hormones in lettuce and tomato plants using hydroponic cultures. For both plants, all hormones were detected in roots. The bioaccumulation factors (BAFs) in lettuce roots were more than 1, which means that these contaminants can be bound to or taken up by the plant roots. Most recently, Adeel et al. (2018) reported the BAFs for E2 and EE2 in lettuce plants were 0.33 and 0.29 at 50 μg L^{-1} initial concentration, indicating the potentials for uptake and bioaccumulation of estrogens in lettuce. Overall, the estrogen contents in lettuce were found to be higher than the toxic level, which is of serious concern to public health. Other studies also reported uptake of estrogens by the freshwater green alga *Nannochloris* (Bai and Acharya, 2019a) and quagga mussel (*Dreissena bugensis*) (Bai and Acharya, 2019b), demonstrating the bioaccumulation potentials of endocrine disrupting chemicals within the aquatic food webs.

Conclusions

Animal manure-borne estrogens and estrogen conjugates are of increasing concern to regulators and the public because of the potential adverse effects on wildlife and human health. To date, many research works have been performed to evaluate the occurrence, fate, and mobility of the contaminants in agricultural lands with manure application; however, further research is still needed to provide a better understanding and recommendation for mitigating the risks. Future studies should focus on the following areas: i) removal efficiencies of estrogens and estrogen conjugates by different manure handling and storage strategies; ii) combined effects of nutrients, organic matter, heavy metals, and other organic compounds on disrupting endocrine systems of organisms; iii) fate and transport of different conjugated estrogens in soil–water systems; iv) uptake and accumulation of estrogens and estrogen conjugates in plants and organisms; and v) comprehensive modeling techniques to predict the fate and risks of estrogenic compounds in the natural environment.

References

Adeel, M., M. Zain, S. Fahad, M. Rizwan, A. Ameen, H. Yi, M.A. Baluch, J.Y. Lee, and Y. Rui. 2018. Natural and synthetic estrogens in leafy vegetable and their risk associated to human health. Environ. Sci. Pollut. Res. Int. 25:36712–36723. doi:10.1007/s11356-018-3588-4

Andaluri, G., R.P.S. Suri, and K. Kumar. 2012. Occurrence of estrogen hormones in biosolids, animal manure and mushroom compost. Environ. Monit. Assess. 184:1197–1205. doi:10.1007/s10661-011-2032-8

Bai, X., and K. Acharya. 2019a. Removal of seven endocrine disrupting chemicals (EDCs) from municipal wastewater effluents by a freshwater green alga. Environ. Pollut. 247:534–540. doi:10.1016/j.envpol.2019.01.075

Bai, X., F.X. Casey, H. Hakk, T.M. DeSutter, P.G. Oduor, and E. Khan. 2015. Sorption and degradation of 17beta-estradiol-17-sulfate in sterilized soil-water systems. Chemosphere 119:1322–1328. doi:10.1016/j.chemosphere.2014.02.016

Bai, X., F.X.M. Casey, H. Hakk, T.M. DeSutter, P.G. Oduor, and E. Khan. 2013. Dissipation and transformation of 17b-estradiol-17-sulfate in soil-water systems. J. Hazard. Mater. 260:733–739. doi:10.1016/j.jhazmat.2013.06.036

Bai, X., A. Lutz, R.W.H. Carroll, K. Keteles, K. Dahlin, M. Murphy, and D. Nguyen. 2018. Occurrence, distribution, and seasonality of emerging contaminants in urban watersheds. Chemosphere 200:133–142. doi:10.1016/j.chemosphere.2018.02.106

Bai, X., S.L. Shrestha, F.X. Casey, H. Hakk, and Z. Fan. 2014. Modeling coupled sorption and transformation of 17beta-estradiol-17-sulfate in soil-water systems. J. Contam. Hydrol. 168:17–24. doi:10.1016/j.jconhyd.2014.09.001

Bai, X.L., and K. Acharya. 2019b. Uptake of endocrine-disrupting chemicals by quagga mussels (Dreissena bugensis) in an urban-impacted aquatic ecosystem. Environ. Sci. Pollut. Res. Int. 26:250–258.

Bircher, S., M.L. Card, G.S. Zhai, Y.P. Chin, and J.L. Schnoor. 2015. Sorption, uptake, and biotransformation of 17-estradiol, 17-ethinylestradiol, zeranol, and trenbolone acetate by hybrid poplar. Environ. Toxicol. Chem. 34:2906–2913. doi:10.1002/etc.3166

Biswas, S., W.L. Kranz, C.A. Shapiro, M. Mamo, and S.L. Bartelt-Hunt. 2013. Use of a surrogate to evaluate the impact of tillage on the transport of steroid hormones from manure-amended agricultural fields. Trans. ASABE 56:1379–1385.

Burkholder, J., B. Libra, P. Weyer, B. Heathcote, D. Koplin, P.S. Thorne, and M. Wichman. 2007. Impacts of waste from concentrated animal feeding operations on water quality. Environ. Health Perspect. 115:308–312. doi:10.1289/ehp.8839

Card, M.L., J.L. Schnoor, and Y.P. Chin. 2012. Uptake of natural and synthetic estrogens by maize seedlings. J. Agric. Food Chem. 60:8264–8271. doi:10.1021/jf3014074

Casey, F.X.M., H. Hakk, J. Simunek, and G.L. Larsen. 2004. Fate and transport of testosterone in agricultural soils. Environ. Sci. Technol. 38:790–798. doi:10.1021/es034822i

Casey, F.X.M., G.L. Larsen, H. Hakk, and J. Simunek. 2003. Fate and transport of 17 beta-estradiol in soil-water systems. Environ. Sci. Technol. 37:2400–2409. doi:10.1021/es026153z

Casey, F.X.M., J. Simunek, J. Lee, G.L. Larsen, and H. Hakk. 2005. Sorption, mobility, and transformation of estrogenic hormones in natural soil. J. Environ. Qual. 34:1372–1379. doi:10.2134/jeq2004.0290

Colucci, M.S., H. Bork, and E. Topp. 2001. Persistence of estrogenic hormones in agricultural soils: I. 17 beta-estradiol and estrone. J. Environ. Qual. 30:2070–2076. doi:10.2134/jeq2001.2070

Colucci, M.S., and E. Topp. 2002. Dissipation of part-per-trillion concentrations of estrogenic hormones from agricultural soils. Can. J. Soil Sci. 82:335–340. doi:10.4141/S01-079

D'Ascenzo, G., A. Di Corcia, A. Gentili, R. Mancini, R. Mastropasqua, M. Nazzari, and R. Samperi. 2003. Fate of natural estrogen conjugates in municipal sewage transport and treatment facilities. Sci. Total Environ. 302:199–209. doi:10.1016/S0048-9697(02)00342-X

Das, B.S., L.S. Lee, P.S.C. Rao, and R.P. Hultgren. 2004. Sorption and degradation of steroid hormones in soils during transport: Column studies and model evaluation. Environ. Sci. Technol. 38:1460–1470. doi:10.1021/es034898e

Desbrow, C., E.J. Routledge, G.C. Brighty, J.P. Sumpter, and M. Waldock. 1998. Identification of estrogenic chemicals in STW effluent. 1. Chemical fractionation and in vitro biological screening. Environ. Sci. Technol. 32:1549–1558. doi:10.1021/es9707973

Dutta, S., S. Inamdar, J. Tso, D.S. Aga, and J.T. Sims. 2010. Free and conjugated estrogen exports in surface-runoff from poultry litter-smended soil. J. Environ. Qual. 39:1688–1698. doi:10.2134/jeq2009.0339

Dutta, S.K., S. Inamdar, J. Tso, and D.S. Aga. 2012. Concentrations of free and conjugated estrogens at different landscape positions in an agricultural watershed receiving poultry litter. Water Air Soil Pollut. 223:2821–2836.

Fan, Z.S., F.X.M. Casey, H. Hakk, and G.L. Larsen. 2007. Persistence and fate of 17 beta-estradiol and tesotosterone in agricultural soils. Chemosphere 67:886–895. doi:10.1016/j.chemosphere.2006.11.040

Fan, Z.S., F.X.M. Casey, H. Hakk, and G.L. Larsen. 2008. Modeling coupled degradation, sorption, and transport of 17 beta-estradiol in undisturbed soil. Water Resour. Res. 44:10. doi:10.1029/2007WR006407

Finlay-Moore, O., P.G. Hartel, and M.L. Cabrera. 2000. 17 beta-estradiol and testosterone in soil and runoff from grasslands amended with broiler litter. J. Environ. Qual. 29:1604–1611. doi:10.2134/jeq2000.00472425002900050030x

Gall, H.E., S.A. Sassman, B. Jenkinson, L.S. Lee, and C.T. Jafvert. 2015. Comparison of export dynamics of nutrients and animal-borne estrogens from a tile-drained Midwestern agroecosystem. Water Res. 72:162–173. doi:10.1016/j.watres.2014.08.041

Goeppert, N., I. Dror, and B. Berkowitz. 2014. Detection, fate and transport of estrogen family hormones in soil. Chemosphere 95:336–345. doi:10.1016/j.chemosphere.2013.09.039

Goeppert, N., I. Dror, and B. Berkowitz. 2015. Fate and transport of free and conjugated estrogens during soil passage. Environ. Pollut. 206:80–87. doi:10.1016/j.envpol.2015.06.024

Gomes, R.L., J.W. Birkett, M.D. Scrimshaw, and J.N. Lester. 2005. Simultaneous determination of natural and synthetic steroid estrogens and their conjugates in aqueous matrices by liquid chromatography/mass spectrometry. Int. J. Environ. Anal. Chem. 85:1–14. doi:10.1080/03067 310512331324745

Gomes, R.L., W. Meredith, C.E. Snape, and M.A. Sephton. 2009a. Analysis of conjugated steroid androgens: Deconjugation, derivatisation and associated issues. J. Pharm. Biomed. Anal. 49:1133–1140. doi:10.1016/j.jpba.2009.01.027

Gomes, R.L., M.D. Scrimshaw, and J.N. Lester. 2009b. Fate of conjugated natural and synthetic steroid estrogens in crude sewage and activated sludge batch studies. Environ. Sci. Technol. 43:3612–3618. doi:10.1021/es801952h

Hanselman, T.A., D.A. Graetz, and A.C. Wilkie. 2003. Manure-borne estrogens as potential environmental contaminants: A review. Environ. Sci. Technol. 37:5471–5478. doi:10.1021/es034410+

He, Z.Q., P.H. Pagliari, and H.M. Waldrip. 2016. Applied and environmental chemistry of animal manure: A review. Pedosphere 26:779–816. doi:10.1016/S1002-0160(15)60087-X

Hobkirk, R. 1985. Steroid sulfotransferases and steroid sulfate sulfatases-characteristics and biological roles. Can. J. Biochem. Cell Biol. 63:1127–1144. doi:10.1139/o85-141

Hutchins, S.R., M.V. White, F.M. Hudson, and D.D. Fine. 2007. Analysis of lagoon samples from different concentrated animal feeding operations for estrogens and estrogen conjugates. Environ. Sci. Technol. 41:738–744. doi:10.1021/es062234+

Ingerslev, F., and B. Halling-Sørensen. 2003: Evaluation of analytical chemical methods for detection of estrogens in the environment. Danish Environmental Protection Agency, Copenhagen, Denmark.

Irwin, L.K., S. Gray, and E. Oberdorster. 2001. Vitellogenin induction in painted turtle, Chrysemys picta, as a biomarker of exposure to environmental levels of estradiol. Aquat. Toxicol. 55:49–60. doi:10.1016/S0166-445X(01)00159-X

Isobe, T., H. Shiraishi, M. Yasuda, A. Shinoda, H. Suzuki, and M. Morita. 2003. Determination of estrogens and their conjugates in water using solid-phase extraction followed by liquid chromatography-tandem mass spectrometry. J. Chromatogr. A 984:195–202. doi:10.1016/ S0021-9673(02)01851-4

Ivie, G.W., R.J. Christopher, C.E. Munger, and C.E. Coppock. 1986. Fate and residues of 4-C-14 estradiol-17-beta after intramuscular injection into Holstein steer calves. J. Anim. Sci. 62:681–690. doi:10.2527/jas1986.623681x

Jacobsen, A.M., A. Lorenzen, R. Chapman, and E. Topp. 2005. Persistence of testosterone and 17beta-estradiol in soils receiving swine manure or municipal biosolids. J. Environ. Qual. 34:861–871. doi:10.2134/jeq2004.0331

Jobling, S., M. Nolan, C.R. Tyler, G. Brighty, and J.P. Sumpter. 1998. Widespread sexual disruption in wild fish. Environ. Sci. Technol. 32:2498–2506. doi:10.1021/es9710870

Johnson, A.C., and J.P. Sumpter. 2001. Removal of endocrine-disrupting chemicals in activated sludge treatment works. Environ. Sci. Technol. 35:4697–4703. doi:10.1021/es010171j

Kellogg, R.L., C.H. Lander, D.C. Moffitt, and N. Gollehon. 2000: Manure nutrients relative to the capacity of cropland and pastureland to assimilate nutrients-spatial and temporal trends for the United States. GSA Publ. nsp00-0579. USDA-NRCS-ERS, Riverside, CA.

Khanal, S.K., B. Xie, M.L. Thompson, S.W. Sung, S.K. Ong, J. Van Leeuwen. 2006. Fate, transport, and biodegradation of natural estrogens in the environment and engineered systems. Environ. Sci. Technol. 40:6537–6546.

Kiang, T.K.L., M.H.H. Ensom, and T.K.H. Chang. 2005. UDP-glucuronosyltransferases and clinical drug-drug interactions. Pharmacol. Ther. 106:97–132. doi:10.1016/j.pharmthera.2004.10.013

Kolodziej, E.P., T. Harter, and D.L. Sedlak. 2004. Dairy wastewater, aquaculture, and spawning fish as sources of steroid hormones in the aquatic environment. Environ. Sci. Technol. 38:6377–6384. doi:10.1021/es049585d

Kolpin, D.W., E.T. Furlong, M.T. Meyer, E.M. Thurman, S.D. Zaugg, L.B. Barber, and H.T. Buxton. 2002. Pharmaceuticals, hormones, and other organic wastewater contaminants in US streams, 1999-2000: A national reconnaissance. Environ. Sci. Technol. 36:1202–1211. doi:10.1021/es011055j

Kumar, V., A.C. Johnson, N. Nakada, N. Yamashita, and H. Tanaka. 2012. De-conjugation behavior of conjugated estrogens in the raw sewage, activated sludge and river water. J. Hazard. Mater. 227-228:49–54. doi:10.1016/j.jhazmat.2012.04.078

Lægdsmand, M., H. Andersen, O.H. Jacobsen, and B. Halling-Sorensen. 2009. Transport and fate of estrogenic hormones in slurry-treated soil monoliths. J. Environ. Qual. 38:955–964. doi:10.2134/jeq2007.0569

Lafrance, P., and E. Caron. 2013. Impact of recent manure applications on natural estrogen concentrations in streams near agricultural fields. Environ. Res. 126:208–210. doi:10.1016/j.envres.2013.05.008

Lai, K.M., K.L. Johnson, M.D. Scrimshaw, and J.N. Lester. 2000. Binding of waterborne steroid estrogens to solid phases in river and estuarine systems. Environ. Sci. Technol. 34:3890–3894. doi:10.1021/es9912729

Lange, I.G., A. Daxenberger, B. Schiffer, H. Witters, D. Ibarreta, and H.H.D. Meyer. 2002. Sex hormones originating from different livestock production systems: Fate and potential disrupting activity in the environment. Anal. Chim. Acta 473:27–37. doi:10.1016/S0003-2670(02)00748-1

Layton, A.C., B.W. Gregory, J.R. Seward, T.W. Schultz, and G.S. Sayler. 2000. Mineralization of steroidal hormones by biosolids in wastewater treatment systems in Tennessee U.S.A. Environ. Sci. Technol. 34:3925–3931. doi:10.1021/es9914487

Lee, L.S., T.J. Strock, A.K. Sarmah, and P.S.C. Rao. 2003. Sorption and dissipation of testosterone, estrogens, and their primary transformation products in soils and sediment. Environ. Sci. Technol. 37:4098–4105. doi:10.1021/es020998t

Lucas, S., and D. Jones. 2006. Biodegradation of estrone and 17 b-estradiol in grassland soils amended with animal wastes. Soil Biol. Biochem. 38:2803–2815. doi:10.1016/j.soilbio.2006.04.033

Ma, L., and S.R. Yates. 2017. Degradation and metabolite formation of estrogen conjugates in an agricultural soil. J. Pharm. Biomed. Anal. 145:634–640. doi:10.1016/j.jpba.2017.07.058

Ma, L., and S.R. Yates. 2018. Degradation and metabolite formation of 17ss-estradiol-3-glucuronide and 17ss-estradiol-3-sulphate in river water and sediment. Water Res. 139:1–9. doi:10.1016/j.watres.2018.03.071

Maier, R.M., I.L. Pepper, and C.P. Gerba. 2000. Terrestrial environment In: R.M. Maier, I.L. Pepper, and C.P. Gerba, editors, Environmental microbiology. Academic Press, Waltham, MA.

Mashtare, M.L., D.A. Green, and L.S. Lee. 2013. Biotransformation of 17 alpha- and 17 beta-estradiol in aerobic soils. Chemosphere 90:647–652. doi:10.1016/j.chemosphere.2012.09.032

Matthiessen, P., D. Arnold, A.C. Johnson, T.J. Pepper, T.G. Pottinger, and K.G.T. Pulman. 2006. Contamination of headwater streams in the United Kingdom by oestrogenic hormones from livestock farms. Sci. Total Environ. 367:616–630. doi:10.1016/j.scitotenv.2006.02.007

Mina, O., H.E. Gall, L.S. Saporito, and P.J.A. Kleinman. 2016. Estrogen transport in surface runoff from agricultural fields treated with two application methods of dairy manure. J. Environ. Qual. 45:2007–2015. doi:10.2134/jeq2016.05.0173

Moore, A.B., G.D. Bottoms, G.L. Coppoc, R.C. Pohland, and O.F. Roesel. 1982. Metabolism of estrogens in the gastrointestinal tract of swine. 1. Instilled estradiol. J. Anim. Sci. 55:124–134. doi:10.2527/jas1982.551124x

Nichols, D.J., T.C. Daniel, P.A. Moore, D.R. Edwards, and D.H. Pote. 1997. Runoff of estrogen hormone 17 beta-estradiol from poultry litter applied to pasture. J. Environ. Qual. 26:1002–1006. doi:10.2134/jeq1997.00472425002600040011x

Olsen, P., K. Bach, H.C. Barlebo, F. Ingerslev, M. Hansen, and B. Halling. 2007. Leaching of estrogenic hormones from manure-treated structured soils. Environ. Sci. Technol. 41:3911–3917. doi:10.1021/es0627747

Orozco-Hernández, L., A.A. Gutierrez-Gomez, N. SanJuan-Reyes, et al. 2018. 17 beta-Estradiol induces cyto-genotoxicity on blood cells of common carp (Cyprinus carpio). Chemosphere 191:118–127. doi:10.1016/j.chemosphere.2017.10.030

Palme, R., P. Fischer, H. Schildorfer, and M.N. Ismail. 1996. Excretion of infused C-14-steroid hormones via faeces and urine in domestic livestock. Anim. Reprod. Sci. 43:43–63. doi:10.1016/0378-4320(95)01458-6

Panter, G.H., R.S. Thompson, and J.P. Sumpter. 1998. Adverse reproductive effects in male fathead minnows (Pimephales promelas) exposed to environmentally relevant concentrations of the natural oestrogens, oestradiol and oestrone. Aquat. Toxicol. 42:243–253. doi:10.1016/S0166-445X(98)00038-1

Panter, G.H., R.S. Thompson, and J.P. Sumpter. 2000. Intermittent exposure of fish to estradiol. Environ. Sci. Technol. 34:2756–2760. doi:10.1021/es991117u

Peterson, E.W., R.K. Davis, and H.A. Orndorff. 2000. 17 beta-estradiol as an indicator of animal waste contamination in mantled karst aquifers. J. Environ. Qual. 29:826–834. doi:10.2134/jeq2000.00472425002900030019x

Pinheiro, A., R.M.R. Albano, T.C. Alves, V. Kaufmann, and M.R. da Silva. 2013. Veterinary antibiotics and hormones in water from application of pig slurry to soil. Agric. Water Manage. 129:1–8. doi:10.1016/j.agwat.2013.06.019

Pollard, A.T., and M.J. Morra. 2017. Estrogens: Properties, behaviors, and fate in dairy manure-amended soils. Environ. Rev. 25:452–462. doi:10.1139/er-2017-0005

Purdom, C.E., P.A. Hardiman, V.J. Bye, N.C. Eno, C.R. Tyler, and J.P. Sumpter. 1994. Estrogenic effects of effluents from sewage treatment works. Chem. Ecol. 8:275–285. doi:10.1080/02757549408038554

Qi, Y., and T.C. Zhang. 2016. Transport of manure-borne testosterone in soils affected by artificial rainfall events. Water Res. 93:265–275. doi:10.1016/j.watres.2016.01.052

Routledge, E.J., D. Sheahan, C. Desbrow, G.C. Brighty, M. Waldock, and J.P. Sumpter. 1998. Identification of estrogenic chemicals in STW effluent. 2. In vivo responses in trout and roach. Environ. Sci. Technol. 32:1559–1565. doi:10.1021/es970796a

Runarsson, T.P., and X. Yao. 2000. Stochastic ranking for constrained evolutionary optimization. IEEE Trans. Evol. Comput. 4:284–294. doi:10.1109/4235.873238

Sabourin, L., P. Duenk, S. Bonte-Gelok, M. Payne, D.R. Lapen, and E. Topp. 2012. Uptake of pharmaceuticals, hormones and parabens into vegetables grown in soil fertilized with municipal biosolids. Sci. Total Environ. 431:233–236. doi:10.1016/j.scitotenv.2012.05.017

Scherr, F.F., A.K. Sarmah, H.J. Di, and K.C. Cameron. 2008. Modeling degradation and metabolite formation kinetics of estrone-3-sulfate in agricultural soils. Environ. Sci. Technol. 42:8388–8394. doi:10.1021/es801850a

Schlüsener, M.P., and K. Bester. 2005. Determination of steroid hormones, hormone conjugates and macrolide antibiotics in influents and effluents of sewage treatment plants utilising high-performance liquid chromatography/tandem mass spectrometry with electrospray and atmospheric pressure chemical ionisation. Rapid Commun. Mass Spectrom. 19:3269–3278. doi:10.1002/rcm.2189

Schuh, M.C., F.X.M. Casey, H. Hakk, T.M. DeSutter, K.G. Richards, E. Khan, and P.G. Oduor. 2011. Effects of field-manure applications on stratified 17 beta-estradiol concentrations. J. Hazard. Mater. 192:748–752. doi:10.1016/j.jhazmat.2011.05.080

Shargil, D., Z. Gerstl, P. Fine, I. Nitsan, and D. Kurtzman. 2015. Impact of biosolids and wastewater effluent application to agricultural land on steroidal hormone content in lettuce plants. Sci. Total Environ. 505:357–366. doi:10.1016/j.scitotenv.2014.09.100

Sheng, G.D., C. Xu, L. Xu, Y.P. Qiu, and H.Y. Zhou. 2009. Abiotic oxidation of 17 beta-estradiol by soil manganese oxides. Environ. Pollut. 157:2710–2715. doi:10.1016/j.envpol.2009.04.030

Shore, L.S., M. Gurevitz, and M. Shemesh. 1993. Estrogen as an environmental pollutant. Bull. Environ. Contam. Toxicol. 51:361–366. doi:10.1007/BF00201753

Shrestha, S.L., X. Bai, D.J. Smith, H. Hakk, F.X.M. Casey, G.L. Larsen, G. Padmanabhan. 2011. Synthesis and characterization of radiolabeled 17b-estradiol conjugates. J. Labelled Comp. Radiopharm. 54:267–271. doi:10.1002/jlcr.1864

Shrestha, S.L., F.X.M. Casey, H. Hakk, D.J. Smith, and G. Padmanabhan. 2012. Fate and transformation of an estrogen conjugate and its metabolites in agricultural soils. Environ. Sci. Technol. 46:11047–11053. doi:10.1021/es3021765

Sosienski, T.A. 2017: The occurrence and fate of steroid hormones from manure amended agriculture fields. Crop and Soil Environmental Sciences, Virginia Polytechnic Institute and State University, Blacksburg, VA.

Steiner, L.D., V.J. Bidwell, H.J. Di, K.C. Cameron, and G.L. Northcott. 2010. Transport and modeling of estrogenic hormones in a dairy farm effluent through undisturbed soil lysimeters. Environ. Sci. Technol. 44:2341–2347. doi:10.1021/es9031216

Thorne, P.S. 2007. Environmental health impacts of concentrated animal feeding operations: Anticipating hazards- Searching for solutions. Environ. Health Perspect. 115:296–297. doi:10.1289/ehp.8831

Tremblay, L.A., J.B. Gadd, and G.L. Northcott. 2018. Steroid estrogens and estrogenic activity are ubiquitous in dairy farm watersheds regardless of effluent management practices. Agric. Ecosyst. Environ. 253:48–54. doi:10.1016/j.agee.2017.10.012

van Donk, S.J., S. Biswas, W.L. Kranz, D.D. Snow, and S.L. Bartelt-Hunt. 2013. Transport of steroid hormones in the vadose zone after land application of beef cattle manure. Trans. ASABE 56:1327–1338.

Verderame, M., E. Limatola, and R. Scudiero. 2016. Estrogenic contamination by manure fertilizer in organic farming: A case study with the lizard Podarcis sicula. Ecotoxicology 25:105–114. doi:10.1007/s10646-015-1571-0

Zhang, F.S., Y.F. Xie, X.W. Li, D.Y. Wang, L.S. Yang, and Z.Q. Nie. 2015. Accumulation of steroid hormones in soil and its adjacent aquatic environment from a typical intensive vegetable cultivation of North China. Sci. Total Environ. 538:423–430. doi:10.1016/j.scitotenv.2015.08.067

Zhang, H., J.H. Shi, X.W. Liu, X.M. Zhan, and Q.C. Chen. 2014a. Occurrence and removal of free estrogens, conjugated estrogens, and bisphenol A in manure treatment facilities in East China. Water Res. 58:248–257. doi:10.1016/j.watres.2014.03.074

Zhang, X.L., Y.X. Li, B. Liu, W. Jing, and C.H. Feng. 2014b. The effects of estrone and 17 beta-estradiol on microbial activity and bacterial diversity in an agricultural soil: Sulfamethoxazole as a co-pollutant. Ecotoxicol. Environ. Saf. 107:313–320. doi:10.1016/j.ecoenv.2014.06.010

Zhao, S., P.F. Zhang, M.E. Melcer, and J.F. Molina. 2010. Estrogens in streams associated with a concentrated animal feeding operation in upstate New York, USA. Chemosphere 79:420–425. doi:10.1016/j.chemosphere.2010.01.060

Zhao, X.M., and W.S. Lung. 2018. Tracking the fate and transport of estrogens following rainfall events. Water Sci. Technol. 77:2474–2481. doi:10.2166/wst.2018.204

Zheng, W., K.N. Wiles, and L. Dodgen. 2016: Uptake and accumulation of pharmaceuticals and hormones in vegetables after irrigation with reuse water. Illinois Sustainable Technology Center, Champaign, IL.

Zheng, W., Y.H. Zou, X.L. Li, and M.L. Machesky. 2013. Fate of estrogen conjugate 17 alpha-estradiol-3-sulfate in dairy wastewater: Comparison of aerobic and anaerobic degradation and metabolite formation. J. Hazard. Mater. 258–259:109–115. doi:10.1016/j.jhazmat.2013.04.038

Managing Animal Manure to Minimize Phosphorus Losses from Land to Water

Peter J.A. Kleinman,* Sheri Spiegal, Jian Liu, Mike Holly, Clint Church and John Ramirez-Avila

Abstract

Given the expansion of eutrophication in water bodies around the world, the improved management of manure to mitigate phosphorus (P) losses to water has become a global concern. This chapter seeks to frame manure management strategies and practices to minimize P losses to water, with a focus on manure properties, land application, farmstead infrastructure, and farming systems. Although many options exist to better manage manure, and, more specifically, P in manure, doing so requires comprehensive approaches that consider factors far beyond the direct handling of manure and require decisions that may compete or conflict with other priorities, most notably profits and time management.

Eutrophication is the most pervasive concern to freshwater and estuarine water bodies worldwide, with phosphorus (P) pollution continuing to expand the extent of eutrophication, its impacts on aquatic life and its disruption of the benefits of ecosystems to humankind. No sector in agriculture has been more closely tied to the accelerated eutrophication of aquatic systems than animal agriculture, principally as a result of the generation and management of manure. Although P contributions from other sources, agricultural and non-agricultural alike, contribute to the spread of eutrophication globally, the economics and structure of animal production systems, both in modern times and historically, the concentration of P in manure and around animal production zones, and the vagaries of precisely and efficiently using manure resources all compound into myriad opportunities for P to be lost from animal production systems and transferred to the aquatic environment.

Successful manure management strategies must consider aspects of animal production systems that, at first glance, do not appear to connect directly with on-farm manure management practices such as handling, storage, and application. Just as eutrophication manifests itself in water bodies as small as farm ponds and as large as the Baltic Sea, P mitigation opportunities must consider manure management scales that extend well beyond the farm gate. Indeed, successful manure management strategies must be viewed through a comprehensive lens that considers long-term factors as well as the immediate consequences of management decisions. Options to managers are often difficult to implement, because manure

P.J.A. Kleinman, USDA Agricultural Research Service, Pasture Systems and Watershed Management Research Unit, University Park, PA 16802; S. Spiegal, USDA Agricultural Research Service, Range Management Research Unit, Las Cruces, NM 88003; J. Liu, University of Saskatchewan, School of Environment and Sustainability, Saskatoon, SK S7N 5B5; M. Holly, University Wisconsin-Green Bay, Department of Natural and Applied Sciences, Green Bay, WI 54311; C. Church, USDA Agricultural Research Service, Pasture Systems and Watershed Management Research Unit, University Park, PA 16802; J. Ramirez Avila, Mississippi State University, Civil and Environmental Engineering Dept, Mississippi State, MS 39762.

*Corresponding author (peter.kleinman@ars.usda.gov)

doi:10.2134/asaspecpub67.c12

Animal Manure: Production, Characteristics, Environmental Concerns and Management. ASA Special Publication 67. Heidi M. Waldrip, Paulo H. Pagliari, and Zhongqi He, editors.
© 2019. ASA and SSSA, 5585 Guilford Rd., Madison, WI 53711, USA.

SYSTEMIC FACTORS

Maintaining a P balance within a farm; Manure export, brokering, and redistribution; Regulatory programs; Predictive tools for decision support (e.g., P Indices)

INFRASTRUCTURE

Improvement and modernization of barnyards; Stormwater management (both quantity and quality); Adequate manure handling and storage systems to recover P

ANIMAL AND EXCRETA

Animal feed management (precision feeding; adjusting the composition of diets; adding phytase); Manure treatment (adding metal salts)

LAND APPLICATION

"4R" strategies:
Right source; Right form;
Right placement; and Right time

Fig. 1. Managing animal manure across scales to minimize phosphorus losses to water. These scales and their interactions are explored as themes of this chapter.

is a byproduct and not the intended product of animal production and its value, including its potential liabilities (odor, pathogen source, stigma), require management solutions that not only consider social and economic constraints, but take advantage of motivations other than P mitigation to affect change.

This chapter seeks to frame and illustrate the diversity of strategies and practices for manure management that constitute the state of knowledge and the state of the art of managing manures to minimize P losses to water, with a focus on manure properties, land application, farmstead infrastructure, farming systems (Fig. 1). Although examples of manure P management are considered from around the world, there is a noteworthy bias toward manure management in North America (United States and Canada). Further, this chapter does not contain a complete litany of available technologies, even as it strives to highlight recent advances and feature the many dimensions of sustainable manure management. Through the insertion of case studies, we hope to illustrate the opportunities and barriers to manure management options that mitigate P losses to water.

Systemic Factors Affecting Manure Phosphorus, within and among Farms

From the standpoint of manure P, it is appropriate to begin a review of manure management at the level of the animal production system, including not only the farms producing the manure, but the larger networks that provide animal producers with P, be it in fertilizers, grains or forages. A systemic perspective on P cycling helps to differentiate factors that are under the immediate control of producers and those factors outside of a farmer's capacity to manage. It is well-established that the specialization and intensification of modern production systems, vertical integration of certain animal industries (esp. poultry and swine), and consolidation of operations promoted by economies of scale, has wrought great efficiencies in resource flows (MacDonald and McBride, 2009). These economic efficiencies have promoted the uncoupling of nutrient cycles, promoting the export of nutrients from regions with a comparative advantage in producing feeds and forages, and the import of nutrients to areas with a comparative advantage in the production of animals (Lanyon, 2005). Approximately half of the grain produced in the United States is consumed by animals. Despite major improvements in balancing nutrients in feeding regimes, the majority of P fed to animals is not metabolized and therefore destined for manure. As a result, hot spots of manure P are regularly found in regions with high concentrations of animal production (Fig. 2).

Although the benefits of land-applying manure are manifold, it is often difficult to directly substitute manure for synthetic, commercial fertilizers. It is bulky in nature, low in nutrient density, tending toward stoichiometries that are out of balance with crop requirements, unpredictable in nutrient availability to crops, malodorous, a source of pathogens, difficult to store and difficult to handle (Ribaudo et al., 2003; Kleinman et al., 2012). Further, there is a growing number of nutrient management standards that mandate additional paperwork, practices and effort to use manure as a substitute for commercial fertilizers. As a result, manure nutrients are generally undervalued in comparison with commercial nutrient

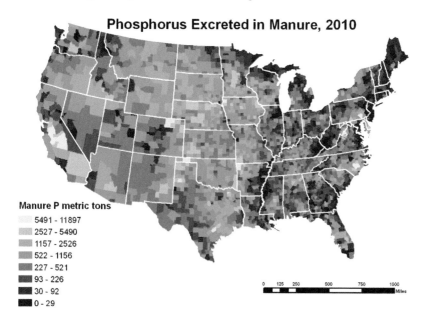

Fig. 2. Estimate of annual excretion of phosphorus in manure in 2010, by county, in the United States. Adapted from Jarvie et al. (2015), based on data from NuGIS (IPNI, 2012).

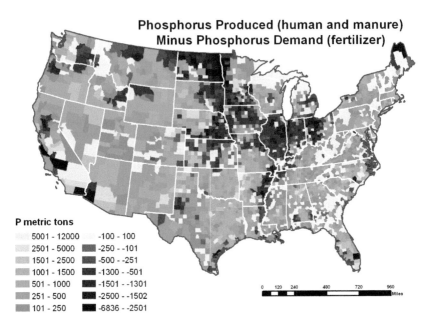

Phosphorus Produced (human and manure)
Minus Phosphorus Demand (fertilizer)

P metric tons

5001 - 12000	-100 - 100
2501 - 5000	-250 - -101
1501 - 2500	-500 - -251
1001 - 1500	-1300 - -501
501 - 1000	-1501 - -1301
251 - 500	-2500 - -1502
101 - 250	-6836 - -2501

Fig. 3. Comparison of manure and biosolid phosphorus (P) with fertilizer P demand in 2010, by county, in the United States. Adapted from Jarvie et al. (2015).

fertilizers, even if their benefits to crop production are nominally appreciated by farmers and well-documented by science (Magdoff et al., 1997; Beegle, 2014). Unsurprisingly, most manure travels short distances from the barns where it is generated to its point of land application, with ranges of 5 to 20 km often cited as a practical limit (Bartelt and Bland, 2007; Hadrich et al., 2010). Well-constructed manure transport programs can move some manures (e.g., poultry litter) much further, up to 250 kg (Herron et al., 2012). If manure's liabilities could be minimized, many opportunities can be found to substitute manure for purchased fertilizer P (Bosch and Napit, 1992; Carreira et al., 2007). Unfortunately, regions with the most productive cropland, such as the mid-western United States, manifest annual deficits in P on agricultural lands that are currently met by fertilizers, even as other regions reveal theoretical surpluses in P through local generation of manure and its human counterpart, wastewater-treatment-plant biosolids (Fig. 3).

The connection of system-level variables to manure management is particularly evident in farm P budgeting, including connections between manure management and feed use efficiency, fertilizer management, even cropping practices (Rotz et al., 2002; Tarkalson and Mikkelsen, 2003; Plaizier et al., 2014). As illustrated in Fig. 4 for a variety of dairy farming systems in the United States, enterprise P budgets can range widely, even within a single industry. The large accumulations of manure P on some dairy operations reflect a combination of factors, from excess P in dairy rations (e.g., Knowlton et al., 2010), to lesser reliance on forages raised on the farm (Ghebremichael et al., 2008), to unnecessary purchases of commercial P fertilizers (Ketterings et al., 2011), to an inability to export manure from the farm (Van Horn et al., 1994). Notably, from a whole farm budgeting perspective, P losses in runoff, often the driving factor behind manure management regulations, are a minor component of the overall enterprise P balances and almost negligible in dairies of arid regions, even though the impact of these small P losses on the water quality can be great (Sharpley et al., 2003a).

Maintaining a P balance within a farming operation is essential to preventing the accumulation of manure P in farm soils–now referred to as "legacy P" due to its role as a long-term source of P to agricultural runoff–one of the most profound environmental management problems facing agriculture (Sharpley et al., 2013). Increasingly, farm P balances have been promoted for guiding strategic decisions (e.g., Chaperon et al., 2007; Soberon et al., 2015), differentiating between strategies that are suited to enterprises with net P surpluses (export manures, acquire more cropland to use the excess manure nutrients, adjust animal numbers- hence manure generation) and enterprises with net P deficits [targeting manure application to fields with low soil P and potassium (K), avoiding manure application to legumes that don't require the nitrogen (N)]. These farmgate assessments serve as an initial check on the potential for additional manure management practices, such as those described below, to meet production and environmental objectives.

For farming systems in which a substantial P surplus exists, manure export is often seen as an essential solution to on-farm accumulation of P, particularly if costs can be absorbed through subsidy or through value added processes that promote the transport of manure off-farm and, preferably, away from local hot spots. Increasingly, in vertically integrated industries with a high degree of strategic coordination (e.g., poultry and swine), new operations are not even designed to land apply manure that is produced and manure export may be required in a contract. Unsurprisingly, manure export programs tend to involve dry manures, either in the form of litters or processed manure solids, which are easier to transport due to their nutrient density (see also discussion below on manure treatment systems). Programs to promote manure export have been very difficult to implement, particularly community programs that involve farm collectives, and, once implemented, equally difficult to sustain (Kleinman et al., 2012). Further, there

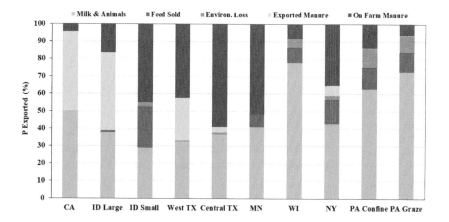

Fig. 4. Enterprise budgets for phosphorus (P) in select dairy farming systems of the United States, showing P distribution in milk and animals sold, feed sold, environmental losses, an on-farm manure accumulation (soil build up), as a percentage of the total quantity of P exported from the farm. Adapted from Holly et al. (2018).

Table 1. Key properties of manures surveyed in the Mid-Atlantic region of the United States. Mean values are followed with standard deviations in parentheses. Nutrient concentrations are expressed on a dry weight equivalent basis. Adapted from Liu et al. (2018a).

Manure type	No. of samples	pH†	Dry matter	Total P	WEP‡	WEP/Total P	Total Ca	Total Mg
						---- % ----		
Cattle, dairy								
Manure – All	275	7.53	9.8 (6.8)	0.60 (0.26)	0.28 (0.13)	46.7 (20.8)	2.96 (2.42)	0.84 (0.36)
– Liquid	29	7.39	2.3 (1.0)	0.91 (0.40)	0.31 (0.13)	34.1 (19.3)	3.95 (1.85)	1.31 (0.43)
– Slurry	179	7.49	7.9 (1.4)	0.60 (0.20)	0.30 (0.12)	50.0 (21.1)	2.94 (2.06)	0.84 (0.30)
– Semi-solid	46	7.58	13.1 (2.35)	0.49 (0.17)	0.23 (0.15)	46.9 (21.4)	3.15 (3.87)	0.70 (0.26)
– Solid	21	7.71	29.5 (6.90)	0.35 (0.10)	0.16 (0.06)	45.7 (12.7)	1.40 (0.56)	0.41 (0.11)
Cattle, beef								
Manure – All	12	n.d.	23.1 (5.4)	0.52 (0.19)	0.21 (0.13)	40.4 (15.2)	1.51 (0.55)	0.52 (0.24)
– Semi-solid	3	n.d.	16.8 (3.9)	0.59 (0.25)	0.34 (0.18)	57.6 (13.0)	1.21 (0.16)	0.49 (0.22)
– Solid	9	n.d.	25.2 (4.1)	0.50 (0.18)	0.17 (0.08)	34.0 (11.5)	1.61 (0.61)	0.53 (0.25)
Compost	8	n.d.	39.1 (21.6)	0.46 (0.19)	0.11 (0.04)	23.9 (12.3)	2.29 (0.89)	0.62 (0.31)
Swine, farrow to wean or farrow to feeder								
Manure – All	36	6.63	3.0 (2.9)	2.68 (0.93)	0.56 (0.30)	20.9 (11.6)	4.27 (2.15)	1.75 (0.64)
– Liquid	31	6.63	2.1 (0.9)	2.51 (0.80)	0.57 (0.33)	22.7 (12.0)	4.02 (1.50)	1.68 (0.58)
Swine, wean to finish or grow to finish								
Manure – All	35	7.87	4.5 (5.9)	2.37 (0.96)	0.41 (0.19)	17.3 (14.8)	2.62 (0.95)	1.61 (0.52)
– Liquid	24	7.98	2.0 (0.8)	2.15 (1.02)	0.42 (0.22)	19.5 (16.5)	2.44 (0.78)	1.54 (0.59)
Swine, mixed stages								
Manure – All	153	7.14	7.5 (5.5)	3.58 (1.38)	0.50 (0.33)	14.0 (15.6)	7.84 (5.62)	2.27 (1.06)
– Liquid	71	7.09	2.2 (1.1)	2.51 (0.70)	0.70 (0.40)	27.9 (16.8)	3.13 (2.56)	1.44 (0.63)
– Slurry	21	7.66	8.0 (2.3)	3.97 (1.33)	0.38 (0.08)	9.6 (7.4)	8.01 (5.48)	2.44 (0.97)
– Semi-solid	60	n.d.	13.2 (1.9)	4.59 (1.09)	0.32 (0.07)	7.0 (4.2)	12.78 (3.4)	3.09 (0.75)
Chicken, layer								
Solid	30	n.d.	80.3 (11.6)	1.12 (0.66)	0.30 (0.18)	26.8 (15.4)	9.11 (2.91)	0.76 (0.49)
Compost	2	n.d.	73.7 (9.4)	2.42 (0.25)	0.51 (0.03)	21.1 (3.6)	9.57 (7.92)	0.99 (0.13)
Chicken, broiler								
Solid	7	7.73	71.5 (8.3)	1.43 (0.36)	0.35 (0.19)	24.5 (11.5)	2.44 (0.68)	0.60 (0.14)
Compost	4	7.64	67.9 (2.3)	1.82 (0.17)	0.46 (0.08)	25.3 (6.7)	2.52 (0.19)	0.80 (0.13)
Turkey								
Solid	10	n.d.	66.8 (6.0)	1.94 (0.61)	0.52 (0.13)	26.8 (6.4)	2.99 (0.99)	0.62 (0.21)
Horse								
Solid	17	n.d.	41.6 (14.2)	0.49 (0.19)	0.27 (0.17)	55.1 (17.8)	1.76 (2.02)	0.43 (0.25)
Compost	7	n.d.	46.8 (15.1)	0.54 (0.33)	0.19 (0.14)	35.2 (22.0)	1.96 (1.63)	0.58 (0.41)

† n.d. = no data

‡ WEP, water extractable phosphorus.

is a need to ensure that exported manure is used prudently to ensure that the exported manure is used in a sustainable fashion (Liu et al., 2016).

Examples of community programs to promote manure transport in the United States, many of which have faced and sometimes succumbed to existential challenges, include a poultry litter pelletizing, now composting, facility in Delaware (Pipkin, 2017), manure brokering within Chesapeake Bay watershed states (Dance, 2017), a manure-to-energy turkey litter incineration plant in Minnesota (MacDonald, 2006), poultry litter baling and brokering programs in Arkansas (O'Keefe, 2011; Herron et al., 2012), a large scale composting facility in Pennsylvania (Torres, 2010), and a community anaerobic digester for dairies in Wisconsin (USEPA, 2015). Internationally, hopeful futures in community manure processing programs may be found in manure-to-energy for Finland's dairy industry (Uutiset, 2018) and nutrient recovery systems being trialed by Sweden's solid waste industry (EasyMining, 2018). In all cases, regulatory programs, such as those promoting circular economies (Nesme and Withers, 2016), are required to ensure that costs associated with exporting manures from farms and processing them at community facilities do not undermine investment, development and maintenance.

Manure Properties Affecting Phosphorus Availability to Runoff

The connection between agricultural P and manure is profound, with ties to the history of modern agriculture and science alike. Phosphorus was the first element discovered by modern science 350 yr ago, obtained from urine, albeit 5500 L of human urine (Sharpley et al., 2018). Bird and bat guano were once the major sources of fertilizer P in the 19th Century (Giaimo, 2015), as were, briefly, deposits of fossilized dung (Schwarcz, 2017). The relative inefficiency of vertebrates in metabolizing P from food sources, combined with the reactive nature of P, enriches excreta with relatively high concentrations of the element. When these excreta are combined with bedding, water, and other materials to form manure, it is not surprising that manure characteristics vary widely (He et al., 2016).

The nutrient composition of manures (e.g., Total P, Total Ca, Total Mg in Table 1) can be influenced by the biology of the animal as well as by management factors. In a survey of 140 manure samples, Kleinman et al. (2005) reported distinctive differences in total P by animal species when normalized on a dry weight basis: from a low for the major ruminant livestock (0.5% for beef cattle and 0.7% for dairy cattle), to higher concentrations in monogastric species (1.6% for broiler chickens, 2.6% for layer chickens, 2.4% for turkey, and 2.9% for swine). Nutrient compositions in manures are also influenced by animal diets, as well as by manure storage and handling processes (Rotz, 2004). For example, Toor et al. (2005b) reported a 40% decrease in manure P concentrations after reducing total P in dairy diets from 5.1 to 3.6 g kg^{-1}. Eghball (2000) and Preusch et al. (2002) found that composting significantly lowered the N to P ratio in manures, as a result of volatilization of N at the same time that P was conserved.

As factors such as diet and manure management have changed over time, so too have the properties of animal manures. In fact, Liu et al. (2018a) compared properties of manures from the earlier survey of Kleinman et al. (2005) with properties of manures from the same region approximately one decade later (Table 1). They found that while total P in manure of some animals (beef cattle, broiler chickens and turkeys) remained similar at the two points in time, total P in the manure from dairy cattle declined significantly (- 30%), as did total P in layer chicken manure (- 39%) and swine manure (- 46%). In part, these changes are a sign of the recognition of the need for parsimony in dietary P to curtail P in manure (e.g., National Research Council, 2001), in no small part due to efforts to educate animal producers (Applegate and Angel, 2008). Strategies to manipulate animal diets to affect the quantity of P include precisely managing P supplements, adjusting the composition of diets (Dou et al., 2002; Toor et al. 2005a, 2005b), as well as adding phytase to poultry and swine feed or otherwise breaking down organic P forms in feed (Humer et al., 2015). As monogastric animals, poultry and swine lack the phytase generating bacteria found in the anaerobic environment of the rumen in cattle, sheep and goats (Yanke et al., 1998). Phytin, or phytic-acid, is the dominant form of P found in corn and soybeans and must be hydrolyzed to be metabolized. The advent of commercial phytase in the late 1990s led to significant changes in feed formulations in poultry and swine diets, with well documented reductions in the total P content of manure as long as the phytase amendment was tied to a reduction in P supplementation (Smith et al., 2004; Applegate and Angel, 2008).

While total P in manures is an essential metric for P budgets, it is not necessarily a good indicator of the immediate availability of P in land-applied manure to the environment (Moore et al., 1999). Water extractable P (WEP; Kleinman et al., 2007) has been widely used to predict the potential for land-applied manure, particularly manure applied to the soil surface, to directly transfer dissolved forms of P to runoff water (Withers et al., 2001; Kleinman et al., 2002; Brandt and Elliott, 2003). In some areas, WEP is used to adjust environmental recommendations for land application of manure (Elliott et al., 2006), and WEP is an important variable in the most sophisticated computational models for P runoff (Vadas et al., 2009). Just as total P varies widely across animal species, so too does WEP. In the survey of Liu et al. (2018a), WEP ranged by an order of magnitude, with the highest average concentrations found in manures from turkey and swine manure, followed by chickens (layers and broilers), followed by dairy, beef, and horse manures (Table 1). In general, composted manures had lesser WEP concentrations. Water extractable P comprised 11 to 58% of total P in these manures and composts. Notably, Liu et al. (2018a) found WEP to be negatively related to the dry matter content of manures, and positively related to estimates of P sorption saturation (total P/[total Ca + Mg + Fe + Al + Mn]). There was also an indication that increasing pH of dairy and swine manures (from 6.6 to 8.3) reduced WEP. All these relationships point to opportunities to lower manure WEP concentrations through practices that affect manure physical and chemical traits.

A large body of literature documents the potential to reduce manure WEP using various metal salts, borrowing from processes in wastewater treatment plants. Perhaps the best accepted amendment for manures is alum ($Al_2(SO_4)_3$), which is used in the production of over 1 billion broiler chickens in the United States alone. Notably, alum's adoption by the broiler industry reflects its contribution to ammonia conservation and documented improvements in the growth and health of housed birds (Choi and Moore, 2008). Other salts that have been trialed include ferric chloride ($FeCl_3$), ferric sulfate ($FeSO_4$), aluminum chloride ($AlCl_3$), and gypsum ($CaSO_4$), all of which show varying potential to lower WEP in manure but none of which have gained widespread acceptance (Xin et al., 2011). Generally, a reduction in the solubility of P in manure with salt addition does not significantly impact P availability to crops, although the added P sorption capacity from salt amendments in manure can also reduce P solubility in soils, as demonstrated by long-term studies with alum-treated poultry litter (Huang et al., 2016). To address cost, particularly when an amendment does not offer ancillary benefits that would improve profitability of animal production (alum has a benefit to cost ratio of 2 due to aforementioned improvement in chicken health; Moore et al., 2000), waste materials have often been considered (water treatment residuals rich in Al and Fe, Elliott et al., 2002; Ca-, Al- and Fe-rich by-products from coal-fired plants, Dou et al., 2003; Al and Fe residues from treatment of acid mine drainage, Adler and Sibrell, 2003). Notably, in liquid manures, there are acute exposure concerns associated with the use of amendments that include S, due to the potential generation of hydrogen sulfide (H_2S) gas under anaerobic conditions (Fabian-Wheeler et al., 2017).

Farmstead Infrastructure

The management of manure in and around animal production facilities represents one of the greatest opportunities to control acute losses of P from animal production systems. Indeed, modern barns, manure handling and storage systems, and standards for barnyards and drainage system are intended to prevent discharges of manure and its constituents with stormwater. At the core of farmstead manure management strategies is the principal of efficiently isolating and containing manure by preventing the interaction of farmstead stormwater with surfaces and structures where manure is found.

Catastrophic discharges of manure can occur when infrastructure is improperly sited, insufficiently maintained, overwhelmed by extreme storm events, or simply fails (Ogejo, 2009). As climate change accelerates, there is a need for continuous reconsideration of how manure infrastructure is designed and managed (Wright et al., 2013). For instance, 33 swine lagoons discharged manure and six lagoons suffered structural damage in North Carolina as a result of rain and flooding from Hurricane Florence of 2018, despite major changes to the siting, design, and management of lagoons following in the aftermath of Hurricane Floyd in 1999 (NCDEQ, 2018).

Efforts to mitigate manure P losses to the environment due to farm infrastructure are regularly hindered by historical inertia, supported by cultural factors that resist change, as well as the cost of most infrastructure projects. Around the world, past standards (or a lack there-of) and traditional development practices often did not consider water quality implications, and even relied on periodic flooding or direct discharge to waterways as a form of manure management. As late as 2000, an estimated 30 to 70% of China's animal manure was directly discharged to rivers (Strokal et al., 2016). Even when direct discharge to waterways is not a primary factor, inadequate farmstead infrastructure can undermine watershed efforts to mitigate P loss. In a classic case study on this matter, Meals (1993) reported that barnyard improvement strategies on small, New England dairy farms were the most effective practice in mitigating watershed P losses and were essential to the success of watershed management activities around Vermont's LaPlatte River. Similarly, modernization of barnyards on small dairy farms was a principal feature of New York City's successful remediation of P pollution in watersheds serving as the source of its drinking water (USEPA, 2007).

In addition to the modernization of barnyards, a variety of filtration approaches have been used to treat and detain and/or retain stormwater discharge. Traditional practices to remove P from stormwater include vegetated filter strips, detention or

Washington State University on-farm struvite extraction plant
https://Puyallup.wsu.edu/lnm/struvite-extraction/

60-80% P recovery efficiency

USDA-ARS Nutrient Extraction Plants for Swine Operations
Vanotti et al. (2018)

99+ % P removal efficiency

P recovered as calcium phosphate and biosolids

USDA-ARS/Penn State mobile, manure P extraction system
Church et al. (2018)

99% P removal efficiency
P recovered as organic solids

Fig. 5. Examples of on-farm and mobile treatment systems developed in the United States to recover phosphorus from liquid dairy and swine manures.

retention basins, and constructed wetlands (Young et al., 1980; Walker, 1987; Schwer and Clausen, 1989; House et al., 1994; Drizo et al., 1997). Invariably, these filters are overwhelmed by large volumes of stormwater flow and concentrations of P in runoff from farmsteads. While traditional filtration practices can be very effective in controlling particulate P losses from stormwater, assuming they are not undermined by preferential flows (Kim et al., 2006), the capacity of traditional practices to bind dissolved forms of P in runoff from heavily manured sites tends to be overwhelmed by the process of P sorption saturation (Dillaha et al., 1988, 1989; Kleinman, 2017). Indeed, due to the reversibility of sorption processes, heavily P-saturated filters can become sources of dissolved P to runoff. In an extreme example, a constructed wetland used to treat feedlot runoff in Manitoba, Canada eventually became P saturated such that total P concentrations in wetland discharge were 30% greater than concentrations in the feedlot runoff (Pries and McGarry, 2002).

Due to concerns related to the efficacy of traditional runoff treatment practices in mitigating P loss, options to filter stormwater from farmstead areas and other concentrated discharges (e.g., tile and ditch drainage) are growing. Most experience with P filtration out of stormwater with these alternatives remains in the research realm, but a myriad of opportunities exist to implement these systems around older farmsteads that were not developed with modern stormwater management standards (Kleinman et al., 2012). For instance, Bird and Drizo (2010) describe the design of systems to treat milkhouse waste based on residual slag materials, with dissolved P filtration efficiencies up to 76%, with a maximum filter media replacement period of five to six months. In a review of alternative P filtration systems for nonpoint source runoff, Buda et al. (2012) emphasize the extremely high filtration efficiencies (> 60%) of such systems, while Penn et al. (2017) highlight retention time and the sorption properties of the filter media as critical factors in determining filtration efficacy. To compensate for short retention times in stormwater treatment systems, Qin et al. (2018) recommend filter media based on Fe and Al sorption, which is more rapid than precipitation of dissolved P with Ca. Due to cost, most P filters proposed for agricultural settings employ byproduct materials from various industrial processes, although natural P sorbing materials have also been investigated (McDowell et al., 2008; Vohla et al., 2011; Buda et al., 2012), and there is interest in promoting the use of more expensive P sorbing media that would enable the recovery of captured P as a fertilizer product.

Another key consideration in leveraging farmstead infrastructure to minimize P losses to water are manure handling and storage systems. These systems are a principal determinants of manure nutrient properties, particularly when explaining differences within individual species (Table 1). From a qualitative perspective, even manure management strategies aimed at N conservation (e.g., separating urine and feces in dairy barns) have implications to P management as they can substantially alter N/P, and therefore rates of land application, which are frequently driven by crop N demand. Further, systems that generate readily-transportable manures (i.e., dry) or value-added materials (e.g., composts, vermicomposts) are generally needed to promote long-distance transport of manures from their source. However, because P is conserved across the manure handling and storage process, as opposed to N which may suffer significant volatile losses, farm P balances are seldom impacted by handling and storage systems without conscientious effort to engage in manure export. Although potentially self-evident, this is an important distinction because technologies such as anaerobic digesters and solid separators are regularly promoted in watershed mitigation programs (e.g., Kleinman et al., 2012), but their impact on watershed P losses is difficult to enumerate without explicit ties to activities in the field or at the farmgate.

Although P recovery methods are common in municipal wastewater treatment systems, fewer methods exist for livestock manure. A variety of systems have been developed, but their economic viability is still in question (Fig. 5). One of the most common approaches to P recovery from animal manures is to 'force' the formation of struvite (NH_4MgPO_4; Burns and Moody, 2002) by amending the manure with Mg, typically the limiting ion for struvite formation in manures and then removing the struvite. The formation of struvite is sensitive to the ratios of Mg, NH_4, PO_4, and pH, but in liquid swine manures with low solids content (< 2%), the struvite formed can be readily recovered using gravity settling. In one laboratory demonstration, recovery of P in struvite from swine manure slurries was as high as 90% (Burns and Moody, 2002). However, the higher solids content of dairy manures interferes with the gravity settling of struvite, and so, pretreatment to remove manure solids is required. One approach that has been tested uses a fluidized bed reactor to extract P in the form of struvite from anaerobically digested dairy manure liquor. Reported extraction efficiencies for this approach range from 60 to 80% (Fig. 5) (Washington State University, 2018).

Other manure treatment systems have been proposed that recover Ca-P compounds. For instance, in South Carolina, Vanotti et al. (2005, 2010) developed a full-scale, two-stage system for a swine production facility that can treat raw swine manure (Fig. 5). This system uses a nitrification bioreactor to reduce carbonate and ammonium buffers, followed by the addition of $Ca(OH)_2$ to precipitate Ca–P (Vanotti et al., 2005). Manure solids containing about 95% of the P are collected after the treatment process and are exported for end uses in compost or low-solubility fertilizer. Other benefits of this system are the removal of odors and pathogens. Notably, many Ca-based P recovery systems and anaerobic digesters generally volatilize much of the manure N content as they adjust pH to favor Ca-P precipitation unless recovery systems are included to capture NH_3 off-gassed (Karunanithi et al., 2015; Garcia-Gonzales et al., 2016).

Because a substantial amount of P is found in manure solids, liquid–solid separation systems can be used to recover P, albeit in bulk forms that do not readily substitute for commercial fertilizers. Given the costs of bedding, recovery of manure solids, and composting is a common practice on many large dairies, recycling the associated P within the barn rather than applying it to land. Church et al. (2016, 2017) demonstrated that up to 60% of P could be recovered from dairy slurry with liquid–solid separation technologies (screw press followed by centrifugation). When these technologies were combined with chemical treatment, up to 99% of manure P was recovered in various solids, leaving a liquid effluent that preserves more than 90% of the N in the manure (Digested Organics, 2018; Livestock Water Recycling, 2018; Church et al., 2016, 2017, 2018). A mobile version of the solid–separation and chemical treatment system, MAnure PHosphorus EXtraction (MAPHEX), was developed to treat manure slurries on small dairies (Fig. 5) (Church et al., 2016, 2017, 2018).

Land Application of Manure

Globally, animal manure accounts for over 60% of the N and P applied to land each year (Potter et al., 2010), a poignant statistic that, in its magnitude, highlights both the potential for manure to substitute for commercial fertilizer and the imperative that manure be used sustainably in global food production (Magdoff et al., 1997). The tenets of agronomic manure management can be found in the evolving principals of "4 R" nutrient management stewardship (Right *rate*, Right *timing*, Right *placement*, Right *form*) as well as in its predecessor, critical source area management (Sharpley et al., 2003b; Ehmke, 2014). These principals seek to elevate consideration of P use efficiency by crops as well as to curtail the environmental impact of land-applied manure P. While easily stated, crop P use efficiency and environmental protection can be extremely difficult to achieve, largely due to the complex, site-specific interactions of management and environmental factors, but also because P management is but one component of farming and therefore must accommodate other priorities that impinge on the availability of finances, time, technology and land (Syers et al., 2008). Therefore, despite decades of continual improvements in manure management options, it remains the rule that

Table 2. Summary of the effects of different manure application methods on runoff and water quality relative to broadcast application. Adapted from Maguire et al. (2011).†

Method	Runoff volume		Erosion		Total phosphorus load		Dissolved phosphorus load	
	Row cropland	Pasture/ grassland	Row cropland	Pasture/ grassland	Row cropland	Pasture/ grassland	Row cropland	Pasture/ grassland
Chisel Injection								
Chisel with sweep	--	--	--	--	--	--	--	--
Spike/knife	--	--	--	--	94% less	--	--	--
Disk Injection								
Shallow disk	--	3-35% lower	0%	68% less	0-91% less	84% less	71-94% less	--
Tandem disk	--	--	--	--	--	--	--	--
Aerator		0-81% lower		28% more to 69% less	94% less	0-88% less	96% less	13-90% less
Tillage								
by moldboard plow	9-56% lower	--	--	--	90% less	--	84% less	--
by chisel plow	14-66% lower	--	0-97% more	--	90% more to 81% less	--	0-68% less	
by double disk	20% lower	--	--	--	--	--	--	--

† Data obtained from Ross et al. (1979), Laflen and Tabatai (1984), Andraski et al. (2003), Davarede et al. (2004), Shah et al. (2004), van Vliet et al. (2006), Butler et al. (2008), Franklin et al. (2007), Burcham et al. (2008), Johnson et al. (2011)

land-applied manure P is inefficiently used in crop production, and that historical attitudes toward manure management, while evolving, diminish the value of this important resource and therefore serve as disincentive for changing manure management practices. Thus, as much as the knowledge and management options discussed below can contribute to sustainable management of manure P, education is ultimately at the heart of advancing the agronomic management of manure P (Genskow, 2012).

To understand how different manure application practices affect P fate in the environment, it is useful to characterize manure P losses on the basis of the processes and pathways of P transfer from the soils where it is land applied to the waters where it accelerates eutrophication. Land applied manure contributes P to runoff directly, through the "incidental transfer" or "wash-off" of P in manure, or indirectly, through soils that have accumulated P from past manure applications and now lose that P to runoff through the processes of erosion, desorption, reductive dissolution and mineralization (Buda et al., 2013). The wash-off of manure P from soil is a short-term concern, involving transfers of recently applied manure P during the season in which it is applied and often characterized by severely elevated P concentrations in runoff water (surface and subsurface), whereas soil-mediated losses are persistent, albeit generally lower in concentration. Most notably, the accumulation of P in soil creates so-called "legacy" sources of P that can continue to enrich runoff for years, even decades, after manure application ceases (Kleinman et al., 2011; Sharpley et al., 2013).

Manure Application Methods

Surface application of manure by broadcasting, such as with flails, rotors and splash plates, continues to be the most common form of land application. Because surface application of manure leaves ammoniacal forms of manure N vulnerable to

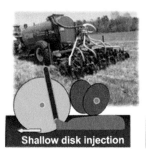

Fig. 6. Examples of low-disturbance manure applicators that improve incorporation of manure into the soil: shallow disk injection, aeration and manure banding, and high pressure injection. The cartoons below the photographs depict the discrete nature of manure (in blue) incorporation into soil.

volatilization (e.g., Dell et al., 2012), it tends to exacerbate N/P imbalances as well as to leave manure P vulnerable to wash-off by runoff. Consequently, a substantial body of research explores alternative methods of application that incorporate manure into soil where chemi-sorption and immobilization of P can be promoted, preferably with minimal disturbance to soils to prevent adverse impacts on soil erosion caused by tillage and to permit manure application to perennial forage crops (Table 2).

Incorporation of manure into soil can be achieved via a variety of low-disturbance technologies, many of which are compatible with soil conservation objectives, some of which are suited to perennial forage crops and all with trade-offs that affect their adoption. A few examples of technologies compared by Johnson et al. (2011) under no-till corn production, as well as the spatial pattern of manure incorporation, are illustrated in Fig. 6. In general, the more manure is removed from the soil surface, the less manure P is available to wash-off processes over the short-term and, where tillage is not used, the less soil P is vertically stratified over the long-term. However, applicators, such as aerators, may also impact rainfall infiltration and therefore runoff volumes, providing dramatic, short-term improvements in infiltration (Table 2).

While both agronomic and environmental benefits can be quantified with these types of technologies (Waldrip et al., 2010; Liu et al., 2016), their adoption by farmers and contract manure applicators can be hampered by a variety of factors, real and perceived. Broadcast applicators tend to employ well-established technologies that are relatively easy to maintain. In contrast, low disturbance manure incorporation technologies can be expensive relative to broadcast technologies, all of the examples in Fig. 6 would replace a simple splash-plate used in broadcasting liquid manure. Injectors and aerators all suffer damage from contact with stones, and their performance can be hampered by antecedent soil conditions (e.g., moisture) as well as a slope. Further, broadcasting applies manure over wide swaths quickly compared with alternative manure application methods that tend to slow down progress toward emptying barns and manure storages–a priority when manure spreading periods are brief. Other factors affecting adoption include interaction of low disturbance incorporation technologies with residue (e.g., the high-pressure system in Fig. 6 tended to rake corn residue and had to be cleaned and reset regularly), impacts on soil surface topography (e.g., mounding of soil by steel ground engagement units), and perceived adverse impacts on crop growth (e.g., potential for salt-burning by concentrated bands of injected manure, although this has not been documented).

Manure Application Rate

Changing manure application rates impacts both the short-term wash-off of manure P and the long-term accumulation of legacy soil P. When manure is applied to the soil surface, there are clear relationships between the rate of application of a particular manure and the loss of dissolved P in runoff in the first few events following application (Kleinman and Sharpley, 2003). This relationship diminishes with time and is modified by factors such as manure properties (concentration of WEP, moisture content) and the rainfall-to-runoff-ratio (Vadas et al., 2004). Wash-off of manure P and the influence of manure application rate are less of a concern when manure is fully incorporated into soil (Maguire et al., 2011).

While the punctuated losses of P in the first runoff events following the surface application of manure can be dramatic, the gradual build-up of soil P that comes with applying manure in excess of crop P uptake can be a more insidious water quality concern over the long-term. Left unchecked, the accumulation of manure P in soils can build legacy P reserves that may require many decades to reverse, creating conditions that undermine watershed mitigation programs (Sharpley et al., 2013). Today, unchecked application of manure is less common than several decades ago, but it is still a concern. For instance, a recent study of over 15,000 soil samples from the top 20 poultry production counties in Mississippi found that average soil test P levels (Mehlich-3) increased from 66 mg kg^{-1} in 2002 to 187 mg kg^{-1} in 2012 (Ramirez-Avila et al., 2017). Given the relatively low N/P of most manures, variable rates of manure nutrient mineralization, and general need to purge manure storages in an expedient fashion, persuading farmers to apply manure at lower rates, especially those corresponding to crop P requirement, can be difficult. As a result, many areas allow manure application rates calculated to meet crop P needs across a longer rotation (e.g., three year rotation of multiple crops), trading off short-term risks of wash-off with long-term balances of soil P.

Manure Application over Subsurface Drainage

Managing manure application over tile drains is a complex subject, fraught with trade-offs that sometimes upend useful generalizations. Subsurface P transport to tile drains is greatest when preferential flow pathways such as biopores and cracks minimize contact between infiltrating water and soil, thereby minimizing dissolved P sorption from leachate. These macropores may also transmit particulate P. When manure is broadcast, tillage prior to manure application may disrupt P leaching via preferential flow pathways, but tillage following manure application may translocate manure to lower depths where it comes in contact with macropores that serve as conduits for preferential flow (Kleinman et al., 2009). Similarly, injection of manure may place highly soluble forms of manure in contact with macropores, although the action of an injector may smear or otherwise destroy macropores around the injection zone (Cooley et al., 2013; Feyereisen et al., 2010). Soil properties can substantially influence P leaching losses, sometimes counterintuitively. Liu et al. (2012) found that incorporation of swine slurry reduced leaching of dissolved P by 64% in a well-structured clay loam soil where P transport was dominated by preferential flow pathways, but had little effect in a sandy soil where P loss was dominated by matrix flow. Since macropores serving as preferential flow pathways to tile drains are only found within a few meters of the tile drain, strategies to avoid manure application in these areas have been proposed, as have strategies that would isolate tillage to areas in closest proximity to drains (Ruark et al., 2009).

Such approaches require knowledge of drainage system layout, technology to perform precision management, and, most importantly, an impetus to instigate efforts that can be time-consuming and costly.

Winter Manure Management

While manure is ideally applied in warm seasons when nutrients can be best used by crops (Ehmke, 2014; IPNI, 2014), manure application in cool seasons of late fall, winter, and early spring is not uncommon in many areas of the U.S. and Canada, due to practical reasons such as the lack of long-term manure storage facilities (Dou et al., 2001), and the unavailability of labors and/or unfavorable (wet) field conditions for manure applications in spring (Lewis and Makarewicz, 2009; Liu et al., 2018b). This, as a result, leads to an elevated risk of P losses, for which winter manure applications on frozen, snow-covered, and water-saturated soils have become a particular concern because of coincidence of manure nutrient sources with active transport pathways (Srinivasan et al., 2006; Williams et al., 2011; Vadas et al., 2017). Furthermore, there are fewer management options that can be implemented to minimize the impacts of nutrient losses associated with winter manure applications. Although the practice of frost injection has been developed, it has only been used at local scales.

Phosphorus runoff associated with winter manure applications is greatly influenced by environmental (e.g., weather, soil, and hydrology) and management (e.g., manure, soil, and crop) factors that interactively determine nutrient runoff characteristics (Steenhuis et al., 1981; Fleming and Fraser, 2000; Srinivasan et al., 2006; Williams et al., 2011, 2012; Liu et al., 2017; Vadas et al., 2017, 2018; Stock et al., 2019). In particular, P runoff is largely dependent on the extent of manure nutrient infiltration in soil prior to and during runoff events, which can be determined by both long-term climatic patterns (Vadas et al., 2017) and short-term weather conditions (Collick et al., 2016). Due to the complex influential factors, studies have reported conflicting nutrient loss trends, with many observing elevated risks of N and P losses following manure application to frozen, snow-covered, or water-saturated soils in winter (Midgley and Dunklee, 1945; Klausner et al., 1976; Young and Mutchler, 1976; Phillips et al., 1981; Maule and Elliott, 2006; Komiskey et al., 2011) but some reporting no impact of winter manure application on water quality (e.g., Young and Holt, 1977; Ginting et al., 1998). To minimize potential P losses, however, mitigation strategies are needed in place wherever winter manure applications are conducted. In a recent study model simulating the efficiencies of a large range of management options on P runoff, Liu et al. (2017) found that although winter and fall manure applications increased P runoff as compared with spring applications, targeting manure applications to low-risk fields (gently sloped fields away from streams) in winter and fall or use of a cover crop could reduce P runoff losses to levels below those resulting from spring application if low-risk fields were not targeted.

Due to the water quality concerns associated with winter manure applications, many countries in the world have developed mandatory regulations or voluntary guidelines to guide manure management in this critical season (Liu et al., 2018b). In the United States and Canada, there are a diverse set of winter manure management directives that vary among states and provinces. The directives range from complete prohibition of manure application during winter months to conditional restriction by law on when, where, how, and what manure is applied to voluntary adoption of conservation practices associated with winter manure applications to no specific directives at all (Liu et al., 2018b). It should be noted that a total of 26 states in the U.S. have stricter directives for large-sized animal operations (i.e., Animal Feeding Operations and Concentrated Animal Feeding Operations) than for small operations as of December 2014, which is due to a greater concern over manure nutrient losses from large operations with higher animal densities than from small operations (Liu et al., 2018b). In general, the policies have been designed to take into account several factors such as their climatic conditions, extents of animal production, and considerations of water quality sensitivity to nutrients. In particular, the sensitivity of water quality to nutrients seems to be the first factor considered when deciding the regulations/guidelines by many. In the Chesapeake Bay region in the United States, for example, regulations and guidelines are found to be more restrictive in Delaware (Delaware Department of Agriculture, 2003) and Maryland (Maryland Department of Agriculture, 2012) that are in close proximity to the Bay, in spite of the milder winter conditions compared with the more distant, northern states of New York (New York Natural Resources Conservation Service, 2013) and Pennsylvania (Pennsylvania Code, 2005; Pennsylvania Department of Environmental Protection, 2011). Similarly, in the Lake Winnipeg basin in Canada, although Saskatchewan and Manitoba have similar climates and production settings, while Saskatchewan allows winter manure applications, Manitoba has banned winter manure applications for many years (for large animal farms since 1998 (Manitoba Environment Act, 1998) and for all animal farms since 2013 (Manitoba Sustainable Development, 2017) due to its close proximity to Lake Winnipeg.

In-field Stacking of Manure

Stacking of dry manure is a common, short-term management practice in which dry manure is staged in the field prior to land application. Although it is recommended that manure should be stored under a permanently roofed structure that prevents litter from exposure to rain (Rasnake et al., 1995; Sylvester, 2007; Ogejo and Collins, 2009; Cunningham et al., 2012), it is still common to stack poultry litter in crop fields in anticipation of spreading that manure during a short time frame. To minimize nutrient runoff from the manure stacked in the fields, most U.S. states and Canadian provinces have developed relevant management guidelines, which require or recommend covering field stacks with plastic sheeting or with a reinforced, ultraviolet-resistant cover (Ogejo and Collins, 2009), having a set-back from environmentally-sensitive areas (streams, water wells, drainage ditches, and sinkholes) (Delaware Code Online, 2015; Pennsylvania Department of Environmental Protection, 2011), and siting the stacks on lands with gentle slopes (Pennsylvania Department of Environmental Protection, 2011). Covering manure stacks with a plastic sheet can be a cost-effective strategy to reduce P and N runoff while better preserving the N value in the manure. While some studies have pointed out concerns related to covering a manure stack under certain circumstances, such as, elevated nutrient leaching by water generated by microorganism respiration under aerobic conditions (Dewes, 1995), most of the studies highlight the benefits of covering the manure stacks (Zebarth et al., 1999; Sommer, 2001; Felton et al., 2007; Nicholson et al., 2010; Doody et al., 2012; Liu et al., 2015). In a field study conducted on poultry litter stacks, for example, Liu et al. (2015) found that although P losses from both covered and uncovered stacks were generally low due to their high capacity to hold precipitation water, covering stacks reduced leachate total P losses by 25 to 100 times such that leachate P losses from covered stacks were similar to that in the controls with no manure stacking. It should be noted, however, that manure stacking could result in the development of P "hotspots" in fields (Doody et al., 2013; Liu et al., 2015). After two years of poultry litter stacking, for example, Liu et al. (2015) observed WEP concentrations in upper 5-cm soils were as high as 120 to 240 mg kg⁻¹ under the covered stack and 140 to 250 mg kg⁻¹ under the uncovered stack. Therefore, relocating stack sites is needed when field stacking of manure is practiced.

Predictive Tools for Decision Support

A large number of decision support tools have been developed to address P loss from land applied manure, most intending to differentiate between sites and practices that will exacerbate runoff losses of manure P, and many intending to identify manure management options. Decision support tools range from the P Index, to complex watershed models that predict the outcome of different scenarios, to short-term risk forecasts (Kleinman et al., 2017). Most of these tools include input and process information related to source and transport factors (the central construct of the critical source area paradigm), but they differ in their representation of P cycling, fate and transport processes, as well as in their temporal and spatial scales of inference (Radcliffe et al., 2015). These differences are tied to the amount of information needed to set up a tool, its computational demands (processing time and capability), and the required expertise of end-users (Bolster et al., 2017).

No decision support tool for managing manure P has received more attention than the P Index, which has been applied to various conditions around the world to assess site potential for P loss (i.e., fields with greatest vulnerability to P loss in runoff), helping to guide manure and fertilizer management, among other things. The P Index originated in the United States (Lemunyon and Gilbert, 1993), prompted by federal policy, and was rapidly modified into different versions to address the needs of individual states, with up to 47 states identifying this tool as the basis for their P-based management guideline (Sharpley et al., 2003b). While most versions of the P Index focus solely on P losses via surface runoff from agricultural fields, a substantial number of modern P Indices now consider vulnerability to subsurface P loss, principally in relationship to artificial drainage (Shober et al., 2017). With nearly two decades of experience in using the P Index as a site assessment tool, a variety of concerns have been raised, from inconsistency and inaccuracy in its assessment of P loss potential to ineffectiveness in promoting change in manure management (Sharpley et al., 2017). Even so, the P Index remains the best-accepted site assessment tool for P-based manure management, with continuing efforts to improve its performance and utility (Ketterings et al., 2017; Kleinman et al., 2017).

The P Index has contributed to the development of other tools intended to help animal producer and advisors to comply with state and national nutrient management regulations. Tools such as the Pennsylvania Nutrient Balance Worksheet and the Manure Management Planner (MMP), assist animal producers or advisors in assembling a comprehensive nutrient management plan by entering information about the operation's fields, crops, storage, animals, and application equipment (Effland, 2010). In addition to the P Index, other, more complex, tools have regularly been advanced and advocated for site assessment and manure management guidance by farmers and their advisors, as well as for support of nutrient management and conservation programs. The Annual Phosphorus Loss Estimator (APLE) and its derivative, APLE for Cattle Lots (APLE-Lots), are spreadsheet models that employ empirical algorithms to predict the consequences of manure management options on annual and sediment-bound P loss in surface runoff. The AGricultural Non-Point Source Pollution Model (AGNPS), the Agricultural Policy/Environmental eXtender (APEX) and the Soil and Water

Assessment Tool (SWAT) are highly calibrated, process-based simulation models that predict daily surface and subsurface P losses at field and watershed scales. Although these models apply to different scales of inference, many of them have been adapted to site assessment, either to bolster the P Index, or to replace the P Index (e.g., White et al., 2010). Efforts to further advance the applicability of these models, and other comparable models, to manure management decisions remain an area of active research.

All the tools presented in this discussion have been widely evaluated under different management scenarios, including the use of different manure sources and application methods (Bolster et al., 2017; Collick et al., 2016; Nelson et al., 2017; Osmond et al., 2017; Ramirez-Avila et al., 2012; Yuan et al., 2013). Efforts to corroborate the performance of these decision support tools have demonstrated that the P Index can perform as well as the other, more complex water quality models when assessing field-scale P loss vulnerability (Osmond et al., 2017). Notably, while APEX, SWAT, and AGNPS have been repeatedly used to satisfactorily predict nutrient transport processes, their use in site assessment requires careful calibration to ensure accuracy of their P loss prediction (Baffaut et al., 2017; Ramirez-Avila et al., 2017). Furthermore, none of these models were developed to simulate the unique processes of subsurface P transport (Radcliffe et al., 2015).

The site assessment tools described above all provide strategic support for manure management, as opposed to short-term support for operational decisions that must adapt to weather, changing management priorities, and other dynamic factors. In response to a perceived need for short-term decision support, a new class of tools has been developed that combine short-term weather forecasts with hydrologic modeling techniques to identify when and where manure should be applied. Some tools, such as Wisconsin's Manure Advisory System have been adopted by local nutrient management programs while others remain in development (Easton et al., 2017).

Conclusions

Manure management has profound implications to the biogeochemical cycling of P, both as a result of the specialization and intensification of modern agricultural production systems, and as a result of the many ways in which P can be lost from animal production operations, with little economic consequence to operators. Many options exist to better manage manure, and, more specifically, P in manure, to minimize their impact on water quality. However, doing so requires comprehensive approaches that consider factors far beyond the direct handling of manure and require decisions that may compete or conflict with other priorities, most notably profits and time management.

Increasingly, the connection has been made between manure management and long-term food security, recognizing the finite supply of easily-mined P sources that minimize its cost and therefore the cost of food (Jarvie et al., 2015). Given the large proportion of grain crops, in addition to forages, that support animal production, increases in agronomic use efficiency of manure P should not only be seen as a benefit to the farmer, but to the whole of modern society. It is in this context that the imperative exists to remove the many barriers to P-based manure management.

References

Adler, P.R., and P.L. Sibrell. 2003. Sequestration of phosphorus by acid mine drainage floc. J. Environ. Qual. 32:1122–1129. doi:10.2134/jeq2003.1122

Andraski, T.W., L.G. Bundy, and K.C. Kilian. 2003. Manure history and long-term tillage effects on phosphorus losses in runoff. J. Environ. Qual. 32:1782–1789. doi:10.2134/jeq2003.1782

Applegate, T.J., and R. Angel. 2008. Phosphorus Requirements for Poultry, Purdue Extension, Anim. Sci.-583-W. https://www.extension.purdue.edu/extmedia/as/as-583-w.pdf (Verified 9 Dec. 2018).

Baffaut, C., N.O. Nelson, J.A. Lory, G.M.M.M.A. Senaviratne, A.B. Bhandari, R.P. Udawatta, D.W. Sweeney, M.J. Helmers, M.W. Van Liew, A.P. Mallarino, and C.S. Wortmann. 2017. Multisite Evaluation of APEX for water quality: I. Best professional judgment parameterization. J. Environ. Qual. 46:1323–1331. doi:10.2134/jeq2016.06.0226

Bartelt, K.D., and W.L. Bland. 2007. Theoretical analysis of manure transport distance as a function of herd size and landscape fragmentation. J. Soil Water Conserv. 62:345–352.

Beegle, D. 2014. Nutrient management and the Chesapeake Bay. J. Contemp. Water Res. Educ. 151:3–8. doi:10.1111/j.1936-704X.2013.03146.x

Bird, S.C. and A. Drizo. 2010. EAF steel slag filters for phosphorus removal from milk parlor effluent: The effects of solids loading, alternate feeding regimes and in-series design. Water 2: 484-499. doi:10.3390/w2030484

Bolster, C.H., A. Forsberg, A. Mittelstet, D. Radcliffe, D. Storm, J. Ramirez-Avila, A.N. Sharpley, and D. Osmond. 2017. Comparing an annual and daily time-step model for predicting field-scale phosphorus loss. J. Environ. Qual. 46:1314–1322. doi:10.2134/jeq2016.04.0159

Bosch, D.J. and K.B. Napit. 1992. Economics of transporting poultry litter to achieve more efficient use as fertilizer. J Soil Water Conserv 47: 342-346.

Brandt, R.C., and H.A. Elliott. 2003. Phosphorus runoff losses from surface-applied biosolids and dairy manure. Joint Residuals and Biosolids Management Conference. Water Environmental Federation, Alexandria, VA. doi:10.2175/193864703790898477

Buda, A.R., G.F. Koopmans, R.B. Bryant, and W.J. Chardon. 2012. Emerging technologies for removing nonpoint phosphorus from surface water and groundwater: Introduction. J. Environ. Qual. 41:621–627. doi:10.2134/jeq2012.0080

Buda, A.R., P.J.A. Kleinman, R.B. Bryant, G.W. Feyereisen, D.A. Miller, P.G. Knight, and P.J. Drohan. 2013. Forecasting runoff from Pennsylvania landscapes. J. Soil Water Conserv. 68(3):185–198. doi:10.2489/jswc.68.3.185

Burcham, M., R.O. Maguire, M. Alley, W. Thomason, and B. Jones. 2008. Injecting dairy manure to decrease nitrogen and phosphorous in runoff. SERA17 Annual Conference, Grasonville, MD.

Burns, R.T., and L.B. Moody. 2002. Phosphorus recovery from animal manures using optimized struvite precipitation. Proceedings of Coagulants and Flocculants: Global Market and Technical Opportunities for Water Treatment Chemicals, Chicago, IL. 22–24 May 2002. University of Tennessee, Knoxville, TN. http://wastemgmt.ag.utk.edu/Pubs/floc%20conf%20paper.pdf (Accessed 10 Dec. 2018).

Butler, D.M., D.H. Franklin, M.L. Cabrera, A.S. Tasistro, K. Xia, and L.T. West. 2008. Evaluating aeration techniques for decreasing phosphorus export from grasslands receiving manure. J. Environ. Qual. 37:1279–1287. doi:10.2134/jeq2007.0289

Carreira, R.I., K.B. Young, H.L. Goodwin, Jr., and E.J. Wailes. 2007. How far can poultry litter go? A new technology for litter transport. J. Agric. Appl. Econ. 39:611–623. doi:10.1017/S1074070800023300

Chaperon, C.O.I., V. Girard, and Y. Chorfi. 2007. On-farm phosphorus budget: Model to predict yearly phosphorus contents in manure of dairy herds. Can. Vet. J. 87:407–411.

Choi, I.H., and P.A. Moore, Jr. 2008. Effect of various litter amendments on ammonia volatilization and nitrogen content of poultry litter. J. Appl. Poult. Res. 17:454–462. doi:10.3382/japr.2008-00012

Church, C.D., A.N. Hristov, R.B. Bryant, P.J.A. Kleinman, and S.K. Fishel. 2016. A novel treatment system to remove phosphorus from liquid manure. Appl. Eng. Agric. 32:103–112. doi:10.13031/aea.32.10999

Church, C.D., A. Hristov, R.B. Bryant, and P.J.A. Kleinman. 2017. Processes and treatment systems for treating high phosphorus containing fluids. US Patent 9790.110B2. Date issued 21 Feb.

Church, C.D., A.N. Hristov, P.J.A. Kleinman, S.K. Fishel, M.R. Reiner, and R.B. Bryant. 2018. Versatility of the MAnure PHosphorus Extraction (MAPHEX) System in removing phosphorus, odor, microbes, and alkalinity from dairy manures: A four-farm case study. Appl. Eng. Agric. 34:567–572. doi:10.13031/aea.12632

Collick, A.S., T.L. Veith, D.R. Fuka, P.J.A. Kleinman, A.R. Buda, J.L. Weld, R.B. Bryant, P.A. Vadas, M.J. White, R.D. Harmel, and Z.M. Easton. 2016. Improved simulation of edaphic and manure phosphorus loss in SWAT. J. Environ. Qual. 45:1215–1225. doi:10.2134/jeq2015.03.0135.

Cooley, E.T., M.D. Ruark, and J.C. Panuska. 2013. Tile drainage in Wisconsin: Managing tile-drained landscapes to prevent nutrient loss. Univ. Wisconsin Cooperative Extension publication GWQ064. https://fyi.extension.wisc.edu/drainage/files/2012/06/3-Managing-Tile-Drained-Landscapes-to-Prevent-Nutrient-Loss-DF.pdf (Accessed 28 Jan. 2019).

Cunningham, D.L., C.W. Ritz, and W.C. Merka. 2012. Best management practices for storing and applying poultry litter. The University of Georgia Cooperative Extension Bulletin 1230, Athens, GA. http://extension.uga.edu/publications/detail.cfm?number=B1230 (Accessed 8 Jan. 2019).

Dance, S. 2017. Maryland spends $1M a year to transport chicken litter, to the benefit of the Chesapeake and poultry companies. https://www.baltimoresun.com/features/green/blog/bs-md-manure-export-20170115-story.html (Accessed 11 Jan. 2019).

Daverede, I.C., A.N. Kravchenko, R.G. Hoeft, E.D. Nafziger, D.G. Bullock, J.J. Warren, and L.C. Gonzini. 2004. Phosphorus runoff from incorporated and surface applied liquid swine manure and phosphorus fertilizer. J. Environ. Qual. 33:1535–1544. doi:10.2134/jeq2004.1535

Delaware Code Online. 2015. Title 3–Agriculture. http://delcode.delaware.gov/title3/title3.pdf (Accessed 8 Jan. 2019).

Delaware Department of Agriculture. 2003. Nutrient management regulations. https://agriculture.delaware.gov/nutrient-management/laws-regulations/ (Accessed 27 Jan. 2019).

Dell, C.J., P.J.A. Kleinman, J.P. Schmidt, and D.B. Beegle. 2012. Low disturbance manure incorporation effects on ammonia and nitrate loss. J. Environ. Qual. 41:928–937. doi:10.2134/jeq2011.0327

Dewes, T. 1995. Nitrogen losses from manure heaps. Biol. Agric. Hortic. 11:309–317. doi:10.1080/01448765.1995.9754715

Digested Organics. 2018. Nutrient concentration and water reclamation system. Digested Organics, Farmington Hills, MI. https://www.digestedorganics.com/wp-content/uploads/2015/02/Digested-Organics-NCWR-Flow-Diagram.pdf?0ddc6c&0ddc6c (Accessed 10 Dec. 2018).

Dillaha, T.A., J.H. Sherrard, D. Lee, S. Mostaghimi, and V.O. Shanholtz. 1988. Evalution of vegetative filter strips as a best management practice for feed lots. J. Water Pollut. Control Fed. 60:1231–1238.

Dillaha, T.A., R.B. Reneau, S. Mostaghimi, and D. Lee. 1989. Vegetative filter strips for agricultural nonpoint source pollution control. Tran ASABE 32: 513-519. doi:10.13031/2013.31033.

Doody, D.G., J.S. Bailey, and C.J. Watson. 2013. Evaluating the evidence-base for the Nitrate Directive regulations controlling the storage of manure in field heaps. Environ. Sci. Policy 29:137–146. doi:10.1016/j.envsci.2013.01.009

Doody, D.G., R.H. Foy, J.S. Bailey, and D. Matthews. 2012. Minimising nutrient transfers from poultry litter field heaps. Nutr. Cycling Agroecosyst. 92:79–90. doi:10.1007/s10705-011-9473-x

Dou, Z., D.T. Galligan, C.F. Ramberg, C. Meadows, and J.D. Ferguson. 2001. A survey of dairy farming in Pennsylvania: Nutrient management practices and implications. J. Dairy Sci. 84:966–973. doi:10.3168/jds.S0022-0302(01)74555-9

Dou, Z., K.F. Knowlton, R.A. Kohn, Z. Wu, L.D. Satter, G. Zhang, J.D. Toth, and J.D. Ferguson. 2002. Phosphorus characteristics of dairy feces affected by diets. J. Environ. Qual. 31:2058–2065. doi:10.2134/jeq2002.2058

Dou, Z., G.Y. Zhang, W.L. Stout, J.D. Toth, and J.D. Ferguson. 2003. Efficacy of alum and coal combustion by-products in stabilizing manure phosphorus. J. Environ. Qual. 32:1490–1497. doi:10.2134/jeq2003.1490

Drizo, A., A.C. Frost, K.A. Smith, and J. Grace. 1997. The use of constructed wetlands in phosphate and ammonium removal from wastewater. Water Sci. Technol. 35:95–102. doi:10.2166/wst.1997.0173

Easton, Z.M., P.J.A. Kleinman, A.R. Buda, M.T. Walter, N. Emberston, S. Reed, D. Goehring, P.J. Drohan, J. Lory, and A. Sharpley. 2017. Short-term forecasting tools for agricultural nutrient management. J. Environ. Qual. 46: 1257–1269. doi:10.2134/jeq2016.09.0377

EasyMining. 2018. Ash2Phos. http://www.easymining.se/our-technologies/ash2phos/ash2phos-information/ (Verified 7 Dec. 2018).

Effland, W. 2010. Report on soil information systems of the USDA Natural Resources Conservation Service. USDA-NRCS, Washington, D.C. http://www.fftc.agnet.org/library.php?func=view&id=20110809091556 (Accessed Feb 9, 2019).

Eghball, B. 2000. Nitrogen mineralization from field-applied beef cattle feedlot manure or compost. Soil Sci. Soc. Am. J. 64:2024–2030. doi:10.2136/sssaj2000.6462024x

Ehmke, T. 2014. The 4Rs of nutrient management. Crops & Soils 47(5):4–10. doi:10.2134/cs2014-47-5-1

Elliott, H.A., G.A. O'Connor, P. Lu, and S. Brinton. 2002. Influence of water treatment residuals on phosphorus solubility and leaching. J. Environ. Qual. 31:1362–1369. doi:10.2134/jeq2002.1362

Elliott, H.A., R.C. Brandt, P.J.A. Kleinman, A.N. Sharpley, and D.B. Beegle. 2006. Estimating source coefficients for phosphorus site indices. J. Environ. Qual. 35:2195–2201. doi:10.2134/jeq2006.0014

Fabian-Wheeler, E.E., M.L. Hile, D.J. Murphy, D.E. Hill, R. Meinen, H.A. Elliott, and D. Hofstetter. 2017. Operator exposure to hydrogen sulfide from dairy manure storages containing gypsum bedding. J. Agric. Saf. Health 23:9–22. doi:10.13031/jash.11563

Felton, G.K., L.E. Carr, and M.J. Habersack. 2007. Nutrient fate and transport in surface runoff from poultry litter stock piles. Trans. ASABE 50:183–192. doi:10.13031/2013.22399

Feyereisen, G.W., P.J.A. Kleinman, G.J. Folmar, L.S. Saporito, C.D. Church, T.R. Way, and A.L. Allen. 2010. Effect of direct incorporation of poultry litter on phosphorus leaching from Coastal Plain soils. J. Soil Water Conserv. 65:243–251. doi:10.2489/jswc.65.4.243

Fleming, R., and H. Fraser. 2000. Impacts of winter spreading of manure on water quality: Literature review. Ridgetown College, University of Guelph, Ridgetown, Ontario, Canada.

Franklin, D.H., M.L. Cabrera, L.T. West, V.H. Calvert, and J.A. Rema. 2007. Aerating grasslands: Effects of runoff and phosphorus losses from applied broiler manure. J. Environ. Qual. 36:208–215. doi:10.2134/jeq2006.0012

Garcia-Gonzalez, M.C., M. B. Vanotti, and A. A. Szogi. 2016. Recovery of ammonia from anaerobically digested manure using gas-permeable membranes. Scientia Agricola 73. doi:10.1590/0103-9016-2015-0159 (Accessed 11 Dec. 2018).

Genskow, K.D. 2012. Taking stock of voluntary nutrient management: Measuring and tracking change. J. Soil Water Conserv. 67:51–58. doi:10.2489/jswc.67.1.51

Ghebremichael, L.T., T.L. Veith, J.M. Hamlett, and W.J. Gburek. 2008. Precision feeding and forage management effects on phosphorus loss modeled at a watershed scale. J. Soil Water Conserv. 63:280–291. doi:10.2489/jswc.63.5.280

Giaimo, C. 2015. When the western world ran on guano. Atlas Obscura, Brooklyn, N.Y. https://www.atlasobscura.com/articles/when-the-western-world-ran-on-guano (Verified 9 Dec. 2018).

Ginting, D., J.F. Moncrief, S.C. Gupta, and S.D. Evans. 1998. Corn yield, runoff, and sediment losses from manure and tillage systems. J. Environ. Qual. 27:1396–1402. doi:10.2134/jeq1998.00472425002700060x

Hadrich, J.C., T.M. Harrigan, and C.A. Wolf. 2010. Economic comparison of liquid manure transport and land application. Appl. Eng. Agric. 26:743–758. doi:10.13031/2013.34939

He, Z., P.H. Pagliari, and H.M. Waldrip. 2016. Applied and environmental chemistry of animal manure: A review. Pedosphere 26:779–816. doi:10.1016/S1002-0160(15)60087-X

Herron, S.L., A.N. Sharpley, S. Watkins, and M. Daniels. 2012. Poultry litter management in the Illinois River Watershed of Arkansas and Oklahoma. Fact Sheet FSA 9535. Cooperative Extension Service, Division of Agriculture, University of Arkansas, Fayetteville, AR. http://www.uaex.edu/Other_Areas/publications/PDF/FSA-9535.pdf (Accessed 26 Apr. 2019).

Holly, M., P. Kleinman, J. Baker, D. Bjorneberg, M. Boggess, R. Bryant, R. Chintala, G. Feyereisen, J. Gamble, A. Leytem, K. Reed, A. Rotz, P. Vadas, and H. Waldrip. 2018. Nutrient management challenges and opportunities across U.S. dairy farms. J. Dairy Sci. 101:6632–6641. doi:10.3168/jds.2017-13819

House, C.H., S.W. Broome, and M.T. Hoover. 1994. Treatment of nitrogen and phosphorus by a constructed upland-wetland wastewater treatment system. Water Sci. Technol. 29:177–184. doi:10.2166/wst.1994.0185

Huang, L., P.A. Moore, Jr., P.J.A. Kleinman, K.R. Elkin, M. Savin, D. Pote, and D. Edwards. 2016. Reducing phosphorus runoff and leaching from poultry litter with alum: Twenty year small plot and paired-watershed studies. J. Environ. Qual. 45:1413–1420. doi:10.2134/jeq2015.09.0482

Humer, E., C. Schwarz, and K. Schedle. 2015. Phytate in pig and poultry nutrition. J. Anim. Physiol. Anim. Nutr. 99:605–625. doi:10.1111/jpn.12258

International Plant Nutrition Institute. 2012. A Nutrient Use Information System (NuGIS) for the U.S. IPNI, Peachtree Corners, GA. http://nugis.ipni.net/About%20NuGIS/ (Verified 6 Dec. 2018).

International Plant Nutrition Institute. 2014. History of the "4Rs". Int. Plant Nutr. Inst., Peachtree Corners, GA. http://www.ipni.net/article/IPNI-3284 (Accessed 5 Jan. 2016).

Jarvie, H., A.N. Sharpley, D. Flaten, P.J.A. Kleinman, A. Jenkins, and T. Simmons. 2015. The pivotal and paradoxical role of phosphorus in a resilient water-energy-food security nexus. J. Environ. Qual. 44:1049–1062. doi:10.2134/jeq2015.01.0030

Johnson, K.N., P.J.A. Kleinman, D.B. Beegle, and H.A. Elliott. 2011. Effect of dairy manure slurry application in a no-till system on phosphorus runoff. Nutr. Cycling Agroecosyst. 90:201–212. doi:10.1007/s10705-011-9422-8

Karunanithi, R., A.A. Szogi, N. Bolan, R. Naidu, P. Loganathan, P.G. Hunt, M.B. Vanotti, C.P. Saint, Y.S. Ok, and S. Krishnamoorthy. 2015. Phosphorus recovery and reuse from waste streams. Adv. Agron. 131:173–250. doi:10.1016/bs.agron.2014.12.005

Ketterings, Q.M., K.J. Czymmek, and S.N. Swink. 2011. Evaluation methods for a combined research and extension program used to address starter phosphorus fertilizer use for corn in New York. Can. J. Soil Sci. 91:467–477. doi:10.4141/cjss10001

Ketterings, Q.M., S. Cela, A.S. Collick, S.J. Crittenden, and K.J. Czymmek. 2017. Restructuring the P Index to better address P management in New York. J. Environ. Qual. 46:1372–1379. doi:10.2134/jeq2016.05.0185

Kim, Y.J., L.D. Geohring, J.H. Jeon, A.S. Collick, S.K. Giri, and T.S. Steenhuis. 2006. Evaluation of the effectiveness of vegetative filter strips for phosphorus removal with the use of a tracer. J. Soil Water Conserv. 61:293–302.

Klausner, S.D., P.J. Zwerman, and D.F. Ellis. 1976. Nitrogen and phosphorus losses from winter disposal of dairy manure. J. Environ. Qual. 5:47–49. doi:10.2134/jeq1976.00472425000500010010x

Kleinman, P.J.A., A.M. Wolf, A.N. Sharpley, D.B. Beegle and L.S. Saporito. 2005. Survey of water extractable phosphorus in manures. Soil Sci. Soc. Am. J. 67. 701-708. doi:10.2136/sssaj2004.0099

Kleinman, P.J.A. 2017. The persistent environmental relevance of soil phosphorus sorption saturation. Curr. Pollut. Rep. 3(2):141–150. doi:10.1007/s40726-017-0058-4.

Kleinman, P.J.A., and A.N. Sharpley. 2003. Effect of broadcast manure on runoff phosphorus concentrations over successive rainfall events. J. Environ. Qual. 32:1072–1081. doi:10.2134/jeq2003.1072

Kleinman, P.J.A., A.N. Sharpley, B.G. Moyer, and G.F. Elwinger. 2002. Effect of mineral and manure phosphorus sources on runoff phosphorus. J. Environ. Qual. 31:2026–2033. doi:10.2134/jeq2002.2026

Kleinman, P., D. Sullivan, A. Wolf, R. Brandt, Z. Dou, H. Elliott, J. Kovar, A. Leytem, R. Maguire, P. Moore, L. Saporito, A. Sharpley, A. Shober, T. Sims, J. Toth, G. Toor, H. Zhang, and T. Zhang. 2007. Selection of a water extractable phosphorus test for manures and biosolids as an indicator of runoff loss potential. J. Environ. Qual. 36:1357–1367. doi:10.2134/jeq2006.0450

Kleinman, P.J.A., A.N. Sharpley, L.S. Saporito, A.R. Buda, and R.B. Bryant. 2009. Application of manure to no-till soils: Phosphorus losses by sub-surface and surface pathways. Nutr. Cycling Agroecosyst. 84:215–227. doi:10.1007/s10705-008-9238-3

Kleinman, P.J.A., A.N. Sharpley, A.R. Buda, R.W. McDowell, and A.L. Allen. 2011. Soil controls of phosphorus runoff: Management barriers and opportunities. Can. J. Soil Sci. 91:329–338. doi:10.4141/cjss09106

Kleinman, P., D. Beegle, K. Saacke Blunk, K. Czymmek, R. Bryant, T. Sims, J. Shortle, J. McGrath, D. Dostie, R. Maguire, R. Meinen, Q. Ketterings, F. Coale, M. Dubin, A. Allen, K. O'Neill, M. Davis, B. Clark, K. Sellner, M. Smith, L. Garber, and L. Saporito. 2012. Managing manure for sustainable livestock production in the Chesapeake Bay Watershed. J. Soil Water Conserv. 67:54A–61A. doi:10.2489/jswc.67.2.54A

Kleinman, P.J.A., A.N. Sharpley, A.R. Buda, Z.M. Easton, J.A. Lory, D.L. Osmond, D.E. Radcliffe, N.O. Nelson, T.L. Veith, and D.G. Doody. 2017. The promise, practice, and state of planning tools to assess site vulnerability to runoff phosphorus loss. J. Environ. Qual. 46:1243–1249. doi:10.2134/jeq2017.10.0395

Knowlton, K.F., D.K. Beede, and E. Kebreab. 2010. Phosphorus and calcium utilization and requirements in farm animals. In: D. Vitti and E. Kebraib, editors, Phosphorus and calcium requirements of ruminants. CAB International. doi:10.1079/9781845936266.0112

Komiskey, M.J., T.D. Stuntebeck, D.R. Frame, and F.W. Madison. 2011. Nutrients and sediment in frozen-ground runoff from no-till fields receiving liquid-dairy and solid-beef manures. J. Soil Water Conserv. 66:303–312. doi:10.2489/jswc.66.5.303

Laflen, J.M., and M.A. Tabatabai. 1984. Nitrogen and phosphorus losses from corn-soybean rotations as affected by tillage practices. Trans. ASAE 27:58–63. doi:10.13031/2013.32735'

Lanyon, L.E. 2005. Phosphorus, animal nutrition and feeding: Overview. In: J.T. Sims and A.N. Sharpley, editors, Phosphorus: Agriculture and the environment. Agronomy Monograph. American Society of Agronomy, Madison, WI. p. 561–586.

Lemunyon, J.L., and R.G. Gilbert. 1993. The concept and need for a phosphorus assessment tool. J. Prod. Agric. 6:483–486. doi:10.2134/jpa1993.0483

Lewis, T.W., and J.C. Makarewicz. 2009. Winter application of manure on an agricultural watershed and its impact on downstream nutrient fluxes. J. Great Lakes Res. 35:43–49. doi:10.1016/j.jglr.2008.08.003

Liu, J., H. Aronsson, L. Bergström, and A.N. Sharpley. 2012. Phosphorus leaching from loamy sand and clay loam topsoils after application of pig slurry. Springerplus 1:53. doi:10.1186/2193-1801-1-53

Liu, J., P.J.A. Kleinman, D.B. Beegle, J.L. Weld, A.N. Sharpley, L.S. Saporito, and J.P. Schmidt. 2015. Phosphorus and nitrogen losses from poultry litter stacks and leaching through soils. Nutr. Cycling Agroecosyst. 103:101–114. doi:10.1007/s10705-015-9724-3

Liu, J., P.J.A. Kleinman, D.B. Beegle, C.J. Dell, T.L. Veith, L.S. Saporito, K. Hun, D.H. Pote, and R.B. Bryant. 2016. Subsurface application enhances benefits of manure redistribution. Agric. Environ. Lett. 1:150003. doi:10.2134/ael2015.09.0003

Liu, J., T.L. Veith, A.S. Collick, P.J.A. Kleinman, D.B. Beegle, and R.B. Bryant. 2017. Seasonal manure application timing and storage effects on field- and watershed-level phosphorus losses. J. Environ. Qual. 46:1403–1412. doi:10.2134/jeq2017.04.0150

Liu, J., J.T. Spargo, P.J.A. Kleinman, R. Meinen, P.A. Moore, Jr., and D.B. Beegle. 2018a. Water extractable phosphorus in livestock manures and composts: Quantities, characteristics, and temporal changes. J. Environ. Qual. 47:471–479. doi:10.2134/jeq2017.12.0467

Liu, J., P.J.A. Kleinman, H. Aronsson, D. Flaten, R.W. McDowell, M. Bechmann, D.B. Beegle, T.P. Robinson, R.B. Bryant, H.B. Liu, A.N. Sharpley, and T.L. Veith. 2018b. A review of regulations and guidelines related to winter manure application. Ambio 47:657–670. doi:10.1007/s13280-018-1012-4

Livestock Water Recycling. 2018. https://www.livestockwaterrecycling.com/the-system.html (Accessed 10 Dec. 2018).

MacDonald, P. 2006. Poultry litter to power. Manure Manager, Simcoe, ON. https://www.manuremanager.com/poultry/poultry-litter-to-power-1219 (Verified 7 Dec. 2018).

MacDonald, J.M., and W.D. McBride. 2009. The transformation of U.S. livestock agriculture. Scale efficiency, and risks. U.S. Department of Agriculture Economic Research Service, Washington, D.C. www.ers.usda.giv/publications/eib43 (Accessed 28 Jan. 2019).

Magdoff, F., L. Lanyon, and W. Lieghardt. 1997. Nutrient cycling, transformations, and flows: Implications for a more sustainable agriculture. Adv. Agron. 60:1–73. doi:10.1016/S0065-2113(08)60600-8

Maguire, R.O., P.J.A. Kleinman, C. Dell, D.B. Beegle, R.C. Brandt, J.M. McGrath, and Q.M. Ketterings. 2011. Manure management in reduced tillage and grassland systems: A review. J. Environ. Qual. 40:292–301. doi:10.2134/jeq2009.0228

Manitoba Environment Act. 1998. Livestock manure and mortalities management regulation. Government of Manitoba, Winnipeg, MB. https://web2.gov.mb.ca/laws/regs/current/_pdf-regs.php?reg=42/98 (accessed 28 Jan. 2019).

Manitoba Sustainable Development. 2017. Restriction on winter application of nutrients. Province of Manitoba, Winnipeg, MB. https://www.gov.mb.ca/sd/waterstewardship/wqmz/pdf/vary_spring_nutrient_application_dates.pdf (accessed 28 Jan. 2019).

Maryland Department of Agriculture. 2012. Nutrient application requirements. Maryland Nutrient Management Manual Section 1. Nutrient recommendations. Maryland Department of Agriculture, Annapolis, MD. http://mda.maryland.gov/resource_conservation/Documents/nm_manual/1-D1-1-1D1-6.pdf (Accessed 28 Jan. 2019).

Maule, C., and J. Elliott. 2006. Effect of hog manure injection upon soil productivity and water quality. Part II. Elstow site 2000-2005. ADF Project 98000094. Saskatoon, Canada.

McDowell, R.W., A.N. Sharpley, and W. Bourke. 2008. Treatment of drainage water with industrial by-products to prevent phosphorus loss from tile-drained land. J. Environ. Qual. 37:1575–1582. doi:10.2134/jeq2007.0454

Meals, D.W. 1993. Assessing nonpoint phosphorus control in the LaPlatte River watershed. Lake Reservoir Manage. 7:197–207. doi:10.1080/07438149309354271

Midgley, A.R., and D.E. Dunklee. 1945. Fertility runoff losses from manure spread during the winter. University of Vermont, Agricultural Experiment Station Bulletin 523:1–19.

Moore, P.A., Jr., T.C. Daniel, and D.R. Edwards. 1999. Reducing phosphorus runoff and improving poultry production with alum. Poult. Sci. 78:692–698. doi:10.1093/ps/78.5.692

Moore, P.A., Jr., T.C. Daniel, and D.R. Edwards. 2000. Reducing phosphorus runoff and inhibiting ammonia loss from poultry production with alum. J. Environ. Qual. 29:37–49. doi:10.2134/jeq2000.00472425002900010006x

National Research Council. 2001. Nutrient requirements of dairy cattle: Seventh revised edition. The National Academies Press, Washington, D.C. doi:10.17226/9825

Nelson, N.O., C. Baffaut, J.A. Lory, G.M.M.M. Anomaa Senaviratne, A.B. Bhandari, R.P. Udawatta, D.W. Sweeney, M.J. Helmers, M.W. Van Liew, A.P. Mallarino, and C.S. Wortmann. 2017. Multisite evaluation of APEX for water quality: II. Regional parameterization. J. Environ. Qual. 46:1349–1356. doi:10.2134/jeq2016.07.0254

Nesme, T., and P.J.A. Withers. 2016. Sustainable strategies towards a phosphorus circular economy. Nutr. Cycling Agroecosyst. 104:259–264. doi:10.1007/s10705-016-9774-1

New York Natural Resources Conservation Service. 2013. Conservation practice standard: Nutrient management (Ac.) Code 590. http://www.nrcs.usda.gov/Internet/FSE_DOCUMENTS/nrcs144p2_027006.pdf (Accessed 28 Jan. 2019).

Nicholson, F., A. Rollett, and B. Chambers. 2010. Review of pollutant losses from solid manures stored in temporary field heaps. Project WT1006 Report. UK Department of Environment, Food and Rural Affairs (DEFRA), London, U.K. http://www.defra.gov.uk (Accessed 8 Jan. 2019).

North Carolina Department of Environmental Quality. 2018. DEQ Dashboard, Animal Operations-Swine Lagoons. https://deq.nc.gov/news/deq-dashboard#animal-operations—swine-lagoon-facilities (Verified 12 Dec. 2018).

Ogejo, J.A. 2009. Poultry and livestock manure storage: Management and safety. Virginia Cooperative Extension Publication 442-308, Blacksburg, VA. http://pubs.ext.vt.edu/content/dam/pubs_ext_vt_edu/442/442-308/442-308_pdf.pdf (Verified 12 Dec. 2018).

Ogejo, J.A., and E.R. Collins. 2009. Storing and handling poultry litter. Publication 442-054. Virginia Cooperative Extension, Blacksburg, VA. http://pubs.ext.vt.edu/442/442-054/442-054.html (Accessed 8 Jan. 2019).

O'Keefe, T. 2011. Baling makes poultry litter portable. WattAgNet, Rockford, IL. https://www.wattagnet.com/articles/8486-baling-makes-poultry-litter-portable (Accessed 7 Dec. 2018).

Osmond, D., C. Bolster, A. Sharpley, M. Cabrera, S. Feagley, A. Forsberg, C. Mitchell, R. Mylavarapu, J. Larry Oldham, D.E. Radcliffe, J.J. Ramirez-Alva, D.E. Storm, F. Walker, and H. Zhang. 2017. Southern Phosphorus Indices, water quality data, and modeling (APEX, APLE, and TBET) results: A comparison. J. Environ. Qual. doi:10.2134/jeq2016.05.0200

Penn, C.J., I. Chagas, A. Klimeski, and G. Lyngsie. 2017. A review of phosphorus removal structures: How to assess and compare their performance. Water 9:583. doi:10.3390/w9080583

Pennsylvania Code. 2005. Act 38 Nutrient Management Regulations. Penn State Extension, College Station, PA. http://extension.psu.edu/plants/nutrient-management/act-38/act-38-nutrient-management-regulations (Accessed 28 Jan. 2019).

Pennsylvania Department of Environmental Protection. 2011. Land application of manure: A supplementary to manure management for environmental protection. Manure Management Plan Guidance 361-0300-002. http://www.elibrary.dep.state.pa.us/dsweb/Get/Document-86014/361-0300-002%20combined.pdf (Accessed 8 Jan. 2019).

Phillips, P.A., J.L.B. Culley, F.R. Hore, and N.K. Patni. 1981. Pollution potential and corn yields from selected rates and timing of liquid manure applications. Trans. ASAE 24:139–144. doi:10.13031/2013.34213

Pipkin, W. 2017. Perdue turns to composting to get more poultry waste off farm fields. Ches. Bay J. https://www.bayjournal.com/article/perdue_builds_compost_facility_to_get_more_poultry_waste_off_farm_fields (Accessed 7 Dec. 2018).

Plaizier, J.C., G. Legesse, K. Ominiski, and D. Flaten. 2014. Whole-farm budgets of phosphorus and potassium on dairy farms in Manitoba. Can. J. Anim. Sci. 94:119–128. doi:10.4141/cjas2013-089

Potter, P., N. Ramankutty, E.M. Bennett, S.D. Donner. 2010. Characterizing the spatial patterns of global fertilizer application and manure production. Earth Interactions 14: 1-21. doi:10.1175/2009EI288.1

Preusch, P.L., P.R. Adler, L.J. Sikora, and T.J. Tworkoski. 2002. Nitrogen and phosphorus availability in composted and uncomposted poultry litter. J. Environ. Qual. 31:2051–2057. doi:10.2134/jeq2002.2051

Pries, J., and P. McGarry. 2002. Feedlot stormwater runoff treatment using constructed wetlands. Archive of Agri-Environmental Programs in Ontario, Ottawa, ON. http://agrienvarchive.ca/bioenergy/download/WEAO_2001_Pries.pdf (Verified 20 Dec. 2018).

Qin, Z., A. Shober, K.G. Scheckel, C.J. Penn and K.C. Turner. 2018. Mechanisms of phosphorus removal by phosphorus sorbing materials. J. Environ Qual 47: doi:10.2134/jeq2018.02.0064

Radcliffe, D.E., D.K. Reid, K. Blomback, C.H. Bolster, A.S. Collick, Z.M. Easton, W. Francesconi, D.R. Fuka, H. Johnsson, K. King, M. Larsbo, M.A. Youssef, A.S. Mulkey, N.O. Nelson, K. Persson, J.J. Ramirez-Alva, F. Schmieder, and D.R. Smith. 2015. Applicability of models to predict phosphorus losses in drainage fields: A review. J. Environ. Qual. 44:614–628. doi:10.2134/jeq2014.05.0220

Ramirez-Avila, J.J., S.L. Ortega-Achury, and W.H. McAnally. 2012. Application of the APEX model to determine water quality assessments in agricultural fields in the Mississippi Delta. Research Report. Geosyst. Res. Inst., Mississippi State Univ. Starkville, MS.

Ramirez-Avila, J.J., D.E. Radcliffe, D. Osmond, C. Bolster, A. Sharpley, S.L. Ortega-Achury, A. Forsberg, and J.L. Oldham. 2017. Evaluation of the APEX model to simulate runoff quality from agricultural fields in the southern region of the United States. J. Environ. Qual. 46:1357–1364. doi:10.2134/jeq2017.07.0258

Rasnake, M., T. Pescatore, and D. Overhults. 1995. Proper handling and storage of poultry litter. University of Kentucky Poultry Extension. http://www2.ca.uky.edu/agc/pubs/id/id117/id117. htm (Accessed 9 Jan. 2019).

Ribaudo, M., N. Gollehon, M. Aillery, J. Kaplan, R. Agapoff, L. Christensen, V. Breneman, and M. Peters. 2003. Manure management for water quality: Costs of animal feeding operations of applying nutrients to land. Agricultural Economic Report 824. USDA Economic Research Service, Washington, D.C.

Ross, I.J., J.P. Sizemore, J.P. Bowden, and C.T. Haan. 1979. Quality of runoff from land receiving surface application and injection of liquid dairy manure. Trans. ASAE 22:1055–1062. doi:10.2134/jeq2015.09.0498

Rotz, C.A. 2004. Management to reduce nitrogen losses in animal production. J. Anim. Sci. 82(E. Suppl.): E119–E137. doi:10.2527/2004.8213_supplE119x

Rotz, C.A., A.N. Sharpley, L.D. Satter, W.J. Gburek, and M.A. Sanderson. 2002. Production and feeding strategies for phosphorus management on dairy farms. J. Dairy Sci. 85:3142–3153. doi:10.3168/jds.S0022-0302(02)74402-0

Ruark, M.D., J.C. Panuska, E.T. Cooley, and J. Pagel. 2009. Tile drainage in Wisconsin: Understanding and locating tile drainage systems. University of Wisconsin Extension publication GWQ054. University of Wisconsin-Madison, Madison, WI. http://learningstore.uwex.edu/Assets/pdfs/ GWQ054.pdf (Accessed 28 Jan. 2019).

Schwarcz, J. 2017. England's coprolite rush had big effect on world. McGill University Office for Science and Society Newsletter, Montreal, QC. https://www.mcgill.ca/oss/article/environment-health-history-quirky-science/joe-schwarcz-englands-coprolite-rush-had-big-effect-world (Verified 8 Dec. 2018).

Schwer, C.B., and J.C. Clausen. 1989. Vegetative filter treatment of dairy milkhouse waste water. J. Environ. Qual. 18:446–451. doi:10.2134/jeq1989.00472425001800040008x

Shah, S.B., J.L. Miller, and T.J. Basden. 2004. Mechanical aeration and liquid dairy manure application impacts on grassland runoff water quality and yield. Trans. ASAE 47:777–788. doi:10.13031/2013.16109

Sharpley, A.N., T. Daniel, T. Sims, J. Lemunyon, R. Stevens, and R. Parry. 2003a. Agricultural phosphorus and eutrophication, 2nd ed. ARS–149. U.S. Department of Agriculture, Agricultural Research Service, Washington, D.C.

Sharpley, A.N., J.L. Weld, D.B. Beegle, P.J.A. Kleinman, W.J. Gburek, P.A. Moore, Jr., and G. Mullins. 2003b. Development of phosphorus indices for nutrient management planning strategies in the U.S. J. Soil Water Conserv. 58:137–152.

Sharpley, A., H.P. Jarvie, A. Buda, L. May, and P. Kleinman. 2013. Phosphorus legacy: Overcoming the effects of past management practices to mitigate future water quality impairment. J. Environ. Qual. 42:1308–1326. doi:10.2134/jeq2013.03.0098

Sharpley, A., P. Kleinman, C. Baffaut, D. Beegle, C. Bolster, A. Collick, Z. Easton, J. Lory, N. Nelson, D. Osmond, D. Radcliffe, T. Veith, and J. Weld. 2017. Evaluation of phosphorus site assessment tools: Lessons from the USA. J. Environ. Qual. 46:1250–1256. doi:10.2134/jeq2016.11.0427

Sharpley, A., H. Jarvie, D. Flaten, and P. Kleinman. 2018. Celebrating the 350th anniversary of phosphorus discovery: A conundrum of deficiency and excess. J. Environ. Qual. 47:774–777. doi:10.2134/jeq2018.05.0170

Shober, A.L., A.R. Buda, K.C. Turner, N.M. Fiorellino, A.S. Andres, J.M. McGrath, and J.T. Sims. 2017. Assessing coastal plain risk indices for subsurface phosphorus loss. J. Environ. Qual. 46:1270–1286. doi:10.2134/jeq2017.03.0102

Smith, D.R., P.A. Moore, Jr., C.V. Maxwell, B.E. Haggard, and T.C. Daniel. 2004. Reducing phosphorus runoff from swine manure with dietary phytase and aluminum chloride. J. Environ. Qual. 33:1048–1054. doi:10.2134/jeq2004.1048

Soberon, M., Q. Ketterings, K. Czymmek, S. Cela, and C. Rasmussen. 2015. Whole farm nutrient mass balance calculator for New York dairy farms. Cornell Field Crops newsletter. http://blogs. cornell.edu/whatscroppingup/2015/03/25/whole-farm-nutrient-mass-balance-calculator-for-new-york-dairy-farms/ (Verified 6 Dec. 2018).

Sommer, S.G. 2001. Effect of composting on nutrient loss and nitrogen availability of cattle deep litter. Eur. J. Agron. 14:123–133. doi:10.1016/S1161-0301(00)00087-3

Srinivasan, M.S., R.B. Bryant, M.P. Callahan, and J.L. Weld. 2006. Manure management and nutrient loss under winter conditions: A literature review. J. Soil Water Conserv. 61:200–209.

Steenhuis, T.S., G.D. Bubenzer, J.C. Converse, and M.F. Walter. 1981. Winter-spread manure nitrogen loss. Trans. ASAE 24(2):436–441. doi:10.13031/2013.34270

Stock, M.N., F.J. Arriaga, P.V. Vadas, and K.G. Karthikeyan. 2019. Manure application timing drives energy absorption for snowmelt on an agricultural soil. J. Hydrol. 569:51–60. doi:10.1016/j.jhydrol.2018.11.028

Strokal, M., L. Ma, Z. Bai, S. Luan, C. Kroeze, O. Oenema, G. Velthof, and F. Zhang. 2016. Alarming nutrient pollution of Chinese rivers as a result of agricultural transitions. Environ. Res. Lett. 11:024014. doi:10.1088/1748-9326/11/2/024014

Syers, J.K., A.E. Johnston, and D. Curtin. 2008. Efficiency of soil and fertilizer phosphorus use: Reconciling changing concepts of soil phosphorus behaviour with agronomic information. U.N. Food and Agriculture Organization Plant Nutrition Bulletin 18. http://www.fao.org/3/a-a1595e.pdf (Verified 10 Jan. 2019).

Sylvester, P. 2007. Poultry manure field stockpiling. University of Delaware Kent County Cooperative Extension, Dover, DE. http://extension.udel.edu/kentagextension/2007/11/27/poultry-manure-field-stockpiling/ (Accessed 8 Jan. 2019).

Tarkalson, D.D., and R.L. Mikkelsen. 2003. A phosphorus budget of a poultry farm and a dairy farm in the Southeastern U.S. and the potential impacts of diet alterations. Nutr. Cycl. Agroecosyst. 66:295–303. doi:10.1023/A:1024435909139

Toor, G.S., B.J. Cade-Menun, and J.T. Sims. 2005a. Establishing a linkage between phosphorus forms in dairy diets, feces, and manures. J. Environ. Qual. 34:1380–1391. doi:10.2134/jeq2004.0232

Toor, G.S., J.T. Sims, and Z. Dou. 2005b. Reducing phosphorus in dairy diets improves farm nutrient balances and decreases the risk of nonpoint pollution of surface and ground waters. Agric. Ecosyst. Environ. 105:401–411. doi:10.1016/j.agee.2004.06.003

Torres, C. 2010. Oregon dairy organics opens its doors to the public. Lancaster Farmer, Ephrata, PA. https://www.lancasterfarming.com/news/main_edition/oregon-dairy-organics-opens-its-doors-to-the-public/article_02bc642b-5d12-53c8-ad38-bbd907cf70b4.html (Verified 7 Dec. 2018).

U.S. Environmental Protection Agency. 2007. Restoration and protection activities in the Upper Branch of the Delaware River protects New York City's drinking water supply. Section 319 nonpoint source protection success story: New York. U.S. Environmental Protection Agency, Washington, D.C. https://www.epa.gov/sites/production/files/2015-11/documents/ny_wbde.pdf (Verified 19 Dec. 2018).

U.S. Environmental Protection Agency. 2015. Dane County Community Digester–Waunakee, WI. U.S. Environmental Protection Agency, Washington, D.C. https://www.epa.gov/sites/production/files/2016-05/documents/dane_county_agstar_site_profile_final_508_093015.pdf (Verified 7 Dec. 2018).

Uutiset. 2018. Finland to probe potential of cattle manure for energy production. Uutiset, 10 July. https://yle.fi/uutiset/osasto/news/finland_to_probe_potential_of_cattle_manure_for_energy_production/10297989 (Verified 7 Dec. 2018).

Vadas, P.A., P.J.A. Kleinman, and A.N. Sharpley. 2004. A simple method to predict dissolved phosphorus in runoff from surface applied manures. J. Environ. Qual. 33:749–756. doi:10.2134/jeq2004.7490

Vadas, P.A., L.W. Good, P.A. Moore, Jr., and N. Widman. 2009. Estimating phosphorus loss in runoff from manure and fertilizer for a phosphorus loss quantification tool. J. Environ. Qual. 38:1645–1653. doi:10.2134/jeq2008.0337

Vadas, P.A., L.W. Good, W.E. Jokela, K.G. Karthikeyan, F.J. Arriaga, and M. Stock. 2017. Quantifying the impact of seasonal and short-term manure application decisions on phosphorus loss in surface runoff. J. Environ. Qual. 46:1395–1402. doi:10.2134/jeq2016.06.0220

Vadas, P.A., M.N. Stock, G.W. Feyereisen, F.J. Arriaga, L.W. Good, and K.G. Karthikeyan. 2018. Temperature and manure placement in a snowpack affect nutrient release from dairy manure during snowmelt. J. Environ. Qual. 47:848–855. doi:10.2134/jeq2017.12.0464

Van Horn, H.H., A.C. Wilkie, W.J. Powers, and R.A. Nordstedt. 1994. Components of dairy manure management systems. J. Dairy Sci. 77:2008–2030. doi:10.3168/jds.S0022-0302(94)77147-2

Van Vliet, L.J.P., S. Bittman, G. Derksen, and C.G. Kowalenko. 2006. Aerating grassland before manure application reduces runoff nutrient loads in a high rainfall environment. J. Environ. Qual. 35:903–911. doi:10.2134/jeq2005.0266

Vanotti, M.B., A.A. Szogi, and P.G. Hunt. 2005. Wastewater treatment system. U.S. Patent 6893,567 B1. Date issued: 17 May.

Vanotti, M.B., A.A. Szogi, and L.M. Fetterman. 2010. Wastewater treatment system with simultaneous separation of phosphorus and manure solids. U.S. Patent 7674,379 B2. Issued 25 Dec.

Vohla, C., M. Koiv, H.J. Bavor, F. Chazarenc, and U. Mander. 2011. Filter materials for phosphorus removal from wastewater in treatment wetlands– A review. Ecol. Eng. 37:70–89. doi:10.1016/j.ecoleng.2009.08.003

Waldrip, H.M., Z. He, and S. Erich. 2010. Effects of poultry manure amendment on phosphorus uptake by ryegrass, soil phosphorus fractions and phosphatase activity. Biol. Fertil. Soils. doi:10.1007/s00374-011-0546-4

Walker, W.W., Jr. 1987. Phosphorus removal by urban runoff detention basins. Lake Reserv. Manage. 3:314–326. doi:10.1080/07438148709354787

Washington State University. 2018. Livestock nutrient management: Struvite extraction. Washington State University, Pullman, WA. https://puyallup.wsu.edu/lnm/struvite-extraction/ (Accessed 11 Dec. 2018).

White, M.J., D.E. Storm, P. Busteed, M.D. Smolen, H. Zhang, and G. Fox. 2010. A quantitative phosphorus loss assessment tool for agricultural fields. Environ. Model. Softw. 25:1121–1129. doi:10.1016/j.envsoft.2010.03.017

Williams, M.R., G.W. Feyereisen, D.B. Beegle, and R.D. Shannon. 2012. Soil temperature regulates phosphorus loss from lysimeters following fall and winter manure application. Trans. ASABE 55:871–880. doi:10.13031/2013.41529

Williams, M.R., G.W. Feyereisen, D.B. Beegle, R.D. Shannon, G.J. Folmar, and R.B. Bryant. 2011. Manure application under winter conditions: Nutrient runoff and leaching losses. Trans. ASABE 54:891–899. doi:10.13031/2013.37114

Withers, P.J.A., S.D. Clay, and V.G. Breeze. 2001. Phosphorus transfer in runoff following application of fertilizer, manure and sewage sludge. J. Environ. Qual. 30:180–188. doi:10.2134/jeq2001.301180x

Wright, P., J.L. Rennells, A.T. DeGaetano, and C. Gooch. 2013. Impacts of climate change in the Northeast on manure storage. Waste to worth: Spreading science and solutions, Denver, CO. 1-5 Apr. 2013. Livestock and Poultry Environmental Learning Community, University of Nebraska, Lincoln, NE. https://articles.extension.org/pages/67643/impacts-of-changing-climate-in-the-northeast-on-manure-storage (Verified 12 Dec. 2018).

Xin, H., R.S. Gates, A.R. Green, F.M. Mitloehner, P.A. Moore, Jr., and C.M. Wathes. 2011. Environmental impacts and sustainability of egg production systems. Poult. Sci. 90:263–277. doi:10.3382/ps.2010-00877

Yanke, L.J., H.D. Bae, L.B. Selinger, and K.J. Cheng. 1998. Phytase activity of anaerobic ruminal bacteria. Microbiology 144:1565–1573. doi:10.1099/00221287-144-6-1565

Young, R.A., and C.K. Mutchler. 1976. Pollution potential of manure spread on frozen ground. J. Environ. Qual. 5:174–179. doi:10.2134/jeq1976.00472425000500020013x

Young, R.A., and R.F. Holt. 1977. Winter applied manure: Effects of annual runoff, erosion, and nutrient movement. J. Soil Water Conserv. 32:219–222.

Young, R.A., T. Hutrods, and W. Anderson. 1980. Effectiveness of vegetative buffer strips in controlling pollution from feedlot runoff. J. Environ. Qual. 9:483–487. doi:10.2134/jeq1980.00472425000900030032x

Yuan, Y., M.A. Locke, R.L. Bingner, and R.A. Rebich. 2013. Phosphorus losses from agricultural watersheds in the MS delta. J. Environ. Manage. 115:14–20. doi:10.1016/j.jenvman.2012.10.028

Zebarth, B.J., J.W. Paul, and K. Chipperfield. 1999. Nutrient losses to soil from field storage of solid poultry manure. Can. J. Soil Sci. 79:183–190. doi:10.4141/S98-050

Organomineral Fertilizers and Their Application to Field Crops

William B. Smith, Melissa Wilson, and Paulo Pagliari*

Abstract

The combination of animal manure and mineral fertilizer to produce organomineral fertilizers is a new concept in animal waste management. Compared with animal manure, organomineral fertilizers have increased nutrient concentration so that lower application rates can be used. The production of organomineral fertilizers produces a more stable, balanced, and uniform product with predictable nutrient availability and nutrient release. The advantage of organomineral fertilizers over mineral fertilizers is the supply of a range of macro- and micronutrients in addition to organic matter. Although some limited information exists on the use of organomineral fertilizers in grain crops, most research has been conducted on vegetables, fruits, and specialty crops. Overall, research has shown that organomineral fertilizers can improve plant growth parameters such as yield and nutrient uptake to a greater degree than when manure or fertilizers are used alone. This chapter provides a review of the current literature on the use of organomineral fertilizers for food production.

Animal manure and mineral fertilizers have been widely used for centuries to provide macro- and micronutrients for crop production. Manure has the added benefit of providing organic matter when applied to soil, and when properly used, can provide conditions for crop production that surpass those provided by mineral fertilizer alone (Reddy et al., 2000; Fernandes et al., 2003; Steiner et al., 2007; Ayinla et al., 2018). For example, Steiner et al. (2007) reported that application of chicken manure led to significantly greater rice (*Oryza sativa* L.) and sorghum (*Sorghum bicolor* L.) yield than mineral fertilizer and higher soil pH, and extractable P, Ca, and Mg, as well. However, in many cases crop yields are similar between plots receiving manure or chemical fertilizer (Eghball and Power, 1999) suggesting that both are excellent nutrient sources.

Despite being widely used, there are several known drawbacks to the use of manure and mineral fertilizers. For example, to meet crop nutritional needs, mineral fertilizers are often needed to balance the nutrients in manure, which often has a low N to P ratio. Application of manure to meet the N needs of crops will in most cases result in over application of P, leading to environmental issues over time. Chemical tests should always be performed on manures before they are land applied so that known amounts of nutrients are applied. Another issue related to manure management is the high cost associated with transportation of manure to far distances due to its high moisture content. Although mineral fertilizers have the benefit of being able to be custom blended to meet all crop

W.B. Smith, Department of Animal Science & Veterinary Technology. Tarleton State University, 309A Joe W. Autry Agriculture Building. Box T-0070. Stephenville, TX 76402; M. Wilson, Department of Soil, Water, and Climate, University of Minnesota. 439 Borlaug Hall, 1991 Upper Buford Circle, Saint Paul, MN 55108; P. Pagliari, Department of Soil, Water, and Climate, University of Minnesota. Southwest Research and Outreach Center. 23669 130th St. Lamberton, MN 56152. *Corresponding author (pagli005@umn.edu).

doi:10.2134/asaspecpub67.c18

Animal Manure: Production, Characteristics, Environmental Concerns and Management. ASA Special Publication 67. Heidi M. Waldrip, Paulo H. Pagliari, and Zhongqi He, editors.
© 2019. ASA and SSSA, 5585 Guilford Rd., Madison, WI 53711, USA.

nutritional needs, there are still environmental issues associated with overapplication. More recently, there is an increasing shortage of raw materials used for fertilizer production as natural resources become depleted. An additional problem with mineral fertilizers is their cost, which is one of the most limiting factors for their use in developing countries such as in Africa, Asia, and South America. In light of the negative problems associated with each of these nutrient sources, there have been calls for options that are more environmentally and economically sound (Ojo et al., 2014; Sakurada et al., 2016; Ayinla et al., 2018).

Combining mineral fertilizers with manure or other waste products through industrial processes to produce a new product, organomineral fertilizers, is one alternative that has promise. The nutrient concentration is increased in this process so lower application rates can be used, and a more stable, balanced, and uniform product is created. As this becomes an industrialized nutrient source, it will also be packaged and well labeled with the nutrients composition clearly written which could be used to calculate the amount of nutrients required for any crop. An added benefit is that nutrient availability rates are more easily predicted (Sakurada et al., 2016). Furthermore, the use of organomineral fertilizer also has the advantage of supplying a range of macro- and micronutrients in addition to organic matter. The combination of animal manure and mineral fertilizer provides a great possibility to improve soil conditions of degraded agricultural soils and also maintain productivity in soils that are not yet degraded. For example, it has been reported that combining manure and chemical fertilizer provides a priming effect where the mixed material can outyield the manure or chemical fertilizer when applied alone (Reddy et al., 2000; Mandal et al., 2007; Eifediyi and Remison, 2010; Ayinla et al., 2018).

The objectives of this chapter are to: i) summarize the production process and chemical composition of organomineral fertilizers, and ii) evaluate the effectiveness of organomineral fertilizer as a plant nutrient source.

Manure as a Nutrient Source and its Limitations

In its raw form, animal manure it is primarily water (from 6 to 90%) and organic carbon (from 18 to 47%). Nutrients such as N account for 1.4 to 5.5%, P accounts for 0.2 to 4.8%, and K accounts for 1.0 to 7.8%, on a dry matter basis (Pagliari and Laboski, 2012; He et al., 2016). Nitrogen, P, and sulfur (S) are usually present in two forms in manure, inorganic minerals and also organic compounds. The inorganic mineral behaves just as inorganic fertilizer when applied to soils, while the organic compounds must first be mineralized before the nutrients can be utilized by plants and microbes (Pagliari and Laboski, 2013, 2014; Pagliari, 2014). This mixture of organic and inorganic nutrients in manure allows manure to act as a slow release fertilizer (from organic compounds) and also as a quick release fertilizer (from inorganic minerals) (Tejada et al., 2005; Makinde et al., 2007; Pagliari and Laboski, 2013, 2014).

Among the different animal species, monogastric animals have the highest concentration of nutrients in the manure. This is primarily because their manure has less moisture and also the feed has a higher concentration of nutrients, primarily N and P, which is in excess of what is needed by the animals. For example, average total N in manures from monogastric animals is 4.7% and from ruminants is 2.6%; average total P from monogastric animals is 2.6% and from ruminants is 0.9%, while the average for total K in manures from monogastric animals is 3.2% and from ruminants is 2.5% (Pagliari and Laboski, 2012; He et al., 2016). The higher nutrient concentration in manure from monogastric animals typically results in lower application rates to soils.

While animal manure is a great source of nutrients for plants, there are limitations with regards to land application due to environmental and health concerns. For instance, one of the main drawbacks of using manure as a nutrient source is the fact the nutrient concentrations are not in the same range of that required by most crops. In most cases the ratio between N and P_2O_5 in manure (about 2:1 to 1:1) is much lower than most crops requirement (about 4:1). Application of manure or organic wastes with low N/P ratio will result in the overapplication of P. Parham et al. (2002) reported that manure application in excess of crop needs has led to an increase in soil P ranging from 8 to 40 kg P ha^{-1} yr^{-1}; suggesting that in some cases manure is applied by almost twice its requirements for P. Overapplication of P increases the labile soil P pools (bioavailable) and the non-labile (more recalcitrant) pools in the soils (Waldrip et al., 2015). This is problematic since it has been reported that the higher the soil test P level the higher the potential for nonpoint source pollution (Bundy et al., 2001; Andraski and Bundy, 2003). Therefore, overapplication of manure (in excess of crop needs) poses a direct threat to aquatic ecosystems.

In the 2017 national water quality inventory reported to U.S. Congress, the EPA reported that over 46% of U.S. fresh water had some level of pollution and received the lowest ranking in its ability to support life (USEPA, 2017). In most cases agriculture is a contributor to nonpoint source pollutants, in particular N and P. Phosphorus is the most limiting nutrient in fresh water systems and even small amounts of added P can have significant negative impact on the system's eutrophication status. On the other hand, N is the most limiting nutrient in salt water systems. Manure is a strong contributor of P and N to aquatic systems primarily due to poor manure management strategies and lack of strong regulations. However, the cost associated with proper manure handling is extremely high as manure in its raw form can only be economically transported to a few miles away from the point of production. As a result, manure has been for many decades overapplied to agricultural lands near the point of production. This practice has led to P build up in the soils to levels that are considered excessive by any index utilized to assess soil P status (Waldrip et al., 2015).

At a more localized scale, manure is a potential source of pathogens to both water systems and food production systems (Solomon et al., 2002; Sivapalasingam et al., 2004). Islam et al. (2004) reported that *Samonella enterica* can remain in the soil for up to 231 d after the pathogen was introduced to soils by means of composted poultry manure. In addition, radishes and carrots grown in the contaminated soils showed the presence of *S. enterica* in tissue for up to 203 d after seeds were sown (Islam et al., 2004). The authors also tested the effects of adding contaminated water and found the same results as those observed for when contaminated composted manure was applied. Solomon et al. (2002) reported that *E. coli* introduced by contaminated water can enter lettuce plants through the roots and remain in the tissue protected against sanitation by virtue of its inaccessibility. Similarly, Islam et al. (2004) reported that *E. coli* remained in the soil for up to 217 d after dairy and poultry manure compost application. The authors reported that lettuce and parsley had detectable numbers of *E. coli* for as long as 177 d after seeds were sown (Islam et al., 2004). Although manure can be a source of pathogens, safety measures are available. Guan and Holley (2003) reported that

Fig. 1. Pelletized chicken manure mixed with chemical fertilizer monoammonium phosphate, potassium chloride, and calcogran (Tancal–Comeìrcio e Industria de Cal, 20% Ca and 10% Mg) used in the study of Sakurada L. et al. (2016). Photo credit to Marcelo A. Batista.

keeping manure at 25 °C for 90 d will render manure free from most pathogens. Aorigele and Simujide (2008) reported that temperatures between 30 and 60 °C are ideal to eliminate *E. coli* from cattle manure. There are also bacterial inhibitors ($CaCN_2$) which can be used to safely eliminate *E. coli* from manure during composting when temperatures are not favorable such as during the winter in temperate climate environments (Aorigele and Simujide, 2008). No research has shown grain crops being contaminated with pathogens from animal manure.

As evidenced, the application of excessive amounts of nutrients, organic or inorganic, above what a crop can remove will lead to environmental problems such as surface water eutrophication and water table contamination (Tejada et al., 2005). In addition, there are significant concerns regarding pathogens transferring from manure to our water and food systems. Alternatives are needed to curb these issues and the technology for creating organomineral fertilizers has promise.

Complexes of Manure and Mineral Fertilizer
Processes of Organomineral Creation

Many materials, such as food waste, sewage, and other industrial wastes, have been used for the production of organomineral fertilizers (Chassapis and Roulia, 2008; Rady, 2012; Kominko et al., 2017), but this chapter will focus on organomineral fertilizers produced with animal manure. Table 1 provides a list of peer reviewed research where different forms of organominerals fertilizers were used in different crops. In particular, manure from poultry animals have been the most widely used

Table 1. Selected reference list for peer-reviewed research using various forms of organomineral fertilizers.

Materials Used	Product Form	Final Nutrient Composition (N-P-K)	Test Crop	Reference
House hold waste + inorganic fertilizer		3.5– 2.5– 4.0	Maize	Ayeni et al., 2012
		3.5– 2.5– 4.0	Tomato	Ayeni and Ezeh 2017
Cow dung + house hold waste + inorganic fertilizer		2.9– 1.1– 0.7	Yam	Oshunsanya and Akinrinola, 2014
		2.9– 1.1– 0.7	Maize	Babalola et al., 2007
Cow dung + inorganic fertilizer	Fused Organomineral fertilizer	3.5– 2.5– 4.0	Cabbage	Olaniyi and Ojetayo 2011
		Unknown	Egusi Melon	Makinde et al., 2007
Various mixes of composted pig slurry + poultry litter + spent mushroom compost + cocoa husks + moistened shredded paper + blood and feather meal		Various composition ranging from 10– 3– 6 to 3– 5– 10	grasslands	Rao et al., 2007
Poultry manure + inorganic fertilizer		3– 15– 2	Corn	Sakurada et al. 2016
Sewage Sludge		10– 10– 10	No test crop	Kominko et al., 2017
Pine wood savings + urban waste compost + chicken bedding + inorganic fertilizer	Fused Organomineral fertilizer liquid and solid	1.5– 3.5– 1.0	Melon	Fernandes et al., 2003
Kola pod husk (product 1) Cow dung + city waste (product 2)	Ground materials	Unknown	*Amaranthus cruentus*	Makinde et al., 2011
Compost amended with inorganic fertilizer	Not reported	5.1– 4.4– 1.1	Watermelon	Ojo et al., 2014
Granulated poultry manure	Granulated manure mixed with fertilizer	5– 20– 2	Corn	Sakurada et al. 2016
Solid poultry + solid dairy + liquid hog manure + inorganic fertilizer	Granulated manure mixed with fertilizer	7– 4– 4	Potato	Zebarth et al., 2005

Fig. 2. Chicken manure fused with monoammonium phosphate into a single granule used in the study of Sakurada L. et al. (2016). Photo credit to Marcelo A. Batista.

in the production of organomineral fertilizers because of the inherently higher nutrient density in this type of manure. This facilitates the production of a more balanced organomineral fertilizer using lower amounts of inorganic fertilizer.

In general, two types of organomineral fertilizers are reported in the scientific literature. One in which pelletized or granulated manure and inorganic fertilizer granules are mixed together (from herein after called "granulated manure mixed with fertilizer"), and one in which pelletized manure and fertilizer are fused together in a process called granulation to produce one product (from herein after called "fused organomineral fertilizer"). In the first process, two distinct products are visible, the pelletized manure and the fertilizer granules (Fig. 1). In the second, only one distinct granule is visible, the fused organomineral fertilizer (Fig. 2). Different nutrient sources and concentrations can be added to the pelletized or granulated manure or to the mix during pelletization or granulation (for fused organomineral fertilizers), which will lead to the production of materials with different final composition. The reader should note that throughout the chapter, the term organomineral fertilizer will also be used to refer to either type of product, especially when reports did not report whether they tested granulated manure mixed with fertilizer or fused organomineral fertilizer. See Chapter 24 for a review on pelletized manure for on- and off-farm use.

Rao et al. (2007) has reported the most detailed process for the creation of organomineral fertilizers, but aside from this study the only information on how organomineral fertilizers are fabricated is presented in patents. For example, Kazemzadeh (1998) describes the production of odorless and sterilized manure pellets. The process involves mixing of dried manure with a dry binder agent material, followed by the addition of steam and water or other sludge type materials. The mixture is then added to an extruder which homogenizes the materials, then at this stage high pressure (100 psi) and high temperature (125 °C) is applied. After the material have dried to about 7% moisture, the pellets can then be mixed with fertilizer granules to produce granulated manure mixed with fertilizer, or fused with fertilizer to produced fused organomineral fertilizer. Varshovi (2005) described a method where wet manure is mixed with dry chemical fertilizer prior to granulation, then placed into a rotating mixer or granulation drum, and dried during the granulation process. Similarly, dried and granulated manure can be mixed with liquid chemical fertilizer, then placed into a rotating mixer or granulation drum, and then the mixed materials dry during the granulation process.

Chemical Properties of Organomineral Fertilizer

Unlike chemical fertilizer, organomineral fertilizers do not have fixed or specific chemical properties; they will vary according to the fabrication process. For example, monoammonium phosphate (MAP) fertilizer has a distinct chemical composition of 11% N and 52% P_2O_5, diammonium phosphate (DAP) fertilizer has a distinct chemical composition of 18% N and 46% P_2O_5, and urea is 46% N. Organomineral fertilizers are produced with varied ratios of nutrients so each product may be different.

Several researchers have reported details on the materials used in production of organomineral fertilizers. Sakurada et al. (2016) reported that the organomineral fertilizers tested in their study had the composition of 5–20–2 (N-P_2O_5-K_2O) and 3–15–2 (N-P_2O_5-K_2O) for fused organomineral fertilizer (Fig. 2) and granulated manure mixed with fertilizer (Fig. 1), respectively. The organomineral fertilizer tested by Sakurada et al. (2016) was produced from chicken manure as the organic source, MAP, KCl, and Calcogran (Tancal–Comèircio e Industria de Cal) (20% Ca and 10% Mg) as the chemical fertilizer sources. The amount of organic carbon remaining in the final product was 41% and 44% and pH was 6.3 and 5.1 in the fused organomineral fertilizer and granulated manure mixed with fertilizer, respectively (Sakurada et al., 2016). Fernandes et al. (2003) reported that the organomineral fertilizer tested in their study was produced by mixing pine wood shavings, urban waste compost, and chicken manure with phospho-gypsum, urea, triple superphosphate, single superphosphate, potassium chloride, a biocatalyst, and a solubilizing catalyst. The process generated a final product containing 2–7-4 ($N-P_2O_5-K_2O$) with 42% organic carbon and a pH of 6.6 (Fernandes et al., 2003). Makinde et al. (2007) tested an organomineral fertilizer produced from cow dung mixed with inorganic N, having a composition of 4–0.5–1 ($N-P_2O_5-K_2O$), 36% organic carbon, and a pH of 6.1.

Other researchers have provided less information on the production of the organomineral fertilizer that was tested but did provide the final chemical composition. For example, Zebarth et al. (2005) tested an organomineral produced from several mixed organic and chemical source with a final chemical composition of 7–4-4 ($N-P_2O_5-K_2O$). Ojo et al. (2014) used two organomineral fertilizers of unknown origin with the chemical properties 5–4-1 ($N-P_2O_5-K_2O$) and 1–0.7–2 ($N-P_2O_5-K_2O$). Ayeni and Ezeh (2017) tested an organomineral fertilizer with chemical composition of 4–3-4 ($N-P_2O_5-K_2O$). As can be seen from the studies reported in this section, the chemical composition of organomineral fertilizers are wide and change dramatically.

Soil Response to Organomineral Fertilizers

Organomineral fertilizers have a wide range of macro- and micronutrients as well as physical properties, most likely due to the variability in the manures and mineral fertilizers used to make them. Research has shown that land application of organominerals can promote changes to physical, chemical, and biological properties of soils. Table 2 reports peer-reviewed studies that reported changes in soil properties due to organomineral fertilizer application.

Soil Physical Responses to Organomineral Fertilizers

Cumulative infiltration rates have been reported to increase after an organomineral fertilizer (62 cm) was applied to soils compared with plots that were untreated (37 cm) (Babalola et al., 2007). Soil loss as a result can also be minimized when organomineral fertilizer is used. Babalola et al. (2007) reported that plots treated with an organomineral fertilizer had a yearly soil loss of 800 kg ha[-1] while in the control it was over 1000 kg ha[-1]. However, Babalola et al. (2007) also reported that under certain conditions runoff in plots receiving organomineral fertilizer is also much greater than the control plots. As a result of higher runoff rates, there were higher loads of nutrient running off, particularly P and N, in organomineral treated plots (Babalola et al., 2007).

Table 2. Effects of granulated organomineral fertilizer or fused organomineral fertilizer on soil physical, chemical, and biological properties.

Reference	Soil Chemical Properties	Soil Physical Properties	Soil Biological Properties
Ayeni et al., 2012	Increased soil pH; increased OM; and increased available P and K		
Ayeni and Ezeh 2017	Increased soil pH; increased OM; increased available P, Ca, Mg, K, Cu, Mn, and Zn		
Tejada et al., 2005	Increased available N, P, K, and Ca		
Ojeniyi et al., 2009	Increased soil pH; increased available N, P, K, and Ca		
Babalola et al., 2007		Improved infiltration rate; minimized soil loss through erosion; potential increased runoff when conditions are right	
Mandal et al., 2007			Increased microbial biomass carbon; higher roots exudates; increased dehydrogenase activity;

Soil Chemical Responses to Organomineral Fertilizers

Nutrient availability throughout the growing season is the key factor determining productivity. A nutrient source that is capable of supplying nutrients for a growing plant during its vegetative and reproductive cycles is desirable. Like raw manures, organomineral fertilizers combine nutrients in organic forms, which must be mineralized before it becomes available for plant uptake, and also in inorganic forms, which are available soon after application (Pagliari and Laboski, 2013, 2014). The stability of the organic and inorganic nutrient forms appears to depend on how the organomineral is formulated. When organic and chemical fertilizers are fused together to form one single granule, nutrient availability seems to be slower but lasts for multiple growing seasons compared with when the nutrient sources are mixed together and two products are distinguishable (Makinde et al., 2011). Others have reported that when organic and chemical fertilizers are fused together, the presence of inorganic minerals seems to increase mineralization of organic nutrients, extending nutrient availability during the growing season (Mandal et al., 2007; Ojo et al., 2014). The rapid solubilization of the minerals provides a rapid supply of available nutrients and the mineralization of organic nutrients provides a steady supply of nutrients during the critical stages in the growing season (Olaniyi and Ojetayo, 2011).

Zebarth et al. (2005) compared the availability of N in organomineral fertilizer with that of chemical fertilizer and found similar availability when applied at similar N rates. Tejada et al. (2005) reported that inorganic N is more readily available for leaching (as much as 10%) when granulated or pelletized manure is mixed with chemical fertilizer compared with applied organomineral fertilizer. Phosphorus leaching was also markedly greater (16 kg ha^{-1}) in granulated manure mixed with mineral fertilizer than in fused organomineral fertilizer (7 kg ha^{-1}). In addition, K and Ca leaching was also greater by 29% and 24%, respectively, when granulated manure was mixed with mineral fertilizer than in organomineral fertilizer. Ojeniyi et al. (2009) reported that plant N, P, and K uptake was improved with organomineral fertilizer, most likely due to an increase of soil nutrient levels, compared with non-fertilized and chemical fertilized plots. Micronutrients, in particular Cu, Mn, and Zn have also been reported to increase in plant tissue after the application of organomineral fertilizer compared with mineral fertilizer (Ayeni and Ezeh, 2017).

Soil pH is another important chemical factor that not only drives nutrient availability but can impact crop growth as well. For instance, most crops thrive in the pH range of 6 to 7.5 and phosphorus is most available in the range from 6 to 7. The effects of organomineral fertilizer on soil pH can be dramatic compared with chemical fertilizer. Ojeniyi et al. (2009) reported that organomineral fertilizer had improved effects on soil pH compared with non-fertilized and chemical fertilized plots. Ayeni and Ezeh (2017) found that soil pH in plots receiving organomineral fertilizer increased from 6.8 (control plots) to 7.9 (organomineral plots); while the use of chemical fertilizer caused a drop in the soil pH (6.1). In a different study, application of organomineral fertilizer increased soil pH from 6.0 in the control plots to 6.2, whereas soil pH after application of chemical fertilizer was 5.9 (Ayeni et al., 2012).

Improving soil organic matter levels by the addition of organic amendments can provide a wide range of benefits ranging from improved soil physical and chemical properties to improved habitat for soil microfauna (Kominko et al., 2017). Organomineral fertilizers have been shown to increase soil organic matter compared with mineral fertilizers alone (Ayeni et al., 2012). Potential benefits of increased organic matter can include improved aggregate stability, water holding capacity, bulk density, infiltration, cation exchange capacity, and also added carbon to fuel microbial activity.

Soil Biological Responses to Organomineral Fertilizers

Microbial biomass carbon has been shown to be significantly higher in plots that are treated with organomineral fertilizer than in plots treated with mineral fertilizers alone (Mandal et al., 2007). Mandal et al. (2007) reported that microbial biomass carbon in plots treated with organomineral was 517 mg kg^{-1}, significantly greater than in the control, 261 mg kg^{-1}, where no nutrient sources were added. The higher microbial biomass carbon was attributed to higher above ground growth leading to higher photosynthesis rates leading to high C being transported to the roots. As a result, higher amounts of roots exudates were excreted which provided a carbon source for microbial communities to flourish (Mandal et al., 2007). Mandal et al. (2007) also reported increased dehydrogenase activity in soils treated with organomineral fertilizer compared with control and chemical fertilizers. Dehydrogenase activity has been shown to be influenced more by the quality than by the quantity of organic carbon added into soil. Thus, the significant effect of organomineral fertilizer on dehydrogenase activity might be related to the more easily decomposable components of crop residues on the metabolism of soil microorganisms resulting from organomineral application. However, not all microbial activity indicators respond to the addition of organomineral or mineral fertilizer. For example, acid phosphatase activity has shown to be unaffected by the addition of organomineral and chemical fertilizers (Mandal et al., 2007).

Plant Response to Organomineral Fertilizer

Yield Response

Possibly the earliest adopters of organomineral fertilizers were the producers of vegetable and specialty crops, which have by far the largest number of research data available in the literature. Table 1 list research articles and the crops being used for tests with orgamineral fertilizer. Ayeni and Ezeh (2017) evaluated the effect of organic, organomineral, and chemical fertilizers on soil properties and tomato (*Solanum lycopersicum* L.) development and growth. Tomato plants receiving organomineral fertilizer had greater plant height (39 cm vs. 26 cm), root dry matter (53 g vs. 30 g), and fruit yield (539 g vs. 430 g) than commercial fertilizer when applied at equivalent rates (Ayeni and Ezeh, 2017). The best application rate for organomineral fertilizer and chemical fertilizer for okra (*Abelmoschus esculentus* [L.] Moench) was tested in Nigeria (Olaniyi et al., 2010). The application of 3 Mg organomineral fertilizer per ha resulted in an equivalent number of okra fruit per ha while fruit yield was increased by approximately 10 kg ha-1 compared with the control (Olaniyi et al., 2010). Furthermore, the authors also reported that combining organomineral fertilizer and chemical fertilizer significantly increased okra yield when compared with either source applied alone; and, along with an increase in fruit yield, plants also exhibited higher leaf nitrogen and greater fruit protein, K, and Fe uptake (Olaniyi et al., 2010). In a separate experiment, the effects of organic fertilizer, organomineral fertilizer, and chemical fertilizer on cabbage (*Brassica oleracea* L.) production parameters were studied (Olaniyi and Ojetayo, 2011). The authors reported that the use of organomineral fertilizer resulted in greater head length, diameter, and yield than the control (Olaniyi and Ojetayo, 2011). Similar effects have been observed in the production of cucumber (*Cucumis sativus* L.; Olaniyi et al., 2009), lettuce (*Lactuca sativa* L.; Olaniyi, 2008), and potato (*Solanum tuberosum* L.; Zebarth et al., 2005). However, application of organomineral fertilizer to sweet potatoes (*Ipomoea batatas* [L.] Lam.) resulted in no change in measurable quality parameters (Kareem, 2013).

Plant response to organomineral fertilizer has shown positive results, encouraging its adoption across a wide range of production crops including grain. When fused organomineral fertilizer was applied at rates of 5 or 10 Mg ha^{-1} to maize, grain yields were increased by 2.11 and 2.13 Mg ha^{-1}, respectively (Ayeni et al., 2012). In addition, application of fused organomineral fertilizer lead to greater corn grain yield than application of chemical fertilizer at comparable rates (Ayeni et al., 2012). Babalola et al. (2007) compared the use of vetiver grass (*Chrysopogon nigritanus* [Benth.] Veldkamp) strip with organomineral fertilizer and their potential to act as fertilizer to improve corn grain yield and also as soil cover to improve soil and water conservation. The authors found that the use of organomineral fertilizer led to higher plant height (about 10% taller plants in plots treated with organomineral fertilizer), and greater grain yields ranging from 40% to 57% in plots treated with organomineral fertilizer than in plots treated with vetiver grass strips when adequate moisture was available. In drier years, when rain events were limited during the flowering stage no significant differences were observed between organomineral fertilizer and vetiver grass strips (Babalola et al., 2007). Sakurada et al. (2016) used a greenhouse study to compare the cumulative effects of fused organomineral fertilizer with pelletized manure mixed with chemical fertilizer and also with chemical fertilizer used alone on corn plants development. The authors reported that for the first two cycles corn biomass in fused organomineral and chemical fertilizer were not different and where both greater than that in the pelletized manure mixed with chemical fertilizer (Sakurada et al., 2016). In the third and fourth cycles, however, fused organomineral fertilizer started to be less effective than chemical fertilizer and pelletized manure mixed with chemical fertilizer (Sakurada et al., 2016). The results of Sakurada et al. (2016) shows that the different methods to fabricate the different organomineral fertilizer will impact their performance and residual effects. Tejada et al. (2005) investigated the effects of fused organomineral fertilizer and organic matter mixed with chemical fertilizer on wheat production parameters and nutrient uptake. Wheat grain yield significantly increased by 115 kg ha^{-1}, protein increased by 0.5%, and number of spikes m^{-2} increased by 11 when treated with fused organomineral fertilizer compared with organic matter mixed with chemical fertilizer (Tejada et al., 2005). The use of two different organomineral fertilizers resulted in plant height and stem girth of grain amaranth (*Amaranthus cruentus* L.) similar to the application of mineral fertilizer and dry shoot weight greater than the application of mineral fertilizer (Olowoake, 2014). Titrated levels of organomineral fertilizer derived from poultry litter resulted in approximately linear increases in plant height, leaf area, tillering, straw yield, and grain yield in rice (*Oryza sativa* L.) (Egbuchua and Enujeke, 2013).

Seldom mentioned is the effect of organomineral fertilizer in the production of forage crops. The use of biosolids treated with urea was used as a fertilizer source in the production of forage crops for silage (Smith et al., 2015). Across multiple years, there was no effect of organomineral fertilizer on dry matter (DM) yield of winter wheat (*Triticum aestivum* L.) or forage maize (*Zea mays* L.) relative to the use of conventional fertilizer (Smith et al., 2015). Similarly, there was no effect of organomineral fertilizer on the silage yield of perennial ryegrass (*Lolium perenne* L. ssp. *perenne*) (Smith et al., 2015). However, in a single year of testing, the use of organomineral fertilizer resulted in a decrease in the yield of spring barley (*Hordeum vulgare* L.) by 20% relative to conventional fertilizer (Smith et al., 2015).

Efficacy of Organomineral Fertilizer and Nutrient Uptake

There is evidence to support the hypothesis that mineral uptake from organomineral fertilizers may be more efficient than from traditional commercial fertilizers. When tomato plants were grown in soil receiving organomineral fertilizer, they exhibited greater K, Ca, Mg, Fe, Mn, and Zn uptake than plants grown in soil receiving inorganic fertilizer (Ayeni and Ezeh, 2017). When grown in the presence of 1.5 Mg ha^{-1} organomineral fertilizer, wheat grain exhibited an increase in N (28.1 vs. 27.3%), K (8.4 vs. 8.1%), and Fe (129.7 vs. 126.3 ppm) when compared with a complex of organic and inorganic fertilizers (Tejada et al., 2005). Nitrogen use efficiency in perennial ryegrass has been shown to exist in the range of 0.33 to 0.37 kg N kg^{-1} DM when organomineral fertilizer is applied (Antille et al., 2014). Varying rates of organomineral fertilizer increased maize concentration of N (2.5, 5, and 10 Mg ha^{-1}), P (10 Mg ha^{-1}), K (2.5, 5, and 10 Mg ha^{-1}), Fe (2.5, 5, and 10 Mg ha^{-1}), Cu (2.5, 5, and 10 Mg ha^{-1}), and Mn (5 and 10 Mg ha^{-1}) compared with the control (Ayeni et al., 2012). Wheat grain N (0.8 g kg^{-1}), K (0.3 g kg^{-1}), and Fe (3.4 mg kg^{-1}) uptake were increased in plots treated with fused organomineral fertilizer compared with organic matter mixed with chemical fertilizer (Tejada et al., 2005).

On the other hand, some studies have shown that nutrient uptake may be enhanced when organomineral fertilizers are combined with commercial fertilizers. Makinde et al. (2011) investigated the effect of chemical fertilizer, kola pod husk, and fused organomineral fertilizer on amaranth growth and nutrient uptake. The authors reported that N, P, K, Ca, and Mg uptake was greater when plants were treated with fused organomineral fertilizer mixed with chemical fertilizer than with fused organomineral applied alone (Makinde et al., 2011). In another study, corn plants grown in a greenhouse were able to take up greater amounts of P when grown in pots treated with granulated organomineral fertilizer or chemical fertilizer applied alone than from plants grown in pots treated with fused organomineral fertilizer (Sakurada et al., 2016).

Other Benefits and Potential Environmental Implications Related to Organomineral Fertilizers

Though overapplication of fertilizer in any form is a concern from an environmental stand point, organomineral fertilizers may provide some solutions. When an organomineral fertilizer was compared with a complex of organic and inorganic nutrients, inorganic N losses were reduced by 16%, P losses were reduced by 55%, and an 11% reduction in the loss of Mg with the use of an organomineral fertilizer (Tejada et al., 2005). Similarly, organomineral fertilizer resulted in an increase in mineralization rate and decrease in N loss relative to mineral fertilizer when applied to perennial rye grass (Antille et al., 2014). In an experiment to test the leaching effect of fertilizer sources, Richards et al. (1993) found that approximately 26% of the NO_3 from organomineral fertilizer could be detected in the leachate versus 95% of that derived from ammonium nitrate. This is supported by the work of Florio et al. (2016) who found similar results in the application of organomineral fertilizers to potted perennial ryegrass. In this same experiment, however, NH_3 emissions were shown to be elevated in pots treated with organomineral fertilizers relative to mineral-treated controls (Florio et al., 2016). As an added environmental selling point, Adewole et al. (2010) found that the application of organomineral fertilizer was successful in promotion of *Helianthus annuus* L. to remove 42, 43, and 57% of Cu, Pb, and Cd, and the promotion of *Tithonia diversifolia* (Hemsl.) A. Gray to remove 23, 44, and 35% of Cu, Pb, and Cd from a soil contaminated with paint effluent.

Conclusion

The use of organomineral fertilizers, in the form of granulated manure mixed with fertilizer or fused organomineral fertilizer, is increasing as the potential benefits are starting to be realized. The ability to design a fertilizer with known nutrient quantity and predictable nutrient release rates that acts as a fast and slow release nutrient source is appealing. Improved soil physical, chemical, and biological properties have been reported as result of organomineral application. Plant responses to organomineral fertilizer have also shown to be positive, as in most cases organomineral fertilizers result in greater plant growth parameters, such as yield and nutrient uptake, than those observed with the use of chemical fertilizer used alone. Limited research has been done to evaluate the potential benefits and drawbacks of using organomineral fertilizer as the sole nutrient source for food production. Future research needs to be developed so that best management practices for row crop and vegetable crop production can be developed for the use of organomineral fertilizers.

References

Adewole, M.B., M.K.C. Sridhar, and G.O. Adeoye. 2010. Removal of heavy metals from soil polluted with effluents from a paint industry using *Helianthus annuus* L. and *Tithonia diversifolia* (Hemsl.) as influenced by fertilizer applications. Biorem. J. 14(4):169–179. doi:10.1080/10889868.2010.514872

Andraski, T.W., and L.G. Bundy. 2003. Relationships between phosphorus levels in soil and in runoff from corn production systems. J. Environ. Qual. 32(1):310–316. doi:10.2134/jeq2003.3100

Antille, D.L., R. Sakrabani, and R.J. Godwin. 2014. Effects of biosolids-derived organomineral fertilizers, urea, and biosolids granules on crop and soil established with ryegrass (*Lolium perenne* L.). Commun. Soil Sci. Plant Anal. 45(12):1605–1621. doi:10.1080/00103624.2013.875205

Aorigele, Y.J., and W.C. Simujide. 2008. Sterilization effects of bacterial inhibitor on Escherichia Coli in cattle manure compost. Agric. Sci. Technol. 9.

Ayeni, L.S., E.O. Adeleye, and J.O. Adejumo. 2012. Comparative effect of organic, organomineral and mineral fertilizers on soil properties, nutrient uptake, growth and yield of maize (*Zea mays*). International Research Journal of Agricultural Science and Soil Science 2(11):493–497. doi:10.13140/RG.2.2.15371.18721

Ayeni, L.S., and O.S. Ezeh. 2017. Comparative effect of NPK 20:10:10, organic and organo-mineral fertilizers on soil chemical properties, nutrient uptake and yield of tomato (*Lycopersicon esculentum*). Applied Tropical Agriculture 22(1):111–116.

Ayinla, A., I.A. Alagbe, B.U. Olayinka, A.R. Lawal, O.O. Aboyeji, and E.O. Etejere. 2018. Effects of organic, inorganic and organo-mineral fertilizer on the growth, yield and nutrient composition of *corchorus olitorious* (L). Ceylon J. Sci., Biol. Sci. 47(1):13–19. doi:10.4038/cjs.v47i1.7482

Babalola, O., S.O. Oshunsanya, and K. Are. 2007. Effects of vetiver grass (*Vetiveria nigritana*) strips, vetiver grass mulch and an organomineral fertilizer on soil, water and nutrient losses and maize (*Zea mays* L.) yields. Soil Tillage Res. 96(1–2):6–18. doi:10.1016/j.still.2007.02.008

Bundy, L.G., T.W. Andraski, and J.M. Powell. 2001. Management practice effects on phosphorus losses in runoff in corn production systems. J. Environ. Qual. 30(5):1822. doi:10.2134/jeq2001.3051822x

Chassapis, K., and M. Roulia. 2008. Evaluation of low-rank coals as raw material for Fe and Ca organomineral fertilizer using a new EDXRF method. Int. J. Coal Geol. 75(3):185–188. doi:10.1016/j.coal.2008.04.006

Egbuchua, C.N., and E.C. Enujeke. 2013. Effects of different levels of organomineral fertilizer on the yield and yield components of rice (Oryza Sativa L.) in a coastal flood plain soil, Nigeria. IOSR J. Agric. Vet. Sci. 4(2): 1–5. www.iosrjournals.org (accessed 26 Sept. 2018).

Eghball, B., and J.F. Power. 1999. Phosphorus- and nitrogen-based manure and compost applications. Soil Sci. Soc. Am. J. 63(4):895. doi:10.2136/sssaj1999.634895x

Eifediyi, E.K., and S.U. Remison. 2010. Growth and yield of cucumber (*Cucumis sativus* L.) as influenced by farmyard manure and inorganic fertilizer. J. Plant Breed. Crop Sci. 2(7):216–220.

Fernandes, A.L.T., G.P. Rodrigues, and R. Testezlaf. 2003. Mineral and organomineral fertirrigation in relation to quality of greenhouse cultivated melon. Sci. Agric. 60(1):149–154. doi:10.1590/S0103-90162003000100022

Florio, A., B. Felici, M. Migliore, M.T. Dell'Abate, and A. Benedetti. 2016. Nitrogen losses, uptake and abundance of ammonia oxidizers in soil under mineral and organo-mineral fertilization regimes. J. Sci. Food Agric. 96(7):2440–2450. doi:10.1002/jsfa.7364

Guan, T.Y., and R.A. Holley. 2003. Pathogen survival in swine manure environments and transmission of human enteric illness–a review. J. Environ. Qual. 32(2):383–392 http://www.ncbi.nlm.nih.gov/pubmed/12708660 (Accessed 2 Sept. 2018). doi:10.2134/jeq2003.3830

He, Z., P.H. Pagliari, and H.M. Waldrip. 2016. Applied and environmental chemistry of animal manure: A review. Pedosphere 26(6):779–816. doi:10.1016/S1002-0160(15)60087-X

Islam, M., M.P. Doyle, S.C. Phatak, P. Millner, and X. Jiang. 2004. Persistence of enterohemorrhagic Escherichia coli O157:H7 in soil and on leaf lettuce and parsley grown in fields treated with contaminated manure composts or irrigation water. J. Food Prot. 67(7):1365–1370 http://www.ncbi.nlm.nih.gov/pubmed/15270487 (Accessed 2 Sept. 2018).

Kareem, I. 2013. Fertilizer treatment effects on yield and quality parameters of sweet potato (Ipomoea batatas). Res. J. Chem. Environ. Sci. 1(3): 40–49. www.aelsindia.com (Accessed 26 Sept. 2018).

Kazemzadeh, M. 1998. Process for producing odorless organic and semi-organic fertilizer. U.S. Patent No. 5,772,721. Date issued 30 June. https://patents.google.com/patent/US5772721A/en (Accessed 8 Sept. 2018).

Kominko, H., K. Gorazda, and Z. Wzorek. 2017. The possibility of organo-mineral fertilizer production from sewage sludge. Waste Biomass Valorization 8(5):1781–1791. doi:10.1007/s12649-016-9805-9

Makinde, E.A., L.S. Ayeni, and S.O. Ojeniyi. 2011. Effects of organic, organomineral and NPK fertilizer treatments on the nutrient uptake of *Amaranthus cruentus* (L.) on two soil types in Lagos, Nigeria. Journal of Central European Agriculture 12(1):114–123. doi:10.5513/JCEA01/12.1.887

Makinde, E.A., O.T. Ayoola, and M.O. Akande. 2007. Effects of an organo– mineral fertilizer application on the growth and yield of maize. Aust. J. Basic Appl. Sci. 3(10):1152–1155.

Mandal, A., A.K. Patra, D. Singh, A. Swarup, and R. Ebhin Masto. 2007. Effect of long-term application of manure and fertilizer on biological and biochemical activities in soil during crop development stages. Bioresour. Technol. 98(18):3585–3592. doi:10.1016/j.biortech.2006.11.027

Ojeniyi, S.O., O. Owolabi, O.M. Akinola, and S.A. Odedina. 2009. Field study of effect of organomineral fertilizer on maize growth, yield soil and plant nutrient composition in Ilesa, Southwest Nigeria. Niger. J. Soil Sci. 19(1):11–16 https://www.cabdirect.org/cabdirect/abstract/20113109797 (Accessed 8 Sept. 2018).

Ojo, J.A., A.A. Olowoake, and A. Obembe. 2014. Efficacy of organomineral fertilizer and un-amended compost on the growth and yield of watermelon (Citrullus lanatus Thumb) in Ilorin Southern Guinea Savanna zone of Nigeria. International Journal of Recycling of Organic Waste in Agriculture 3(4):121–125. doi:10.1007/s40093-014-0073-z

Olaniyi, J.O. 2008. Comparative effects of the source and level of nitrogen on the yield and quality of lettuce. Am. J. Sustain. Agric. 2(3): 225–228. http://go.galegroup.com.ezp1.lib.umn.edu/ps/i.do?id=GALE%7CA215515411&v=2.1&u=umn_wilson&it =r&p=EAIM&sw=w (Accessed 26 Sept. 2018).

Olaniyi, J.O., W.B. Akanbi, O.a. Olaniran, and O.T. Ilupeju. 2010. The effect of organo-mineral and inorganic fertilizers on the growth, fruit yield, quality and chemical compositions of okra. J. Anim. Plant Sci. 9(1):1135–1140.

Olaniyi, J., E. Ogunbiyi, and D. Alagbe. 2009. Effects of organo-mineral fertilizers on growth, yield and mineral nutrients uptake in cucumber. J. Anim. Plant Sci. 5(1):437–442 http://www.m.elewa.org/JAPS/2009/5.1/4.pdf (Accessed 3 Apr. 2019).

Olaniyi, J.O., and A.E. Ojetayo. 2011. Effect of fertilizer types on the growth and yield of two cabbage varieties. J. Anim. Plant Sci. 12:1573–1582 http://www.biosciences.elewa.org/JAPS (Accessed 8 Sept. 2018).

Olowoake, A.A. 2014. Influence of organic, mineral and organomineral fertilizers on growth, yield, and soil properties in grain amaranth (Amaranthus cruentus L.). J. Org. 1(1): 39–47. https://jorganics.files.wordpress.com/2014/10/1150.pdf (Accessed 26 Sept. 2018).

Oshunsanya, S. and T. Akinrinola. 2014. Changes in soil physical properties under yam production on a degraded soil amended with organomineral fertilizers. Global Research Journal 2: 087-093.

Pagliari, P.H. 2014. Variety and solubility of phosphorus forms in animal manure and their effects on soil test phosphorus In: Z. He and H. Zhang, Applied manure and nutrient chemistry for sustainable agriculture and environment. Springer Netherlands, Dordrecht. p. 141–161. doi:10.1007/978-94-017-8807-6_8

Pagliari, P.H., and C.A.M. Laboski. 2012. Investigation of the inorganic and organic phosphorus forms in animal manure. J. Environ. Qual. 41(3):901. doi:10.2134/jeq2011.0451

Pagliari, P.H., and C.A.M. Laboski. 2013. Dairy manure treatment effects on manure phosphorus fractionation and changes in soil test phosphorus. Biol. Fertil. Soils 49(8):987–999. doi:10.1007/s00374-013-0798-2

Pagliari, P.H., and C.A.M. Laboski. 2014. Effects of manure inorganic and enzymatically hydrolyzable phosphorus on soil test phosphorus. Soil Sci. Soc. Am. J. 78(4):1301. doi:10.2136/sssaj2014.03.0104

Parham, J., S. Deng, W. Raun, and G. Johnson. 2002. Long-term cattle manure application in soil. Biol. Fertil. Soils 35(5):328–337. doi:10.1007/s00374-002-0476-2

Rady, M.M. 2012. A novel organo-mineral fertilizer can mitigate salinity stress effects for tomato production on reclaimed saline soil. S. Afr. J. Bot. 81:8–14. doi:10.1016/j.sajb.2012.03.013

Rao, J.R., M. Watabe, T.A. Stewart, B.C. Millar, and J.E. Moore. 2007. Pelleted organo-mineral fertilisers from composted pig slurry solids, animal wastes and spent mushroom compost for amenity grasslands. Waste Manag. 27(9):1117–1128. doi:10.1016/j.wasman.2006.06.010

Reddy, D.D., A.S. Rao, and T. Rupa. 2000. Effects of continuous use of cattle manure and fertilizer phosphorus on crop yields and soil organic phosphorus in a Vertisol. Bioresour. Technol. 75(2):113–118. doi:10.1016/S0960-8524(00)00050-X

Richards, J.E., J.-Y. Daigle, P. LeBlanc, R. Paulin, and I. Ghanem. 1993. Nitrogen availability and nitrate leaching from organo-mineral fertilizers for personal use only. Can. J. Soil Sci. 73:197–208 www.nrcresearchpress.com (Accessed 26 Sept. 2018). doi:10.4141/cjss93-022

Sakurada L.R., L.M.A. Batista, T.T. Inoue, A.S. Muniz, and P.H. Pagliar. 2016. Organomineral phosphate fertilizers: Agronomic efficiency and residual effect on initial corn development. Agron. J. 108(5):2050–2059. doi:10.2134/agronj2015.0543

Sivapalasingam, S., C.R. Friedman, L. Cohen, and R.V. Tauxe. 2004. Fresh produce: A growing cause of outbreaks of foodborne illness in the United States, 1973 through 1997. J. Food Prot. 67(10): 2342–2353. http://foodprotection.org/doi/pdf/10.4315/0362-028X-67.10.2342?code=FOPR-site (Accessed 2 Sept. 2018).

Smith, G.H., K. Chaney, C. Murray, and M.S. Le. 2015. The effect of organo-mineral fertilizer applications on the yield of winter wheat, spring barley, forage maize and grass cut for silage. J. Environ. Prot. (Irvine Calif.) 06(02):103–109. doi:10.4236/jep.2015.62012

Solomon, E.B., S. Yaron, and K.R. Matthews. 2002. Transmission of Escherichia coli O157:H7 from contaminated manure and irrigation water to lettuce plant tissue and its subsequent internalization. Appl. Environ. Microbiol. 68(1):397–400 http://www.ncbi.nlm.nih.gov/pubmed/11772650 (Accessed 2 Sept. 2018). doi:10.1128/AEM.68.1.397-400.2002

Steiner, C., W.G. Teixeira, J. Lehmann, T. Nehls, J.L.V. de Macêdo, W.E.H. Blum, and W. Zech. 2007. Long term effects of manure, charcoal and mineral fertilization on crop production and fertility on a highly weathered Central Amazonian upland soil. Plant Soil 291(1–2):275–290. doi:10.1007/s11104-007-9193-9

Tejada, M., C. Benitez, and J.L. Gonzalez. 2005. Effects of application of two organomineral fertilizers on nutrient leaching losses and wheat crop. Agron. J. 97(3):960–967. doi:10.2134/agronj2004.0092

USEPA. 2017. National water quality inventory: Report to Congress. United States Environmental Protection Agency, Washington, D.C. https://www.epa.gov/waterdata/national-water-quality-inventory-report-congress (Accessed 2 Sept. 2018).

Varshovi, A. 2005. Organic-based fertilizer. U.S. Patent 6,852,142 B2. Date issued: 8 Feb. https://patents.google.com/patent/US6852142B2/en (Accessed 8 Sept. 2018).

Waldrip, H.M., P.H. Pagliari, Z. He, R.D. Harmel, N.A. Cole, and M. Zhang. 2015. Legacy phosphorus in calcareous soils: Effects of long-term poultry litter application. Soil Sci. Soc. Am. J. 79(6):1601. doi:10.2136/sssaj2015.03.0090

Zebarth, B.J., R. Chabot, J. Coulombe, R.R. Simard, J. Douheret, and N. Tremblay. 2005. Pelletized organo-mineral fertilizer product as a nitrogen source for potato production. Can. J. Soil Sci. 85(3):387–395. doi:10.4141/S04-071

Composting with or without Additives

Chryseis Modderman*

Abstract

The intensity of animal agriculture has increased with a growing world population. Therefore, proper animal manure management is crucial for disposing of livestock waste while preserving environmental quality. Composting animal manure has many benefits for both livestock and crop producers as a nutrient source and waste disposal method. Managing compost can be time consuming and labor intensive as there are specific parameters that must be met and maintained such as temperature, moisture, and aeration. Compost additives may be used to alter these parameters to accelerate the composting process or improve the quality of the final product.

Composting manure is a process in which animal feces, urine, and bedding are stacked, turned, and managed in a way that promotes decomposition. The finished product has a soil-like texture and odor, and may be spread onto agricultural land as a crop or pasture nutrient source. Composting has many benefits including destruction of harmful pathogens and weed seeds, and reduction in overall volume and odor. However, proper composting requires daily observation and weekly management by the producer, which may seem too labor-intensive to some. There are biological, organic, and inorganic additives that alter the state of compost and aid in the pace of decomposition or quality of finished product.

Benefits of Composting Manure

Both raw and composted manure may be spread on crop or pasture land and are excellent sources of the macro- and micronutrients needed for plant development. While both forms of manure benefit soil health and crop production, there are benefits to creating and land-applying composted manure over raw manure. One such benefit is that composted manure is more nutrient-dense than raw manure. Raw manure typically contains less than 10% total nutrient by weight, and that extra bulk increases time and transportation costs compared to commercial fertilizers. Composting breaks down bedding and manure into a product that occupies less volume than raw manure. Composted manure is, therefore, cheaper to transport than raw manure, which bolsters farmer profitability (Lory et al., 2006).

Another agronomic benefit of composting manure is that it kills weed seeds in manure. Some weed seeds found in livestock feed can retain their viability when passed through the animal's digestive system and end up in manure. And when that manure is applied to crop or pasture land, those weed seeds can germinate and flourish, causing the cash crop or pasture grasses to compete with the weeds for sunlight, water, and nutrients. When some weeds produce over 100,000 to 1 million seeds per plant, even a small rate of survival in digestion can cause significant weed problems in the field (Katovich et al., 2005). Applying composted

C. Modderman, University of Minnesota Extension. 46352 State Highway 329, Morris, MN 56267. *Corresponding author (cmodderm@umn.edu)

doi:10.2134/asaspecpub67.c19

Animal Manure: Production, Characteristics, Environmental Concerns and Management. ASA Special Publication 67. Heidi M. Waldrip, Paulo H. Pagliari, and Zhongqi He, editors.

rather than raw manure can mitigate these risks, as the high internal heat associated with composting kills most weed seeds. A study conducted in California determined that total weed seed viability was reduced by 99% under proper composting conditions. Killing weed seeds during composting, before they have had the chance to germinate in the field, can help cut down on costs associated with applying extra herbicides to combat these weeds (Cudney et al., 1992).

Composted over raw manure alsohelps protect the environment against nutrient pollution, both directly and indirectly. Directly, the form of nitrogen in composted manure is much more stable than that of raw manure, making it less likely to be lost to the environment (Rynk et al., 1992; Eghball, 2000). Indirectly, composted manure reduces the likelihood of pollution due to over-application on crop or pasture land. Accidental over-application is more likely to occur when the actual nutrient content of the manure is variable, and therefore, unknown; and raw manure has a high degree of inherent variability both spatially and over time. In contrast, composted manure's nutrients are more uniform and stable than raw manure (Lory et al., 2006). Therefore, the farmer has a clearer picture of the nutrients actually being applied when using composted rather than raw manure, which results in more-accurate application calculations and rates; and that results in reduced risk of over-application (Rynk et al., 1992).

Much like the destruction of weed seeds, the heat generated by composting manure also kills human and animal pathogens that are present in raw manure. This can help reduce infection risk for animals and farmers that come in direct contact with the manure; and also for members of society that might encounter manure runoff that has escaped into a waterway or food source (Rynk et al., 1992; Lung et al., 2001). In addition, composted manure has a less-offensive odor than raw manure. In fact, manure compost smells more like soil than feces and urine. This makes composted manure a more palatable option for those farmers with neighbors that might find the smell of raw manure application offensive.

While composting manure has its benefits, its main drawback is that it is more labor-intensive than simply stockpiling manure. There are certain parameters of temperature, moisture, aeration, carbon and nitrogen, pile size, and particle size that must be followed for successful breakdown of manure and bedding (Whichuk and McCartney 2012). The process must be monitored and the pile turned periodically. This can all seem very overwhelming to a farmer who is interested in composting but does not know where to begin. There is a lot that can go wrong with a compost pile, but most of them can be corrected if the farmer has the knowledge and skills.

The Composting Process

Composting uses microbes that already exist in manure to break down organic matter (such as undigested feed and bedding) and requires certain conditions to be successful. Well managed compost may take four to six months to reach completion; but if a producer diligently turns, aerates, and manages the pile, it can reach completion in as little as two months. The basic steps of composting include piling raw product, then turning and mixing that pile to aerate and maintain temperature and moisture so that the breakdown process continues. Knowing when to turn the pile and proper compost management is almost an art, and it takes the right combination of particle size, moisture content, oxygen, temperature, and carbon to nitrogen ratio to work efficiently (Table 1) (Rynk et al., 1992).

The overall size of a compost pile will impact material decomposition and will be determined by the size and needs of an operation. Some may have one large pile, while others may have multiple, smaller piles at different decomposition stages. A compost pile should, at minimum, be 3-feet square by 3-feet deep. Anything smaller than that will be unable to generate the internal heat necessary for composting. The maximum size should not exceed that which available machinery can effectively turn and mix (Rynk et al., 1992).

Another important composting factor is particle size. When building a compost pile, small particle size is desirable as that lends to more overall surface area; which, in turn, allows better material access for the decomposing microbes (Doublet et al., 2011). However, if the particles are too small, they will fit together too closely and limit the amount of oxygen in the pile that the microbes depend on. Particle sizes of 1/8 inch to 2 inches are considered best. Manure is crumbly and can be broken into bits of that size, but coarse bedding such as corn stalks may need to be shredded if they are too large (Rynk et al., 1992).

One of the most important factors that regulates breakdown of a compost pile is temperature. In the life cycle of a compost pile, there are three temperature phases: i) initial activation, ii) thermophilic, and iii) maturation. Throughout all three phases, temperature should be monitored with a 36-inch temperature probe. The amount of heat in a compost pile is a good indicator of how well the process is progressing; and temperatures that are inconsistent with the pile's current phase are usually the first indication of a problem (Bernal et al., 2009).

The first phase is initial activation, also known as the warm up, and is the shortest, only lasting for a few days to a month. It occurs from the time the pile is formed until the internal temperature reaches 105 °F. During this time, the microbes in manure start to become active and begin the breakdown process. The thermophilic phase is second and is also known as the hot composting stage. This phase is where the real work is done by the microbes and can last four to six months when temperatures within the pile cycle from 110 to 150 °F. Maintaining the pile at these temperatures is crucial since the microbes are most productive in this range. If the pile falls below 110 °F, the microbes will become dormant and decomposition will slow or stop. Conversely, if the pile reaches temperatures over 160 °F, microbes will start to die and decomposition will also slow or stop. When temperatures approach 160 °F, the pile should be turned. This will lower the overall temperature and maintain the composting process. If temperatures are too low in this stage, and they do not rise with mixing, it may be a sign that other components are out of balance such as moisture, C/N, or oxygen (Bernal et al., 2009). If the compost is sufficiently broken down, appears soil-like, and temperatures

Table 1. Characteristics of successful composting (Rynk et al. 1992)

Characteristic	Reasonable range	Preferred range
Particle size	1/16– 4 inches	1/8– 2 inches
Temperature	105- 160 °F	110- 150 °F
Moisture	40– 65%	50– 60%
Oxygen	5- 20%	10– 15%
Carbon to nitrogen ratio (C/N)	20:1– 40:1	25:1– 30:1

no longer rise above 110°F, it is a sign that the second phase is over and the final phase is beginning. The final phase, maturation or cool curing, can last up to 4 mo. It is when the compost pile returns to lower temperatures, usually below 105 °F. This phase involves further decomposition by fungus, worms, insects, and other organisms. In some cases, this step may be omitted and the compost used as a nutrient source directly after the thermophilic phase completes. But compost that is allowed to cure will be a more stable and complete product, which is especially valuable if marketing is the intent (Bernal et al., 2009).

Moisture in compost is also crucial for the reactions needed to break down organic materials; and it also helps regulate compost temperature. The optimum level for composting is 50 to 60% moisture, but many of the microbes will still be effective at 40 to 65%. If moisture content is too high, the pore space within the pile will fill up with water, limiting the microbes' oxygen supply and suppressing the composting process. Moisture contents over 60% may hinder composting unless action is taken to turn and aerate the pile more frequently. Conversely, compost breakdown will also be hindered if the moisture content is too low. The microbes necessary for composting need a moist environment, and a dry one will slow or stop microbial processes. Composting can occur at as little as 25% moisture, but it will be very slow; and the destruction of weed seeds will be hampered if moisture is below 35%, even if temperatures are high enough (Rynk et al., 1992). There is a simple way to check compost moisture called the "squeeze test". While wearing gloves, a handful of compost is squeezed. If water drips from the compost, it is too wet; if the compost crumbles, it is too dry. Compost with optimal moisture will feel moist and hold its shape, but without dripping. If moisture is not within the optimal range, there are ways to alter it. Covering the pile with a roof or tarp protects it from getting too wet from excess rainfall, or too dry from sun exposure. Also, if the pile is too wet, additional aeration and turning can help dry it; and a too-dry pile can be moistened with water from a spray hose or bucket.

In addition, aerobic (oxygenated) conditions are necessary for composting, with the greatest need for high oxygen levels at the beginning of the process. A minimum of 5% oxygen in the pore space is necessary, and 10 to 15% is optimal (Rynk et al., 1992). Within the compost pile, oxygen will be most abundant in the outer layers; and the compressed inner core of the pile will have the least pore space, and therefore, the least oxygen. Turning and mixing the compost is important to spread oxygen throughout the pile; bringing the center of the pile to the outside, and the outside to the center.

Another important component of composting manure is the carbon to nitrogen ratio (C/N). If it is unbalanced, it can impede the whole composting process. The carbon sources in composting are typically bedding, and the nitrogen sources are both manure and bedding, depending on the bedding type. The optimal ratio for composting is between 25:1 and 30:1 (Rynk et al., 1992). Determining that the C/N is unbalanced can be difficult without a manure test. If the pile is not breaking down properly, and other factors- such as moisture and temperature- have been ruled out, it may be beneficial to send samples of the compost pile to a manure laboratory to test for C/N. If the C/N is lower than 20:1, nitrogen will be lost to the atmosphere as ammonia, and if the C/N is higher than 40:1, nitrogen will be immobilized by the excess carbon (Rynk et al., 1992).

Troubleshooting Compost Piles

Because of the multi-faceted and complex nature of successful composting, it can be difficult to pinpoint which factor– or combination of factors– described above is causing a problem within a compost pile. Indications that there is a problem include foul odor and not decomposing or heating. Use Table 2 to help identify and fix common composting issues.

Compost Additives

While composting may be successfully done with simply the manure and bedding originally present, other materials may be introduced to aid in decomposition or overall quality. When compost is amended with these additives, components discussed above such as temperature, moisture content, pH, and nutrient availability may be altered. These additives fall into three general categories: i) inorganic, ii) organic, or iii) biological.

Inorganic additives are typically mineral and include chemicals, lime, clay, and industrial waste. Often this type of compost amendment is an industrial byproduct and, therefore, available in abundance and at low cost (Gomes et al., 2016).

Table 2. Common composting problems and their solutions (MPCA, 2011).

Symptom of problem	Cause of symptom	Solution
Rotten egg smell	Not enough oxygen due to compaction	Turn and mix pile to create air pockets. If particle size is too small (< 1 inch), add bulkier particles such as woodchips about 2" in size.
	Excessive moisture (water drips from squeeze test)	Turn and mix pile to aid drying. If particle size is too small, add bulkier particles such as woodchips.
Ammonia smell	Excess nitrogen	Add more carbon sources (straw, leaves, etc.). See section on C:N for more carbon sources.
	Pile too small	Increase pile size to at least 3 ft x 3 ft x 3 ft deep. In the winter it needs to be larger: at least 5 ft x 5 ft.
	Pile too dry– most common problem (use squeeze test to tell moisture content)	While turning and mixing pile, add water with a hose or bucket. Let pile rest for several hours, then retest with the squeeze test. Add more water if necessary.
Pile does not heat up	Not enough nitrogen	Add nitrogen sources (grass clippings, hay, etc.). See section on C:N for more nitrogen sources.
	Not enough oxygen	Turn and mix pile to introduce oxygen. If particles are too small, add bulkier items such as woodchips.
	Cold weather	Make sure the pile is large enough (at least 5 ft x 5 ft). Turn and mix pile less frequently than in warm weather.
	Composting is complete	Compost is complete when it resembles soil and is crumbly.
Attracts insects, millipedes, slugs, etc.	This is normal	To minimize insect problem, keep at the proper moisture level. Make sure the pile is heating to high enough temperatures to kill any eggs laid by the insects.

Table 3. Pounds of bulk material to add per 100 pounds of manure to raise the C/N to 30:1 (Rynk et al. 1992)

Material	Avg. material C/N	Initial manure C/N		
		10:1	15:1	20:1
Straw, general	80:1	295	150	75
Straw, oat	60:1	370	190	95
Straw, wheat	125:1	240	125	65
Sawdust	440:1	195	100	50
Wood chips, hardwood	560:1	190	100	50
Wood chips, softwood	640:1	190	100	50
Newsprint	625:1	190	100	50
Leaves	55:1	215	215	110

Organic additives are usually plant material and may include residual straw from crops, grass clippings, bark, cornstalks, and biochar (Dias et al., 2010). These amendments contain their own nutrients, so when using them, care must be taken to avoid unbalancing the overall C to N ratio of a compost pile (Doublet et al., 2011). In that same line of reasoning, organic additives can be used to remedy an unbalanced C to N ratio. In a pile with a C/N that is too low, carbon sources, such as straw or wood chips, can be added to the pile. Table 3 shows the amount (in pounds) of each carbon source needed to raise the C/N to the appropriate level. If the C to N ratio is too high, nitrogen sources, such as grass clippings or hay, can be added. Table 4 shows how much of each nitrogen source can be added to lower the C to N ratio to the correct level (Rynk et al., 1992).

Biological additives refer to microorganisms that are added to a compost pile. These are generally commercial products that are marketed for aiding decomposition or removing odors and gases. Using microbiological additives comes with inherit uncertainty since, unlike commercial pesticides, private companies are not required to disclose the active ingredients and often describe their product with vague terms such as "over 60 strains of active bacteria" (Wakase et al., 2008). However, DNA assays have shown that many biological additives include the bacterial genera *Alcaligenes*, *Bacillus*, *Clostridium*, *Enterococcus*, and *Lactobacillus* (Dubois et al., 2004; Sasaki et al., 2006).

Effects of Compost Additives

Many marketed compost additives claim to improve compost by stimulating microbial activity, though this parameter is poorly-defined and difficult to measure. Even if their claim is true, simply increasing microbial numbers does not necessarily improve composting. However, there are additives that do improve rate of decomposition and overall quality by prolonging the thermophilic composting stage, improving aeration, regulating moisture content, and reducing odor and greenhouse gas emissions; though not all of these compost improvements are induced through microbial activity (Barthod et al., 2018).

Temperature

Increased temperatures leading to a prolonged thermophilic stage reduces overall composting time, thereby improving efficiency. Many studies have shown an increased temperature and microbial activity when using inorganic, organic, and biological compost additives. For example, the addition of organic additives such as grass clippings, branches, crushed wood pallets, bark, and cornstalks increased overall composting temperatures and thermophilic stage (Doublet et al., 2011). Swine manure compost amended with bamboo biochar at a rate of 6% to 9% took only 1 d to reach the thermophilic phase compared to the control that took 4 d (Chen et al., 2010). The same result was observed with poultry manure as added biochar reduced the time to reach the thermophilic phase by four to 5 d compared to the control (Czekala et al., 2016). Similarly, biochar at a rate of 15% increased temperatures rapidly in the early stages of food waste composting, leading to a significantly shorter composting time of 50 to 60 d (Waqas et al., 2018). However, biochar should not be added at rates higher than 20% as it may impede compost decomposition (Liu et al., 2017).

Inorganic commercial products such as zeolite, kaolinite, chalk, ashes, and sulfates increased the thermophilic stage of biowaste composting from two to 3 wk (Himanen and Hanninen 2009). In addition, zeolite-amended swine manure compost reached the thermophilic stage more quickly, and remained in that phase longer, than the control; thereby accelerating the composting process (Venglovsky et al., 2005). In contrast, calcium superphosphate was found to actually delay the thermophilic phase and suppress temperatures in swine manure composting (Zhang et al., 2017). Bentonite, phosphor-gypsum, and lime were found to have no impact on composting temperature (Gabhane et al., 2012; Li et al., 2012)

Both the biological yeast additive *Pichia kudriavzevii* and the microbial inoculum containing lactic acid bacteria, yeast, and phototrophic bacteria showed increased temperatures and acceleration of food waste composting (Manu et al., 2017; Nakasaki and Hirai, 2017).

Aeration

Though simply turning a compost pile will promote aeration, there are also additives that can increase oxygen levels Due to their natural porosity, organic bulking agents such as biochar, wood chips, pine cones, and crushed branches can increase aeration in a compost pile (Kulcu and Yaldiz, 2007; Czekala et al., 2016). These are useful when compost particle size is too small which impedes the ability of air pockets to form within the pile. Care must be taken when using additives to improve oxygen levels as too much aeration can cause the compost to resist heating to the appropriate level. Also, using additives for aeration is not a substitute for mechanical aeration; and the compost should still be turned and mixed.

Moisture Content

Maintaining compost moisture within the optimal zone is crucial for effective biological activity and decomposition, and there are additives that can help regulate moisture. To reduce moisture in a pile that is too wet, absorptive materials

Table 4. Pounds of bulk material to add per 100 pounds of manure to lower the C/N to 30:1

Material	Avg. material C/N	Initial manure C/N		
		40:1	45:1	50:1
Hay, general	22:1	70	95	115
Hay, legume	16:1	30	40	50
Grass clippings	17:1	35	45	55

such as cornstalks or sawdust. Wheat straw, however, was found to not impact moisture content (Miner et al., 2001). Adding inorganic clay such as bentonite can help retain moisture in a pile that is too dry (Li et al., 2012).

Odor and Gas Emissions

During the composting process, unpleasant odors may be emitted, which can be unpleasant for neighbors. Ammonia and sulfur-containing compounds are the primary sources of foul odor in composting, and there are additives to mitigate those odors. Biochar added to compost can decrease ammonia loss and odor in manure that contains high levels of nitrogen, such as poultry litter (Steiner et al., 2010). However, in general, organic additives that increase aeration will stimulate microbial activity, thereby increasing ammonia loss and odor. Waqas et al. (2018) found that biochar may actually enhance ammonia production and odor due to increased microbial activity; and Jiang et al. (2015) demonstrated the same effect with cornstalks.

Inorganic additives may produce better results since they typically are not bulking agents. Adding zeolite to compost has been shown to reduce both ammonia and sulfurous odors, with the amount of zeolite added directly proportional to the reduction in odor (Lefcourt and Meisinger 2001). Both sodium nitrate and sodium nitrite significantly reduced sulfurous odors in swine manure composting by reducing emissions of dimethyl sulfide and dimethyl disulfide (Zang et al., 2017). Magnesium hydroxide and phosphoric acid, along with other magnesium and phosphate salts, have also shown success in reducing ammonia loss and odor (Jeong and Hwang, 2005; Ren et al., 2010; Zhang et al., 2017).

Conclusions

Though it can be labor intensive, composting manure is a process that has many benefits such as reduced overall volume and destruction of weed seeds and pathogens. Components of composting, such as temperature, moisture, and aeration, take active management to ensure proper decomposition, and there are organic, inorganic, and biological additives that can help regulate those parameters while decreasing odor. However, there may be contradicting affects among additives that improve one aspect while detracting from another. Those positives and negatives should be considered in each case where composting additives are used.

Literature Cited

Barthod, J., C. Rumpel, and M. Dignac. 2018. Composting with additives to improve organic amendments. A review. Agron. Sustain. Dev. 38:17. doi:10.1007/s13593-018-0491-9

Bernal, M., J. Alburquerque, and R. Moral. 2009. Composting of animal manures and chemical criteria for compost maturity assessment. A review. Bioresour. Technol. 100:5444–5453. doi:10.1016/j.biortech.2008.11.027

Chen, Y., X. Huang, Z. Han, X. Huang, B. Hu, D. Shi, and X. Wu. 2010. Effects of bamboo charcoal and bamboo vinegar on nitrogen conservation and heavy metals immobility during pig manure composting. Chemosphere 78(9):1177–1181. doi:10.1016/j.chemosphere.2009.12.029

Cudney, D., S. Wright, T. Schultz, and J. Reints. 1992. Weed seed in dairy manure depends on collection site. Calif. Agric. 46(3):31–32.

Czekala, W., K. Malinska, R. Caceres, D. Janczak, J. Dach, and A. Lewicki. 2016. Co-composting of poultry manure mixtures amended with biochar– The effect of biochar on temperature and C-CO2 emission. Bioresour. Technol. 200:921–927. doi:10.1016/j.biortech.2015.11.019

Dias, B., C. Silva, F. Higashikawa, A. Roig, and M. Sanchez-Monedero. 2010. Use of biochar as bulking agent for the composting of poultry manure: Effect on organic matter degradation and humification. Bioresour. Technol. 101(4):1239–1246. doi:10.1016/j.biortech.2009.09.024

Doublet, J., C. Francou, M. Poitrenaud, and S. Houot. 2011. Influence of bulking agents on organic matter evolution during sewage sludge composting; consequences on compost organic matter stability and N availability. Bioresour. Technol. 102(2):1298–1307. doi:10.1016/j.biortech.2010.08.065

Dubois, J., S. Hill, L. England, T. Edge, L. Masson, J. Trevors, and R. Brousseau. 2004. The development of a DNA microarray-based assay for the characterization of commercially formulated microbial products. J. Microbiol. Methods 58(2):251–262. doi:10.1016/j.mimet.2004.04.011

Eghball, B. 2000. Nitrogen mineralization from field-applied beef cattle feedlot manure or compost. Soil Sci. Soc. Am. J. 64:2024–2030. doi:10.2136/sssaj2000.6462024x

Gabhane, J., S. William, R. Bidyadhar, P. Bhilawe, D. Anand, A. Vaidya, and S. Wate. 2012. Additives aided composting of green waste: Effects on organic matter degradation, compost maturity, and quality of the finished compost. Bioresour. Technol. 114:382–388. doi:10.1016/j.biortech.2012.02.040

Gomes, H., W. Mayes, M. Rogerson, D. Stewart, and I. Burke. 2016. Alkaline residues and the environment: A review of impacts, management practices and opportunities. J. Clean. Prod. 112(4):3571–3582. doi:10.1016/j.jclepro.2015.09.111

Himanen, M., and K. Hanninen. 2009. Effect of commercial mineral-based additives on composting and compost quality. Waste Manag. 29(8):2265–2273. doi:10.1016/j.wasman.2009.03.016

Jeong, Y., and S. Hwang. 2005. Optimum doses of Mg and P salts for precipitating ammonia into struvite crystals in aerobic composting. Bioresour. Technol. 96:1–6. doi:10.1016/j.biortech.2004.05.028

Jiang, T., G. Li, Q. Tang, X. Ma, G. Wang, and F. Schuchardt. 2015. Effects of aeration method and aeration rate on greenhouse gas emissions during composting of pig feces in pilot scale. J. Environ. Sci. (China) 31:124–132. doi:10.1016/j.jes.2014.12.005

Katovich, J., R. Becker, and J. Doll. 2005. Weed seed survival in livestock systems. University of Minnesota Extension, Saint Paul, MN .

Kulcu, R., and O. Yaldiz. 2007. Composting of goat manure and wheat straw using pine cones as a bulking agent. Bioresour. Technol. 98:2700–2704. doi:10.1016/j.biortech.2006.09.025

Lefcourt, A., and J. Meisinger. 2001. Effect of adding alum or zeolite to dairy slurry on ammonia volatilization and chemical composition. J. Dairy Sci. 84:1814–1821. doi:10.3168/jds.S0022-0302(01)74620-6

Li, R., J. Wang, Z. Zhang, F. Shen, G. Zhang, R. Qin, X. Li, and R. Xiao. 2012. Nutrient transformations during composting of pig manure with bentonite. Bioresour. Technol. 121:362–368. doi:10.1016/j.biortech.2012.06.065

Liu, N., J. Zhou, L. Han, S. Ma, X. Sun, and G. Huang. 2017. Role and multi-scale characterization of bamboo biochar during poultry manure aerobic composting. Bioresour. Technol. 241:190–199. doi:10.1016/j.biortech.2017.03.144

Lory, J., R. Massey, and B. Joern. 2006. Using manure as a fertilizer for crop production. United States Environmental Protection Agency, Washington, D.C.

Lung, A., C. Lin, J. Kim, M. Marshall, R. Nordstedt, N. Thompson, and C. Wei. 2001. Destruction of Escherichia coli O157:H7 and Salmonella Enteritidis in cow manure composting. J. Food Prot. 64(9):1309–1314. doi:10.4315/0362-028X-64.9.1309

Manu, M., R. Kumar, and A. Garg. 2017. Performance assessment of improved composting system for food waste with varying aeration and use of microbial inoculum. Bioresour. Technol. 234:167–177. doi:10.1016/j.biortech.2017.03.023

Miner, F., R. Koenig, and B. Miller. 2001. The influence of bulking material type and volume on in-house composting in high-rise, caged layer facilities. Compost Sci. Util. 9:50–59. doi:10.1080/1065657X.2001.10702016

MPCA. 2011. Diagnosing common composting problems. Minnesota Pollution Control Agency, Saint Paul, MN.

Nakasaki, K., and H. Hirai. 2017. Temperature control strategy to enhance the activity of yeast inoculated into compost raw material for accelerated composting. Waste Manag. 65:29–36. doi:10.1016/j.wasman.2017.04.019

Ren, L., F. Schuchardt, Y. Shen, G. Li, and C. Li. 2010. Impact of struvite crystallization on nitrogen losses during composting of pig manure and cornstalk. Waste Manag. 30:885–892. doi:10.1016/j.wasman.2009.08.006

Rynk, R., van de Kamp, M., Wilson, G., Singley, M., Richard, T., Kolega, J., Gouin, F., Laliberty, L., Kay, D., Murphy, D., Hoitink, H., and Brinton, W. 1992. On farm composting. Northeast Regional Agricultural Engineering Service (NRAES-54). Ithaca, NY.

Sasaki, H., O. Kitazume, J. Nonaka, K. Hikosaka, K. Otawa, K. Itoh, and T. Nakai. 2006. Effect of a commercial microbiological additive on beef manure compost in the composting process. Anim. Sci. J. 77:545–548. doi:10.1111/j.1740-0929.2006.00384.x

Steiner, C., K. Das, N. Melear, and D. Lakly. 2010. Reducing nitrogen loss during poultry litter composting using biochar. J. Environ. Qual. 39:1236–1242. doi:10.2134/jeq2009.0337

Venglovsky, J., N. Sasakova, M. Vargova, Z. Pacajova, I. Placha, M. Petrovsky, and D. Harichova. 2005. Evolution of temperature and chemical parameters during composting of the pig slurry solid fraction amended with natural zeolite. Bioresour. Technol. 96(2):181–189. doi:10.1016/j.biortech.2004.05.006

Wakase, S., H. Sasaki, K. Itoh, K. Otawa, O. Kitazume, J. Nonaka, M. Satoh, T. Sasaki, and Y. Nakai. 2008. Investigation of the microbial community in a microbiological additive used in a manure composting process. Bioresour. Technol. 99:2687–2693. doi:10.1016/j.biortech.2007.04.040

Waqas, M., A. Nizami, A. Aburiazaiza, M. Barakat, I. Ismail, and M. Rashid. 2018. Optimization of food waste compost with the use of biochar. J. Environ. Manage. 216:70–81. doi:10.1016/j.jenvman.2017.06.015

Whichuk, K., and D. McCartney. 2012. Animal manure composting: Stability and maturity evaluation. In: Z. He, editor, Applied research of animal manure– Challenges and opportunities beyond the adverse environmental concerns. Nova Science Publishers, NY. p. 203–262.

Zang, B., S. Li, F. Michel, G. Li, D. Zhang, and Y. Li. 2017. Control of dimethyl sulfide and dimethyl disulfide odors during pig manure composting using nitrogen amendment. Bioresour. Technol. 224:419–427. doi:10.1016/j.biortech.2016.11.023

Zhang, D., W. Luo, J. Yuan, G. Li, and Y. Luo. 2017. Effects of woody peat and superphosphate on compost maturity and gaseous emissions during pig manure composting. Waste Manag. 68:56–63. doi:10.1016/j.wasman.2017.05.042

Thermochemical Processing of Animal Manure for Bioenergy and Biochar

Mingxin Guo*, Hong Li, Brian Baldwin, and Jesse Morrison

Abstract

Animal manure containing high organic carbon and mineral nutrients is an available feedstock for bioenergy and biochar. This chapter reviews the thermochemical techniques for processing poultry litter, cattle manure, swine manure, and sewage sludge into biochar and biofuels and summarizes the quality characteristics of these bioproducts. Pyrolysis, gasification, hydrothermal liquefaction, and combustion are the existing thermochemical techniques for converting animal manure into valuable bioenergy and biochar products. Dry manure solids ranging in energy density from 13 to 19 MJ kg^{-1} can be combusted directly as a firewood alternative. The major products from pyrolysis of manure are biochar and bio-oil, from gasification are syngas and biochar, and from hydrothermal liquefaction are biocrude oil and hydrochar. Manure-derived biochar and hydrochar carry significant N, P, K nutrients, and both are a promising soil amendment. Manure-derived bio-oil, biocrude oil, and syngas contain substantial impurities and may pose challenges for subsequent upgrading processes. Up to 100% of the feed N may be lost during thermochemical processing of manure, forming a major concern for such treatment. Low temperature (400–450 °C) pyrolysis was recommended for producing quality biochar from manure. Currently thermochemical conversion of animal manure to bioenergy and biochar is technically feasible, but needs to be improved in economic viability.

Worldwide the animal production industry rears 30.1 billion livestock and poultry animals and generates 13 billion metric tons (~5 billion dry tons) of manure waste per annum (Zhang et al., 2017; FAO, 2018). The annual generation of animal manure in the United States is estimated at 1 billion wet tons (~335 million dry tons), with the recoverable portion reaching 73 million dry tons (Zhang and Schroder, 2014; Gollehon et al., 2016). Animal manure, consisting primarily of feces, urine, and bedding materials (e.g., straw and sawdust), has relatively high contents of organic carbon (OC) and nitrogen (N), phosphorus (P) and potassium (K) nutrients. Solid manures (e.g., cattle manure, swine manure, poultry litter, and sheep manure) generally contain 29.4 to 74.1% water and 25.9 to 60.6% dry matter (Brown, 2013). Of the dry matter, 49.6 to 84.7% is attributed to organic matter (OM) and 15.3 to 50.4% to mineral ash components (Moral et al., 2005). On a dry mass basis, solid manures typically possess 396 to 697 g OC kg^{-1}, 27.0 to 44.7 g N kg^{-1}, 7.7 to 21.8 g P kg^{-1}, and 18.5 to 23.9 g K kg^{-1} (Moral et al., 2005; Brown, 2013).

Abbreviations: AD, anaerobic digestion; BFB, bubbling fluidized bed; CAFO, concentrated animal feeding operation; CEC, cation exchange capacity; EC, electrical conductivity; ER, equivalence ratio; HHV, higher heating value; HR, heating rate; HTC, hydrothermal carbonization; HTT, highest treatment temperature; IBI, International Biochar Initiative; OC, organic carbon; OM, organic matter; PL, poultry litter; PS, particle size; S/C, solid to feed carbon ratio; SBR, steam to biomass ratio; SSA, specific surface area; VRT, vapor residence time; WHC, water holding capacity.

Keywords: Manure, pyrolysis, gasification, hydrothermal liquefaction, biochar, bio-oil, syngas

M. Guo, Department of Agriculture and Natural Resources, Delaware State University, Dover, DE 19901; H. Li, Department of Animal and Food Sciences, University of Delaware, Newark, DE 19716; B. Baldwin and J. Morrison, Department of Plant and Soil Sciences, Mississippi State University, Mississippi State, MS 39762. * Corresponding Author (mguo@desu.edu)

doi:10.2134/asaspecpub67.c21

Animal Manure: Production, Characteristics, Environmental Concerns and Management. ASA Special Publication 67. Heidi M. Waldrip, Paulo H. Pagliari, and Zhongqi He, editors.

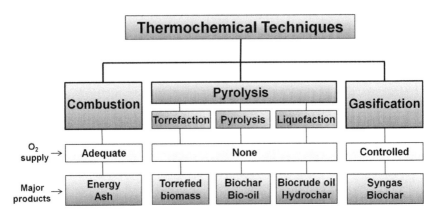

Fig. 1. Common thermochemical biomass conversion techniques.

Animal manure is a valuable source of nutrients, OM, and bioenergy. Traditionally, manure has been applied to cropland as a fertilizer and soil conditioner to promote crop production. Land application of animal manure is an effective, environmentally-benign, and value-added approach to dispose of waste while recovering nutrients and recycling OM. At the global scale, nutrients in land-applied manure account for 14% of N, 25% of P, and 40% of K in the total soil nutrient inputs (Sheldrick et al., 2003). In regions with concentrated animal feeding operations (CAFOs), excess manures are commonly managed through anaerobic digestion (AD) to produce biogas, a gaseous biofuel alternative to natural gas. As a matter of fact, manure is among the best biogas feedstocks given its small particle size, optimal C/N ratio, and high moisture content (Karim et al., 2005). After AD, the digestate can be land-applied to recover the nutrients originally in the manure. In remote, undeveloped areas with limited access to firewood, animal manure is combusted as an alternative fuel for cooking and even used as an additive for construction materials (Teenstra et al., 2014). Combustion is the most efficient thermochemical method to recover bioenergy from biomass. The higher heating values (HHV) of dry solid cattle manure, swine solids, and poultry litter are 18.0, 17.6, and 15.8 MJ kg^{-1}, respectively, slightly lower than wheat straw (18.3 MJ kg^{-1}) and sugarcane bagasse (19.4 MJ kg^{-1}) (Uchimiya and He, 2012). The ash from manure combustion retains most of the original P, K, Ca, and Mg nutrients (most N is lost via volatilization) and is a high quality soil amendment. Pagliari et al. (2009; 2010) examined turkey manure ash for its P, K, S, Zn, and other elemental contents and its promotive effect on alfalfa (*Medicago sativa* L.) and corn (*Zea mays* L.) growth when applied as a soil amendment. Other thermochemical methods that have been investigated in recent years to recover bioenergy from animal manure include pyrolysis (for bio-oil) and gasification (for syngas) (Ro, 2012). Thermochemical processing of animal manure using these two techniques yields a dark, porous solid residue termed "biochar", a promising soil amendment capable of persistently improving soil quality and facilitating plant growth. Biochar is environmentally recalcitrant and water and nutrient retentive, making it desirable in a variety of agricultural and environmental applications (Guo et al., 2016). The yield and quality of biochar and bioenergy from thermochemical processing of animal manure, however, vary with the feedstock type, pre-treatment, and processing operations. This chapter aims to introduce the existing thermochemical techniques for producing bioenergy and biochar from animal manure and elucidate the quality characteristics of thermochemical products derived from different animal manures.

Thermochemical Techniques

Thermochemical techniques have been designed to treat biomass materials under high temperature (commonly 200–900 °C) conditions to generate bioenergy and other valuable products. These techniques are classified into three broad categories, depending on the O_2 availability during treatment: combustion (with adequate O_2 supply), pyrolysis (with little or no O_2 supply), and gasification (with controlled O_2 supply) (Fig. 1).

Combustion is a high temperature-initiated exothermic oxidation reaction in which a substance reacts rapidly with an oxidizing agent (mostly atmospheric O_2) to generate heat and light in the form of fire. The ignition temperature (the lowest temperature for a substance to ignite) of biomass materials is around 190 to 260 °C. When biomass ignites, a portion of the material is transformed into a tarry vapor, and the subsequent burning of the tarry vapor forms flame. The flame temperature of wood is in the range of 600 to 1000 °C. Combustion of high quality coal and charcoal does not generate a flame, as no tarry vapor is formed when a "clean" fuel burns. Biomass materials are transformed into CO_2 and H_2O through combustion, leaving behind a small amount of solid residues as mineral ash. Emissions of NOx and SO_2 are generally minimal from biomass combustion but are dependent on the source material. Incomplete combustion further emits CO and tarry vapor (smoke). Combustion of firewood can be facilitated by increasing the temperature and the O_2 supply and reducing the moisture and bulk density of the wood fuel.

Pyrolysis is heating biomass materials in the absence of O_2 (air) to obtain various valuable products. There are three types of pyrolysis techniques: torrefaction, conventional pyrolysis, and liquefaction (Fig. 1). Torrefaction is mild, partial pyrolysis using low temperature (200–300 °C), low heating rate (typically <1 °C s^{-1}), short treatment time (e.g., 20–40 min), and O_2-limited conditions to transform biomass materials into a dry, hydrophobic, higher energy density, and easy-to-grind brownish product (Koppejan et al., 2012). The technique is usually used to treat fuelwood to enhance combustion performance and storage stability. During torrefaction, 20 to 30% of the fuelwood weight and 10% of the initial energy content are lost in pyrolysis vapors, yet the moisture content of the torrefied biomass is decreased to <3% and the calorific value increased by 15–25% by weight (Tumuluru et al., 2011). Liquefaction, also termed hydrothermal pyrolysis or hydrothermal carbonization (HTC), is the treatment of water slurry composed of 15 to 35 wt% finely-grinded biomass under high temperature (250–374°C), high pressure (4–22 MPa), and O_2-free conditions for 30 to 60 min to produce biocrude oil, a liquid capable of being upgraded to hydrocarbon fuels (Tekin et al., 2014). With addition of catalysts (e.g., Na_2CO_3 or Fe), 9.4 to 32.6% of biomass (e.g., swine manure, sewage sludge, microalgae, and lignocellulosic materials) could be converted to biocrude oil through hydrothermal liquefaction, with products demonstrating a low oxygen content (~16.6%) and high HHV (32.0–34.7 MJ kg^{-1}) (Vardon et al., 2011; Tzanetis et al., 2017). The solid residue from liquefaction is called "hydrochar." Liquefaction also results in a gaseous product (mainly CO_2) which typically accounts for 2 to 15% of the feed biomass (Kalderis et al., 2014). If HTC is operated at supercritical water

Table 1. Classification of pyrolysis techniques (Vamvuka, 2011; He et al., 2016; Jouhara et al., 2018).

Technique	Slow pyrolysis		Fast pyrolysis	
	Slow pyrolysis	Intermediate pyrolysis	Fast pyrolysis	Flash pyrolysis
Common HTT ‡	300–500 °C	500–700 °C	700–900 °C	800–1000 °C
Solid residence time	Hours to days	Minutes	Seconds	<2 s†
Heating rate	<1 °C s^{-1}	5–50 °C s^{-1}	>200 °C s^{-1}	>500 °C s^{-1}
Feed particle size	Up to logs	<1 cm	<2 mm	<1 mm
Reactor type	Fixed bed	Moving bed	Fluidized bed	Fluid bed
Char yield§	30–40%	25–35%	15–30%	10–20%
Bio-oil yield§	25–35%	35–50%	50–70%	65–75%
Syngas yield§	20–35%	20–30%	15–20%	10–20%

† The vapor residence time (from reactor to condenser) is also controlled at < 1 s.
‡ HTT, highest treatment temperature
§ The yield is relative to the ash-free dry mass of feedstock.

conditions (temperature >374°C and pressure >22.1 MPa), H_2-enriched syngas will be generated instead of biocrude oil (Correa and Kruse, 2018). The technology is then termed "supercritical water gasification" or "hydrothermal gasification" and is designed specifically for the production of H_2 from biomass. This technology requires the solid content of the feed slurry <8% (Ro, 2012).

Conventional pyrolysis (referred to as pyrolysis hereafter) has long been practiced for producing charcoal and bio-oil from biomass. In the absence of O_2, pyrolytic decomposition of biomass starts at 190 °C and becomes highly active at >350 °C. Organic matter is transformed into char and tarry vapor. The tarry vapor is emitted from the pyrolysis reactor and the inherent heavy components, including water steam, high molecular weight organic compounds, and tar particulates, are cooled and condense at room temperature to form a brownish liquid called bio-oil. The uncondensable portion of the tarry vapor is called synthesis gas or "syngas," with CO and H_2 as the major constituents (Fig. 2). The char(coal) from pyrolysis is a solid biofuel, a promising soil amendment, and a material with many industrial and environmental uses. Bio-oil is a mixture of over 200 organic compounds (Weldekidan et al., 2018) and can be used as an industrial feedstock for various chemicals or refined to produce biogasoline, renewable biodiesel, green jet fuel, and other drop-in biofuels. Syngas can be burned directly to generate electricity or upgraded to biohydrogen, dimethyl ether, biopropane, and various drop-in biofuels (Guo and Song, 2019). The quality and relative proportion of the three pyrolysis products, however, vary with the feedstock and the pyrolysis conditions. The pyrolysis temperature (highest treatment temperature, HTT) generally proceeds at 300 to 1000 °C. There is little pressure built up in the pyrolysis reactor, allowing thermic biomass decomposition to occur at ambient atmospheric pressure (Guo et al., 2015). Complete pyrolytic composition of biomass yields char(coal) demonstrating high storage stability and combustion performance. Hours to days may be needed to achieve complete pyrolysis as indicated by no further tarry vapors being evolved, especially when the pyrolysis temperature is at the lower end (Guo et al., 2012). According to the residence time of biomass materials in the pyrolysis reactor and the heating rate (the rate for heat to transfer into feed particles and raise their temperature in the core), pyrolysis can be divided into slow pyrolysis (reaction time: minutes to days; heating rate: <50 °C s⁻¹) and fast pyrolysis (reaction time: seconds; heating rate: >200 °C s⁻¹). Further subdivisions can be made into slow pyrolysis (reaction time: hours to days; heating rate: <1 °C s⁻¹), intermediate pyrolysis (reaction time: minutes; heating rate: 5–50 °C s⁻¹), fast pyrolysis (reaction time: seconds; heating rate: >200 °C s⁻¹), and flash pyrolysis (reaction time: < 2 s; heating rate: >500 °C s⁻¹; the vapor residence time is also shortened by vacuum or sweep N_2 gas flow) (Table 1). Plasma pyrolysis is a flash pyrolysis technique using high temperature plasmas generated by an electricity-powered plasma torch consisting of two electric arcs to pyrolyze organic solids at 800 to 1000 °C (Punčochář et al., 2012). Each pyrolysis technique involves specific types of pyrolysis reactors and has particular feed requirements (Table 1). In general, slow (slow to intermediate) pyrolysis systems (e.g., charcoal kilns, retorts, screw/auger pyrolyzers, and rotating drum pyrolyzers) require lower infrastructure investment and are employed to maximize charcoal production, while fast (fast to flash) pyrolysis systems (e.g., fluid bed pyrolyzers, circulating fluidized bed pyrolyzers, and rotating cone/vortex/vacuum reactors) require higher capital costs and are used to maximize bio-oil generation (Guo et al., 2012). In slow pyrolysis, the typical char, bio-oil, and syngas yields are 25 to 40%, 25 to 45%, and 20 to 35% of the ash-free dry feed mass, respectively; in fast pyrolysis, the numbers are 10 to 30%, 50 to 75%, and 10 to 20%, respectively (Table 1). As an endothermic process, pyrolysis requires external energy to maintain the pyrolytic reactions inside the reactor. The heat energy is usually furnished by an external heater, by burning a small portion of the feedstock, or by recycling the syngas and char generated (Fig. 2).

Gasification is a thermochemical process with the objective of maximizing syngas production by heating organic materials at 800 to 1200 °C with controlled O_2 supply. It is a well-developed technology, having been commercially practiced for nearly 100 years (Sikarwar et al., 2017). Gasification takes place in equipment

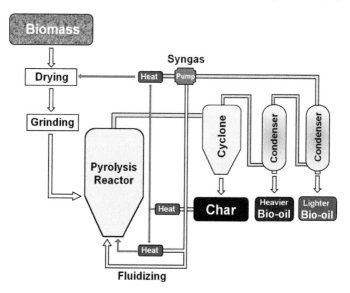

Fig. 2. A typical fluid bed pyrolysis system.

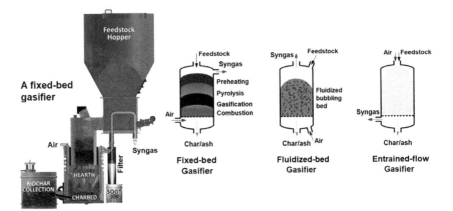

Fig. 3. Common gasification systems for producing syngas.

called a "gasifier," in which the biomass experiences dehydration, pyrolysis, oxidation (combustion and cracking), and reduction, and finally is transformed into uncondensable gases, predominantly CO, H_2, CO_2, and CH_4 and meagerly C_2H_2, C_2H_4, C_2H_6, C_3H_6, C_3H_8, and C_4H_{10} (Lima et al., 2009; Ro et al., 2010; Mondal et al., 2011; Pandey et al., 2016). There are three common types of gasifiers: fixed-bed, fluidized-bed, and entrained-flow gasifiers, with the first two mostly used for biomass gasification (Fig. 3). A plasma gasifier (in plasma gasification) is a fixed-bed gasifier heated by high temperature plasmas of ionized N_2 gas as it passes through high-voltage electric arcs (Janajreh et al., 2013). Similar to pyrolysis, gasification also results in three products: syngas, bio-oil (tar), and char/ash. The yields and chemical composition of the gasification products, however, vary with the feedstock type, moisture content, particle size, and gasification operation parameters such as temperature, gasification (oxidizing) medium, equivalence ratio (ER, the air fuel ratio for complete combustion; normally 0.2–0.5), gasifier pressure, and gasifier configuration (Yao et al., 2018). The gasification efficiency (percentage of feed energy recovered in cold syngas; also termed cold gas efficiency) typically ranges from 65% to 81%. The carbon conversion efficiency (recovery of feed carbon in syngas) varies from 70% to 90% (NREL, 2012). Approximately 5% to 15% (up to 20%) of the feed C is lost in char and/or ash and the balance in bio-oil (tar). The efficiencies can be improved through the addition of catalysts (e.g., nickel-based catalysts, olivine, and dolomite) to gasifier, or using pure O_2 as the oxidation medium (Sikarwar et al., 2017). To achieve satisfactory gasification efficiencies, feed particle size is limited to <51 mm for fixed-bed gasifiers and to <0.15 mm for entrained-flow gasifiers. Larger particles encourage char and tar generation and concurrently reduce syngas yield. Air gasification of wood typically yields 1 to 3 Nm^3 kg^{-1} of dry syngas consisting of 5 to 15% H_2, 20 to 30% CO, 1 to 3% CH_4, and 5 to 15% CO_2 and possessing an HHV of 4 to 7 MJ Nm^{-3} (Sikarwar et al., 2017). Using pure O_2 in place of air reduces the N_2 proportion in syngas and consequently, improves the syngas HHV to 12 to 28 MJ Nm^{-3} (Sikarwar et al., 2017). Steam may be loaded to the charbed area (Fig. 3) to improve the gasification efficiency. Hydrogen-enriched syngas can be obtained if steam is the gasification medium. To maintain an optimal temperature for biomass gasification, the feed moisture content must be <30%. In the moisture range of 0 to 30% a "wetter" feed generally leads to greater proportions of CO_2 and H_2 and less CO in the syngas (Yao et al., 2018). Gasification of char starts at 700 °C. Elevating the gasification temperature generally promotes formation of syngas, but can increase the risk of ash agglomeration. To avoid slagging, low (e.g., <2%) ash biomass materials are preferred. Syngas derived from high ash biomass usually carries many impurities including tar, carbon particulates, NH_3, HCN, COS, CS_2, $(CH_3)_2S$, H_2S, halides, and alkali metals (Sikarwar et al., 2017). A complete gasification system typically includes a syngas purification unit and an energy converter (e.g., an internal combustion engine) in line with the gasifier (Fig. 3).

Combustion of Animal Manure as a Firewood Alternative

In some rural areas of many countries, where fossil fuels are not available and firewood is scare, animal manure has been the primary energy source for households. For centuries, dungs of local ruminant animals have been collected, dried, and used as fuel for cooking and heating (NAS, 1980). In India, for example, it is a routine activity for village women to collect cattle dung and shape them into patties by hand. The dung patties are dried under the sun and piled in the yard for cooking (Fig. 4). Straw and other crop residues may be added to cattle dungs before shaping the material into "cakes." The average size of a dry dung patty is around 20 cm in diameter and weighs approximately 120 g. Dry cattle dung has an HHV of 18.0 MJ kg^{-1}, which is slightly lower than that of firewood (Uchimiya and He, 2012). After burning in earthen ovens, each dung patty can provide 2.1 MJ of heat energy.

It has been estimated that more than 2 billion people in the world use dry animal dung as an energy source (Biofuel, 2010). In India alone, 300 to 400 million tons of wet cattle dung (dry weight 60–80 million tons) are annually consumed as a cooking fuel (NAS, 1980). Manures of ruminant animals are rich in lignocellulosic materials. The OM content of dry ruminant dungs is between 51 to 70% (Moral et al., 2005). The energy content of dry dung is comparable to crop residues. Use of animal manure as fuel helps meet the local energy demand with a cheap, readily available fuel and reduces waste disposal issues. Nevertheless, most of the valuable N in dung is lost to the atmosphere during combustion, though other nutrients including P, K, Ca, and Mg can be recovered if the ash is applied to cropland. Burning dung patties in earthen ovens also releases air pollutants such as fine particulate matter (Mudway et al., 2005). Using animal manure as fuel may generate negative impacts on soil quality over a long term by depriving cropland of OM and some critical nutrients. It is indeed a last option to tackle fuel scarcity. In those regions, highly efficient dung burning stoves should be designed and disseminated.

Fig. 4. Cattle dung patties prepared as firewood for cooking in India.

Table 2. Selected studies on converting manure and sewage sludge into char and bioenergy.

Manure	Feedstock pretreatment	Technology	Technical conditions	Yield¶ Biochar	Bio-oil	Syngas	Reference
Poultry litter	<4 mm granules Moisture 8.3% OM 71.5%	Slow pyrolysis	400 °C HR 20 °C/min Complete pyrolysis	57.6%	30.4%	12.0%	Xiao, 2014
Poultry litter	Oven dried	Slow pyrolysis	350 °C and 450 °C for 30 min HR 1.7 °C/min	69.7% 63.0%	NA NA	NA NA	Domingues et al., 2017
Poultry litter	Moisture 10.2% OM 51.1%	Slow pyrolysis	620 °C HR 13 °C/min 2 h reaction time	46.4%	37.7%	15.9%	Ro et al., 2010
Poultry litter	5×5 mm pellets	Slow pyrolysis	700 °C 1 h reaction time	40.7%	NA	NA	Lima et al., 2009
Poultry litter	Oven dried PS <0.28 mm OM 62.9%	Slow pyrolysis	560 °C and 760 °C HR 3 °C/s	40.4% 24.5%	14.3% 26.6%	45.3% 48.9%	Weldekidan et al., 2018
Poultry litter	Moisture 33.4% PS <0.84 mm OM 78.8%	Fast pyrolysis	BFB reactor 500 °C VRT <5 s	32.9%	43.5%	23.6%	Agblevor et al., 2010
Poultry litter	PS <0.5 mm	Fast pyrolysis	800 °C HR 300 °C/s	22.2%	67.8%	10.0%	Lima et al., 2009
Poultry litter	Moisture 22.1 PS <2.8 mm OM 82.4%	Gasification	800 °C, BFB reactor, air, ER 0.25, mixed with 8 w% lime	23.6%	5.2%	1.39 Nm³ kg⁻¹‡	Pandey et al., 2016
Poultry litter	Moisture 30.2% PS <5 mm OM 80.2%	Gasification	580–680 °C, updraft fixed bed reactor, air, ER 0.15	33.0%	8.7%	58.3%§	Taupe et al., 2016
Chicken manure	Freeze dried OM 84.1%	Liquefaction	10% solid slurry 250 °C for 1 h 300 °C,	54.5% 58.1%	44.5% NA	<1% NA	Heilmann et al., 2014
Cattle manure	Oven-dried PS <2 mm	Slow pyrolysis	400 °C, 500 °C, and 700 °C for 2 h	44.9% 39.8% 37.1%	NA NA NA	NA NA NA	Yue et al., 2017
Cattle manure	Air dried PS <0.5 mm OM 80.3%	Fast pyrolysis	350 °C, 400 °C, 450 °C, 500 °C, and 550 °C BFB reactor	42.0% 34.0% 28.5% 26.0% 20.0%	47.5% 53.8% 51.8% 50.0% 49.0%	10.5% 12.2% 19.7% 24.0% 31.0%	Suttibak, 2013
Cattle manure	Moisture 7.8% PS <1.2 mm OM 76.1%	Gasification	BFB reactor, 800 °C, air and steam, ER 0.20, SBR 0.5	25.2%	NA	30%†	Wu et al., 2012
Cattle manure	Moisture 8.1% PS <1 mm OM 79.8%	Gasification	BFB reactor, 800 °C, steam, S/C 1.5; SBR 0.2	NA	3.5%	0.82 Nm³ kg⁻¹	Schweitzer et al., 2018
Cattle manure	Moisture 10.0% OM 81.4%	Gasification	BFB reactor, 800 °C, air, ER 0.25	20.0%	NA	0.94 Nm³ kg⁻¹#	Engler et al., 2010
Cattle manure	Freeze dried OM 88.0%	Liquefaction	15% solid slurry 260 °C for 1 h	36.9%	62.1%	<1%	Heilmann et al., 2014
Cattle manure	Oven dried PS <0.42 mm OM 75.8%	Liquefaction	200 °C and 280 °C for 4 h	58.3% 43.5%	NA NA	NA NA	Wu et al., 2017
Cattle manure	Moisture 8.2% PS 0.2–0.8 mm	Liquefaction	5% solid slurry 180 °C, 220 °C, and 260 for 30 min 20% solid slurry	84.0% 69.8% 50.8%	13.0% 21.8% 33.8%	3.0% 8.4% 15.4%	Reza et al., 2016
Horse manure	Oven dried OM 92.5%	Slow pyrolysis	400 °C 1 h reaction time HR 10 °C min	34%	NA	NA	Tsai et al., 2015
Yak manure	Oven dried PS <0.42 mm OM 83.9%	Slow pyrolysis	300 °C and 500 °C for 3 h	NA	NA	NA	Zhang et al., 2018
Swine solids	Moisture 12.8% OM 60.6%	Slow pyrolysis	620 °C HR 13 °C min⁻¹ 2 h reaction time	44.8%	33.4%	21.8%	Ro et al., 2010
Swine solids	Moisture 10% PS <1 mm OM 95.4%	Slow pyrolysis	420 °C 20 min reaction time	40.3%	NA	NA	Marchetti et al., 2012
Swine solids	Oven dried PS 0.25–0.6 mm OM 80.1%	Fast pyrolysis	500 °C and 600 °C, BFB reactor, VRT 1.5 s	39.2% 36.4%	45.3% 39.9%	15.5% 23.7%	Azuara et al., 2013
Swine solids	Moisture 12.1% PS <1 mm OM 80.1%	Gasification	BFB reactor 800 °C, steam, S/C 1.5; SBR 0.2	NA	3.5%	0.75 Nm³/kg	Schweitzer et al., 2018
Swine solids	Freeze dried OM 84.9%	Liquefaction	10% solid slurry 225 °C for 2 h 300 °C,	33.9% 56.5%	65.1% 36.5%	<1% 7.0%	Heilmann et al., 2014
Sewage sludge	Moisture 80.8% OM 75.3%	Slow pyrolysis	400 °C, 500 °C, and 600 °C BFB reactor	42.0% 39.1% 36.5%	50.0% 50.7% 52.0%	8.0% 10.2% 11.5%	Xue et al., 2015
Sewage sludge	Oven dried PS <1 mm OM 82.2%	Fast pyrolysis	487.5 °C VRT 2 s BFB reactor	28.7%	47.8%	23.5%	Arazo et al., 2017
Sewage sludge	Oven dried PS 0.25–0.6 mm OM 61.0%	Fast pyrolysis	530 °C SRT 6 min BFB reactor	49.7%	37.8%	12.5%	Martinez et al., 2014
Sewage sludge	Moisture 7.8% PS <2 mm OM 52.5%	Gasification	800 °C S/C 1.5; SBR 0.2	NA	8.6%	0.81 Nm³/kg	Schweitzer et al., 2018
Sewage sludge	Moisture 80.8% OM 75.3%	Liquefaction	170 °C, 220 °C, and 270 °C for 20 min	32.3% 19.6% 9.3%	67.5% 79.4% 88.6%	0.2% 1.0% 2.1%	Xue et al., 2015

† gasification efficiency

‡ gasification efficiency 89.2% and carbon conversion efficiency 73.0%

§ gasification efficiency 26.0% and carbon conversion efficiency 44.0%

¶ on dry weight basis

gasification efficiency 55.6%

†† BFB, bubbling fluidized bed; ER: equivalence ratio; HR, heating rate; NA, not available; OM, organic matter content; PS, particle size; SBR, steam to biomass ratio; S/C, solid (bed material) to feed carbon ratio; SRT, solid residence time; VRT, vapor residence time.

Pyrolysis of Animal Manure for Biochar and Bioenergy

The interest in pyrolytic conversion of animal manure to biochar started in the early 2000s, when the Amazonian *terra preta* soil–implied biochar may be a potential solution to the global soil degradation and climate change (Guo, 2016). "Biochar is fine-grained or granular charcoal made from heating vegetative biomass, bones, manure solids, or other plant-derived organic residues in an oxygen-free or oxygen-limited environment and used as a soil amendment for agricultural and environmental purposes" (Guo et al., 2016). As a soil amendment, biochar shows promise due to its high recalcitrance and high water and nutrient retentive capacity. Compared with other organic amendments such as animal manure, compost, and crop residues that are quickly mineralized within several years following land application, biochar can remain for hundreds to thousands of years. Amending soil with biochar is anticipated to enhance fertility, overall quality, and crop productivity of the treated soils over a long term, as observed in *terra preta* (Guo, 2016). Though highly desirable to crop producers, biochar is currently restricted from large scale application by its high production cost and low availability (Guo et al., 2016). Manufacturing low-cost biochar from freely available agricultural byproducts including animal manure, food processing waste, and crop residues has been explored for feasibility. The co-products bio-oil and syngas add additional value to pyrolytic biochar production. Conversion of manure to biochar and bioenergy ameliorates waste management issues of CAFOs. Manure-derived biochar contains significant amounts of slow-release, plant-available nutrients. It acts better in supplying nutrients than plant-derived biochar, which is generally low in nutrients, and raw manure, which is readily biodegradable and releases nutrients in flush (Li et al., 2018).

Substantial research has been conducted on production, characterization, and potential uses of biochar and bio-oil from different animal manures by adjusting pyrolysis operations (Table 2). The solid fractions of poultry litter, cattle manure, swine manure, sewage sludge, and other manures were collected, dried, ground, and processed using slow pyrolysis and fast pyrolysis techniques at 300 to 800 °C to produce biochar, bio-oil, and syngas. A survey of the literature demonstrates that dry manure is a feasible pyrolysis feedstock for biochar and bioenergy. Manure type and the operation conditions alter the proportion of pyrolysis products. Yields of biochar, bio-oil, and syngas ranged from 20 to 70%, 14 to 68%, and 7 to 49% of the feed mass, respectively (Table 2). Lower pyrolysis temperature and heating rate facilitated biochar production while higher pyrolysis temperature and heating rate favored bio-oil and syngas generation. The heating rate is controlled by reactor temperature, feed characteristics (e.g., particle size, moisture, chemical composition, and density), and feeding rate. Tarry vapors are actively evolved from pyrolytic decomposition of biomass at 350 to 500 °C (Antal and Gronli, 2003). To ensure the essential heating rate for rapid tarry vapor evolution, a smaller feed particle size must be adopted if the reactor temperature is not adequately high. Research trials indicate slow pyrolysis at 300 to 500 °C with the feed particle size at <5 mm was effective to produce biochar from animal manure, while fast pyrolysis at 700 to 800 °C with the feed particle size at <1 mm was efficient to convert manures to bio-oil and syngas (Table 2). Trials claimed to be "fast pyrolysis" (<600 °C using fluidized-bed reactors) actually belonged to intermediate pyrolysis (Table 1) and therefore, demonstrated a relatively low bio-oil yield (~40–50% of feed mass; Table 2). Furthermore, most of

Table 3. General properties of manure-derived biochar and hydrochar ‡.

Feedstock	Production conditions	HHV MJ kg⁻¹	SSA m² g⁻¹	CEC cmol_c kg⁻¹	pH	Ash %	OC %	Nutrient, g kg⁻¹			Reference
								N	P	K	
Poultry litter	350 °C and	NA	NA	41.6	9.7	50.2	31.2	NA	NA	NA	Domingues et al., 2017
	450 °C pyrolysis	NA	NA	40.2	10.2	55.3	27.2	NA	NA	NA	
Poultry litter	400 °C and	13.5	3.9	41.7	10.3	56.6	36.1	26.3	26.3	81.2	Xiao, 2014; Song and Guo, 2012
	600 °C pyrolysis	NA	5.8	29.2	11.5	60.8	32.5	1.2	30.5	91.5	
Poultry litter	620 °C pyrolysis	13.5	NA	NA	NA	53.2	41.5	27.7	17.2	67.7	Ro et al., 2010
Poultry litter	700 °C pyrolysis	NA	238	NA	8.1	NA	NA	NA	36.8	NA	Lima et al., 2009
Poultry litter	635 °C gasification	14.3	NA	NA	NA	54.8	33.1	6.4	29.3	87.8	Taupe et al., 2016
Poultry litter	800 °C gasification	NA	NA	NA	NA	NA	10–34	NA	17–32	54–96	IBI, 2010
Chicken manure	250 °C Liquefaction	NA	NA	NA	7.0	NA	45.1	22.3	32.9	NA	Heilmann et al., 2014
Cattle manure	400 °C pyrolysis	NA	NA	NA	9.0	70.3	NA	13.5	4.4	NA	Singh et al., 2010
Cattle manure	300 °C	NA	5.0	NA	8.3	NA	51.3	17.0	0.14†	1.4†	Yue et al., 2017
	400 °C	NA	10.2	NA	10.1	NA	51.7	15.9	0.06†	2.6†	
	500 °C	NA	2.4	NA	10.6	NA	52.5	14.5	0.06†	3.5†	
	700 °C pyrolysis	NA	2.4	NA	10.5	NA	52.9	10.6	0.02†	4.4†	
Cattle manure	180 °C,	19.6	NA	NA	5.8	18.7	46.6	20	18	4	Reza et al., 2016
	220 °C, and	21.1	NA	NA	4.7	23.0	50.4	21	21	3	
	260 °C liquefaction	22.1	NA	NA	4.4	25.4	54.0	25	24	1	
Cattle manure	250 °C Liquefaction	NA	NA	NA	5.2	NA	56.2	37.6	18.6	NA	Heilmann et al., 2014
Cattle manure	200 °C and	16.0	NA	NA	NA	34.4	39.7	14.6	NA	NA	Wu et al., 2017
	280 °C liquefaction	17.9	NA	NA	NA	40.4	41.6	17.9	NA	NA	
Horse manure	400 °C pyrolysis	22.5	NA	NA	NA	NA	53.1	6.5	NA	NA	Tsai et al., 2015
Yak manure	300 °C and	NA	3.6	NA	7.6	NA	41.6	32	45.2	27.8	Zhang et al., 2018
	500 °C pyrolysis	NA	17.3	NA	10.2	NA	41.3	30	54.1	28.5	
Swine solids	420 °C pyrolysis	NA	NA	NA	9.7	34.5	51.1	21.1	38.5	NA	Marchetti et al., 2012
Swine solids	450 °C and	NA	162	NA	9.9	50.9	33.7	25.7	71.0	20.7	Sun et al., 2018
	600 °C pyrolysis	NA	206	NA	10.2	52.3	35.6	24.6	79.9	23.4	
Swine solids	620 °C pyrolysis	18.3	NA	NA	NA	44.7	50.7	32.6	71.5	25.6	Ro et al., 2010
Swine solids	500 °C and 600 °C fast pyrolysis	17.7	NA	NA	NA	47.8	44.7	18.1	49.3	NA	Azuara et al., 2013
		17.0	NA	NA	NA	52.1	43.9	16.2	52.6	NA	
Swine manure	250 °C liquefaction	NA	NA	NA	6.6	NA	55.0	25.7	39.1	NA	Heilmann et al., 2014
Sewage sludge	400 °C pyrolysis	18.0	NA	NA	NA	NA	NA	NA	NA	NA	Xue et al., 2015
Sewage sludge	487 °C fast pyrolysis	7.4	NA	NA	9.0	65.9	19.7	45.5	NA	NA	Arazo et al., 2017
Sewage sludge	220 °C liquefaction	17.3	NA	NA	NA	NA	NA	NA	NA	NA	Xue et al., 2015

† Water-extractable nutrient contents
‡ If not specified, pyrolysis refers to slow pyrolysis
§ HHV, higher heating value; SSA, specific surface area; CEC, cation exchange capacity; OC, organic carbon content; NA, not available.

the mineral ash components in manure remained in the biochar. Under similar pyrolysis conditions more biochar was generated from feedstocks with an initial higher ash content (Table 2; Wang et al., 2015). Pyrolysis of manure retained P and K nutrients in the biochar, yet a significant portion of the initial N was lost in bio-oil and syngas (Song and Guo, 2012; Weldekidan et al., 2018).

In addition to various mineral nutrients, pyrolysis-derived manure biochars also contain 27 to 53% of OC and have an HHV in the range of 13.5 to 18.3 MJ kg⁻¹ (Table 3). The material is indeed a quality soil amendment and can also serve as a solid fuel. Bio-oil from manure is a mixture of numerous organic acids, alcohols, aldehydes, esters, ketones, phenolics, paraffins, and N-containing organic compounds (Weldekidan et al., 2018) and possesses an HHV of 19.7 to 35.4 MJ kg⁻¹ (Agblevor et al., 2010; Xiao, 2014; Arazo et al., 2017), 50 to 80% of that of crude petroleum oil. Syngas generated from pyrolysis of manure can be recycled by direct combustion to provide heat to sustain the pyrolysis process and dry the manure feed (Fig. 2). Overall, feed quality, reactor temperature and heating rate were the most important parameters that determine the yield and characteristics of manure-derived pyrolytic products (Table 2). Information on the effect of reaction time (solid residence time in the reactor), however, is lacking. Complete pyrolytic decomposition generates biochar with the highest carbon stability. The reaction time to complete pyrolytic decomposition is influenced by pyrolysis temperature, heating rate, and feeding rate. Yue et al. (2017) compared the results of slow pyrolysis of cattle manure in a batch reactor for 0.5 h and 1 h. The shorter reaction time resulted in a greater yield of biochar with lower pH and ash content, but higher N content and water soluble P. In designing pyrolysis operations for converting manure to biochar and bio-oil, the reaction time has to be considered.

Mobile and small pyrolysis units with moving-bed or fluidized bed reactors have been developed to convert manure to biochar and bio-oil in a continuous flow pattern

Fig. 5. A mobile fluidized-bed pyrolyzer (top) and a small fluidized bed fast pyrolysis unit (bottom) for converting animal manure into biochar and bio-oil.

(Fig. 5). In the United States, poultry litter is the most common feed option due to its relatively low moisture content. In China, pyrolysis has been practiced to valorize sewage sludge (Zhang et al., 2016a). Crop residues and other plant materials may be added to increase the suitability of manure for pyrolytic processing (Ro et al., 2010).

Gasification of Animal Manure for Bioenergy

Gasification of animal manure for bioenergy in syngas has been studied and practiced. The solid fractions of chicken, cattle, and swine manures as well as sewage sludge were dried to a moisture content <15%, processed to <5 mm particles, and gasified in fixed-bed gasifiers at 630 to 800 °C with 0.15 to 0.25 ER air supply or 0.20 to 0.88 SBR (steam to biomass ratio) steam provision (Table 2). Gasification efficiency was between 30% to 89% (Table 2). The yield of syngas (excluding N_2) ranged from 0.75 to 1.39 Nm^3 kg^{-1}. The calorific value of raw syngas (including 46–69% N_2) was in the range of 2.8 to 4.7 MJ Nm^{-3} (Wu et al., 2012; Pandey et al., 2016 Taupe et al., 2016). Gasification of manure also resulted in 20 to 33% of the original feed mass in char/ash and 3–9% of the initial feed mass in bio-oil/tar (Table 2). Similar to pyrolysis bio-oil, bio-oil from manure gasification also contains numerous organic compounds and many impurities (Taupe et al., 2016). In general, conversion efficiency increased as the gasification temperature and the ER were elevated in the test ranges. Due to its high ash content, manure demonstrated a conversion efficiency 10 to 50% lower than that of wood pellets under similar gasification conditions. The raw syngas also contained higher concentrations of impurities (e.g., tar, NH_y, H_2S and Cl) (Schweitzer et al., 2018). Optimal manure gasification conditions were identified as 800 °C and 0.25 ER (SBR = 0). Modifications to the raw manure required drying to <10% moisture, the addition of 8% limestone, and a feed particle size of <3 mm. Under the optimal conditions the gasification efficiency may reach 89%. The melting point of manure ash is around 600 to 700 °C (Engler et al., 2010; Acharya et al., 2014; Taupe et al., 2016). At gasification temperature >750 °C, ash slagging may generate a vitreous mass, blocking the gasifier bed (Engler et al., 2010). To avoid bed agglomeration, steam may be used in place of air as the gasification medium. No ash slagging was observed in steam gasification of swine or cattle manure nor sewage sludge at 820 °C (Schweitzer et al., 2018).

In the United States, small gasifiers with fixed-bed or fluidized-bed reactors have been developed and employed by farms to produce syngas from poultry litter, cattle

Litter loading system

Fig. 6. A small fixed-bed gasifier to convert chicken litter into syngas for heating at Frye Poultry Farm (Wardensville, WV).

manure, swine manure, and horse manure (Fig. 6; CoalTec Energy, 2018). The syngas is used to generate electricity or provide heat for greenhouses and chicken barns (Engler et al., 2010). The char and/or ash is applied to cropland as a soil amendment (IBI, 2010).

Hydrothermal Liquefaction of Animal Manure for Biocrude Oil and Hydrochar

Hydrothermal liquefaction has been explored to valorize manure by converting it to biocrude oil and hydrochar. Swine, dairy, and poultry manures (no bedding materials) as well as sewage sludge were processed into 5 to 15% solid–water slurry and heated to 170 to 280 °C in a "pressure cooker" (reactor) for 20 to 120 min (Table 2). The hydrochar and biocrude oil yields were 9.3 to 84.0% and 13.0 to 88.6%, respectively, of the solid feedstock mass. A small fraction (0.2–15.4%) of feed ended up in HTC gas (mainly CO_2). In general, lower HTC temperature, coarser feed particles, higher slurry solid contents, and shorter treatment time facilitated hydrochar generation, while the opposite parameters promoted biocrude oil production. Sewage sludge showed a higher bio-crude oil yield than animal manures, which contain significant amounts of lignocellulosic components (Table 2). Hydrothermal liquefaction of biomass is typically conducted at temperature >250 °C for maximum biocrude oil yield. At 250 °C and 300 °C, the steam pressure of water in the HTC reactor is close to 4.0 and 8.6 MPa, respectively. These studies were aimed at hydrochar production, manure P extraction, or enhanced waste dewatering, and therefore, relatively low temperatures were used (Escala et al., 2013; Heilmann et al., 2014).

Hydrochar has been evaluated for uses as a soil conditioner, contaminant immobilizer, and electrode material (Abel et al., 2013). Manure-derived hydrochars contain N and P nutrients at significant concentrations (Table 3). Nevertheless, a substantial portion of the mineral nutrients originally in animal manure are lost in the process water during HTC treatment (Escala et al., 2013; Heilmann et al., 2014). Compared with the original feed, the OC content and the calorific value of manure-derived hydrochar are significantly greater. Hydrothermal liquefaction at 240 °C increased the HHV of cow manure from 14.1 MJ kg^{-1} to 17.9 MJ kg^{-1} in the hydrochar (Wu et al., 2017). Manure-derived hydrochars demonstrated an HHV in the range of 16.0 to 22.1 MJ kg^{-1} (Table 3) and therefore, could putatively be used as a solid fuel. Analyses indicate hydrothermal processing of sewage sludge and drying the hydrochar consumes less energy than drying sludge for incineration (Escala et al., 2013). Manure-derived biocrude oil is a mixture of various organic compounds including acids, ketones, phenols, aldehydes, and others (Wu et al., 2017). Its OC content (water-free) ranges from 66.6% to 75.2% and HHV from 33.5 to 35.5 MJ kg^{-1} (Skaggs et al., 2018). Hydrothermal liquefaction of manure, sewage sludge, food waste, and waste vegetable oils and fats generated in the United States would produce 22.3 GL yr^{-1} of biocrude oil, containing 23.9% of the nation's current energy demand (Skaggs et al., 2018).

One advantage of HTC is that manure does not need to be dried prior to hydrothermal processing. In addition to harvesting valuable products, HTC provides other benefits including disinfection, odor reduction, and dewaterability improvement of manures (Heilmann et al., 2014). Hydrothermal processing of manure at 130 to 175 °C greatly improves its biodegradability (Stuckey and McCarty, 1984). The technique has been used to pre-treat AD media for improved biogas production and bioenergy recovery. Hydrothermal liquefaction of animal manure for biocrude oil and hydrochar, however, has not been commercially practiced because low economic returns make it inviable. The cost of biocrude oil from hydrothermal liquefaction of sludge was estimated at US $0.67 L^{-1} using the present technology (PNNL, 2017).

Quality Characteristics and Potential Applications of Manure-derived Biochar

Biochars generated from different biomass materials through different thermochemical techniques and processing conditions vary significantly in quality characteristics and therefore, demonstrate varied effects on soil quality, plant growth, and contamination mitigation in field applications. The performance of biochar is further influenced by the application rate, application method, and original soil properties. The following parameters have been recommended for biochar quality evaluation: particle size distribution, specific surface area (SSA), cation exchange capacity (CEC), water holding capacity (WHC), pH, electrical conductivity (EC), mineral ash content, lime equivalency, germination inhibition assay, OC content, carbon stability (or H:OC ratio), total and available N, P, Ca, Mg, and S content, and content of polycyclic aromatic hydrocarbons, dioxins and/or furans, polychlorinated biphenols, and the common toxic elements (Guo et al., 2012; IBI, 2015). The biochar products with great environmental recalcitrance, high nutrient and water retention capacities, and minimal contaminant content are considered to have the greatest desirability.

Biochars generated from pyrolysis and gasification of animal manures contain 44.7 to 70.3% mineral ash and 27.2 to 52.9% OC (Table 3). They are typically alkaline, demonstrating a pH value in the range of 7.6 to 11.5. The products carry significant amounts of N, P, and K (Table 3), but are drastically lower in SSA when compared to activated carbon (Qiu and Guo, 2010). In general, the ash content and pH of biochar increases as the thermochemical processing temperature is elevated, while the OC content gradually decreases. Hydrochars from HTC processing of manure demonstrate a lower pH value (5.2–7.2) and ash content (18.7–40.4%) yet substantially higher OC content (40.4–56.2%). Their N, P, K content is generally lower than biochars produced from low temperature (e.g., <500 °C) pyrolysis (Table 3). The carbon stability and environmental recalcitrance of biochars and hydrochars have not been adequately evaluated. Carbon stability of biochar increases with elevating the pyrolysis temperature (Song and Guo, 2012). Hydrochar does not have the recalcitrance of biochar. Using acid chromate oxidation methods, Naisse et al. (2013) determined ~70% of the OC in biochars generated from 1200 °C gasification of poplar wood, wheat straw, and olive meal was oxidation resistant, while the percentage for hydrochars generated from 230 °C HTC of the same plant materials was <10%. Field soil incubation trials with <2 mm biochar prepared by 750 °C slow pyrolysis of miscanthus biomass and with <2 mm hydrochar produced from 200 °C hydrothermal liquefaction of the same material revealed that the biochar-C changed little while 23 to 30% of the hydrochar-C was mineralized in 19 months (Marco et al., 2016). Evidently, biochar would remain in soil for many years, while hydrochar disappear in a couple of years; nevertheless, both chars can be used as a soil amendment.

Manure-derived biochar possesses significant quantities of labile OC and nutrients (Song and Guo, 2012; Li et al., 2018) and therefore, are useful as a soil conditioner and a minor nutrient source in crop production and land reclamation

(Buyantogtokh, 2013). Poultry litter- and swine manure-derived biochars are relatively rich in P (Table 3) and may be applied to remediate heavy metal-contaminated soils (Lahori et al., 2017). Dairy manure solid–derived biochar has shown to be capable of capturing nutrients from feedlot effluent (Ghezzehei et al., 2014). Poultry litter–derived biochar after acidification treatment was effective to suppress NH_3 emissions in chicken houses (Zhang et al., 2016b). Due to the potential release of nutrients, manure-derived biochar may not qualify as a bioretention medium to treat stormwater (Tian et al., 2016).

Converting Animal Manure to Biochar: A Case Study

We converted poultry litter (PL) to biochar through complete slow pyrolysis at HTT varying from 300 to 600 °C and evaluated the quality of the biochar products. Granular PL (<4 mm, moisture 7.7%) acquired from a local PL processing facility was packed into a metal canister and heated in a muffle furnace at 20 °C min^{-1} to the peak temperatures of 300, 350, 400, 450, 500, 550, and 600 °C, respectively. Peak temperatures were maintained until no visible smoke was emitted from the vent. Once the pyrolysis reaction was complete, the canister was withdrawn and immediately sealed. After cooling to room temperature, the biochar in the canister was transferred into a plastic bag and stored at room temperature prior to quality characterization.

Standard methods were followed to characterize the PL-derived biochars from 300, 350, 400, 450, 500, 550, and 600 °C pyrolysis, referred to hereafter as C300, C350,

Table 4. Quality characteristics of poultry litter (PL) and the derived biochars from different temperature slow pyrolysis†. C300, C350, C400, C450, C500, C550, and C600 refer to PL-derived biochars from pyrolysis at 300, 350, 400, 450, 500, 550, and 600 °C, respectively (Song and Guo, 2012).

Sample	Yield %‡	Ash %	pH¶	OC %	Stable OC, %	N g kg^{-1}	P g kg^{-1}	K g kg^{-1}	CEC cmol$_c$ kg^{-1}	SSA m^2 g^{-1}	WHC g g^{-1}
Raw PL		28.53	7.1	35.51	NA	30.66	15.14	41.77	NA	NA	NA
C300	60.13	47.87	9.5	37.99	13.98	41.71	22.73	69.28	51.1	2.68	0.88
C350	56.17	51.29	10.2	37.65	15.49	32.22	24.02	74.55	41.6	3.42	0.99
C400	51.52	56.62	10.3	36.10	16.35	26.30	26.29	81.16	41.7	3.94	1.01
C450	48.69	58.66	10.4	35.22	24.38	22.25	26.59	85.70	38.2	4.35	1.10
C500	47.57	60.58	10.7	34.47	29.46	12.14	27.87	87.92	35.8	4.77	0.99
C550	46.62	60.65	11.0	33.88	29.16	3.11	29.84	89.69	31.6	5.16	0.98
C600	45.71	60.78	11.5	32.52	28.08	1.18	30.54	91.51	29.2	5.79	0.95

† Values are means of triplicate measurements. The coefficient of variation for the measurements was less than 2%.

‡ % of the dry feed mass

§ measured in 1:5 solid/water slurry

¶ OC, organic carbon content; CEC, cation exchange capacity; SSA, specific surface area measured by BET N_2 adsorption methods; WHC, water holding capacity at -0.1 bar.

C400, C450, C500, C550, and C600, respectively (Song and Guo, 2012). Yield of PL-derived biochar ranged from 45.7 to 60.1% of the dry feed mass, decreasing as the pyrolysis temperature was elevated (Table 4). The ash content was in the range of 47.9 to 60.8%, increasing concurrently with the pyrolysis temperature. The pH value fell between 9.5 and 11.5, higher than that of raw PL (pH 7.1) and increasing with elevating the pyrolysis temperature. The OC content, however, showed a decreasing trend in the range of 32.5 to 38.0% as the pyrolysis temperature increased (Table 4). The stable (oxidation-resistant) portion of OC in C300 was 36.8% and in C600 was 86.3%, increasing steadily as the pyrolysis temperature was increased. Total N content of C300 was 41.7 g kg^{-1}, higher than that of raw PL (30.7 g kg^{-1}). The N level decreased rapidly as the pyrolysis temperature increased and became 1.2 g kg^{-1} for C600. At 400 °C pyrolysis temperature, 55.8% of the original feed N was lost to bio-oil and syngas. At $\geq$500 °C, the N loss was greater than 95.0%. Nearly all the P and K found in raw PL were recovered in biochar after pyrolysis, resulting in much greater P (22.7–30.5 g kg^{-1}) and K (69.3–91.5 g kg^{-1}) concentration for PL-derived biochars (Table 4). The PL-biochars were low in SSA (2.7–5.8 m^2 g^{-1}) but relatively high in CEC (29.2–51.1 cmol$_c$ kg^{-1}). The WHC was in the range of 0.88 to 1.10 g g^{-1}, lower than greenwaste compost, but higher than mineral soil (Table 4). Overall, high quality PL-derived biochar was obtained by low temperature pyrolysis. Considering the N recovery and product stability, 400 to 450 °C was recommended for slow pyrolysis of PL to produce biochar for agricultural uses.

Conclusions

Thermochemical processing of animal manure to produce bioenergy and biochar is a potential approach for valorizing agricultural byproducts. Research has demonstrated that valuable fuel and biochar products can be produced from animal manure solids through the thermochemical techniques pyrolysis, gasification, hydrothermal liquefaction, and direct combustion. The calorific value of dry manure solids is 14 to 19 MJ kg^{-1}. In certain regions of the world, animal manure has long been used as a fuel for heating and cooking. The resulting ash is applied to cropland for mineral nutrient recycling. Pyrolysis and gasification of animal manure generate three products: biochar, bio-oil, and syngas, all possessing energy at 13 to 35 MJ kg^{-1} and capable of serving as a fuel. Hydrothermal liquefaction of manure yields hydrochar and biocrude oil. Manure-derived biochar and hydrochar carry substantial plant nutrients and both are promising as soil amendments. Bio-oil and biocrude oil can be refined to produce various drop-in biofuels. Syngas can be burned directly for farm-heating or electricity generation, or synthesized into liquid biofuels and industrial chemicals. Bio-oil, biocrude oil, and syngas generated from manure contain more impurities than those derived from plant materials and may pose additional challenges in subsequent upgrading processes. Moreover, a major portion to all of the feed N is lost during thermochemical processing of manure, bringing nutrient recovery concerns. Though technically feasible, thermochemical conversion of manure to bioenergy and biochar is currently low in economic viability.

References

Abel, S., A. Peters, S. Trinks, H. Schonsky, M. Facklam, and G. Wessolek. 2013. Impact of biochar and hydrochar addition on water retention and water repellency of sandy soil. Geoderma 202–203:183–191. doi:10.1016/j.geoderma.2013.03.003

Acharya, B., A. Dutta, S. Mahmud, M. Tushar, and M. Leon. 2014. Ash analysis of poultry litter, willow and oats for combustion in boilers. Journal of Biomass to Biofuel 1:16–26.

Agblevor, F.A., S. Beis, S.S. Kim, R. Tarrant, and N.O. Mante. 2010. Biocrude oils from the fast pyrolysis of poultry litter and hardwood. Waste Manag. 30:298–307. doi:10.1016/j.wasman.2009.09.042

Antal, M.J., and M. Gronli. 2003. The art, science, and technology of charcoal production. Ind. Eng. Chem. Res. 42:1619–1640. doi:10.1021/ie0207919

Arazo, R.O., D.A.D. Genuino, M.D.G. de Luna, and S.C. Capareda. 2017. Bio-oil production from dry sewage sludge by fast pyrolysis in an electrically-heated fluidized bed reactor. Sustainable Environmental Research 27:7–14.

Atienza-Martínez, M., G. Gea, J. Arauzo, S.R.A. Kersten, and A.M.J. Kootstra. 2014. Phosphorus recovery from sewage sludge char ash. Biomass Bioenergy 65:42–50. doi:10.1016/j.biombioe.2014.03.058

Azuara, M., S.R.A. Kersten, and A.M.J. Kootstra. 2013. Recycling phosphorus by fast pyrolysis of pig manure: Concentration and extraction of phosphorus combined with formation of value-added pyrolysis products. Biomass Bioenergy 49:171–180. doi:10.1016/j.biombioe.2012.12.010

Biofuel. 2010. Solid biofuels. http://biofuel.org.uk/solid-biofuels.html (accessed 31 July 2018).

Brown, C. 2013. Available nutrients and value for manure from various livestock types. Factsheet AGDEX 538. Ministry of Agriculture and Food, Ontario, Canada.

Buyantogtokh, U. 2013. Reclamation of abandoned mine land with poultry litter biochar. Graduate thesis. Delaware State University, Dover, DE.

CoalTec Energy. 2018. Manure management. CoalTec Energy, Evansville, IN. http://www.coaltecenergy.com/ (accessed 20 Aug. 2018).

Correa, C.R., and A. Kruse. 2018. Supercritical water gasification of biomass for hydrogen production – Review. J. Supercrit. Fluids 133:573–590. doi:10.1016/j.supflu.2017.09.019

Domingues, R.R., P.F. Trugilho, C.A. Silva, I.C.N.A. de Melo, L.C.A. Melo, Z.A. Magriotis, M.A. Sánchez-Monedero . 2017. Properties of biochar derived from wood and high-nutrient biomasses with the aim of agronomic and environmental benefits. PLoS One 12(5):e0176884. doi:10.1371/journal.pone.0176884

Engler, C., S. Capareda, and S. Mukhtar. 2010. Assembly and testing of an on-farm manure to energy conversion BMP for animal waste pollution control. Technical Report No. 366. Texas Water Resources Institute, College Station, TX.

Escala, M., T. Zumbuhl, C. Koller, R. Junge, and R. Krebs. 2013. Hydrothermal carbonization as an energy-efficient alternative to established drying technologies for sewage sludge: A feasibility study on a laboratory scale. Energy Fuels 27:454–460. doi:10.1021/ef3015266

FAO. 2018. FAOStat. Food and Agriculture Organization, United Nations, Rome, Italy. http://www.fao.org/faostat/en/#data (accessed 24 July 2018).

Ghezzehei, T.A., D.V. Sarkhot, and A.A. Berhe. 2014. Biochar can be used to capture essential nutrients from dairy wastewater and improve soil physico-chemical properties. Solid Earth 5:953–962. doi:10.5194/se-5-953-2014

Gollehon, N.R., R.L. Kellogg, and D.C. Moffitt. 2016. Estimates of recoverable and non-recoverable manure nutrients based on the census of agriculture—2012 results. USDA Natural Resources Conservation Service, Washington, D.C.

Gronwald, M., C. Vos, M. Helfrich, and A. Don. 2016. Stability of pyrochar and hydrochar in agricultural soil - A new field incubation method. Geoderma 284:85–92. doi:10.1016/j.geoderma.2016.08.019

Guo, M. 2016. Pyrogenic carbon in Terra Preta soils. In: M. Guo, Z. He, and S.M. Uchimiya, editors, Agricultural and environmental applications of biochar: Advances and barriers. SSSA Spec. Publ. 63. Soil Science Society of America, Madison, WI. p. 15–27, doi:10.2136/sssaspecpub63.2014.0035.5.

Guo, M., and W. Song. 2019. The growing U.S. bioeconomy: Drivers, development, and constraints. N. N. Biotechnol. 49:48–57. doi: 10.1016/j.nbt.2018.08.005

Guo, M., W. Song, and J. Buhain. 2015. Bioenergy and biofuels: History, status, and perspective. Renew. Sustain. Energy Rev. 42:712–725. doi:10.1016/j.rser.2014.10.013

Guo, M., Y. Shen, and Z. He. 2012. Poultry litter-based biochar: Preparation, characterization, and utilization. In: Z. He, editor, Applied research in animal manure management: Challenges and opportunities beyond the adverse environmental impacts. Nova Science Publishers, Hauppauge, NY. p. 171–202.

Guo, M., Z. He, and S.M. Uchimiya. 2016. Introduction to biochar as an agricultural and environmental amendment. In: M. Guo, Z. He, and S.M. Uchimiya, editors, Agricultural and environmental applications of biochar: Advances and barriers. SSSA Spec. Publ. 63. SSSA, Madison, WI. p. 1–14, doi:10.2136/sssaspecpub63.2014.0034.

He, Z., S.M. Uchimiya, and M. Guo. 2016. Production and characterization of biochar from agricultural byproducts – Overview and utilization of cotton biomass residues. In: M. Guo, Z. He, and S.M. Uchimiya, editors, Agricultural and environmental applications of biochar: Advances and barriers. SSSA Spec. Publ. 63. SSSA, Madison, WI. p. 63–86. doi:10.2136/sssaspecpub63.2014.0037.5

Heilmann, S.M., J.S. Molde, J.G. Timler, B.M. Wood, A.L. Mikula, G.V. Vozhdayev, E.C. Colosky, K.A. Spokas, and K.J. Valentas. 2014. Phosphorus reclamation through hydrothermal carbonization of animal manures. Environ. Sci. Technol. 48:10323–10329. doi:10.1021/es501872k

IBI. 2010. Practitioner's profile: using chicken litter for biochar. International Biochar Initiative, Vector, NY. https://www.biochar-international.org/poultry/ (accessed 2 Aug. 2018).

IBI. 2015. Standardized product definition and product testing guidelines for biochar that is used in soil–Version 2.1. International Biochar Initiative, Vector, NY. http://www.biochar-international.org/characterizationstandard (accessed 9 Aug. 2018).

Janajreh, I., S.S. Raza, and A.S. Valmundsson. 2013. Plasma gasification process: Modeling, simulation and comparison with conventional air gasification. Energy Convers. Manage. 65:801–809. doi:10.1016/j.enconman.2012.03.010

Jouhara, H., D. Ahmad, I. den Boogaert, E. Katsou, S. Simons, and N. Spencer. 2018. Pyrolysis of domestic based feedstock at temperature up to 3000 °C. Thermal Science and Engineering Progress 5:117–143. doi:10.1016/j.tsep.2017.11.007

Kalderis, D., M.S. Kotti, A. Mendez, and G. Gasco. 2014. Characterization of hydrochars produced by hydrothermal carbonization of rice husk. Solid Earth 5:477–483. doi:10.5194/se-5-477-2014

Karim, K., R. Hoffmann, K.T. Klasson, and M.H. Al-Dahhan. 2005. Anaerobic digestion of animal waste: Effect of mode of mixing. Water Res. 39:3597–3606. doi:10.1016/j.watres.2005.06.019

Koppejan, J., S. Sokhansanj, S. Melin, and S. Madrali. 2012. Status overview of torrefaction technologies. IEA Bioenergy Task 32 Report. International Energy Agency, Paris, France.

Lahori, H., Z. Guo, Z. Zhang, R. Li, A. Mahar, M.A. Awasthi, F. Shen, T.A. Sial, F. Kumbhar, P. Wang, and S. Jiang. 2017. Use of biochar as an amendment for remediation of heavy metal-contaminated soils: Prospects and challenges. Pedosphere 27:991–1014. doi:10.1016/S1002-0160(17)60490-9

Li, W., X. Feng, W. Song, and M. Guo. 2018. Transformation of phosphorus in speciation and bioavailability during converting poultry litter to biochar. Front. Sustain. Food Syst. 2:1–10. doi:10.3389/fsufs.2018.00020

Lima, I.M., A.A. Boateng, and K.T. Klasson. 2009. Pyrolysis of broiler manure: Char and product gas characterization. Ind. Eng. Chem. Res. 48:1292–1297. doi:10.1021/ie800989s

Marchetti, R., F. Castelli, A. Orsi, L. Sghedoni, and D. Bochicchio. 2012. Biochar from swine manure solids: Influence on carbon sequestration and Olsen phosphorus and mineral nitrogen dynamics in soil with and without digestate incorporation. Ital. J. Agron. 7:e26. doi:10.4081/ija.2012.e26

Mondal, P., G.S. Dang, and M.O. Garg. 2011. Syngas production through gasification and cleanup for downstream applications — Recent developments. Fuel Process. Technol. 92:1395–1410. doi:10.1016/j.fuproc.2011.03.021

Moral, R., J. Moreno-Caselles, M.D. Perez-Murcia, A. Perez-Espinosa, B. Rufete, and C. Paredes. 2005. Characterization of the organic matter pool in manures. Bioresour. Technol. 96:153–158. doi:10.1016/j.biortech.2004.05.003

Mudway, I.S., S.T. Duggan, C. Venkataraman, G. Habib, F.J. Kelly, et al. 2005. Combustion of dried animal dung as biofuel results in the generation of highly redox active fine particulates. Part. Fibre Toxicol. 2:6. doi:10.1186/1743-8977-2-6

Naisse, C., M. Alexis, A. Plante, K. Wiedner, B. Glaser, et al. 2013. Can biochar and hydrochar stability be assessed with chemical methods? Org. Geochem. 60:40–44. doi:10.1016/j.orggeochem.2013.04.011

NAS. 1980. Firewood crops: Shrubs and tree species for energy production. National Academy of Sciences, Washington, D.C. The National Academies Press, Atlanta, GA.

NREL. 2012. Biomass gasification technology assessment. NREL/SR-5100-57085. National Renewable Energy Laboratory, Golden, CO.

Pandey, D.S., M. Kwapinska, A. Gomez-Barea, A. Horvat, L.E. Fryda, L.P.L.M Rabou, J.J. Leahy, and W. Kwapinski. 2016. Poultry litter gasification in a fluidized bed reactor: Effects of gasifying agent and limestone addition. Energy Fuels 30:3085–3096. doi:10.1021/acs.energyfuels.6b00058

Pagliari, P., C. Rosen, J. Strock, and M. Russelle. 2010. Phosphorus availability and early corn growth response in soil amended with turkey manure ash. Commun. Soil Sci. Plant Anal. 41:1369–1382. doi:10.1080/00103621003759379

Pagliari, P.H., C.J. Rosen, and J.S. Strock. 2009. Turkey manure ash effects on alfalfa yield, tissue elemental composition, and chemical soil properties. Commun. Soil Sci. Plant Anal. 40:2874–2897. doi:10.1080/00103620903173863

PNNL. 2017. Conceptual biorefinery design and research targeted for 2022: Hydrothermal Liquefaction Processing of Wet Waste to Fuels. PNNL-27186. Pacific Northwest National Laboratory, Richland, WA.

Punčochář, M., B. Ruj, and P.K. Chatterjee. 2012. Development of process for disposal of plastic waste using plasma pyrolysis technology and option for energy recovery. Procedia Eng. 42:420–430. doi:10.1016/j.proeng.2012.07.433

Qiu, G., and M. Guo. 2010. Quality of poultry litter-based granular activated carbon. Bioresour. Technol. 101:379–386. doi:10.1016/j.biortech.2009.07.050

Reza, M.T., A. Freitas, X. Yang, S. Hiibel, H. Lin, and C.J. Coronella. 2016. Hydrothermal carbonization (HTC) of cow manure: Carbon and nitrogen distributions in HTC products. Environ. Prog. Sustainable Energy 35:1002–1011. doi:10.1002/ep.12312

Ro, K.S. 2012. Thermochemical conversion technologies for production of renewable energy and value-added char from animal manures. In: Z. He, editor, Applied research of animal manure: Challenges and opportunities beyond the adverse environmental concerns. Nova Science Publishers, Hauppauge, NY. p. 63–82.

Ro, K.S., K.B. Cantrell, and P.G. Hunt. 2010. High-temperature pyrolysis of blended animal manures for producing renewable energy and value-added biochar. Ind. Eng. Chem. Res. 49:10125–10131. doi:10.1021/ie101155m

Schweitzer, D., A. Gredinger, M. Schmid, G. Waizmann, M. Beirow, R. Spörl, and G. Scheffknecht. 2018. Steam gasification of wood pellets, sewage sludge and manure: Gasification performance and concentration of impurities. Biomass Bioenergy 111:308–319. doi:10.1016/j.biombioe.2017.02.002

Sheldrick, W., J.K. Syers, and J.L. Lyngaard. 2003. Contribution of livestock excreta to nutrient balances. Nutr. Cycling Agroecosyst. 66:119–131. doi:10.1023/A:1023944131188

Sikarwar, V.S., M. Zhao, P.S. Fennell, N. Shah, and E.J. Anthony. 2017. Progress in biofuel production from gasification. Pror. Energy Combust. Sci. 61:189–248. doi:10.1016/j.pecs.2017.04.001

Singh, B., B.P. Singh, and A.L. Cowie. 2010. Characterization and evaluation of biochars for their application as a soil amendment. Soil Res. 48:516–525. doi:10.1071/SR10058

Skaggs, R.L., A.M. Coleman, T.E. Seiple, and A.R. Milbrandt. 2018. Waste-to-energy biofuel production potential for selected feedstocks in the conterminous United States. Renew. Sustain. Energy Rev. 82:2640–2651. doi:10.1016/j.rser.2017.09.107

Song, W., and M. Guo. 2012. Quality variations of poultry litter biochars generated at different pyrolysis temperatures. J. Anal. Appl. Pyrolysis 94:138–145. doi:10.1016/j.jaap.2011.11.018

Stuckey, D.C., and P.L. McCarty. 1984. The effect of thermal pretreatment on the anaerobic biodegradability and toxicity of waste activated sludge. Water Res. 18:1343–1353. doi:10.1016/0043-1354(84)90002-2

Sun, K., M. Qiu, L. Han, J. Jin, Z. Wang, Z. Pan, and B. Xing. 2018. Speciation of phosphorus in plant- and manure-derived biochars and its dissolution under various aqueous conditions. Sci. Total Environ. 634:1300–1307. doi:10.1016/j.scitotenv.2018.04.099

Suttibak, S. 2013. Influence of pyrolysis temperature on yields of bio-oil produced from fast pyrolysis of cow manure. Proc. Sci. Eng. 2013:338–343. 4th International Science, Social Science, Engineering and Energy Conference, Bangkok, Thailand.

Taupe, N.C., D. Lynch, R. Wnetrzak, M. Kwapinska, W. Kwapinski, and J.J. Leahy. 2016. Updraft gasification of poultry litter at farm-scale – A case study. Waste Manag. 50:324–333. doi:10.1016/j.wasman.2016.02.036

Teenstra, E., T. Vellinga, N. Aektasaeng, W. Amatayakul, A. Ndambi, D. Pelster, L. Germer, A. Jenet, C. Opio, and K. Andeweg. 2014. Global assessment of manure management policies and practices. Livestock Research Report 844. Wageningen UR Livestock Research, Wageningen, The Netherlands.

Tekin, K., S. Karagoa, and S. Bektas. 2014. A review of hydrothermal biomass processing. Renew. Sustain. Energy Rev. 40:673–687. doi:10.1016/j.rser.2014.07.216

Tian, J., V. Miller, P.C. Chiu, J.A. Maresca, M. Guo, and P.T. Imhoff. 2016. Nutrient release and ammonium sorption of poultry litter and wood biochars in stormwater treatment. Sci. Total Environ. 553:596–606. doi:10.1016/j.scitotenv.2016.02.129

Tsai, W., C. Huang, H. Chen, and H. Cheng. 2015. Pyrolytic conversion of horse manure into biochar and its thermochemical and physical properties. Waste Biomss Valorization 6:975–981. doi:10.1007/s12649-015-9376-1

Tumuluru, J.S., S. Sokhansanj, J.R. Hess, C.T. Wright, and R.D. Boardman. 2011. A review on biomass torrefaction process and product properties for energy applications. Ind. Biotechnol. (New Rochelle N.Y.) 7:384–401.

Tzanetis, K.F., J.A. Posada, and A. Ramirez. 2017. Analysis of biomass hydrothermal liquefaction and biocrude-oil upgrading for renewable jet fuel production: The impact of reaction conditions on production costs and GHG emissions performance. Renew. Energy 113:1388–1398. doi:10.1016/j.renene.2017.06.104

Uchimiya, M., and Z. He. 2012. Calorific values and combustion chemistry of animal manure. In: Z. He, editor, Applied research of animal manure: Challenges and opportunities beyond the adverse environmental concerns. Nova Science Publishers, Hauppauge, NY. p. 45–62.

Vamvuka, D. 2011. Bio-oil, solid and gaseous biofuels from biomass pyrolysis processes—An overview. Int. J. Energy Res. 35:835–862. doi:10.1002/er.1804

Vardon, D.R., B.K. Sharma, J. Scott, G. Yu, Z. Wang, L. Schideman, Y. Zhang, and T.J. Strathmann. 2011. Chemical properties of biocrude oil from the hydrothermal liquefaction of Spirulina algae, swine manure, and digested anaerobic sludge. Bioresour. Technol. 102:8295–8303. doi:10.1016/j.biortech.2011.06.041

Wang, Y., Y. Lin, P. Chiu, P. Imhoff, and M. Guo. 2015. Phosphorus release behaviors of poultry litter biochar as a soil amendment. Sci. Total Environ. 512–513:454–463. doi:10.1016/j.scitotenv.2015.01.093

Weldekidan, H., V. Strezov, T. Kan, and G. Town. 2018. Waste to energy conversion of chicken litter through a solar-driven pyrolysis process. Energy Fuels 32:4341–4349. doi:10.1021/acs.energyfuels.7b02977

Wu, H., M.A. Hanna, and D.D. Jones. 2012. Fluidized-bed gasification of dairy manure by Box–Behnken design. Waste Manage. Res. 30:506–511. doi:10.1177/0734242X11426173

Wu, K., Y. Gao, G. Zhu, J. Zhu, Q. Yuan, Y. Chen, M. Cai, and L. Feng. 2017. Characterization of dairy manure hydrochar and aqueous phase products generated by hydrothermal carbonization at different temperatures. J. Anal. Appl. Pyrolysis 127:335–342. doi:10.1016/j.jaap.2017.07.017

Xiao, P. 2014. Characterization and upgrading of bio-oil from slow pyrolysis of organic waste. Graduate thesis. Delaware State University: Dover, DE.

Xue, X., D. Chen, X. Song, and X. Dai. 2015. Hydrothermal and pyrolysis treatment for sewage sludge: Choice from product and from energy benefit. Energy Procedia 66:301–304. doi:10.1016/j.egypro.2015.02.064

Yao, Z., S. You, T. Ge, and C. Wang. 2018. Biomass gasification for syngas and biochar co-production: Energy application and economic evaluation. Appl. Energy 209:43–55. doi:10.1016/j.apenergy.2017.10.077

Yue, Y., Q. Lin, Y. Xu, G. Li, and X. Zhao. 2017. Slow pyrolysis as a measure for rapidly treating cow manure and the biochar characteristics. J. Anal. Appl. Pyrolysis 124:355–361. doi:10.1016/j.jaap.2017.01.008

Zhang, B., H. Tian, C. Lu, S.R.S. Dangal, J. Yang, and S. Pan. 2017. Global manure nitrogen production and application in cropland during 1860-2014: a 5 arcmin gridded global dataset for Earth system modeling. Earth System Science Data 9:667–678. doi:10.5194/essd-9-667-2017

Zhang, C., W.D. Alexis, H. Li, M. Guo, and J. Moyle. 2016b. Using poultry litter-derived biochar as litter amendment to control ammonia emissions. ASABE 2016 Annual International Meeting, Orlando, Florida. American Society of Agricultural and Biological Engineers, St. Joseph, MI.

Zhang, H., and J. Schroder. 2014. Animal manure production and utilization in the US. In: Z. He and H. Zhang, editors, Applied manure and nutrient chemistry for sustainable agriculture and environment. Springer, Dordrecht, The Netherlands. p. 1–21. doi:10.1007/978-94-017-8807-6_1

Zhang, J., B. Huang, L. Chen, Y. Li, W. Li, and Z. Luo. 2018. Characteristics of biochar produced from yak manure at different pyrolysis temperatures and its effects on the yield and growth of highland barley. Chem. Spec. Bioavailab. 30(1):57–67. doi:10.1080/09542299.2018.1487774

Zhang, X., Y. Luo, K. Muller, J. Chen, Q. Lin, J. Xu, Y. Tian, H. Cong, and H. Wang. 2016a. Research and application of biochar in China. In: M. Guo, Z. He, and S.M. Uchimiya, editors, Agricultural and environmental applications of biochar: Advances and barriers. SSSA Spec. Publ. 63. Soil Science Society of America, Madison, WI. p. 377–408. doi:10.2136/sssaspecpub63.2014.0049

Removing and Recovering Nitrogen and Phosphorus from Animal Manure

M.B. Vanotti,* M.C. García-González, A.A. Szögi, J.H. Harrison, W.B. Smith, and R. Moral

Abstract

Many areas in the United States produce more manure nutrients than available cropland can assimilate. Among all nutrients in manure, nitrogen (N) and phosphorus (P) cause the greatest environmental concern. Treatment technologies can play an important role in the management of animal manures by providing a more flexible approach to land application and acreage limitations. Development of technologies for nutrient-reuse was identified as one of the five main challenges in waste management within a circular economy. Manure treatment can be accomplished with the use of biological, chemical, and physical methodologies as part of treatment systems that are integrated with the needs of the land and create additional value through nutrient concentration and recycling, water quality credits, energy production, greenhouse gas reductions, and other beneficial by-products. This chapter reviews the alternative approaches for removing and recovering N and P from animal manure using both on-farm and off-farm management. Major components include diet manipulation, solid-liquid separation, and anaerobic digestion. The resulting materials are subsequently treated using a variety of technologies developed for solid or liquid fractions that include: composting, thermochemical processes, extraction of phosphate concentrates, recovery of the ammonia, biological nutrient removal processes, and vegetative or algal nutrient removal systems. Desirable properties of the recovered products and agronomic utilization are also discussed.

Animal production in the United States has increasingly moved to very large-scale operations. Since the 1950s, the animal production has more than doubled while the number of operations has decreased by 80% (Graham and Nachman, 2010). Most dairy cows, poultry, pigs, and beef cattle are now housed in high-density, confined spaces. In addition, the larger scale operations are often in agglomerated geographic distribution. Because of this intensification, expansion and agglomeration, there is a net import of nutrients as feed in some areas and limited capacity of the nearby land to utilize the nutrients. Therefore, many areas in the United States produce more manure nutrients than available cropland can assimilate (Gollehon et al., 2001). Among all nutrients in manure, nitrogen (N) and phosphorus (P) cause the greatest

Abbreviations: ACP, amorphous calcium phosphate; AD, anaerobic digestion; anammox, anaerobic ammonium oxidation; C, carbon; Ca, calcium; CO_2, carbon dioxide; COD, chemical oxygen demand; DAP, di-ammonium phosphate; DM, dry matter; EBPR, enhanced biological phosphorus removal process; $FeCl_2$, ferric chloride; HPNS, high-performance nitrifying sludge; HRT, hydraulic retention time; MAP, monoammonium phosphate; MLE, modified Ludzack-Ettinger process; N, nitrogen; NDN, nitrification–denitrification process; NF, nanofiltration; NH_3, ammonia; NH_4–N, ammonium nitrogen; NO_2–N, nitrate nitrogen; P, phosphorus; PO_4^{3-}, phosphate; PAM polyacrylamide; PAOs, polyphosphate accumulating microorganisms; PVA, polyvinyl alcohol; TAN, total ammoniacal nitrogen; TSP, triple superphosphate; QW, quick wash process; RO, reverse osmosis; TKN, total Kjeldahl N; TSS, total suspended solids; VFA, volatile fatty acids; VTA, vegetative treatment area.

M.B. Vanotti and A.A. Szögi, USDA-ARS, Coastal Plains Soil, Water and Plant Research Center, Florence, SC; M.C. Garcia-Gonzalez, Agriculture Technological Institute of Castilla and Leon (ITACyL), Valladolid, Spain; J.H. Harrison, Washington State University, Department of Animal Sciences, Puyallup, WA; W.B. Smith, Tarleton State University, Department of Animal Science and Veterinary Technology, Stephenville, TX; R. Moral, Miguel Hernandez University, Research Group on Agrochemistry and Environment, Orihuela, Spain. *Corresponding author (matias.vanotti@ars.usda.gov)

doi:10.2134/asaspecpub67.c22

Animal Manure: Production, Characteristics, Environmental Concerns and Management. ASA Special Publication 67. Heidi M. Waldrip, Paulo H. Pagliari, and Zhongqi He, editors.
© 2019. ASA and SSSA, 5585 Guilford Rd., Madison, WI 53711, USA.

environmental concern. It was estimated that 36% of the 170,000 animal feeding operations (AFO) in 2012 could not land-apply all the manure produced on the farm, resulting in farm-level excess manure containing about 868,000 metric tons of farm-level excess manure N and 388,000 metric tons of excess manure P, equal to 59% of the total recoverable manure nutrients produced on the farms (Gollehon et al., 2016). More sustainable techniques using P recovery for both solid and liquid waste are important to close the P cycle in modern human society and address future scarcity (Desmidt et al., 2015; Keyzer, 2010). The largest source of ammonia (NH_3) emissions in the USA is livestock farming, contributing 2.5 million tons per year (EPA, 2014). Treatment technologies can play an important role in the management of animal manures by providing a more flexible approach to land application and acreage limitations and by solving specific problems associated with excess nutrients such as surface and groundwater pollution, ammonia emissions, and P contamination of soils. In addition, fertilizer prices have escalated in recent years (USDA-ERS, 2014), thus there is renewed interest on developing technologies to recover and recycle nutrients from manure. Development of technologies for nutrient-reuse was identified as one of the five main challenges in waste management within a circular economy (Bernal, 2017).

An overview is provided in this chapter on existing alternative technologies for removing and recovering nitrogen and phosphorus from animal manure. The chapter is organized in seven sections as follows: a first section provides a flow diagram on alternative approaches to traditional land application of manures; a second section shows desirable characteristics of the recovered products; a third section shows solids separation and dewatering methods to concentrate nutrients; a fourth section reviews technologies for separation of phosphate concentrates; a fifth section shows technologies for recovery of the nitrogen; a sixth section describes biological N and P removal processes; and the last section describes agronomic utilization of the recovered nutrients.

Approach to Nutrient Removal and Recovery

Treatment can be enhanced with the use of biological, chemical, and physical methodologies, especially in combination as part of holistic systems that 1) are integrated with the needs of the land and other agri-food activities, and 2) maximize the value of manure through energy production, nutrient concentration and recycling, greenhouse gas (GHG) reductions and environmental credits, and other beneficial by-products (Vanotti et al., 2009a). Figure 1 illustrate alternatives approaches to direct land application of manure using both on-farm and off-farm management. While some technologies are in the research stage, many are now in the commercialization stage after years of extensive on-farm testing. The examples show a shift from municipal treatment methods in the near past to a new body of knowledge with methods adapted to the specific characteristics of these wastes and a different purpose for treatment. The flow diagram includes i) liquid systems, where manure leaving the animal production facility is a liquid or slurry material, and ii) dry systems, where all or part of the manure leaving a production facility is directly handled as a solid. The diagram shows three principal components in alternative manure management: diet manipulation, solid-liquid separation, and anaerobic digestion. Diet manipulation can alter both the composition of available phosphorus (i.e., low phytate P soybean in monogastric feed) and limiting amino acids in the feed that changes the amount of

N and P excreted by the animals and contained in the manure. The second principal component is the solid-liquid separation where organic nutrients are separated from the manure and transported and/or treated with a variety of technologies to generate value-added products. Solid-liquid separation of the manure increases the capacity of decision making and opportunities for treatment. The separation up-front in a treatment train allows recovery of the organic compounds, which can be used for the manufacture of compost materials and other value-added products or energy production. These products may include stabilized peat substitutes, humus, bio-chars, bio-oils, organic fertilizers, soil amendments, energy, quick-wash phosphorus, and proteins/amino-acids. The remaining liquid contains mostly soluble nutrients (ammonium and phosphates). A variety of biological, physical or chemical processes can be used on the farm to further remove and/or recover soluble nutrients to achieve specific nutrient management and environmental standards including application to cropland and water reuse. The third principal component in Fig. 1 is the anaerobic digestion (AD) to recover methane from the carbon in the liquid manure. The biogas recovery systems collect methane from the manure to generate electricity or heat. Production of biogas from manure using anaerobic digesters (covered lagoons, plug flow, and complete mix reactors) is projected to be important worldwide. Since AD per se is not a nutrient pollution control practice, nutrient recovery technologies need to be developed in conjunction with AD to address surplus nitrogen and phosphorus. When additional volume of organic substrate feedstock is added to the manure to make the AD process more economical (co-digestion), additional nutrients are also added to the effluents and the need for nutrient recovery technologies is more critical. For example, a national study (Informa Economics, 2013) showed that without nutrient recovery technology (N and P), widespread implementation of digesters in all large dairies in the USA (2647 dairies) would double the land area needed for safe

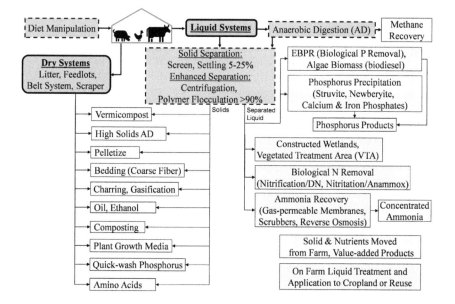

Fig. 1. Schematic diagram of nutrient removal and recovery of nutrients and value-added products from manure.

effluent disposal, from 4.26 to 8.06 million ha. However, if nutrient recovery technologies are provided, the land area for effluent disposal is reduced to 3.09 million ha.

Desirable Properties of Products

Several manure byproducts show significant potential to provide safe sources of nutrients (N, P, and K) and constitute an alternative for synthetic fertilizers. To ensure high quality products and that their use does not lead to overall negative environmental impacts or human health risks, nutrient recovered products should meet the following basic requirements of: i) agronomic efficacy; and ii) limits on contaminants and pathogens.

The agronomic efficacy of recovered nutrient materials for use as plant fertilizers depends on the plant nutrient availability of the product. The plant nutrient availability can be determined using either chemical or bioassay tests (Bauer et al., 2007). For instance, chemical tests for plant available P are based on specific extractant solutions such as water and neutral ammonium citrate that are used to extract plant available P fractions (water-soluble plus citrate-soluble P) and non-available P (citrate-insoluble) from the fertilizer material (AOAC International, 2000). Bioassay tests are based on the plant response to an amendment under greenhouse conditions or in field plot trials. The bioassay trials are the most consistent tests for predicting plant nutrient availability, however they are more time-intensive than chemical tests.

With respect to contaminant limits, any recovered nutrient material should not be a major source of toxic pollutants for soils and plants such as heavy metals (Loganathan et al., 2003), and of nutrient losses to surface runoff and eutrophication (Chien et al., 2011). An advantage of recovering nutrients from animal manure is their lower heavy metal concentrations than other P-rich waste streams such as industrial and municipal biosolids, reducing the risks of exceeding regulatory limits for land application and deemed environmentally safer (Vaccari, 2011, Ehmann et al., 2017, Szögi et al., 2018). An added advantage of manure recovered nutrients such as magnesium and calcium phosphates is that they are sparingly soluble in water and dissolved phosphate ions quickly form bonds with soil particles which reduces P loss by leaching or surface runoff (Rahman et al., 2011; Szögi et al., 2012). The reduction of pathogen along with the recovery of phosphates was reported by Vanotti et al. (2005a). They found that nitrification/denitrification treatment after solids separation was very effective in reducing pathogens in liquid swine manure and that the N removal treatment followed by a P removal step via alkaline calcium precipitation produced both a concentrated P material and a pathogen free, sanitized effluent.

Furthermore, desirable properties of manure byproducts are linked to the specific nutrient recovery technology and production process conditions. In the USA, a Nutrient Recycling Challenge (NRC) that searched affordable technologies that recycle nutrients from livestock manure and create value added products, established criteria for such technologies (Ziobro, 2017). The NRC criteria was developed with inputs from swine and dairy producers, EPA, USDA, WWF, WRF, and various environmental, business and academic groups. Aside from cost effectiveness/affordability, desirable characteristics of the nutrient recovery technologies are (Ziobro, 2017):

· Recovers and concentrates nutrients.

· Produces materials with predictable nutrient concentrations.

- Byproducts can be used directly with no need of further industrial processing other than the recovery process.
- Ability to produce low-nutrient effluent from liquid manure stream.
- Compatible with existing animal production facilities and manure management systems.
- Yields additional multiple benefits such as odor and pathogen reduction, reduces greenhouse gases, generates energy and water for on-farm reuse, and provides other ecosystem benefits such as protection and restoration of water quality and water quality credits.

Solids Separation and Dewatering to Concentrate Nutrients

Solid-Liquid Separation

Solid-liquid separation is a processing technology used to divide the liquid and solid fractions of manure using gravity, mechanical, and/or chemical processes. Mechanical and gravity solid-liquid separation has been traditionally used to reduce organic loading on a treatment lagoon or other holding pond, to improve pumping characteristics, or to recover solids from lagoon sludge (Chastain, 2013). The separation efficiency is higher when the manure is fresh and it decreases with storage time due to degradation of the organic particles (Kunz et al., 2009). Average mass separation efficiencies of total solids, total N and total P in dairy manure by mechanical separation using screw presses with screen sizes 0.5 to 2.25 mm were approximately 45%, 18%, and 21%, respectively (Aguirre-Villegas et al., 2017). Although this coarse separation recovers little of the nutrients in the manure, it produces a fibrous material that has a value. For example, separated fibrous solids from AD digestate has been used as a bedding material for cows to replace purchased sawdust or straw (Jensen et al,. 2016), and as a substrate for production of quality composts for peat substitution in the horticultural industry (Hummel et al., 2014).

However, new advances over the last 15 yr in equipment and flocculant applications for chemically enhancing solid-liquid separation treatment have improved removal efficiency of solids and specific plant nutrients such as N and P (Hjorth et al., 2010; Chastain, 2013). Centrifugation generate a centrifugal force that enhances solids and nutrient separation, especially the P that is usually contained in very fine particles. Average separation indexes at centrifugation were identified by Hjorth et al. (2010) as: 14% volume, 61% dry matter, 28% total N, and 71% total P. Performance of all the mechanical and gravity options can be enhanced significantly by addition of flocculants before separation as described later (subsection on separation with flocculants). High-rate solids separation facilitates the use of further methods to recover nutrients that would otherwise be unsuited for use with raw wet manure (Fig. 1).

Physical and Mechanical Case

A common practice on dairy farms is to remove large particle solids prior to storage of the liquid in a lagoon. The advantage of this practice is to remove solids that can potentially limit liquid storage, produce odors, and clog irrigation systems. Limited

Table 1. Nutrient content and information of separated anaerobically digested dairy manure and pre-consumer food wastes (Ma et al., 2017).

Item	Screw press separator (2009)	Rotating drum separator (2011)	Rotating drum separator (2012)
m³ of liquid manure and pre-consumer food wastes	303.8	504.3	445.6
kg Solids in effluent	4743	9082	11,395
kg wet separated solids	3834	10,214	5743
% DM of separated solids	41.0	25.4	25
kg separated solids, DM basis	1585	2474	1727
% separated solids	25.0	21.3	13.2
kg separated N	79.4	63.5	37.2
kg liquid N	544	1010	833
% separated N	12.9	6.3	4.3
kg separated P	17.2	10.7	7.7
kg liquid P	62.6	111.1	62.6
% separated P	21.5	9.23	11.1

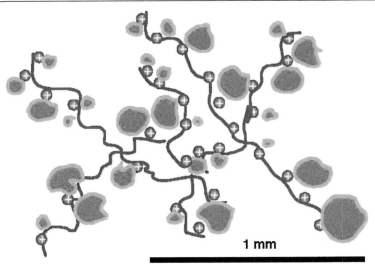

1 mm

Fig. 2. Schematic of a floc formed by a cationic polyacrylamide polymer in liquid swine manure (Vanotti et al., 2002).

mass-balance data is available to document the partition of solids and nutrients with liquid-solids separators at a farm scale. The case shows nutrient and solids partition after mechanical liquid-solid separation using of anaerobically digested dairy manure and pre-consumer food wastes. Data in Table 1 was collected from a farm that utilized an anaerobic digester which received daily ~ 227 m³ of dairy manure and ~ 114 m³ of pre-consumer food wastes. Two different liquid–solids separators were evaluated, a screw press separator (Eys, Daritech Inc., Lynden, WA), and rotating drum screen separator (DT-360, Daritech Inc., Lynden, WA). The data (Table 1) on nutrient partitioning after liquid–solids separation indicate a range: in solids separation of 13 to 25%, in N separation of 4.3 to 12.9%, and in P separation of 9.2 to 21.5%.

The EYS screw press liquid–solids separator system resulted in greater removal of solids, N, and P, but was observed to require greater maintenance and had a lower liquid throughput rate. The overall observation of significance is that the majority (> 75%) of solids and nutrients remains with the liquid fraction (Ma et al., 2017).

Enhanced Separation with Flocculants

Organic polymers such as polyacrylamide (PAM) are useful to increase separation of suspended solids and carbon compounds from liquid swine and dairy manure (Vanotti and Hunt, 1999; Zhang and Lei, 1998). Along with the solids, there is a significant separation of organic nutrient elements contained in small suspended particles typical of these wastes. Vanotti et al. (2002) showed that 80.4% of the total suspended solids (TSS), 78% of the nitrogen (N), and 93% of the phosphorus (P) fractions in flushed swine manure that are potentially removable by phase separation were contained in particles less than 0.3 mm in size.

Polyacrylamides are moderate to high molecular weight, long-chained, water-soluble organic polymers. The long polymer molecules destabilize suspended, charged particles by adsorbing onto them and building bridges between several suspended particles (Fig. 2). With flocculation, the effective particle size is increased by agglomeration of small particles into a larger particle, or floc, which separates from the liquid and dewater more readily. This larger size can significantly enhance manure solids and nutrient retention by screens and separation of colloidal particles by settling (Vanotti and Hunt, 1999; Vanotti et al., 2002). Polyacrylamides have varied characteristics such as molecular weight and charge type (+, 0,-), density distribution of charge (0% to 100%), and chain structure that provide them with a variety of chemical performance characteristics and uses. For example, polyacrylamides have been used to enhance separation and thickening processes in the food industry, remove solids from municipal wastewater, clarify potable water, enhance screening and settling of manure, and as a soil conditioner to reduce irrigation water erosion (Chastain et al., 2001; Sojka et al., 1998; Vanotti et al., 2018a). For manures, the most efficient flocculants tested had these common characteristics: linear cationic PAMs with high molecular weight and medium charge density (20–40%) (Vanotti and Hunt, 1999; Hjorth et al., 2010). Polyacrylamides have been used to enhance the performance of a wide range of mechanical and physical manure separators such as incline screens, rotary screens, screw presses, filter presses, rotary presses, and dissolved air flotation devices. Compared with metal coagulants (aluminum sulfate and ferric chloride), the

Table 2. Removal of phosphorus and nitrogen from flushed swine manure by flocculation and screening (data from Vanotti et al., 2002).

	TP		Organic P		TKN		Organic N		
Polymer rate	Concentration	Removal efficiency	Concentration	Removal efficiency	Concentration	Removal efficiency	Concentration	Removal efficiency	N/P Ratio
mg L^{-1}		(%)	(mg L^{-1})	(%)	(mg L^{-1})	(%)	(mg L^{-1})	(%)	
			Initial Liquid Manure Before Passing 1-mm Screen Separator						
0	270	–	223	–	1293	–	569	–	4.79
			Liquid Fraction After 1-mm Screen Separator						
0	243	10	200	10	1200	7	497	13	4.94
140	71	74	18	92	841	35	85	85	11.85

optimal application rate for organic polymers is significantly lower, typically about 150 mg L^{-1} in swine manure and about 300 mg L^{-1} in dairy manure (Chastain et al., 2001; Vanotti et al., 2002; Szögi et al., 2018) vs. 200 to 4000 mg L^{-1} for metal coagulants (Barrow et al., 1997; Kirk et al., 2003). In a study in a feeder-to-finish swine operation in NC, separation by screening alone was not effective; efficiencies were 15% TSS, and < 15% N and P (Table 2). Flocculation treatment substantially increased retention of the small manure particles even with a relatively large screen (1-mm) with separation efficiency of 95% TSS. The polymer flocculation and screening treatment effectively removed organic nutrients (92% P and 85% N) but had no effect on the dissolved ammonia and phosphate fractions. The selective separation of the nutrients (organic vs. dissolved) increased the N/P ratio of the effluent (from 4.8 to 12.1), resulting in a more balanced effluent for nutrient needs of crops. This helps solve problems of excess phosphorus accumulation in soils of wastewater spray fields. The same study showed that the polymer use efficiency (g TSS separated per g polymer) is significantly affected by the solids strength of the manure, and as a result, the chemical cost of treating diluted manure was seven times more expensive than the cost of treating higher-strength waste. Therefore, with very diluted manure streams typical of flushing systems, a preconcentration step is recommended before solid–liquid separation with flocculants to increase the polymer use efficiency. In a third-generation system, Vanotti et al. (2014) used a decanting tank (gravity settling) to preconcentrate the manure before solid–liquid separation with flocculants and rotary press separator in a 1200-sow farrow-to-feeder operation that used a flushing system and generated 140 m³ of diluted manure per day. The decanting tank removed 60 to 70% of the TSS in the flushed effluent and concentrated the manure 15 times (from 0.3% to 4.5% TSS). This preconcentration strategy increased polymer use efficiency 5.4 times (from 52 to 279 g g^{-1}), and reduced the total manure volume processed by the solid separator press by 98 m³ per day. This lower volume was one of the major advances of the third-generation project because it increased solid separator press capacity and lowered operating expenses when flocculation is adapted to flushing systems.

Natural flocculants may have an important role in waste management because of increased cost of energy and renewed interest on organic farming systems. Garcia et al. (2009) indicated that naturally-occurring flocculants such as chitosan can be as effective as synthetic polymers for the separation of solids and nutrients from concentrated dairy manure effluents. Chitosan is a natural, biodegradable polycationic polymer that is the deacetylated form of chitin, a polymer found in certain fungi and in the exoskeleton of arthropods such as shrimp and crab shell waste. Garcia et al. (2009) used various rates of chitosan to flocculate mixtures of dairy manure and the lagoon supernatant used to flush freestall alleys. The flocculated manure was dewatered using 1-mm and 0.25-mm screens. The results showed that separation by screening alone (1-mm screen) was not effective (average efficiencies were 60% for TSS, 22% for TKN, and 25% for TP), and that mixing with chitosan before screening substantially increased separation efficiency: at optimum rates, separation efficiencies were 97% for TSS, 79% for TKN, and 58% for TP.

Composting of Separated Solids

Composting is the aerobic microbial breakdown and stabilization of organic matter involving mineralization and partial humification of the organic matter, producing a stabilized and sanitized material. Composting of animal manure should be seen as a technology that adds value, producing a high-quality product focused on specific agricultural markets such as soilless media for nursery crops, orchard mulching, and organic farming (Bernal et al., 2009). The quality criteria are established in terms of nutrient content, humified and stabilized organic matter, the maturity degree, hygienisation, and the absence of phytotoxic compounds (Bernal et al., 2009, Moral et al., 2009). The advantages of composting animal manures compared with direct application to soil of untreated manure are sanitation, reduction of volume and moisture, odor removal, safe storage, and a more uniform, easier-to-transport byproduct than untreated manure (Millner et al., 2014). Composting is a well-suited technology for on-farm management of agricultural residues, but it is not an environmentally sustainable technology for treating manure alone. Certain chemical and physical characteristics of animal manures are not adequate for composting and could limit the efficiency of the process: excess of moisture, low porosity, high N concentration versus organic C (which gives a low C/N ratio), and in some cases, high pH values (Bernal et al., 2009). When composting animal manure, significant N loss can occur through NH_3 volatilization, which is the main pathway of N loss during this process (Pardo et al., 2015). Therefore, different aeration strategies, substrate conditioning-feedstock formulation, bulking agents, and amendments have been used in manure composting to control the process to reduce composting time and costs, enhance the quality of compost, and ultimately avoid NH_3 volatilization losses (Pardo et al., 2015). For instance, a full-scale study in a centralized solids processing facility evaluated the combined effect of feedstock formulation and bulking agent to optimize the composting process of pig manure (Vanotti et al., 2006). The feedstock formulation included polymer separated swine manure solids (SS) combined with cotton gin waste (CGT) in a mixture 1SS:2CGT (volume basis). The compost product recovered 56% of the original material weight, 25% of the material volume, 61% of the C, 95% of the N, and 100% of the P. The product also met strict EPA Class A bio-solids quality standards due to low pathogen levels. In another study, using swine manure with zeolite or alum amendments, NH_3 emissions were reduced 85 to 92% with the amended swine manure compost retaining three times more N than the unamended control; the addition of these amendments did not appear to significantly affect the composting process (Bautista et al., 2011).

Thermochemical Processes

Thermochemical treatment processes are waste-to-energy technologies that involve high temperatures to convert organic feedstocks into gases, hydrocarbon fuels, and char or ash residues (Cantrell et al., 2008). The operating temperature and oxygen content can vary with different thermochemical methods. These methods include torrefaction (200–300 °C), pyrolysis (300–800 °C and no oxygen), gasification (> 800°C and some oxygen), and combustion (800–1,000 °C with oxygen). At lower process temperatures and in absence of oxygen, biochar (charcoal) is generated as a solid byproduct of pyrolysis. At high combustion temperatures, manure is converted into P-rich ash and other byproducts of the process (particulate matter, carbon monoxide and dioxide, dioxins, nitrogen and sulfur oxides, hydrochloric acid, heavy metals, and polycyclic aromatic carbon) that require additional control methods to reduce their emissions into the environment (Stingone and Wing, 2011). Thermochemical conversion of manure using heat alters nutrient contents. Whereas combustion of manure produces a P-concentrated ash but very low in N (Table 3), biochar generated from manure retains most of the P and up to one third of the N content in the original manure

Table 3. Thermal processes of animal manure and nutrient concentration in byproducts.

Thermal Process	Temperature	Manure Feedstock	Solid Product	N	P	K	Reference
	°C				g kg^{-1}		
Combustion	> 800	Poultry litter	Ash	–	53.0	3.9	Codling et al., 2002
	800–1000	Poultry litter	Ash	–	76.2	53.0	Acharya et al., 2014
Pyrolysis	350	Dairy	Biochar	2.6	10.0	14.3	
	700	Dairy	Biochar	1.5	16.9	23.1	
	350	Poultry litter	Biochar	4.5	20.8	48.5	Cantrell et al., 2012
	700	Poultry litter	Biochar	2.1	31.2	74.0	
	350	Swine	Biochar	3.5	38.9	17.8	
	700	Swine	Biochar	2.6	59.0	25.7	
Hydrothermal carbonization	160–240	Swine (wet)	Hydrochar	28.5	26.0	8.0	Song et al., 2017

feedstock (Ro et al., 2010). Among these different thermochemical conversion technologies, pyrolysis has attracted a lot of interest for animal manure management because the biochar produced can be used as a soil amendment for carbon sequestration, improvement of water infiltration, and as a source of plant nutrients (Hunt et al., 2013; Sigua et al., 2016; Schulz and Glaser, 2013; Novak et al., 2016). For many manure types, N concentration increases at process temperature of 350 °C due to the organic mass lost as carbon dioxide (Table 3), but N eventually decreases with pyrolysis at temperatures as high as 700 °C (Cantrell et al., 2012). However, pyrolysis of wet animal manure alone is less energetically viable because of its low energy output versus the large energy required to evaporate moisture from the manure feedstock (Lentz et al., 2017). Therefore, copyrolysis of animal manures with high-energy density feedstocks can make the total pyrolysis process energetically sustainable. Pyrolysis of wet swine manure blended with spent plastic mulch wastes produced a solid biochar and combustible gas with a heating value higher than natural gas (Ro et al., 2014). According to this report, pyrolysis of the swine solids could be energetically sustainable by copyrolyzing a feedstock consisting of dewatered swine solids (75% moisture) with just 10% plastic mulch waste. On the other hand, hydrothermal carbonization methods can pyrolyze a manure slurry directly. Hydrothermal carbonization is a relatively low temperature (180–350 °C) process that can treat wet manure under pressurized liquid water and produce a valuable solid char called "hydrochar" (Libra et al., 2011). Because evaporation of water is avoided, hydrothermal carbonization requires much less energy input than pyrolysis. Aside from their potential for improving soil fertility (Ro et al., 2016), the surface properties of manure-based hydrochars make them useful as an environmental sorbent for pollutants such as endocrine disrupting chemicals, herbicides, and polyaromatic hydrocarbons (Sun et al., 2011, 2012; Han et al., 2016). Despite its favorable energetics, hydrothermal carbonization of animal manure is still an emerging technology as the full-scale implementation of this technology for manure treatment and beneficial uses of its byproducts need further research and development (Lentz et al., 2017).

Technologies for Recovery of Phosphorus

Magnesium Phosphates (Struvites)

Phosphorus can be harvested from liquid dairy and swine manure by precipitating magnesium ammonium phosphate ($MgNH_4PO_4$) crystals, as a compound called struvite (Bouropoulos and Koutsoukos, 2000). Struvite can be recovered from manure that has been processed with or without the use of an anaerobic digester, however more efficient phosphorus recovery is achieved when inorganic levels of P are greatest. Soil incubations showed that struvite provided the lesser P mobility away from the fertilizer granules compared to conventional fertilizers MAP and TSP (Nascimento et al., 2018). Phosphorus recovered as struvite can be used as an efficient, slow-release fertilizer for a multitude of plants and crops (Münch and Barr, 2001; Nelson et al., 2003; Gonzalez-Poncer and Garcia-Lopez, 2007; Yetilmezsoy and Zengin, 2009; Hilt et al., 2016).

Struvite is effectively precipitated out of wastewater at a pH between 7.0 and 7.5 and when the conditions involving the molar ratio of $Mg:PO_4:NH_3$ are met. The chemical formula for struvite is $MgNH_4PO_4$ with six water molecules (Chirmuley, 1994). While wastewaters, particularly manure, have naturally high levels of phosphate (PO_4^{3-}) and ammonium (NH_4^+), they often do not contain sufficient levels of magnesium to meet the required molar ratio for struvite precipitation. As a result, a source of magnesium must be added to the wastewater before struvite precipitation can occur, and studies have found magnesium chloride ($MgCl$) to be the best source of magnesium as it assists with ammonium nitrogen (NH_4–N) removal (Rahman et al., 2011). It has been shown that increasing the molar ratio to 1.2:1:1 for Mg^{2+}, NH_4^+, and PO_4^{3-} ions respectively yields the maximum amount of struvite precipitation (Kozik et al., 2011).

The pH of the wastewater is a critical factor for struvite production. Struvite solubility decreases with increasing pH (Ohlinger et al., 1999; Celen and Turker, 2001; Stratful et al., 2001; Battistoni et al., 2001; Dastur, 2001; Adnan et al., 2003), and crystal growth rate increases with increasing pH. Calcium can impact the optimal pH for the formation of struvite. Struvite precipitation occurred at a pH of 7.5 to 9 in wastewater with low calcium levels compared with a pH of 7.0 to 7.5 in wastewater with high calcium levels (Hao et al., 2008). The pH of wastewater is normally below 7.0, but can be increased either by aeration or by the addition of pH modifiers. Aeration results in carbon dioxide (CO_2) evolving out of solution. As CO_2 is stripped from the solution, the pH of the wastewater increases, making the solution suitable for struvite formation. A study by Liu et al. (2011) found that struvite formation is proportional to the aeration rate, and that maximum struvite production was achieved when the aeration rate was 0.73 L min^{-1}. However, in some reactors (such as fluidized bed reactors) aeration is not necessary for struvite formation, and chemicals can be used to increase the pH. In cone-shaped fluidized bed reactors, the wastewater is pumped up through the reactor so that the up-flow of wastewater mixes with a bed of struvite crystals (Benyahia et al., 2000; Bowers and Westerman, 2005). In this system a basic chemical, such as ammonia or sodium hydroxide is pumped into the reactor to increase the pH.

Research has shown that pure struvite (99.7%) is formed between the pH of 7.0 to 7.5, while struvite formed in the pH range of 8.0 to 9.0 is only 30 to 70% pure struvite, and struvite formed at a pH of 9.5 or above contains less than 30% struvite (Hao et al., 2013). The reason for these impurities is that as the pH increases, other compounds such as calcium phosphate and magnesium phosphate are able to precipitate out of solution. Thus, maintaining wastewater pH in the 7.0 to 7.5 range is critical for maximal precipitation of pure struvite when competing ions are present.

The presence of competing ions, specifically Ca, can greatly diminish the amount of P recovered as struvite. Calcium in the wastewater binds with phosphate ions and precipitates the P out of solution as calcium phosphate (Le Corre et al., 2005). If the P is bound to calcium, then it is unavailable for struvite production (Doyle and Parsons, 2002). To achieve adequate struvite precipitation, the wastewater needs to have a low calcium concentration or a calcium to phosphate ($Ca^{2+}:PO_4^{3-}$) ratio that is less than 0.5 (Moerman et al., 2009). This problem occurs in wastewaters that are high in calcium or have a large $Ca^{2+}:PO_4^{3-}$ molar ratio, such as municipal water and dairy manure. Dairy manure has high concentrations of Ca due to the animals' diet being high in calcium to sustain milk production and prevent disease (Block, 1984). For wastewaters high in Ca, an extra step is required to disassociate the calcium phosphate bonds. The two methods that have been studied to accomplish this are the addition of a chelating agent (Ethylenediaminetetraacetic acid; EDTA) and the addition of an acid. Ethylenediaminetetraacetic acid preferentially binds with calcium leaving the phosphate available in solution, and the acid decreases the pH to increase the solubility of the calcium phosphate and put the phosphate ions back into solution (Zhao et al., 2010). Without the addition of a chelating agent or an acid, total P removal as struvite from dairy manure rarely exceeds 15% (Harris et al., 2008; Zhao et al., 2010). By adding EDTA in high concentrations, it has been shown that a total P removal rate as struvite to be about 80% (Zhao et al., 2010), while adding acid has resulted in up to 70 to 75% total P removal (Bowers, 2006; Zhang and Jahng, 2010).

One way to manage the Ca issue is the use of oxalic acid. The most valuable property of oxalic acid is its ability to form complexes with metals, which is used in removing heavy metal contaminants from wastewaters and a variety of solid materials (Arnott, 1995; Gadd, 1999). Calcium oxalate is quite insoluble (Gadd et al., 2014), making oxalic acid the perfect choice for both decreasing the pH of dairy manure and precipitating the calcium out of solution.

Struvite recovery via a fluidized bed was initially developed for swine manure, and studies have found around 80% total P removal as struvite (Bowers and Westerman, 2005). Poultry manure has also shown high rates of phosphate removal, ranging from 90 to 98% (Demirer and Yilmazel, 2013). Both manure types exhibit high rates of P removal by struvite because the animals are not fed diets high in calcium compared to dairy cows.

Calcium Phosphates

Calcium Phosphate Recovery from Animal Wastewater

A technology was developed to remove soluble P from animal wastewater and other high-ammonia strength effluents as calcium phosphate and had the advantage of requiring minimal chemical addition and producing a valuable byproduct (Vanotti et al., 2003, 2005b). It is based on the distinct chemical equilibrium between phosphorus and calcium ions when the carbonate and ammonium interferences are substantially eliminated before precipitating phosphate as calcium phosphate by the addition of lime. Animal wastewater is a mixture of urine, water, and feces. Livestock urine usually contains more than 55% of the excreted N, of which more than 70% is in the form of urea (Sommer and Husted, 1995). Hydrolysis of urea by the enzyme urease produces ammonium (NH_4^+) and carbonate according to the following reaction:

$$CO(NH_2)_2 + 2H_2O \rightarrow 2NH_4^+ + CO_3^{2-} \tag{1}$$

Therefore, a substantial part of the inorganic carbon in liquid manure is produced during decomposition of organic compounds. Carbonate alkalinity and NH_4^+ are the most important chemical components in liquid manure, contributing to the buffering capacity in the alkaline pH range (Fordham and Schwertmann, 1977; Sommer and Husted, 1995). Alkaline pH is necessary to form a P precipitate with calcium and magnesium compounds (House, 1999). When a calcium or magnesium hydroxide is added to liquid manure, the hydroxide reacts with the existing bicarbonate to form carbonate, with NH_4^+ to form ammonia (NH_3), and with phosphate to form phosphate precipitate compounds (Vanotti et al., 2003). For instance, using calcium hydroxide as an example, the following equations define the reactions:

$$Ca(OH)_2 + Ca(HCO_3)_2 \rightarrow 2CaCO_3 \downarrow + 2H_2O \tag{2}$$

$$5\,Ca^{2+} + 4OH^- + 3HPO_4^{=} \rightarrow Ca_5OH(PO_4)_3 \downarrow + 3H_2O \tag{3}$$

The reaction in Eq. [2] is complete at pH $\geq$ 9.5, whereas that in Eq. [3] starts at pH > 7.0, but the reaction is very slow at pH $\leq$ 9.0. As the pH value of the wastewater increases beyond 9.0, excess calcium ions will then react with the phosphate, to precipitate as calcium phosphate (Eq. [3]). Thus, the presence of bicarbonate in wastewater is an interference in the production of high-grade phosphates. Not expressed in Eq. [2] is the fact that in wastewater containing high NH_4^+ concentration, large amounts of hydrated lime are required to elevate the pH to required values because the NH_4^+ reaction tends to neutralize the hydroxyl ions, according to Eq. [4]:

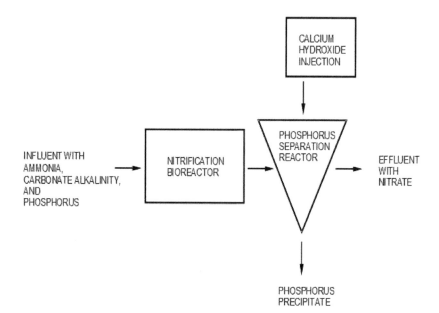

Fig. 3. Schematic showing the basic configuration of a P removal process using calcium phosphate (Vanotti et al., 2005b).

Table 4. Performance of field prototype used to remove and recover phosphorus from lagoon swine wastewater in Duplin County, N.C., using nitrification treatment and chemical phosphorus precipitation (data from Vanotti et al. 2003)

Ca(OH)$_2$ Applied mg L^{-1}	pH	Ca:P Molar Ratio	Alkalinity mg L^{-1}	NH$_4$-N mg L^{-1}	Total N mg L^{-1}	Total P mg L^{-1}	N:P Ratio
Influent (Lagoon liquid)							
–	7.7	–	1738	278	320	71.9	4.45
Effluent after Nitrification Pre-treatment							
–	7.7	–	532	10	300	65.4	4.59
Effluent after Phosphorus Precipitation							
141	9.0	0.82	511	8	303	25.5	11.9
271	9.5	1.58	570	8	301	11.1	27.1
344	10.0	2.00	557	6	299	3.3	90.6
433	10.5	2.52	545	5	299	1.6	186.9

$$Ca(OH)_2 + 2NH_4^+ \rightarrow 2NH_3 + Ca^{2+} + 2H_2O \qquad [4]$$

Consequently, precipitation of P in animal wastewater using an alkaline compound such as lime is very difficult due to the inherent high buffering capacity of liquid manure (NH$_4$–N $\geq$ 200 mg L^{-1} and alkalinity $\geq$ 1200 mg L^{-1}). This buffering effect prevents rapid changes in pH. However, this problem is solved using a prenitrification step that reduces the concentration of both NH$_4^+$ (Eq. [5]) and bicarbonate alkalinity (Eq. [6]) (Vanotti et al., 2003, 2005b):

$$NH_4^+ + 2O_2 \rightarrow NO_3^- + 2H^+ + H_2O \qquad [5]$$

$$HCO_3^- + H^+ \rightarrow CO_2 + H_2O \qquad [6]$$

The buffering effect of NH$_4^+$ (Eq. [4]) is reduced by biological nitrification of the NH$_4^+$ (Eq. [5]). Simultaneously, the interference and buffering effect of bicarbonate (Eq. [2]) is greatly reduced with the acid produced during nitrification (Eq. [6]). These two simultaneous reactions leave a less buffered liquid. In this way, lower rates of lime are added to recover the phosphate as calcium phosphate (Eq. [3]) (Vanotti et al., 2005b). Also, the elimination of inorganic carbon (HCO$_3^-$) by the preceding biological nitrification prevents calcium carbonate formation (Eq. [2]). This leads to production of higher grade calcium phosphates.

The basic process configuration is shown in Fig. 3, comprising: (a) providing wastewater having at least reduced levels of carbonate and ammonia buffers, and at least reduced levels of suspended solids to a reactor vessel, and (b) adding an alkaline earth base to said wastewater to precipitate soluble P (Vanotti et al., 2005b). The process has been used to remove P in various effluents: swine anaerobic lagoon effluents (Vanotti et al., 2003; Szögi and Vanotti, 2009), AD effluents from upflow anaerobic sludge blanket reactors (UASB) treating swine manure (Suzin et al., 2018) and potato processing industry wastewater (Monballiu et al., 2019), and as components of treatment systems for swine manure without lagoons (Vanotti and Szögi, 2008; Vanotti et al., 2018a). In the configuration treating anaerobic lagoon effluents (or any anaerobic digester effluent), the supernatant liquid, rich in NH$_4$–N and carbonate alkalinity, is nitrified and P is subsequently removed by adding hydrated lime. A bench experiment confirmed the theoretical background in Eqs. [1–6]. It compared P removal using various calcium hydroxide rates (1 to 10 moles Ca added/ mol P) applied to swine lagoon wastewater with nitrification and a control without nitrification (Vanotti et al., 2003). The chemical analyses of the control liquid were: $PO_4–P = 63$ mg L^{-1}, $NH_4–N = 300$ mg L^{-1}, alkalinity = 1890 mg L^{-1} and pH = 8.05. After 16-h nitrification, the corresponding analyses were $PO_4–P = 63$ mg L^{-1}, $NH_4–N = 61$ mg L^{-1}, alkalinity = 63 mg L^{-1} and pH = 6.06. Phosphorus removal rates were low (< 34%) in the control but increased to about 100% in the effluent with reduced levels of alkalinity and ammonia. The effectiveness of the technology was tested in a pilot field study obtaining P removal efficiencies of 95% and 98% obtained with Ca/P molar ratios of 2 and 2.5, respectively (Table 4). In practice, the level of treatment added will depend on the degree of P removal desired and should be preferably added in the minimum quantity necessary to balance the N/P ratio of crops or to remediate sprayfields. Since $NH_4–N$ has been converted to nitrate, increased pH does not result in significant gaseous N loss. Therefore, the amount of P removed, and consequently the N/P ratio of the effluent, can also be adjusted in this process to match the N/P ratio needed by the growing crop to which it will be applied. For example, a final N/P ratio of 10.7:1 and 13.4:1 would be needed to match wheat and coastal bermudagrass specific nutrient uptake needs (Edwards and Daniel, 1992), respectively, which can be delivered with about one Ca/P molar ratio treatment (Table 4). Higher N/P ratios (N/P > 30) would be prescribed to clean phosphorus-polluted sprayfields by a negative mass phosphorus balance between P applied and P removed by harvestable plant materials.

In new systems without lagoons, raw liquid manure is first treated through a high-rate solid–liquid separation process to remove most of the carbonaceous material from the wastewater. The separated water is then treated with the nitrification and soluble P removal sequence. A denitrification tank is incorporated into the treatment system to provide total N removal in addition to the P removal. The previous treatment steps (solid–liquid separation and NDN) substantially reduce concentrations of suspended organic particles, alkalinity and the NH_4^+ in the liquid manure (Vanotti and Szögi, 2008), which create new conditions in the liquid that facilitate precipitation and recovery of P using $Ca(OH)_2$. With this process, the use of a crystallizer or fluidized bed reactor with seeding to induce P nucleation is not required because the calcium phosphate produced is mostly amorphous calcium phosphate (ACP). In practice, the soluble P is removed from prenitrified liquid swine manure by adding hydrated lime slurry in a small mixing chamber with 20 min retention time to form a fine precipitate. Depending on the desired level of P removal and effluent disinfection, a target process pH is set in the P-module that is fitted with a pH probe and controller linked to the lime injection pump. Using a process pH of 10.5, the soluble P in the manure was reduced 94.7% during a two-year period that encompassed five consecutives pig production cycles (Vanotti and Szögi, 2008). In another full-scale study using a process pH of 9.5 in the P-module, the soluble P was reduced 79.8% (Vanotti et al., 2018a). The treated effluent is poorly buffered, and the pH decreases readily from 10.5 to 8.5 by short-term aeration (2.5 h) or by natural aeration during storage of the treated effluent. The fine P precipitate is separated by gravity in a settling tank and P-sludge can be dewatered effectively using filter bags (Teknobag-Draimad, Aero-Mod, Inc., Manhattan, KS) in conjunction with an anionic polymer, recovering 99% of total P with respect to unfiltered material (Szögi et al., 2006). To save costs, the settled P-sludge can also be mixed with the raw manure and separated simultaneously in a single separator (Garcia et al., 2007). The recovered P precipitate solid obtained at full-scale was amorphous calcium phosphate with a concentration grade of $24.4 \pm 4.5\%$ P_2O_5 and $> 99\%$ plant availability based on a standard citrate P analysis used by the fertilizer industry.

The process above (Vanotti et al., 2005b) used a natural biological (microbial) acidification with nitrifiers as a way to destroy carbonate alkalinity (Eq. [5] and [6]) and demonstrated the importance of eliminating the inorganic carbon (carbonate) interference for precipitating phosphates from wastewater using phosphorus precipitating compounds calcium and magnesium. There are three other methods for phosphate recovery subsequently developed in which some type of acidification is also used to remove the inhibitory effect of inorganic carbon and also increase dissolution of the P, conditions that result in an effective phosphorus recovery from manure and municipal waste streams and produce high-grade phosphates using an alkaline earth metal: i) one method uses chemical acidification: it adds mineral or organic acids to acidify the manure before the P precipitation step with lime (Quick Wash process described in the following section); ii) another method also uses a biological acidification: it precipitates the P right after the acidification phase of anaerobic digestion in a multiphase anaerobic digestion process (Barack, 2013); iii) the last method uses physical acidification: it adds the P precipitating compounds (calcium or magnesium) to the wastewater after the carbonate alkalinity and ammonia are substantially reduced using gas-permeable membranes (Vanotti et al., 2018b). It was found that, when magnesium is added to wastewater with both the alkalinity and ammonia removed, the phosphates produced are very high grade (46% P_2O_5, $> 98\%$ available), similar to the composition of the biomineral newberyite ($MgHPO_4 \cdot 3H_2O$) found in guano deposits (Vanotti et al., 2017).

Calcium Phosphate Recovery from Raw Manure Solids and Lagoon Sludges Using Quick Wash Process

A novel chemical P recovery process called "Quick Wash" (QW) uses a combination of acid, hydrated lime, and organic polyelectrolytes to selectively extract and recover P (but not the N) from manure solids (Fig. 4). The QW is a patented treatment process (U.S. Patent 8673,046; first filed Feb. 5, 2008) developed for rapid acid wet extraction of P from solid manure and P recovery in a solid concentrated form (Szögi et al., 2014a).

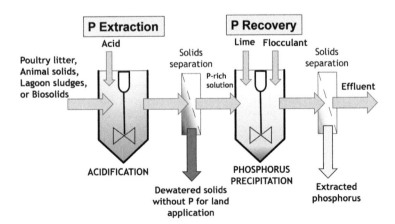

Fig. 4. Schematic diagram of the Quick Wash process to recover calcium phosphate from solid wastes (Szögi et al., 2014a).

It has been used to effectively separate calcium phosphate from a variety of organic wastes: raw poultry litter (Szögi et al., 2008), fresh swine manure solids (Szögi et al., 2015), swine lagoon sludges (Szögi et al., 2018), and municipal biosolids. The QW process produces two materials: a P–depleted, washed residue with a balanced N/P ratio that is more environmentally safe for land application and use in crop production, and a concentrated P material with fertilizer value that can be transported more economically long distances (Szögi et al., 2008). The process include a P extraction step and a P recovery step (Fig. 4). During the P extraction, approximately 80 to 90% of the initial total P in raw animal and poultry waste is selectively extracted by rapid hydrolysis reactions using mineral or organic acids in a pH range of about 3 to 5. The washed residue is further dewatered to prevent unnecessary C and N oxidation and digestion. This first step produces a liquid extract containing low suspended solids and extracted soluble P. In the succeeding P recovery step, the P is precipitated from the liquid extract by lime addition to increase the pH to 9.0 to 11.0 forming a Ca-containing P product. An organic anionic polymer enhances the recovery and P concentration of the product. The precipitated P-rich solid is dewatered while the liquid is recycled back into the quick wash system or land applied. The advantage of this process with respect to thermal pre-treatment is the low content of heavy metals in the recovered P product and conservation of C and N in the acid washed residue.

Szögi et al. (2018) combined the QW process and geotextile dewatering in a system to extract and recover P from swine lagoon sludge. About 83% of total initial P was extracted when the lagoon sludge was acidified to pH 3, and about 79% of total initial P was recovered as a P precipitate with lime addition to pH 10. The calcium phosphate product was identified as amorphous calcium phosphate (ACP) with P grades of 33.2 to 35.5% P_2O_5, higher than rock phosphate, with the advantage that there is no need for additional chemical processing for its use as fertilizer.

Iron and Aluminum Phosphates

Precipitation of phosphorus (P) by metal salts is a main commercial process to remove phosphorus from municipal wastewater. The process has been evaluated for the removal of P from manure using different doses and metal salts (Table 5). Chemicals that have been considered for manure treatment include aluminum sulfate or alum (Al_2SO_4), aluminum chloride ($AlCl_3$), ferric sulfate [$Fe_2(SO_4)_3$] and ferric chloride ($FeCl_3$), alone or in combination with polymers; and applied both in wet and dry manure systems. The most common are alum due to its lower cost, and ferric chloride because it is effective over a wide range of pH (4.0 to 12) (Chastain, 2013). These metal salts can reduce P solubility but also destabilize colloidal particles through coagulation chemical process. Application of these metal salts to precipitate phosphate from manure usually involve injection, mixing, and separation by sedimentation. When Fe and Al salts are added to manure these ions react with hydroxyl ions (OH^-) to form settleable particles of aluminum or ferric hydroxide, in addition the Fe or Al ions react with soluble phosphorus (ortho-P, PO_4^{3-}) to form settleable particles of $AlPO_4$ and $FePO_4$. The P in the recovered aluminum and iron phosphates is generally considered to have low availability to plants, and according to Moore et al. (1998), precipitated $AlPO_4$ in separated solids fraction will not have any fertilizer value at normal ranges of soil pH. The use of metal salts to precipitate P also increase the volume of settled solids, approximately 0.5 to 1.0 g L^{-1} solids increase per g per liter of chemical used for alum, which foretells subsequent handling and disposal problems (Vanotti and Hunt, 1999). The response to increasing the dose of chemicals

Table 5. Removal of phosphorus from swine and dairy manure using different doses of iron and aluminum salts (data obtained from Chastain, 2013).

Type of manure	Type of metal salt	Dose (mg L⁻¹)	TP removal (%)	Reference
Dairy manure	Ferric chloride	201.8	78.1	Barrow et al., 1997†
		403.7	80.8	
		605.6	83.8	
		807.4	87.8	
	Ferric sulfate	193	68.9	
		386	72.6	
		579	79.0	
		772	84.0	
Dairy manure	Ferric chloride	546	81.9	Sherman et al., 2000‡
		1092	88.8	
	Aluminum sulfate	164.8	68.5	
		336.1	72.4	
		500.9	73.1	
		672.2	76.8	
		1008.2	82.5	
		2010.1	96.9	
Swine manure	Aluminum sulfate	0	42	Ndegwa et al., 2001§
		1500	78	
		2000	65	
	Ferric chloride	1500	86	
		2000	45	
Dairy manure	Aluminum sulfate	0	60	Kirk et al., 2003¶
		800	76	
		2000	85	
		4000	90	
		6000	88	
		8000	101	
	Ferric chloride	800	56	
		2000	57	
Swine manure	Aluminum sulfate	0	23	Vanotti and Hunt, 1999#
		1430	90	

† settling time 20 min. Ferric chloride concentration = 13.9%; Ferric sulfate concentration = 10.78%. Total solids content of treated manure = 1%.

‡ settling time 20 min. Total solids content of treated manure = 1%.

§ settling time 4 h. Total solids content of treated manure = 1%.

¶ settling time 24 h. Ferric chloride concentration = 40%, Aluminum sulfate concentration = 40%. Total solids content of treated manure = 2.85%.

settling time 1 h. Flushed swine manure, total solids = 0.18%. Al:P molar ratio = 2.5

-is not linear, and even an excess of chemical can reduce P removal (Table 5, work of Ndegwa et al., 2001). Some authors have observed that when $FeCl_3$ or Alum is applied at high doses, the solids produced tend to float and not to settle due to CO_2 gassing reaction with the chemical (Kirk et al., 2003; Chastain et al., 2001). In these situations, a skimmer would be more appropriate for effluent clarification.

Technologies for Recovery of Nitrogen
Scrubbers and Air Stripping

Air in Barns

One of the strategies for reducing or minimizing NH_3 emissions from livestock production is the use of chemical and/or biological air scrubbers and biofilters, in which exhausted air from the animal houses is led through a wet packed bed to remove ammonia and other water soluble components (Van der Heyden et al., 2015). Full-scale spray scrubbers have been developed for ammonia recovery at poultry facilities with a removal efficiency of 71 to 81% (Hadlocon et al., 2015), and also for deep-pit swine finishing facilities with an average ammonia removal efficiency of 88% (Hadlocon et al., 2014). Melse and Ogink (2005) reported ammonia removal efficiencies of 91 to 99% for acid scrubbers and from 35% to > 90% in biotrickling biofilters. The manure removal system used in the livestock buildings affects the NH_3 emissions available for capture and the design of the scrubbers and biofilters. For example, frequent manure removal, flushing, and separating urine from feces by scrapers or belts have been shown to significantly reduce the NH_3 releases from the buildings by about 50% (Lim et al., 2004; Philippe et al., 2011).

Air scrubbers are applied at mechanically ventilated housing facilities, operating as follows: a packed tower air scrubber is a reactor that has been filled with an inert or inorganic packing material. The packing material usually has a large porosity, or void volume, and a large specific area. A washing liquid is sprayed on top of the packed bed and consequently wetted. Air from the housing facility is introduced, either cross-current or counter-current, resulting in intensive contact between air and water, and enabling mass transfer from gas to liquid phase. A fraction of the trickling water is continuously recirculated; another fraction is discharged and replaced by fresh water (Melse and Ogink, 2005).

In a chemical air scrubber, an acid, usually sulfuric acid, is added to the washing water to keep the pH below 4, shifting the equilibrium NH_3/NH_4^+ toward NH_4^+ as the dissolved NH_3 is captured by the acid, forming an ammonium salt (Eq. [7]). Acid salts, such as aluminum sulfate (alum), sodium bisulfate, potassium bisulfate, ferric chloride and ferric sulfate were found to work as well as strong acids (hydrochloric, phosphoric and sulfuric) for capturing NH_3 (Moore et al., 2018).

$$NH_3 + H^+ \leftrightarrow NH_4^+ \tag{7}$$

In a biological air scrubber, NH_3 captured in the washing water is oxidized by bacteria (nitrification process) according to Eq. [8] (nitritation by ammonium oxidizing bacteria, AOB) and Eq. [9] (nitratation by nitrite oxidizing bacteria, NOB). In a biotrickling filter these bacteria are immobilized in a biofilm on the packing material, and in a bioscrubber, bacteria are contained in a separated bioreactor where nitrification is performed. In the case of biofilters, they consist of a humid filter bed of organic material (but not completed wet) for bacterial growth.

$$NH_3 + 1.5\ O_2 \rightarrow NO_2^- + H^+ + H_2O \tag{8}$$

$$NO_2^- + 0.5\ O_2 \rightarrow NO_3^- \tag{9}$$

Process design and control of air scrubbers are of major importance for their optimal performance. Therefore, packing dimensions and material, air and liquid flow configuration, water flow rate, water discharge, pH control, and inoculation with bacteria should be optimized and controlled (Van der Heyden et al., 2015). To improve capture of other components that are less water soluble such as methane or odor components, addition of organic solvents to the water phase of biological air scrubbers can be used to increase mass transfer and therefore availability for bacteria (Van Groenestijn and Lake, 1999).

Liquid Manures

Ammonia stripping by air or steam is a process that can be applied for efficient ammonia recovery from liquid manure. Both are similar gas-liquid mass transfer processes, steam stripping is essentially a distillation process that takes place at higher temperatures than air stripping (Zeng et al., 2006). In the air stripping process, ammonia is transferred from the waste steam into the air, then absorbed from the air into a strong acid solution generating an ammonium salt (Bonmatí and Flotats, 2003, Jiang et al., 2014).

Nitrogen in liquid manure is mainly present as ammonium ions (NH_4^+) and aqueous ammonia (NH_3) which is a volatile form (Eq. [7]). The equilibrium between both species is strongly dependent on pH and temperature; hence NH_4^+ is more abundant when pH is below 7, regardless of temperature; and NH_3 increases as the pH raises shifting the equilibrium (Eq. [10]). Similarly, the amount of NH_4^+ decreases with a temperature increase which favors NH_3 formation. Therefore, high values of pH and temperature will favor the ammonia stripping process.

$$NH_4^+ + OH^- \leftrightarrow NH_3 + H_2O \qquad\qquad\qquad [10]$$

The amount of NH_3 that can be recovered form a liquid manure, or absorbed in the acidic solution is dependent on two equilibria: NH_3 gas/liquid equilibrium and NH_3 dissociation equilibrium in the liquid (Bonmatí and Flotats, 2003). Therefore, the efficiency of air stripping depends on: pH, temperature, ratio of air to liquid volume, and liquid characteristics. This process has been developed at industrial scale, and it involves the use of stripping towers, compressors, or pumps used to introduce air or steam into the liquid phase. When air stripping is used, absorption or adsorption towers are needed to recover NH_3 from the gas phase; for steam stripping, condensation or absorption equipment is needed to recover NH_3, and no further post-treatment of exhaust gases are required (Zeng et al., 2006). A combined stripper and absorber plant operates by heating the wastewater fed to the plant to around 45 °C and adding NaOH or lime ($Ca[OH]_2$) until a pH of 10.5 to 11 is reached. The use of lime will precipitate carbonates and phosphorous salts before the water is fed to the stripper tower (Zarebska et al., 2015), although other authors have indicated that sodium and potassium hydroxide are more efficient to raise manure pH (Zhang and Jahng, 2010).

A compromise between pH and temperate should be achieved. According to Zarebska et al. (2015), elevated temperatures intensify odor and increase operation costs as well as it enhances water evaporation, causing a decrease of the ammonia concentration in the effluent. However, raising initial pH of fresh pig slurry is essential for air stripping (Bonmatí and Flotats, 2003), and according to Liao et al. (1995) when swine wastewater pH is below 10.5, temperature has more influence on ammonia removal than when pH is above 10.5. Table 6 shows results of ammonia stripping from animal wastes.

Gas-permeable Membranes

The gas-permeable membrane process includes the passage of gaseous ammonia (NH_3) through a microporous hydrophobic membrane and subsequent capture and concentration in an acidic stripping solution on the other side of the membrane. The process can be used for removing and recovering nitrogen from liquid manures in storage tanks (Vanotti and Szögi, 2015), and from the air of poultry and animal barns (Szögi et al., 2014b). For liquid manure applications, the membrane manifolds are submerged in the liquid manure and the NH_3 is removed from the liquid before it escapes into the air; the NH_3 permeates through the membrane pores reaching the acidic solution on the other side (Fig. 5). Once in the acidic solution, NH_3 combines with free protons to form nonvolatile ammonium (NH_4^+) ions that are converted into a valuable NH_4^+ salt fertilizer. Acid is added to a concentrator tank that contains the acidic solutions to an endpoint of pH < 1 whenever the pH of the acidic solution increased to 2.

The concept was successfully tested using digested and raw swine and dairy manures containing a wide range of NH_4–N concentrations (140 to 5000 mg N L^{-1}). Gas-permeable membranes have been shown to effectively recover more than 97% of NH_4^+ from swine wastewater (García-González and Vanotti, 2015; Dube et al., 2016). By using the same stripping solution in 10 consecutive batches treating raw swine manure, the recovered N was concentrated in a clear solution containing 53 g NH_4–N L^{-1} (Vanotti and Szögi, 2015). The rate of N recovery by gas-permeable membranes is higher with increased waste strength. The removal of NH_3 by the gas-permeable membrane increases the acidity in the liquid manure as represented in Fig. 5 (left). The process is responsive to increased pH through addition of alkali chemicals to pH > 9 (García-González and Vanotti, 2015), which leads to an increased release of NH_3 from the manure and capture by the membrane. Vanotti and Szögi (2015) proposed the use of aeration to raise the pH instead of adding alkali chemicals to enhance the removal and recovery of NH_4^+ from livestock effluents using gas-permeable

Table 6. Operational parameters and ammonia removal from liquid manures in different studies using ammonia stripping (data obtained from Zarebska et al., 2015).

Type of manure	NH_4–N content in animal wastes	pH	Time	Temperature	Air flow rate	Ammonia removal	Reference
	(g L^{-1})		(h)	(°C)	(L min^{-1})	(%)	
Swine wastewater	0.8	9.5	55	22	45	90	Liao et al., 1995
	0.8	11.5	7	22	90	90	
Anaerobic digestion effluent (Swine manure and kitchen garbage)	1.5		12	15	10	95.3	Lei et al., 2007
Anaerobic digestion effluent (swine slurry and other organic material)	2.2	10	2	70	-	92.2	Gustin and Marinsek-Logar, 2011
	2.2	10	2	30	-	80	
Swine wastewater	5	10	24	37	0.5	31–70	Zhang and Jahng, 2010
Swine slurry	3.4	9.5	4	80	-	69	Bonmatí and Flotats, 2003
	3.4	11.5	4	80	-	98.8	
Digested swine slurry	3.7	9.5	4	80	-	96	
	3.7	11.5	4	80	-	96	

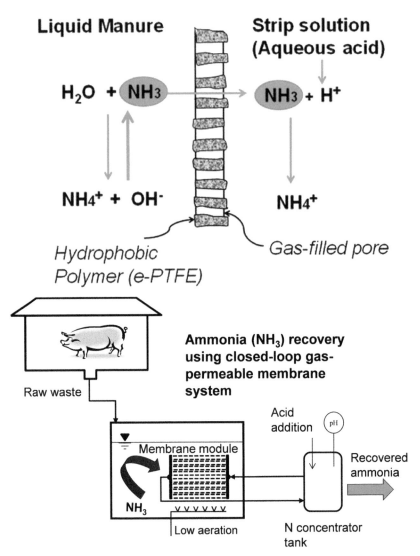

Fig. 5. Ammonia recovery from liquid manure with hydrophobic gas-permeable membranes. Above: Cross-sectional diagram of the process (Dube et al., 2016). Below: Schematic showing a submerged membrane module and an enhanced process configuration using low-rate aeration to increase manure pH and ammonia uptake (Vanotti and Szögi, 2015).

membranes. Low-rate aeration applied to stored livestock effluents results in a pH increase of about 1 unit and increases NH_3 release. García-González et al. (2015) showed that the positive effect of the low-rate aeration on the NH_4^+ recovery rate by the gas-permeable membrane process was equivalent to adding 2.14 g NaOH per L of manure. Dube et al. (2016) showed that the low-rate aeration resulted in a higher pH along with five to six times as fast recovery compared with the same system without aeration. During aeration of the manure, carbonate alkalinity is consumed and OH^- is instantly released, enhancing both the formation of NH_3 and the NH_3 uptake via the gas-permeable membrane. The removal of NH_3 by the gas-permeable membrane increases the acidity in the liquid manure (Vanotti et al., 2017). Using manures of various origins, approximately 4.1 g of carbonate alkalinity is consumed per g of NH_4–N removed by the enhanced process (Daguerre-Martini et al., 2018).

For air applications, the gas-permeable membrane manifolds are exposed to air containing gaseous ammonia (Szögi et al., 2014b). For example, in poultry facilities they are suspended above the litter, and the NH_3 is removed inside the barns close to the litter (Fig. 6). The method was developed for NH_3 capture from the air near the source without intense air movement using gas-permeable membranes and has the potential to reduce ventilation and energy needs along with lowering NH_3 in the air of poultry barns and composting systems (Szögi et al., 2014b). The air application also includes the passage of gaseous NH_3 through a microporous hydrophobic membrane, capture with a circulating diluted acid on the other side of the membrane in a closed loop, and production of a concentrated ammonium salt. In a study using membrane modules made of expanded polytetrafluoroethylene (ePTFE) membranes, about 96% of the NH_3 lost from poultry litter was captured and recovered (Rothrock et al., 2011). These high NH_3 recoveries were obtained regardless of the positioning of the membrane system above or below the litter surface, providing flexibility in design and implementation of the recovery system in a poultry house. The speed of NH_3 recovery by the membranes can be increased by more than 10 times by increasing the pH of the litter with hydrated lime, with the additional benefit of disinfection of the manure (Rothrock et al., 2011). On another study with poultry litter but using

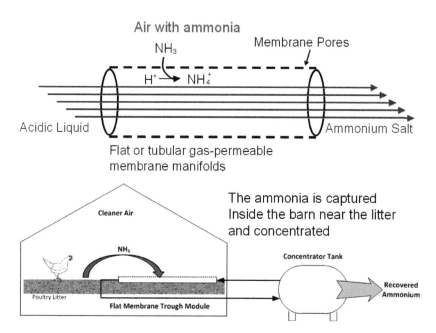

Fig. 6. Recovery of gaseous ammonia from the air of poultry houses using gas-permeable membranes. Top: Cross-sectional diagram of the process producing a concentrated ammonium salt (Rothrock et al., 2010). Bottom: Schematic showing the gaseous ammonia being removed inside the barns using membrane manifolds placed close to the litter surface (Rothrock et al., 2013).

flat gas-permeable membranes, the headspace NH_3 concentration was reduced from 70 to 97% and the recovery of the NH_3 volatilized from poultry litter was 88 to 100% (Rothrock et al., 2013). The potential benefits of this technology include cleaner air inside poultry houses, reduced ventilation costs, and a concentrated liquid ammonium salt that can be used as a plant nutrient solution.

Nanofiltration

Membrane filtration represents a suitable technology for nutrient concentrate production (Massé et al., 2007). Specifically, nanofiltration (NF) membranes can retain molecules with a molecular weight cutoff in the range of 150 to 300 Da, with a very high rejection (> 99%) of multivalent ions, high rejection of organic molecules larger than the molecular weight cutoff (> 90%) and lower rejection of monovalent ions (between 0 and 70%) (Zarebska et al., 2015).

In manure, ammonia (NH_3) exists in equilibrium with dissolved ammonium ions (NH_4^+), being the total ammonia nitrogen (TAN) the sum of both species. This equilibrium is mainly affected by pH and temperature in such a way that high pH or temperature increase NH_3 formation. Ammonium ions in manure usually combine with anions as HCO_3^-, PO_4^{3-} and volatile fatty acids (Massé et al., 2008), being retained by the NF membranes. Therefore, NF membranes can capture NH_3 or retain NH_4^+ depending on the manure pH and on the objective to be achieved: low pH favors NH_4^+ retention and high pH shifts the equilibrium toward NH_3 formation that permeates through the NF membrane (Zarebska et al., 2015).

The performance of nanofiltration membranes is affected by fouling which can be mitigated with a cleaning procedure. According to Zarebska et al. (2015), a combination of pretreatments as centrifugation, followed by ultrafiltration to remove suspended solids ensures an effective NF operation with reduced fouling.

Reverse Osmosis

Reverse osmosis (RO) is a filtration process that forces a solvent through a membrane with a pore size around 0.0001 microns using pressure (10–100 bar). The process retains dissolved solids and salts (concentrate stream) and allows the pure solvent to pass through (permeate stream). The permeate is clean water that can be discharged directly to the environment or reused in the production facility for process water or animal drinking water (Zarebska et al., 2015). Reverse osmosis is usually combined with other solids and nutrient recovery-removal processes to obtain clean water, generally placed at the end of a treatment train so that the effluent has low suspended solids before their use (Ledda et al., 2013; Vaneeckhaute et al., 2017). It can be useful for concentration of the total ammoniacal nitrogen (TAN) captured by acid scrubbers with TAN retentions up to 98.1% (Fu et al., 2011). Compared with the volatile free ammonia (NH_3) that readily diffuses through RO membranes with low retention efficiency (10 to 40%), the ionized molecule (NH_4^+) is more easily retained by membranes because it complexes with anions in manure (Massé et al., 2008); for this reason, an acidified manure at pH < 6.5 is needed to obtain high (> 99%) TAN retention. Reverse osmosis has been used to separate N and potassium (K) in swine manure digestates following solid–liquid separation and ultrafiltration (Ledda et al. (2013): the NH_4–N concentration after ultrafiltration (1852 mg kg^{-1}) was reduced to 72 mg kg^{-1} in the RO permeate and increased to 7263 mg kg^{-1} in the RO concentrate stream. Similarly,

K concentration (2230 mg kg^{-1}) was reduced to 41 mg kg^{-1} in the RO permeate and increased to 7685 mg kg^{-1} in the concentrate stream.

Ion Exchange–Zeolites

Zeolites are crystalline, hydrated aluminosilicates of alkali and alkaline earth cations, having infinite, 3-D structures. They possess various properties (adsorption, cation-exchange, dehydration–rehydration, and catalysis properties), which contribute to a high variety of applications (Mumpton, 1999). For example, zeolite has been used in water and wastewater treatment as well as in animal waste treatment due to its high affinity and selectivity for NH_4^+ ions. Addition of 6.25% zeolite to dairy slurry resulted in a 50% reduction in NH_4^+ volatilization (Lefcourt and Meisinger, 2001), and application of zeolite during storage plus soil application of swine manure reduced NH_4^+ volatilization between 65 and 71% (Portejoie et al., 2003). Regarding poultry manure, application of a layer of zeolite placed on the surface of composting poultry litter reduced NH_4^+ emissions by 44% (Kithome et al., 1999), and Li et al. (2006) in laboratory experiments reported a reduction of 81% of NH_4^+ emission from stored poultry manure with the application of zeolite at a 5% (w/w) rate. Other authors studied addition of zeolite to cattle manure (2% rates) to counteract the inhibitory effect of NH_3 during AD (Borja et al., 1996), and as air scrubber packing material and as a filtration agent in deep-bedded cattle housing (Milan et al., 1999).

Biological Nutrient Removal Processes

Nitrification–Denitrification

The aim of the nitrification–denitrification process is to transform NH_4^+ into innocuous N gas (N_2). Nitrification is a very limiting process in animal waste treatment but a necessary process to be able to remove large amounts of reactive N using biological nitrification–denitrification systems. The effectiveness of the biological nitrification–denitrification process depends on the ability of nitrifying organisms to oxidize ammonium ions (NH_4^+) to nitrite (NO_2^-) and nitrate (NO_3^-). Subsequent reduction to molecular N, denitrification, is rapid with available carbonaceous substrate and an anaerobic environment. Nitrifying bacteria needs oxygen, low organic carbon, favorable temperature, and a growth phase before sufficient numbers are present for effective nitrification (Vanotti and Hunt, 2000). In the absence of enriched nitrifying populations, aerobic treatment of manure can potentially add to environmental problems by stripping ammonia into the atmosphere. For example, when enriched nitrifying populations were added to swine lagoon wastewater, complete NH_4 conversion to oxidized forms was achieved in 14 h with 0% loss of N. However, aeration without addition of nitrifiers resulted in delayed start of nitrification (10 d) and 70% loss of the N (Vanotti and Hunt, 2000). Low nitrification rates during cold weather are also a problem for adoption of biological treatment of NH_4 in livestock effluents because treatment needs to be provided year around and external heating of large tanks in farms is usually unaffordable. This problem has been circumvented by the discovery of a high-performance nitrifying sludge (HPNS) for effective NH_4 removal performance during cold weather conditions (Vanotti et al., 2013a). The unique microbial community composition that contained NH_4 oxidizers, cold tolerant, and floc-forming microorganisms provided a nitrifying sludge capable of very high rates of nitrification at temperatures as low as 5 °C (Ducey et al., 2010). The

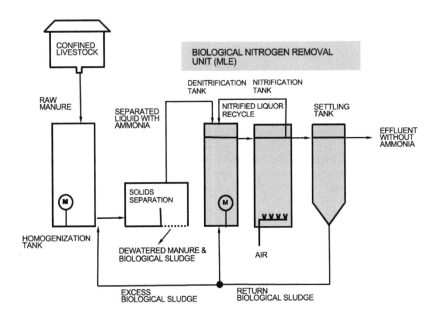

Fig. 7. Biological nitrogen removal after solid-liquid separation using a modified Ludzack-Ettinger (MLE) process with a pre-anoxic configuration that uses endogenous manure carbon for denitrification though an internal recycle that feeds nitrate to the denitrification tank directly from the aerobic zone (adapted from Vanotti et al., 2009b).

HPNS was used for rapid start-up of full-scale plants for swine manure treatment. For example, a 230 m^3 nitrification tank seeded with 1 L of HPNS reached, in 40 d, the target removal rate (95 kg NH_4–N per day) needed to nitrify the ammonia generated by 5200 pigs (Vanotti et al., 2018a). A denitrification tank can be incorporated into the treatment system in fluid connection with a nitrifying tank to provide total N removal. Carbon internal requirements during denitrification are of approximately 6.0 kg COD-manure per kg NO_3–N and 3.5 kg COD-manure per kg NO_2–N. A modified Ludzack-Ettinger (MLE) process (Tchobanoglous et al., 2003) is effective utilizing the endogenous carbon contained in the manure for effective denitrification without external carbon addition. In this process, the nitrified wastewater is continually recycled into the denitrification tank using a pre-denitrification configuration (Fig. 7). In the denitrification tank, denitrification bacteria use the soluble carbon in the separated manure liquid to transform NO_2 and NO_3 into N_2 gas. The MLE process removed 87 to 94% of COD, 95% of TKN and 90 to 98% of NH_4–N in full-scale systems treating separated swine manure in North Carolina (Vanotti et al. (2014) and 2018a). Another way to perform both nitrification and denitrification is to use just one tank and intermittent aeration. Riaño and García-González used the intermittent aeration approach (80 min with aeration and 40 min without aeration) in a full-scale, on-farm treatment plant in Castilla y Leon region, Spain also treating separated manure liquid. This nitrification–denitrification approach was also effective: it removed 84.5% of COD, 95.9% of TKN, and 98.0% of NH_4–N. Another low-cost way to achieve effective denitrification is to recirculate nitrified effluent into the barns so that it reacts with the carbon in fresh manure in the barn pits (Kunz et al., 2012; Vanotti et al., 2007). Besides manure carbon, plants can also provide sufficient carbon for denitrification in farm settings. Poach et al. (2003) combined nitrification of swine lagoon wastewater with constructed wetlands planted with soft-stem bulrush (*Schoenoplectus tabernaemontani*) and obtained total N removals of 85%.

Deammonification (Anammox)

The discovery of the anaerobic ammonium oxidation (anammox) as a new pathway in the N cycle to biologically transform NH_4^+ into N_2 (Mulder et al., 1995) has created great expectations for the treatment of livestock wastewater because it significantly reduces aeration needs and cost of treatment. Compared with the conventional nitrification–denitrification (NDN) process using for biological N removal, the anammox pathway can save more than 50% of the oxygen supply and 100% of the external organic C source for denitrification (Fig. 8).

The anammox process is especially suitable for the removal of N from wastewaters containing high ammonium and low biodegradable organic carbon, such as digestate effluents after waste-to-energy conversion by anaerobic digestion (Magrí et al., 2013). A C/N ratio of the effluent below 0.8 favors the anammox process over traditional denitrification process (Zhang and Jahng, 2010. The anammox process consists of a chemolithoautotrophic bioconversion mediated by Planctomycetes-like bacteria that under anoxic conditions oxidize NH_4^+ using nitrite (NO_2^-) as the electron acceptor (Fig. 8). According to the anammox reaction proposed by Strous et al. (1998) (Eq. [11]), NH_4^+ and NO_2^- are converted to N_2 and nitrate (NO_3^-) under stoichiometric molar ratios of 1.00:1.32:0.26:1.02 for NH_4^+ consumption, NO_2^- consumption, NO_3^- production and N_2 production, respectively.

$$\text{Anammox: } 1\,NH_4^+ + 1.32\,NO_2^- + 0.066\,HCO_3^- + 0.13H^+ \rightarrow 1.02\,N_2 +$$
$$0.26NO_3^- + 0.066\,CH_2O_{0.5}N_{0.15} + 2.03H_2O \qquad [11]$$

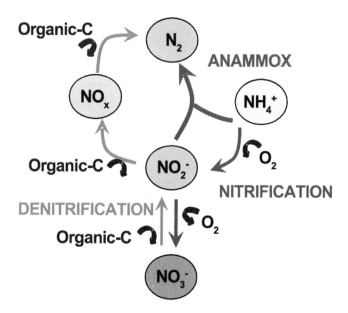

Fig. 8. Diagram showing a new shortcut for the biological removal of nitrogen using partial nitrification and anaerobic ammonium oxidation (anammox).

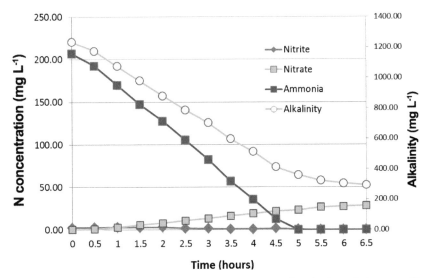

Fig. 9. Biological N removal of digested swine wastewater using single-tank deammonification process combining partial nitritation and anammox. Stoichiometry obtained was:
$NH_4^+ + 0.87\ O_2 \rightarrow 0.45\ N_2 + 0.11\ NO_3^- + 1.41\ H_2O + 1.18\ H^+$ (data from Vanotti et al., 2012).

Partial aerobic oxidation of some NH_4^+ to NO_2^- (partial nitritation) is needed
to produce a suitable influent for the anammox process. According to Eq. [11],
partial nitritation with a conversion efficiency of about 57% (Eq. [12]) is required
to make NO_2^- available to anammox in the right proportion.

> Partial nitritation: $1\ NH_4^+ + 0.85\ O_2 \rightarrow 0.43\ NH_4^+ + 0.57\ NO_2^- + 0.57$
> $H_2O + 1.14\ H^+$ (12)

Deammonification process combines partial nitritation process with anammox process, both working together in a double stage configuration using two
reactors (Furukawa et al., 2009; Magrí et al., 2012b; Qiao et al., 2010) or a single-stage
configuration with partial nitritation and anammox occurring in a single reactor
(Vanotti et al., 2012, De Prá et al., 2016). The overall nitrogen removal reaction by
deammonification process (partial nitritation + anammox) is described by Eq. [13].

> Deammonification: $1\ NH_4^+ + 0.85\ O_2 \rightarrow 0.44\ N_2 + 0.11\ NO_3^- + 1.43\ H_2O$
> $+ 1.1\ H^+$ (13)

In the United States, a novel anammox bacterium strain was discovered
(Candidatus *Brocadia caroliniensis*) that oxidizes NH_4^+ and releases N_2 gas under
anaerobic conditions at the stoichiometric ratios summarized as $NH_4^+ + 1.30\ NO_2^-$
$\rightarrow 1.06\ N_2 + 0.18\ NO_3^-$ (Vanotti et al., 2013b). This anammox bacterium was isolated
from livestock manure sludge and can be preserved long-term via subzero
freezing and lyophilization and reactivated rapidly to facilitate plant start-up
(Rothrock et al., 2011) or immobilized in PVA gel carriers to facilitate biomass
retention inside the reactor (Magrí et al., 2012a). Figure 9 shows results obtained
with digested swine wastewater (anaerobic lagoon effluent) using deammonification process in a single reactor. Deammonification process consumed similar
carbonate alkalinity than the conventional NDN process, at a rate of 4.2 kg alkalinity per kg N removed (vs 3.6 for NDN), but significantly reduced the oxygen

needs by 57% (0.87 mol O_2 per mol NH_4^+ vs. 2.0 for NDN) and eliminated the organic carbon requirement by 100%. This leads to the development of new anammox-based treatment for livestock wastewater that are more energy efficient with a significant decrease in operational costs.

Enhanced Biological Phosphorus Removal

An efficient biological method that has been successfully used to remove P from wastewater is the *enhanced biological phosphorus removal* (EBPR) process. This process has been mainly applied to municipal wastewaters (He et al., 2008; Brdjanovic et al., 1996) but there are some examples of application to removal of P from a variety of animal manures, such as dairy manure (Liu et al., 2014), pre-fermented dairy manure (Yanosek et al., 2003; Güngör et al., 2009), swine wastewater (Tilche et al., 1999; Ra et al., 2000) and digested swine manure (Obaja et al., 2003).

The EBPR process promotes the removal of P from wastewater without the need for chemical precipitants; the P is removed by a group of microorganisms that are known as the polyphosphate accumulating organisms (PAOs). These bacteria store P, removing it from the liquid fraction of the wastewater.

The P uptake and poly-P storage in PAO organisms occur under a particular set of operational conditions where intermittent aeration promote P sequestering and release depending on the medium conditions. Under anaerobic stage a significant release of phosphate is observed, while in the following aerobic stage an even greater amount of phosphate is taken up by the organisms, thereby also removing the phosphate in the incoming wastewater (Blackall et al., 2002), which is called luxury uptake of P. Biomass is then separated from the wastewater resulting in a lower P concentration in the liquid fraction and higher P concentration in the solid (sludge) fraction (Yanosek et al., 2003).

The PAOs can take up carbon sources such as volatile fatty acids (VFAs) under anaerobic conditions and store them intracellularly as poly-β-hydroxyalkanoates (PHAs) (Oehmen et al., 2007). The energy for this PHA accumulation is mainly generated by the cleavage of poly-P and release of phosphate from the cell, and reducing power, that is also required for PHA formation, is produced through the glycolysis of internally stored glycogen (Mino et al., 1998). Therefore, intracellular PHA levels increase in parallel with the assimilation of VFAs and

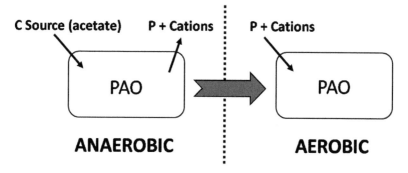

Fig. 10. Changes thought to take place during the aerobic and anaerobic stages of the enhanced bioprocess P removal (EBPR). Polyphosphate accumulating organisms (PAOs) take up organic carbon (C source) and release P and cations in the anaerobic environment. In turn, PAOs take up soluble P and cations under aerobic conditions (Karunanithi et al., 2015).

the release of P to the wastewater (Fig. 10). Under aerobic conditions PAOs are able to use their stored PHA as the energy source for biomass growth, glycogen formation, P uptake and polyphosphate storage (Oehmen et al., 2007) (Fig. 10).

The EBPR process must be monitored for optimal performance. It may suffer deterioration in performance and even failures due to external disturbances, operational factors of the process, and/or microbial competition between PAOs and a group of organisms known as the glycogen accumulating organisms (GAOs) that compete for VFA during the aerobic process. External disturbances causing poor performance include excessive rainfall, shortage of potassium (Brdjanovic et al., 1996), excessive aeration (Brdjanovic et al., 1998), high nitrate loading to the anaerobic zone (Kuba et al., 1994) and nutrient limitation. Operational factors such as pH (Filipe et al., 2001), temperature (Panswad et al., 2003), organic loading rate (OLR) (Ahn et al., 2007) and anaerobic-aerobic contact time (Wang and Park, 2001) are reported.

Vegetative and Algal Nutrient Removal Systems

Constructed Wetlands

When land and demographic conditions are limiting for land application of liquid manure effluents, constructed wetlands are an alternative to management of livestock wastes that could reduce the contamination hazard of soil, air, and water resources. Constructed wetlands, as a component of an on-farm total waste management system, are less land intensive than soil application of manure slurries or livestock wastewater (Humenik et al., 1999; Harrington and McInnes, 2009; Vidal et al., 2018). Although natural wetland systems are used for municipal treatment (Kadlec and Wallace, 2008), they are not considered to be constructed wetlands and cannot be used for livestock manure treatment. Nevertheless, constructed wetlands have become widespread worldwide as a cost-effective alternative for wastewater treatment (Stone et al., 2002; Kadlec and Wallace, 2008). Several types of constructed wetlands can be used to treat animal wastewater and feedlot runoff: free water surface flow, horizontal subsurface flow, free-floating, and vertical-flow systems (NRCS, 2009). The two most prevalent types of treatment wetlands are free water surface and horizontal subsurface flow wetlands (Kadlec, 2009). Surface flow constructed wetlands are the most commonly used wetlands for treating liquid animal manure and the type that the USDA Natural Resources Conservation Service currently recommends in the technical requirements of Conservation Practice Standard 656 for constructed wetland (NRCS, 2016). Surface flow wetlands have the following advantages: (i) ability to efficiently treat the high-strength manure associated with the discharge from animal lagoons and other pretreatment facilities; (ii) relatively low construction costs compared with subsurface systems; (iii) relative ease of management; and (iv) ease of repair and maintenance if problems occur. Subsurface flow constructed wetlands contain gravel or soil media to form a porous bed through which the wastewater passes in a horizontal direction, below the surface of the porous bed. In cold climates, subsurface flow wetlands have the advantage of providing insulation from the cold air because treatment occurs below the porous bed's surface. In addition, these wetlands have virtually no odors and mosquitoes. Although subsurface flow wetlands efficiently remove biodegradable pollutants from wastewater, a major disadvantage is their potential for plugging when used for treatment of livestock wastewater, which usually contain high solids concentrations.

A significant understanding exists on the role of plants, soil, water, and microbial processes that affect nutrient removal from livestock wastewater in wetland treatment systems (Hunt and Poach, 2001; Szögi et al., 2003; Novak et al., 2008; Reddy et al., 2016). For N treatment, in constructed wetlands, the role of the soil substrate is important to the overall constructed wetland function of enhancing water quality since the soil is the supporting medium or substrate for vegetation, habitat for microbes involved in N cycling, and transitional storage of organic and inorganic N (Szögi et al., 2004a). When N-rich livestock wastewater is applied to constructed wetlands, the major expected removal mechanism of N is nitrification–denitrification and to a lesser extent, plant uptake and ammonia volatilization (Hunt et al., 2009; Szögi et al., 2003; Poach et al., 2004). However, the magnitude of these mechanisms is affected by wetland operational parameters such as N form and loads, water depth, retention time, and environmental conditions such as temperature and reduced soil conditions (Kadlec and Wallace, 2008). For instance, high NH^+_4–N concentrations (> 200 mg L^{-1}) can negatively affect plant growth and effectiveness of wetlands built to treat livestock wastewater (Clarke and Baldwin, 2002; Wang et al., 2016). Earlier research on constructed wetlands for dairy wastewater treatment reported by Newman et al. (1999) indicated that N was removed mainly by sedimentation but very little by denitrification, possibly because of solids and N overloading along with ammonia accumulation. To avoid overloading the wetland system with organic C, N, and solids the livestock manure effluents usually require pretreatment by lagoons, vegetative filters, or solids separators. However, Hunt et al. (2002) reported that 87% of the N was removed from constructed wetlands operated at shallow water depth (< 10 cm) receiving anaerobic swine lagoon liquid at N loading rates as high as 25 kg ha^{-1} d^{-1}. According to Hunt et al. (2003) denitrification was likely the major mechanism to reduce N in these constructed wetlands systems that treated swine lagoon wastewater since only 7 to 15% ammonia volatilization occurred during summer conditions (Poach et al., 2002).

Phosphorus is one of the nutrients most difficult to remove from livestock wastewater. Constructed wetlands have the capacity to retain P but its retention in wetland systems is often temporary. Removal of inorganic P from the water column depends on plant uptake, sedimentation, microbial immobilization, soil substrate sorption capacity, and pH and oxidation–reduction potential conditions of the flooded soil substrate (Kadlec and Wallace, 2008). In constructed wetlands, flooded soils at pH < 7, inorganic P is retained in phosphate form by ferric-iron oxides, but oxidation–reduction conditions affect their retention capacity (Szögi et al., 2004a). Under reducing conditions, the ferric iron oxides solubilize releasing P into the water column. Although low P removal rates were reported for constructed wetlands treating livestock effluents (Stone et al., 2004; Szögi et al., 2000; 2004a), a number of investigators reported high P removal rates > 50% are possible in treatment wetlands using a variety of substrates with high P sorption capacity (Westholm, 2006; Vohla et al., 2011).

Constructed wetlands alone cannot remove sufficient N and P from liquid raw manure to meet both the farm nutrient management requirements and land area reduction for safe terminal application. However, placing one or more pretreatment components such as anaerobic lagoon, vegetative filters, or solids separators prior to the wetlands can minimize solids deposition and increase the N and P removal efficiency of the wetlands. Long term, effective removal of P may

Table 7. Removal of nitrogen and phosphorus from manures using microalgae

Type of animal waste	TN initial concentration	TN removed	TP initial concentration	TP removed	Main microalgae	Reference
	(mg L⁻¹)	(%)	(mg L⁻¹)	(%)		
Digested swine manure	303–495†	64–94				González-Fernández et al., 2011
Raw swine manure	324–569†	58–95				González-Fernández et al., 2011
Treated effluent of swine manure	33‡	83	47§	58	*Chlorella sorokiniana*	Hernández et al., 2013
Digested swine manure	132–689†	88–99	5.3–30.4§	54–80	*Oocystis* sp., *Clorella* sp. and *Protderma* sp.	Molinuevo-Salces et al., 2010
Digested manure	100–240	76–83	15–30	63–75	*Chlorella* sp.	Wang et al., 2010
Raw swine manure	59–370‡	55–90			Protoderma sp, Achnanthes sp, Nitzschia sp, Oocystis sp.	de Godos et al., 2009
Digested dairy manure	225	39	25	51	Benthic algae	Wilkie & Mulbry, 2002
Raw dairy manure	1210	60	303	93	Benthic algae	Wilkie & Mulbry, 2002
Dairy industry wastewater	36.3	> 90	112	20–55	*Scenedesmus dimorphus*	González et al., 1997
Dairy industry wastewater	3–36	30–95	112	20–55	*Chlorella vulgaris*	González et al. 1997

† Initial concentration and removal of N-NH4+ (mg L⁻¹)

‡ Initial concentration and removal of total Kjeldahl nitrogen (mg L⁻¹).

§ Initial concentration and removal of soluble P (mg L⁻¹).

require the utilization of soil substrates that are low cost and have high P sorption capacity under flooded conditions.

Vegetative Filters

A Vegetative Treatment Area (VTA) is a strip of perennial herbaceous vegetation established to which agricultural wastes and wastewaters are applied as sheet flow. VTAs are standard components of planned agriculture waste management systems to manage runoff from open lot livestock facilities, usually used after a settling basin (USDA/NRCS, 2008, 2015). Such combination treats runoff by settling, infiltration, and nutrient use. Biological nitrification/denitrification and plant uptake are important processes involved in reduction of N in VTAs. The P in feedlot runoff is adsorbed to solid particles and therefore its removal is related to solid removal efficiencies (Koelsch et al., 2006). Mass removal of N and P can be significant. A summary of VTA performance treating beef feedlot runoff (*n* = 5) and milking and dairy feedlot runoff (*n* = 2) reported by Koelsch et al. (2006) shows total N and total P mass removals averaging approximately 73 ± 32% and 65 ± 35%, respectively.

Vegetative Treatment Area can also be used to remove N from swine lagoon effluents. They can remove large amounts of N per unit area. Szögi et al. (2004b) tested an overland flow system using a 20-cm depth layer of loamy sand soil lined with plastic that was planted with a mixture of fescue (*Festuca arundinacea* Schreb.), Bermuda grass *(Cynodon dactylon* [L.] Pers.) *and* reed canarygrass (*Phalaris arundinacea* L.). The system received very high N application rates per hectare, approximately 64 to 99 kg N ha⁻¹ d⁻¹. Under these conditions, total N removal efficiencies were 36 to

42% providing an annual N removal capacity of about 8400 to 15,100 kg N per hectare mostly due to nitrification and denitrification processes taking place in the soil.

Algae

Microalgae are microscopic photosynthetic organisms that inhabit the illuminated zone of both marine and fresh water bodies, including polluted environments with high concentration of nutrients such as nitrogen (N) and phosphorous (P). In a natural environment, microalgae can liberate oxygen to the water that is used by heterotrophic bacteria to breakdown organic matter, assimilating other nutrients as N and P for biomass production. These microalgal-based systems have been studied to eliminate pollutants from water, especially from urban wastewater, although they have also been used to treat animal waste as dairy manure (Wilkie and Mulbry, 2002; Mulbry et al., 2008), anaerobically digested dairy manure (Mulbry et al., 2008), swine manure (González-Fernández et al., 2011; de Godos et al., 2009), anaerobically digested swine manure (González-Fernández et al., 2011) and the liquid treated fraction of swine manure (Hernández et al., 2013).

In these symbiotic systems, biodegradable organic carbon is removed by bacteria (producing CO_2 and H_2O) and inorganic carbon is mainly removed by microalgae. In the case of N, it is mainly assimilated into algal-bacteria biomass forming proteins, but also by stripping if pH increases in the system (which is very common). According to Oswald (1988), complete N assimilation can be achieved at C/N/P ratios of 100/18/2. Phosphorous removal is also assimilated in form of biomass, but it can also precipitate due to the increase in pH (as a result of CO_2 removal) and the presence of some ions like Ca^{2+}, forming hydroxyapatite (Ruiz-Martínez et al., 2015).

In algal-bacterial systems, biomass growth and pollutant removal will depend on a combination of environmental parameters as temperature, pH, irradiance, concentration of CO_2 and O_2, and inhibitory compounds (Muñoz and Guieysse, 2006); but also on operational parameters such as mixing, HRT, and light penetration. Table 7 shows some results of nutrients removal when treating animal waste.

Agronomic Utilization of Recovered Nutrients

Aside from the technologies necessary for isolation, extraction, and recovery of manure nutrients, one must also consider the response of agronomically important plants to those minerals. Recovery of phosphorus in the form of salts or phosphate crystals has been a major area of focus for recovery of manure nutrients. Recovered magnesium and calcium phosphates are sparingly soluble in water such that dissolved phosphate quickly form bonds with soil particles reducing soluble P loss in surface runoff (Rahman et al., 2011; Szögi et al., 2012). Although recovered P materials have lower solubility in water than commercial P fertilizers, the relative fertilizer effectiveness of recovered phosphates can be similar to commercial fertilizers in terms of plant uptake (Cabeza et al., 2011). An experiment by Szögi et al. (2010), evaluated the biomass production of ryegrass (*Lolium multiflorum* Lam.) of calcium phosphate recovered from poultry litter relative to commercial triple superphosphate (TSP), and raw broiler litter, at increasing P application rates of 0, 22, 44, 88, and 176 mg P per kg soil. Only at the 176 mg P kg^{-1} rate the TSP treatment had higher biomass production than both recovered P and broiler litter. Otherwise, the differences in ryegrass biomass production were not significant for all three P sources. In a similar bioassay experiment, Bauer et al. (2007) evaluated the effectiveness of calcium phosphate recovered from liquid swine manure (Vanotti et al., 2007) compared with commercial TSP for ryegrass P uptake. The recovered calcium phosphate contained 26.2% P_2O_5 and was 0.57% soluble in water and 99.3% citrate-soluble. The recovered calcium phosphate was grounded and pelletized in two particle sizes commonly used in fertilizer applications: 0.5 to 1.0 mm and 2.0 to 4.0 mm. They were compared with commercial TSP at increasing P application rates of 0, 22, 44, 88, and 176 mg P per kg soil. Although the total P uptake increased linearly with application rate for all three fertilizer materials, the recovered phosphate with the smaller particle size resulted in 69% the P uptake of TSP, and the material with the larger particle size resulted in 6% P uptake equivalent after seven weeks of ryegrass growth. Therefore, where a readily available P fertilizer is needed, the finer recovered calcium phosphate material (0.5–1.0 mm) is the best choice. On the other hand, coarse pellets (2.0–4.0 mm) are better suited for applications that require slow release fertilizer. In another study reported by Massey et al. (2009), when struvite recovered from processing of dairy manure was applied to hard red spring wheat (*Triticum aestivum* L., 'Zeke') at a rate of 90 kg ha[-1], plant P concentration was increased in comparison to the negative control, but not when applied at 45 kg ha[-1]. Similarly, wheat subjected to 90 kg ha[-1] struvite or 45 or 90 kg ha[-1] of a mixed dairy product containing fluorapatite, magnesium phosphate, and sand had DM yield greater than the negative control and like those observed with the application of triple super phosphate (Massey et al., 2009). Further studies by Ehmann et al. (2017) evaluated the recovery of phosphorus as phosphate salts from the liquid fraction of swine manure obtained from the BioEcoSIM process. When subjected to up to 0.5 g of phosphate salts from swine manure (up to 200% P requirement), germination of cress (gardencress pepperweed; *Lepidium sativum* L.) was lowered by 27%, but germination of barley (*Hordeum vulgare* L., 'Grace') was increased by 30% (Ehmann et al., 2017). However, DM yield of both cress and barley was increased when grown in the presence of up to 1.5 g phosphate salt (Ehmann et al., 2017).

Biosolids, treated from municipal waste treatment facilities, can have application in the field of agronomy. When tall fescue (*Schedonorus arundinaceus* [Schreb.] Dumort., nom. cons., 'AU Triumph') was grown in the presence of heat-dried or dewatered biosolids at either 622 or 933 kg N ha[-1], DM yield was similar to or greater than the application of 34–0-0 at 202 kg N ha[-1] (Cogger et al., 1999). However, application of 933 kg N ha[-1] was required to see similar results in the application of biosolids to perennial ryegrass (*Lolium perenne* L. ssp. *perenne* 'Prana'; Cogger et al., 1999). Growth of wheat, when measured as plant height or DM weight, was approximately half when treated with struvite-precipitated urine than compared with controls, indicating a slower release of fertilizer nutrients (Ganrot et al., 2007).

Of interest in recent years has been the use of biochar for agronomic applications. When hen compost (poultry manure) was subjected to pyrolysis, the resulting char contained 2.0% N, 13.9% P_2O_5, and 11.0% K_2O, while the ash contained 2.9% N, 13.4% P_2O_5, and 11.2% K_2O (Kaneko et al., 2014). Komatsuna (tender green or Japanese mustard spinach; *Brassica perviridis* [L.H. Bailey] L.H. Bailey) grown in the presence of hen compost ash and char had similar plant heights (21.6 and 24.8 cm, respectively) when compared with mineral fertilizer (22.7 cm), while DM yield was decreased in the presence of ash (Kaneko et al., 2014). Application of biochar from animal manure has been shown to result in biomass increases in corn of up to 43% (Rajkovich et al.,

2012). Hunt et al. (2013) showed that manure biochars (chicken, dairy, beef, turkey, and swine) produced ryegrass DM yields similar to or greater than chemical fertilizer ($CaHPO_4$) at similar P application rates (50 mg P_2O_5 kg^{-1} soil). They found that manure biochars created using 350 °C process temperatures had higher P uptake than those created at 700 °C. Therefore, the processing temperature affected the fertilizer effectiveness of the biochar. An additional advantage of this process is the destruction of pathogens due to high temperatures.

Plant uptake of recovered nutrients can differ from similar application of a commercial mineral fertilizer. Barley's uptake of nitrogen (53.5%) and potassium (41.7%) from phosphate salts was greater than that from mineral fertilizer (38.1 and 23.6%, respectively) when grown on clay soils, though phosphorus uptake was decreased (10.2 versus 15.5%; Ehmann et al., 2017). Phosphorus uptake from algal turf scrubbers from dairy manure were greater than from mineral fertilizer when evaluated in cucumber (*Cucumis sativus* L.) and corn seedlings (*Zea mays* L.) (Mulbry et al., 2005). Nitrogen uptake in tall fescue (*Festuca arundinacea* Schreb.) was greater from the application of heat-dried or dewatered biosolids when compared with the application of 34–0-0 (Cogger et al., 1999).

Conclusions

Solid–liquid separation of the raw or digested manure improves the range of opportunities for removal and recovery of nitrogen (N) and phosphorus (P). Solid-liquid separation of liquid manure has the potential to capture most of the organic N and P pools that are typically contained in small suspended particles. The separation efficiency is greatly increased using centrifugation, or flocculation with natural or synthetic polymers in combination with mechanical and gravity processes. Composting of the manure solids mixed with other residues can produce a high-quality product focused on specific agricultural markets such as soilless media for nursery crops and organic farming. The quality criteria are established based on nutrient content, humified organic matter, the maturity degree, hygienisation, and the absence of phytotoxic compounds. Thermochemical processes can also be applied to manure to concentrate nutrients in biochar, hydrochar, and ash residues. Phosphorus can be harvested from liquid dairy and swine manure by precipitating magnesium ammonium phosphate crystals (struvite). More efficient P recovery is achieved when inorganic levels of P are greatest, such as after anaerobic digestion, and when the interference of competing ions such as calcium is taken care using chelating agents or oxalic acid. Methods have been also developed to efficiently recover P from manures as calcium phosphates. A critical step in these methods is to first remove the carbonate interference before P precipitation with lime. This can be accomplished by addition of acids, or taking advantage of acid produced by microbes, or physical acidification. Iron and aluminum salts can also be used to remove P from wastewater, but with limitations of high dosage rates and low plant availability of the products. For nitrogen, ammonia emissions can be captured from the air in the animal barns using chemical and biological scrubbers. Stripping processes are also available for removing the ammonia from liquid manures absorbed in strong acids favored by high temperatures and pH. Gas-permeable membrane processes have been used to remove and recover N from liquid manure by submerging membrane manifolds in the liquid manure and capturing the ammonia before it escapes into the air. When carbonate alkalinity is present, low-rate aeration results in a higher pH of the manure and higher rate of N recovery by the membrane process. For air applications in poultry barns, the speed of ammonia recovery by gas-permeable membranes can be increased significantly with hydrated lime, that also provides litter disinfection. Reverse osmosis and nanofiltration processes have been used in combination with other solids and nutrient removal processes to obtain cleaner water from manure for reuse; they are placed at the end of a treatment train so that the effluent has low suspended solids before their use. Biological N removal requires effective nitrification that is achieved with enriched nitrifying populations adapted to manure environments. Total N removal is best accomplished utilizing the endogenous carbon contained in the manure for effective denitrification using an MLE process. For biological N removal in digestate effluents, the autotrophic deammonification process that combines partial nitritation and anammox process is especially suitable due to the low biodegradable organic carbon left after waste-to-energy conversion by AD. Polyphosphate accumulating microorganisms (PAOs) have been used in an enhanced biological P removal process to remove P from liquid manures without the need for chemical precipitants; but the process must be closely monitored for optimal performance. Finally, vegetative and algal nutrient removal systems can be used as components of an on-farm total waste management system for polishing the manure effluents and improve water quality. For example, constructed wetlands technology has been used for treatment of N in swine lagoon effluents using much higher N loading rates per surface area compared with traditional land application. Vegetative treatment areas can provide significant mass removal of N and P from the runoff after settling basins in open lot livestock facilities. In both systems, microbial denitrification in the soil is the main process that removes the N. Algal-based systems have been used to remove N and P from digested manures by assimilation of the nutrients into the biomass or precipitation. Aside from the technologies necessary for isolation, extraction, and recovery of manure nutrients, the agronomic efficacy of recovered nutrient materials for use as plant fertilizers is an important consideration in the selection of the best approach for N and P removal and recovery from manure. Other important considerations in this selection process are the predictability of nutrient concentration in the products, their compatibility with existing animal production facilities, and any additional ecosystem benefit their implementation could generate such as carbon credits and water quality credits.

References

Acharya, B., A. Dutta, S. Mahmud, M. Tushar, and M. Leon. 2014. Ash analysis of poultry litter, willow and oats for combustion in boilers. J. Biomass Biofuel. 1:16–26.

Adnan, A., F.A. Koch, and D.S. Mavinic. 2003. Pilot-scale study of phosphorus recovery through struvite crystallisation-II: Applying in-reactor supersaturation ratio as a process control parameter. J. Environ. Eng. Sci. 2:473–483. doi:10.1139/s03-048

Aguirre-Villegas, H., R.C. Larson, and M.D. Ruark. 2017. Solid-liquid separation of manure and effects on greenhouse gas and ammonia emissions. Univ. of Wisconsin-Extension Publication No: UWEX A4131-04 GWQ 076. University of Wisconsin-Madison, Madison, WI.

Ahn, C.H., H.D. Park, and J.K. Park. 2007. Enhanced biological phosphorus removal performance and microbial population changes at high organic loading rates. J. Environ. Eng. 133(10):962–969. doi:10.1061/(ASCE)0733-9372(2007)133:10(962)

AOAC International. 2000. Official methods of analysis. 17th ed. AOAC, Gaithersburg, MD.

Arnott, H.J. 1995. Calcium oxalate in fungi. In: S.R. Khan, editor, Calcium oxalate in biological systems. CRC Press, Boca Raton, FL. p. 73–111.

Barack, P. 2013. Phosphate recovery from acid phase anaerobic digesters. U.S. Patent No. 8568,590 B2. U.S. Patent & Trademark Office, Washington, D.C.

Barrow, J.T., H.H. Van Horn, D.L. Anderson, and R.A. Nordstedt. 1997. Effects of Fe and Ca additions to dairy wastewaters on solids and nutrient removal by sedimentation. Appl. Eng. Agric. 13(2):259–267. doi:10.13031/2013.21598

Battistoni, P., A. De Angelis, P. Pavan, M. Prisciandaro, and F. Cecchi. 2001. Phosphorus removal from a real anaerobic supernatant by struvite crystallization. Water Res. 35:2167–2178. doi:10.1016/S0043-1354(00)00498-X

Bauer, P.J., A.A. Szögi, and M.B. Vanotti. 2007. Agronomic effectiveness of calcium phosphate recovered from liquid swine manure. Agron. J. 99:1352–1356. doi:10.2134/agronj2006.0354

Bautista, J.M., H. Kim, D.H. Ahn, R. Zhang, and Y.S. Oh. 2011. Changes in physicochemical properties and gaseous emissions of composting swine manure amended with alum and zeolite. Korean J. Chem. Eng. 28(1):189–194. doi:10.1007/s11814-010-0312-6

Benyahia, S., H. Arastoopour, T.M. Knowlton, and H. Massah. 2000. Simulation of particles and gas flow behavior in the riser section of a circulating fluidized bed using the kinetic theory approach for the particulate phase. Powder Technol. 112(1-2):24–33. doi:10.1016/S0032-5910(99)00302-2

Bernal, M.P. 2017. Grand challenges in waste management in agroecosystems. Frontiers in Sustainable Food Systems 1:1–4. doi:10.3389/fsufs.2017.00001

Bernal, M.P., J.A. Alburquerque, and R. Moral. 2009. Composting of animal manures and chemical criteria for compost maturity assessment. A review. Bioresour. Technol. 100:5444–5453. doi:10.1016/j.biortech.2008.11.027

Blackall, L.L., G.R. Crocetti, A.M. Saunders, and P.L. Bond. 2002. A review and update of the microbiology of enhanced biological phosphorus removal in wastewater treatment plants. Antonie van Leeuwenhoek 81:681–691. doi:10.1023/A:1020538429009

Block, E. 1984. Manipulating dietary anions and cations for prepartum dairy cows to reduce incidence of milk fever. J. Dairy Sci. 67(12):2939–2948. doi:10.3168/jds.S0022-0302(84)81657-4

Bonmatí, A., and X. Flotats. 2003. Air stripping of ammonia from pig slurry: Characterisation and feasibility as a pre- or post-treatment to mesophilic anaerobic digestion. Waste Manag. 23:261–272. doi:10.1016/S0956-053X(02)00144-7

Borja, R., E. Sanchez, and M.M. Duran. 1996. Effect of the clay mineral zeolite on ammonia inhibition of anaerobic thermophilic reactors treating cattle manure. J. Environ. Sci. Health A 31:479–500.

Bouropoulos, N.C., and P.G. Koutsoukos. 2000. Spontaneous precipitation of struvite from aqueous solutions. J. Cryst. Growth 213:381–388. doi:10.1016/S0022-0248(00)00351-1

Bowers, K. 2006. A new concept for removal of phosphorus from dairy farm waste. Project Report to Small Business Innovation Research Program. Multiform Harvest Inc., Seattle, WA.

Bowers, K.E., and P.W. Westerman. 2005. Performance of cone-shaped fluidized bed struvite crystallizers in removing phosphorus from wastewater. Trans. ASAE 48(3):1227–1234. doi:10.13031/2013.18523

Brdjanovic, D., C.M. Hooijmans, M.C.M. Van Loosdrecht, G.J. Alaerts, and J.J. Heijnen. 1996. The dynamic effects of potassium limitation on biological phosphorus removal. Water Res. 30(10):2323–2328. doi:10.1016/0043-1354(96)00121-2

Brdjanovic, D., A. Slamet, M.C.M. Van Loosdrecht, C.M. Hooijmans, G.J. Alaerts, and J.J. Heijnen. 1998. Impact of excessive aeration on biological phosphorus removal from wastewater. Water Res. 32(1):200–208. doi:10.1016/S0043-1354(97)00183-8

Cabeza, R., B. Steingrobe, W. Römer, and N. Claassen. 2011. Effectiveness of recycled P products as P fertilizers, as evaluated in pot experiments. Nutr. Cycl. Agroecosyst. 91:173. doi:10.1007/s10705-011-9454-0

Cantrell, K.B., T. Ducey, K.S. Ro, and P.G. Hunt. 2008. Livestock waste-to-bioenergy generation opportunities. Bioresour. Technol. 99:7941–7953. doi:10.1016/j.biortech.2008.02.061

Cantrell, K.B., P.G. Hunt, M. Uchimiya, J.M. Novak, and K.S. Ro. 2012. Impact of pyrolysis temperature and manure source on physicochemical characteristics of biochar. Bioresour. Technol. 107:419–428. doi:10.1016/j.biortech.2011.11.084

Celen, I., and M. Turker. 2001. Recovery of ammonia as struvite from anaerobic digester effluents. Environ. Technol. 22:1263–1272. doi:10.1080/09593332208618192

Chastain, J.P. 2013. Solid-liquid separation alternatives for manure handling and treatment. USDA Natural Resources Conservation Service, Washington, D.C.

Chastain, J.P., M.B. Vanotti, and M.M. Wingfield. 2001. Effectiveness of liquid-solid separation for treatment of flushed dairy manure: A case study. Appl. Eng. Agric. 17(3):343–354. doi:10.13031/2013.6210

Chien, S., L. Prochnow, S. Tu, and C. Snyder. 2011. Agronomic and environmental aspects of phosphate fertilizers varying in source and solubility: An update review. Nutr. Cycling Agroecosyst. 89:229–255. doi:10.1007/s10705-010-9390-4

Chirmuley, D.G. 1994. Struvite precipitation in WWTPs: Causes and solutions. Water (Journal of the Australian Water Association) December: 21–23.

Clarke, E., and A.H. Baldwin. 2002. Responses of wetland plants to ammonia and water level. Ecol. Eng. 18:257–264. doi:10.1016/S0925-8574(01)00080-5

Codling, E.E., R.L. Chaney, and J. Sherwell. 2002. Poultry litter ash as a potential phosphorus source for agricultural crops. J. Environ. Qual. 31:954–961. doi:10.2134/jeq2002.9540

Cogger, C.G., D.M. Sullivan, A.I. Bary, and S.C. Fransen. 1999. Nitrogen recovery from heat-dried and dewatered biosolids applied to forage grasses. J. Environ. Qual. 28:754–759. doi:10.2134/jeq1999.00472425002800030004x

Daguerre-Martini, S., M.B. Vanotti, M. Rodriguez-Pastor, A. Rosal, and R. Moral. 2018. Nitrogen recovery from wastewater using gas-permeable membranes: Impact of inorganic carbon content and natural organic matter. Water Res. 137:201–210. doi:10.1016/j.watres.2018.03.013

Dastur, M.B. 2001. Investigation into the factors affecting controlled struvite crystallization at the bench-scale. M.S. Thesis, Department of Civil Engineering, University of British Columbia, Vancouver, B.C.

de Godos I., S. Blanco, P.A. García-Encina, E. Becares, R. Muñoz. 2009. Long-term operation of high rate algal ponds for the bioremediation of piggery wastewaters at high loading rates. Bioresour. Technol. 100:4332–4339. doi:10.1016/j.biortech.2009.04.016

De Prá, M.C., A. Kunz, M. Bortoli, L.A. Scussiato, A. Coldebella, M.B. Vanotti, and H.M. Soares. 2016. Kinetic models for nitrogen inhibition in ANAMMOX and nitrification process on deammonification system at room temperature. Bioresour. Technol. 202:33–41. doi:10.1016/j.biortech.2015.11.048

Desmidt, E., K. Ghyselbrecht, Y. Zhang, L. Pinoy, B. Van der Bruggen, W. Verstraete, K. Rabaey, and B. Meesschaert. 2015. Global phosphorus scarcity and full-scale P-Recovery techniques: A review. Crit. Rev. Environ. Sci. Technol. 45(4):336–384. doi:10.1080/10643389.2013.866531

do Nascimento, C.A.C., P.H. Pagliari, L.A. Faria, and G.C. Vitti. 2018. Phosphorus mobility and behavior in soils treated with calcium, ammonium, and magnesium phosphates. Soil Sci. Soc. Am. J. 82:622–631. doi:10.2136/sssaj2017.06.0211

Doyle, J.D., and S.A. Parsons. 2002. Struvite formation, control and recovery. Water Res. 36:3925–3940. doi:10.1016/S0043-1354(02)00126-4

Dube, P.J., M.B. Vanotti, A.A. Szögi, and M.C. García-González. 2016. Enhancing recovery of ammonia from swine manure anaerobic digester effluent using gas-permeable membrane technology. Waste Manag. 49:372–377. doi:10.1016/j.wasman.2015.12.011

Ducey, T.F., M.B. Vanotti, A.D. Shriner, A.A. Szögi, and A.Q. Ellison. 2010. Characterization of a microbial community capable of nitrification at cold temperature. Bioresour. Technol. 101(2):491–500. doi:10.1016/j.biortech.2009.07.091

Edwards, D.R., and T.C. Daniel. 1992. Environmental impacts of on-farm poultry waste disposal- A review. Bioresour. Technol. 41(1):9–33. doi:10.1016/0960-8524(92)90094-E

Ehmann, A., I.M. Bach, S. Laopeamthong, J. Bilbao, and I. Lewandowski. 2017. Can phosphate salts recovered from manure replace conventional phosphate fertilizer? Agriculture (Basel, Switz.) 7(1):1 doi:10.3390/agriculture7010001

EPA. 2014. National emission inventory-ammonia emissions from animal husbandry operations. U.S. Environmental Protection Agency, Washington, D.C. https://www3.epa.gov/ttnchie1/ap42/ch09/related/nh3inventorydraft_jan2004.pdf (Accessed 26 Mar. 2019).

Filipe, C.D., G.T. Daigger, and C.P. Grady. 2001. Stoichiometry and kinetics of acetate uptake under anaerobic conditions by an enriched culture of phosphorus-accumulating organisms at different pHs. Biotechnol. Bioeng. 76(1):32–43. doi:10.1002/bit.1023

Fordham, A.W., and U. Schwertmann. 1977. Composition and reactions of liquid manure (gulle), with particular reference to phosphate: III. pH-buffering capacity and organic components. J. Environ. Qual. 6(2):140–144. doi:10.2134/jeq1977.00472425000600020008x

Fu, G.M., T. Cai, and Y.B. Li. 2011. Concentration of ammoniacal nitrogen in effluent from wet scrubbers using reverse osmosis membrane. Biosystems Eng. 109:235–240. doi:10.1016/j. biosystemseng.2011.04.005

Furukawa, K., Y. Inatomi, S. Qiao, L. Quan, T. Yamamoto, K. Isaka, and T. Sumino. 2009. Innovative treatment system for digester liquor using anammox process. Bioresour. Technol. 100:5437–5443. doi:10.1016/j.biortech.2008.11.055

Gadd, G.M. 1999. Fungal production of citric and oxalic acid: Importance in metal speciation, physiology and biogeochemical processes. Adv. Microb. Physiol. 41:47–92. doi:10.1016/S0065-2911(08)60165-4

Gadd, G.M., J. Bahri-Esfahani, Q. Li, Y.J. Rhee, Z. Wei, M. Fomina, and X. Liang. 2014. Oxalate production by fungi: Significance in geomycology, biodeterioration, and bioremediation. Fungal Biol. Rev. 28:36–55. doi:10.1016/j.fbr.2014.05.001

Ganrot, Z., G. Dave, E. Nilsson, and B. Li. 2007. Plant availability of nutrients recovered as solids from human urine tested in climate chamber on Triticum aestivum L. Bioresour. Technol. 98:3122–3129. doi:10.1016/j.biortech.2007.01.003

Garcia, M.C., A.A. Szögi, M.B. Vanotti, J.P. Chastain, and P.D. Millner. 2009. Enhanced solid-liquid separation of dairy manure with natural flocculants. Bioresour. Technol. 100:5417–5423. doi:10.1016/j.biortech.2008.11.012

Garcia, M.C., M.B. Vanotti, and A.A. Szögi. 2007. Simultaneous separation of phosphorus sludge and manure solids with polymers. Trans. ASABE 50(6):2205–2215. doi:10.13031/2013.24096

García-González, M.C., and M.B. Vanotti. 2015. Recovery of ammonia from swine manure using gas-permeable membranes: Effect of waste strength and pH. Waste Manag. 38:455–461. doi:10.1016/j.wasman.2015.01.021

García-González, M.C., M.B. Vanotti, and A.A. Szögi. 2015. Recovery of ammonia from swine manure using gas-permeable membranes: Effect of aeration. J. Environ. Manage. 152:19–26. doi:10.1016/j.jenvman.2015.01.013

Gollehon, N.R., M. Caswell, M. Ribaudo, R. Kellogg, C. Lander, and D. Letson. 2001. Confined animal production and manure nutrients. Agriculture Information Bulletin No. (AIB-771). Washington, United States Department of Agriculture, Economic Research Service. https://www.ers.usda.gov/webdocs/publications/42398/17786_aib771_1_.pdf?v=0 (Accessed 27 Mar. 2019).

Gollehon, N.R., R.L. Kellogg, and D.C. Moffitt. 2016. Estimates of recoverable and non-recoverable manure nutrients based on the census of agriculture—2012 Results. USDA Natural Resources Conservation Services, Washington, D.C.

González, L.E., R.O. Cañizares, and S. Baena. 1997. Efficiency of ammonia and phosphorus removal from a Colombian agroindustrial wastewater by the microalgae Chlorella vulgaris and Scenedesmus dimorphus. Bioresour. Technol. 60:259–262. doi:10.1016/S0960-8524(97)00029-1

González-Fernández, C., B. Molinuevo-Salces, and M.C. García-González. 2011. Nitrogen transformations under different conditions in open ponds by means of microalgae-bacteria consortium treating pig slurry. Bioresour. Technol. 102(2):960–966. doi:10.1016/j.biortech.2010.09.052

Gonzalez Poncer, R., and D.M.E. Garcialopez. 2007. Evaluation of struvite as a fertilizer: A comparison with traditional P sources. Agrochimica 51:301–308.

Graham, J.P., and K.E. Nachman. 2010. Managing waste from confined animal feeding operations in the United States: The need for sanitary reform. J. Water Health 8:646–670. doi:10.2166/wh.2010.075

Güngör, K., M.B. Müftügil, J. Arogo, K.F. Knowlton, and N.G. Love. 2009. Prefermentation of liquid dairy manure to support biological nutrient removal. Bioresour. Technol. 100:2124–2129. doi:10.1016/j.biortech.2008.10.052

Guštin, S., and R. Marinsek-Logar. 2011. Effect of pH, temperature and air flow rate on the continuous ammonia stripping of the anaerobic digestion effluent. Process Saf. Environ. Prot. 89:61–66. doi:10.1016/j.psep.2010.11.001

Hadlocon, L.J., L. Zhao, R. Manuzon, and I. Elbatawi. 2014. An Acid Spray Scrubber for Recovery of Ammonia Emissions from a Deep-Pit Swine Facility. Trans. ASABE 57(3):949–960.

Hadlocon, L.J.S., R.B. Manuzon, and L. Zhao. 2015. Development and evaluation of a full-scale spray scrubber for ammonia recovery and production of nitrogen fertilizer at poultry facilities. Environ. Technol. 36(4):405–416. doi:10.1080/09593330.2014.950346

Han, L., K.S. Ro, K. Sun, H. Sun, Z. Wang, J.A. Libra, and B. Xing. 2016. New evidence for high sorption capacity of hydrochar for hydrophobic organic pollutants. Environ. Sci. Technol. 50:13274–13282. doi:10.1021/acs.est.6b02401

Hao, X., C. Wang, L.Y. Lan, and M.C.M. van Loosdrecht. 2008. Struvite formation, analytical methods and effects of pH and Ca2+. Water Sci. Technol. 58(8):1687–1692. doi:10.2166/wst.2008.557

Hao, X., C. Wang, M.C.M. van Loosdrecht, and Y. Hu. 2013. Looking beyond struvite for P-recovery. Environ. Sci. Technol. 47:4965–4966. doi:10.1021/es401140s

Harrington, R., and R. McInnes. 2009. Integrated constructed wetlands (ICW) for livestock wastewater management. Bioresour. Technol. 100:5498–5505. doi:10.1016/j.biortech.2009.06.007

Harris, W.G., A.C. Wilkie, X. Cao, and R. Sirengo. 2008. Bench-scale recovery of phosphorus from flushed dairy manure wastewater. Bioresour. Technol. 99:3036–3043. doi:10.1016/j.biortech.2007.06.065

He, S., A.Z. Gu, and K.D. McMahon. 2008. Progress toward understanding the distribution of accumulibacter among full- scale enhanced biological phosphorus removal systems. Microb. Ecol. 55(2):229–236. doi:10.1007/s00248-007-9270-x

Hernández, D., B. Riaño, M. Coca, and M.C. García-González. 2013. Treatment of agro-industrial wastewater using microalgae–bacteria consortium combined with anaerobic digestion of the produced biomass. Bioresour. Technol. 135:598–603. doi:10.1016/j.biortech.2012.09.029

Hilt, K., J. Harrison, K. Bowers, R. Stevens, A. Bary, and K. Harrison. 2016. Agronomic response of crops fertilized with struvite derived from dairy manure. Water Air Soil Pollut. 227:388. doi:10.1007/s11270-016-3093-7

Hjorth, M., K.V. Christensen, M.L. Christensen, and S.G. Sommer. 2010. Solid-liquid separation of animal slurry in theory and practice. A review. Agron. Sustain. Dev. 30:153–180. doi:10.1051/agro/2009010

House, W.A. 1999. The physico-chemical conditions for the precipitation of phosphate with calcium. Environ. Technol. 20(7):727–733. doi:10.1080/09593332008616867

Humenik, F.J., A.A. Szögi, P.G. Hunt, S. Broome, and M. Rice. 1999. Wastewater utilization: A place for managed wetlands. Asian-australas. J. Anim. Sci. 12:629–632. doi:10.5713/ajas.1999.629

Hunt, P.G., K.B. Cantrell, P.J. Bauer, and J.O. Miller. 2013. Phosphorus fertilization of ryegrass with ten precisely prepared manure biochars. Trans. ASABE 56:1317–1324.

Hunt, P.G., T.A. Matheny, and A.A. Szögi. 2003. Denitrification in constructed wetlands used for treatment of swine wastewater. J. Environ. Qual. 32:727–735. doi:10.2134/jeq2003.7270

Hunt, P.G., and M.E. Poach. 2001. State of the art for animal wastewater treatment in constructed wetlands. Water Sci. Technol. 44:19–25. doi:10.2166/wst.2001.0805

Hunt, P.G., K.C. Stone, T.A. Matheny, M.E. Poach, M.B. Vanotti, and T.F. Ducey. 2009. Denitrification of nitrified and non-nitrified swine lagoon wastewater in the suspended sludge layer of treatment wetlands. Ecol. Eng. 35:1514–1522. doi:10.1016/j.ecoleng.2009.07.001

Hunt, P.G., A.A. Szögi, F.J. Humenik, J.M. Rice, T.A. Matheny, and K.C. Stone. 2002. Constructed wetlands for treatment of swine wastewater from an anaerobic lagoon. Trans. ASAE 45:639–647.

Hummel, R., C. Cogger, A. Bary, and R. Riley. 2014. Marigold and pepper growth in container substrates made from biosolids composted with carbon-rich organic wastes. Horttechnology 24:325–333. doi:10.21273/HORTTECH.24.3.325

Informa Economics. 2013. National Market Value of Anaerobic Digester Products. Report prepared for Innovation Center for U.S. Dairy. American Biogas Council, Washginton, D.C. http://americanbiogascouncil.org/pdf/nationalMarketPotentialOfAnaerobicDigesterProducts_dairy.pdf (Accessed 28 Mar. 2019).

Jensen, J., C. Frear, J. Ma, C. Kruger, R. Hummel, and G. Yorgey. 2016. Digested fiber solids—Methods for adding value. WSU Extension Factsheet FS35E. Washington State University, Pullman, WA.

Jiang, A.P., T.X. Zhang, Q.B. Zhao, X. Li, S.L. Chen, and C.S. Frear. 2014. Evaluation of an integrated ammonia stripping, recovery, and biogas scrubbing system for use with anaerobically digested dairy. Biosystems Eng. 119:117–126. doi:10.1016/j.biosystemseng.2013.10.008

Kadlec, R.H. 2009. Comparison of free water and horizontal subsurface treatment wetlands. Ecol. Eng. 35:159–174. doi:10.1016/j.ecoleng.2008.04.008

Kadlec, R.H., and S. Wallace. 2008. Treatment wetlands. CRC press, Boca Raton, FL. doi:10.1201/9781420012514

Kaneko, K., L. Li, T. Shimizu, H. Matsumura, and T. Takarada. 2014. Fuel gas production and plant nutrient recovery from digested poultry manure. Jpn. Poult. Sci. 51:444–450. doi:10.2141/jpsa.0130184

Karunanithi, R., A.A. Szögi, N. Bolan, R. Naidu, P. Loganathan, P.G. Hunt, M.B. Vanotti, C.P. Saint, Y.S. Ok, and S. Krishnamoorthy. 2015. Phosphorus recovery and reuse from waste streams. Adv. Agron. 131:173–250. doi:10.1016/bs.agron.2014.12.005

Keyzer, M. 2010. Towards a closed phosphorus cycle. De Econ. 158:411–425.

Kirk, D.M., W.G. Bickert, S. Hashsham, and S. Davies. 2003. Chemical additions for phosphorus separation from sand free dairy manure. ASAE Paper No. 034122, ASABE, St. Joseph, MI.

Kithome, M., J.W. Paul, and A.A. Bomke. 1999. Reducing nitrogen losses during simulated composting of pultr manure using adsorbents or chemical amendments. J. Environ. Qual. 28(1):194–201. doi:10.2134/jeq1999.00472425002800010023x

Koelsch, R. K., J. C. Lorimor, and K.R. Mankin. 2006. Vegetative treatment systems for management of open lot runoff: Review of literature. University of Nebraska–Lincoln, Lincoln, NE.

Kozik, A., N. Hutnik, A. Matynia, J. Gluzinska, and K. Piotrowski. 2011. Recovery of phosphate (V) ions from liquid waste solutions containing organic impurities. Chemik 65(7):675–686.

Kuba, T., A. Wachtmeister, M. Van Loosdrecht, and J. Heijnen. 1994. Effect of nitrate on phosphorus release in biological phosphorus removal systems. Water Sci. Technol. 30(6):263–269. doi:10.2166/wst.1994.0277

Kunz, A., R. Steinmetz, S. Damasceno, and A. Coldebela. 2012. Nitrogen removal from swine wastewater by combining treated effluent with raw manure. Sci. Agric. 69(6):352–356. doi:10.1590/S0103-90162012000600002

Kunz, A., R.L.R. Steinmetz, M.A. Ramme, and A. Coldebella. 2009. Effect of storage time on swine manure solid separation efficiency by screening. Bioresour. Technol. 100:1815–1818. doi:10.1016/j.biortech.2008.09.022

Le Corre, K.S., E.V. Jones, P. Hobbs, and S.A. Parsons. 2005. Impact of calcium on struvite crystal size, shape and purity. J. Cryst. Growth 283:514–522. doi:10.1016/j.jcrysgro.2005.06.012

Ledda, C., A. Schievano, S. Salati, and F. Adani. 2013. Nitrogen and water recovery from animal slurries by a new integrated ultrafiltration, reverse osmosis and cold stripping process: A case study. Water Res. 47:6157–6166. doi:10.1016/j.watres.2013.07.037

Lefcourt, A.M., and J.J. Meisinger. 2001. Effect of adding alum or zeolite to dairy slurry on ammonia volatilization and chemical composition. J. Dairy Sci. 84(8):1814–1821. doi:10.3168/jds.S0022-0302(01)74620-6

Lei, X.H., N. Sugiura, C.P. Feng, and T. Maekawa. 2007. Pretreatment of anaerobic digestion effluent with ammonia stripping and biogas purification. J. Hazard. Mater. 145:391–397. doi:10.1016/j.jhazmat.2006.11.027

Lentz, Z.A., J. Classen, and P. Kolar. 2017. Thermochemical conversion: A prospective swine manure solution for North Carolina. Trans. ASABE 60:591–600. doi:10.13031/trans.12074

Li, H., H. Xin, R.T. Burns, and Y. Liang. 2006. Reduction of ammonia emission from stored poultry manure using additives: Zeolite, Al+clear, Ferix-3 and PLT. ASAE paper No. 064188, ASABE Annual International Meeting, Portland, OR.

Liao, P.H., A. Chen, and K.V. Lo. 1995. Removal of nitrogen from swine manure wastewaters by ammonia stripping. Bioresour. Technol. 54:17–20. doi:10.1016/0960-8524(95)00105-0

Libra, J.A., K.S. Ro, C. Kammann, A. Funke, N.D. Berge, Y. Neubauer, M.M. Titirici, C. Fühner, O. Bens, J. Kern, and K.H. Emmerich. 2011. Hydrothermal carbonization of biomass residual: A comparative review of the chemistry, processes and applications of wet and dry pyrolysis. Biofuels 2:71–106. doi:10.4155/bfs.10.81

Lim, T.T., A.J. Heber, F.Q. Ni, D.C. Kendall, and B.T. Richert. 2004. Effects of manure removal strategies on odor and gas emissions from swine finishing. Trans. ASAE 47:2041–2050. doi:10.13031/2013.17801

Liu, Z.H., A. Pruden, J.A. Ogejo, and K.F. Knowlton. 2014. Polyphosphate-and glycogen-accumulating organisms in one EBPR system for liquid dairy manure. Water Environ. Res. 86(7):663–671. doi:10.2175/106143014X13975035525302

Liu, Y.H., M.M. Rahman, J.H. Kwag, J.H. Kim, and C.S. Ra. 2011. Eco-friendly production of maize using struvite recovered from swine wastewater as a sustainable fertilizer source. Asian-Aust. J. Anim. Sci. 24:1699–1705.

Loganathan, P., M.J. Hedley, N. Grace, J. Lee, S. Cronin, N.S. Bolan, and J. Zanders. 2003. Fertiliser contaminants in New Zealand grazed pasture with special reference to cadmium and fluorine—a review. Soil Res. 41:501–532. doi:10.1071/SR02126

Ma, G., J. Shannon Neibergs, J.H. Harrison, and E.M. Whitefield. 2017. Nutrient contributions and biogas potential of co-digestion of feedstocks and dairy manure. Waste Manag. 64:88–95. doi:10.1016/j.wasman.2017.03.035

Magrí, A., F. Béline, and P. Dabert. 2013. Feasibility and interest of the anammox process as treatment alternative for anaerobic digester supernatants in manure processing–An overview. J. Environ. Manage. 131:170–184. doi:10.1016/j.jenvman.2013.09.021

Magrí, A., M.B. Vanotti, and A.A. Szögi. 2012a. Anammox sludge immobilized in polyvinyl alcohol (PVA) cryogel carriers. Bioresour. Technol. 114(2):231–240. doi:10.1016/j.biortech.2012.03.077

Magrí, A., M.B. Vanotti, A.A. Szögi, and K.B. Cantrell. 2012b. Partial nitritation of swine wastewater in view of its coupling with the anammox process. J. Environ. Qual. 41(6):1989–2000. doi:10.2134/jeq2012.0092

Massé, L., D.I. Massé, and Y. Pellerin. 2007. The use of membranes for the treatment of manure: A critical literature review. Biosystems Eng. 98:371–380. doi:10.1016/j.biosystemseng.2007.09.003

Massé, L., D.I. Masse, and Y. Pellerin. 2008. The effect of pH on the separation of manure nutrients with reverse osmosis membranes. J. Membr. Sci. 325(2):914–919. doi:10.1016/j.memsci.2008.09.017

Massey, M.S., J.G. Davis, J.A. Ippolito, and R.E. Sheffield. 2009. Effectiveness of recovered magnesium phosphates as fertilizers in neutral and slightly alkaline soils. Agron. J. 101:323–329. doi:10.2134/agronj2008.0144

Melse, R.W., and N.W.M. Ogink. 2005. Air scrubbing techniques for ammonia and odor reduction at livestock operations: Review of on-farm research in the Netherlands. Trans. ASAE 48(6):2303–2313. doi:10.13031/2013.20094

Milan, Z., E. Sanchez, R. Borja, K. Ilangovan, A. Pellon, N. Rovirosa, P. Weiland, and R. Escobedo. 1999. Deep bed filtration of anaerobic cattle manure effluents with natural zeolite. J. Environ. Sci. Health B 34:305–332. doi:10.1080/03601239909373199

Millner, P., D. Ingram, W. Mulbry, and O.A. Arikan. 2014. Pathogen reduction in minimally managed composting of bovine manure. Waste Manag. 34(11):1992–1999. doi:10.1016/j.wasman.2014.07.021

Mino, T., M.C.M. van Loosdrecht, and J.J. Heijnen. 1998. Microbiology and biochemistry of the enhanced biological phosphate removal process. Water Res. 32:3193–3207. doi:10.1016/S0043-1354(98)00129-8

Monballiu, A., E. Desmidt, K. Ghyselbrecht, and B. Meesschaert. 2019. The inhibitory effect of inorganic carbon on phosphate recovery from upflow anaerobic sludge blanket reactor (UASB) effluent as calcium phosphate. Water Sci. Technol. 78(12)2608–2615. doi:10.2166/wst.2019.026.

Moerman, W., M. Carballa, A. Vandekerckhove, D. Derycke, and W. Verstraete. 2009. Phosphate removal in agro-industry: Pilot- and full-scale operational considerations of struvite crystallization. Water Res. 43:1887–1892. doi:10.1016/j.watres.2009.02.007

Molinuevo-Salces, B., G.M.C. García, and C. González-Fernández. 2010. Performance comparison of two photobioreactors configurations (open and closed to the atmosphere) treating anaerobically degraded swine slurry. Bioresour. Technol. 101(14):5144–5149. doi:10.1016/j.biortech.2010.02.006

Moore, P.A., Jr., W.F. Jaynes, and D.M. Miller. 1998. Effect of pH on the solubility of phosphate minerals. In: Proceedings of the 1998 National Poultry Waste Management Symposium. p. 328-333. Auburn Univ. Printing Service, Auburn, AL.

Moore, P.A., Jr., H. Li, R. Burns, D.M. Miles, R. Maguire, J. Ogejo, M. Reiter, M. Buser, and S.L. Trabue. 2018. Development of the ARS air scrubber: A Device for reducing ammonia, dust and odor in exhaust air from animal rearing facilities. Frontiers in Sustainable Food Systems 2:1–23. doi:10.3389/fsufs.2018.00023

Moral, R., C. Paredes, M.A. Bustamante, F. Marhuenda-Egea, and M.P. Bernal. 2009. Utilisation of manure composts by high-value crops: Safety and environmental challenges. Bioresour. Technol. 100:5454–5460. doi:10.1016/j.biortech.2008.12.007

Mulbry, W., E. Kebede-Westhead, C. Pizarro, and L. Sikora. 2005. Recycling of manure nutrients: Use of algal biomass from dairy manure treatment as a slow release fertilizer. Bioresour. Technol. 96:451–458. doi:10.1016/j.ciortech.2004.05.026.

Mulbry, W., S. Kondrad, C. Pizarro, and E. Kebede-Westhead. 2008. Treatment of dairy manure effluent using freshwater algae: Algal productivity and recovery of manure nutrients using pilot-scale algal turf scrubbers. Bioresour. Technol. 99(17):8137–8142. doi:10.1016/j.biortech.2008.03.073

Mulder, A., A.A. van de Graaf, L.A. Robertson, and J.G. Kuenen. 1995. Anaerobic ammonium oxidation discovered in a denitrifying fluidized bed reactor. FEMS Microbiol. Ecol. 16:177–184. doi:10.1111/j.1574-6941.1995.tb00281.x

Mumpton, F.A. 1999. La roca mágica: Uses of natural zeolites in agriculture and industry. Proc. Natl. Acad. Sci. USA 96:3463–3470. doi:10.1073/pnas.96.7.3463

Münch, E.V., and K. Barr. 2001. Controlled struvite crystallisation for removing phosphorus from anaerobic digester sidestreams. Water Res. 35:151–159. doi:10.1016/S0043-1354(00)00236-0

Muñoz, R., and B. Guieysse. 2006. Algal-bacterial processes for the treatment of hazardous contaminants: A review. Water Res. 40:2799–2815. doi:10.1016/j.watres.2006.06.011

Ndegwa, P.M., J. Zhu, and A. Luo. 2001. Effects of solids levels and chemical additives on removal of solids and phosphorus in swine manure. J. Environ. Eng. 127(12):1111–1115. doi:10.1061/(ASCE)0733-9372(2001)127:12(1111)

Nelson, N.O., R.L. Mikkelsen, and D.L. Hesterberg. 2003. Struvite precipitation in anaerobic swine lagoon liquid: Effect of pH and Mg:P ratio and determination of rate constant. Bioresour. Technol. 89:229–236. doi:10.1016/S0960-8524(03)00076-2

Newman, J.M., J.C. Clausen, and J.A. Neafsey. 1999. Seasonal performance of a wetland constructed to process dairy milkhouse wastewater in Connecticut. Ecol. Eng. 14:181–198. doi:10.1016/S0925-8574(99)00028-2

Novak, J., G. Sigua, D. Watts, K. Cantrell, P. Shumaker, A. Szögi, M.G. Johnson, and K. Spokas. 2016. Biochars impact on water infiltration and water quality through a compacted subsoil layer. Chemosphere 142:160–167. doi:10.1016/j.chemosphere.2015.06.038

Novak, J.M., A.A. Szögi, and D.W. Watts. 2008. Copper and zinc accumulations in sandy soils and constructed wetlands receiving pig manure effluent applications. In: P. Schlegel, S. Durosoy, and A.W. Jongbloed, editors, Trace elements in animal production systems. Wageningen Academic Publishers, The Netherlands. p. 45–54.

NRCS. 2009. Chapter 3, Constructed wetlands. In: NRCS, editor, National engineering handbook, Part 637. Natural Resources Conservation Service, Washington, D.C.

NRCS. 2016. Conservation practice standard, Constructed wetland Code 656-CPS-1. NRCS, Washington, D.C.

Obaja, D., S. Macé, J. Costa, C. Sans, and J. Mata-Alvarez. 2003. Nitrification, denitrification and biological phosphorus removal in piggery wastewater using a sequencing batch reactor. Bioresour. Technol. 87(1):103–111. doi:10.1016/S0960-8524(02)00229-8

Oehmen, A., P.C. Lemos, G. Carvalho, Z. Yuan, J. Keller, L.L. Blackall, and M.A.M. Reis. 2007. Advances in enhanced biological phosphorus removal: From micro to macro scale. Water Res. 41(11):2271–2300. doi:10.1016/j.watres.2007.02.030

Ohlinger, K.N., T.M. Young, and E.D. Schroeder. 1999. Kinetics effects on preferential struvite accumulation in wastewater. J. Environ. Eng. 125:730–737. doi:10.1061/(ASCE)0733-9372(1999)125:8(730)

Oswald, W.J. 1988. Micro-algae and waste-water treatment. In: M.A. Borowitzka and L.J. Borowitzka, editors, Micro-algal biotechnology. Cambridge Univ. Press, Cambridge. p. 305–328.

Panswad, T., A. Doungchai, and J. Anotai. 2003. Temperature effect on microbial community of enhanced biological phosphorus removal system. Water Res. 37(2):409–415. doi:10.1016/S0043-1354(02)00286-5

Pardo, G., R. Moral, E. Aguilera, and A. del Prado. 2015. Gaseous emissions from management of solid waste: A systematic review. Glob. Change Biol. 21:1313–1327. doi:10.1111/gcb.12806

Philippe F.X., J.F. Cabaraux, and B. Nicks. 2011. Ammonia emissions from pig houses: Influencing factors and mitigation techniques. Agriculture, Ecosystems and Environment 141: 245-260, doi:10.1016/j.agee.2011.03.012.

Poach, M.E., P.G. Hunt, G.B. Reddy, K.C. Stone, T.A. Matheny, M.H. Johnson, and E.J. Sadler. 2004. Ammonia volatilization from marsh–pond–marsh constructed wetlands treating swine wastewater. J. Environ. Qual. 33:844-851.

Poach, M.E., P.G. Hunt, E.J. Sadler, T.A. Matheny, M.H. Johnson, K.C. Stone, F.J. Humenik, and J.M. Rice. 2002. Ammonia volatilization from constructed wetlands that treat swine wastewater. Trans. ASAE 45:619–627. doi:10.13031/2013.8825

Poach, M.E., P.G. Hunt, M.B. Vanotti, K.C. Stone, T.A. Matheny, M.H. Johnson, and E.J. Sadler. 2003. Improved nitrogen treatment by constructed wetlands receiving partially nitrified liquid swine manure. Ecol. Eng. 20:183–197. doi:10.1016/S0925-8574(03)00024-7

Portejoie, S., J. Martinez, F. Guiziou, and C.M. Coste. 2003. Effect of covering pig slurry stores on the ammonia emission processes. Bioresour. Technol. 87:199–207. doi:10.1016/S0960-8524(02)00260-2

Qiao, S., T. Yamamoto, M. Misaka, K. Isaka, T. Sumino, Z. Bhatti, and K. Furukawa. 2010. High-rate nitrogen removal from livestock manure digester liquor by combined partial nitritation-anammox process. Biodegradation 21:11–20. doi:10.1007/s10532-009-9277-8

Ra, C.S., K.V. Lo, J.S. Shin, J.S. Oh, and B.J. Hong. 2000. Biological nutrient removal with an internal organic carbon source in piggery wastewater treatment. Water Res. 34:965–973. doi:10.1016/S0043-1354(99)00189-X

Rajkovich, S., A. Enders, K. Hanley, C. Hyland, A.R. Zimmerman, and J. Lehmann. 2012. Corn growth and nitrogen nutrition after additions of biochars with varying properties to a temperate soil. Biol. Fertil. Soils 48:271–284. doi:10.1007/s00374-011-0624-7

Rahman, M.M., Y.H. Liu, J.H. Kwag, and C.S. Ra. 2011. Recovery of struvite from animal wastewater and its nutrient leaching loss in soil. J. Hazard. Mater. 186:2026–2030. doi:10.1016/j.jhazmat.2010.12.103

Reddy, G.B., C.W. Raczkowski, J.S. Cyrus, and A.A. Szögi. 2016. Carbon sequestration in a surface flow constructed wetland after 12 years of swine wastewater treatment. Water Sci. Technol. 73:2501–2508. doi:10.2166/wst.2016.112

Ro, K.S., K.B. Cantrell, and P.G. Hunt. 2010. High-temperature pyrolysis of blended animal manures for producing renewable energy and value-added biochar. Ind. Eng. Chem. Res. 49:10125–10131. doi:10.1021/ie101155m

Ro, K.S., P.G. Hunt, M.A. Jackson, D.L. Compton, S.R. Yates, K. Cantrell, and S. Chang. 2014. Co-pyrolysis of swine manure with agricultural plastic waste: Laboratory-scale study. Waste Manag. 34:1520–1528. doi:10.1016/j.wasman.2014.04.001

Ro, K.S., J.M. Novak, M.G. Johnson, A.A. Szögi, J.A. Libra, K.A. Spokas, and S. Bae. 2016. Leachate water quality of soils amended with different swine manure-based amendments. Chemosphere 142:92–99. doi:10.1016/j.chemosphere.2015.05.023

Rothrock, M.J., Jr., A.A. Szögi, and M.B. Vanotti. 2010. Recovery of ammonia from poultry litter using gas-permeable membranes. Trans. ASABE 53(4):1267–1275. doi:10.13031/2013.32591

Rothrock, M.J., Jr., A.A. Szögi, and M.B. Vanotti. 2013. Recovery of ammonia from poultry litter using flat gas permeable membranes. Waste Manag. 33(6):1531–1538. doi:10.1016/j.wasman.2013.03.011

Rothrock, M.J., Jr., M.B. Vanotti, A.A. Szögi, M.C. Garcia, and T. Fujii. 2011. Long-term preservation of anammox bacteria applied microbiology and biotechnology. Appl. Microbiol. Biotechnol. 92(1):147–157. doi:10.1007/s00253-011-3316-1

Ruiz-Martínez, A., J. Serralta, I. Romero, A. Seco, and J. Ferrer. 2015. Effect of intracellular P content on phosphate removal in Scenedesmus sp. Experimental study and kinetic expression. Bioresour. Technol. 175:325–332. doi:10.1016/j.biortech.2014.10.081

Schulz, H., and B. Glaser. 2012. Effects of biochar compared to organic and inorganic fertilizers on soil quality and plant growth in a greenhouse experiment. J. Plant Nutr. Soil Sci. 175:410–422. doi:10.1002/jpln.201100143

Sherman, J.J., H.H. Van Horn, and R.A. Nordstedt. 2000. Use of flocculants in dairy wastewaters to remove phosphorous. Appl. Eng. Agric. 16(4):445–452. doi:10.13031/2013.5222

Sigua, G.C., J.M. Novak, and D.W. Watts. 2016. Ameliorating soil chemical properties of a hard setting subsoil layer in Coastal Plain USA with different designer biochars. Chemosphere 142:168-175. doi:10.1016/j.chemosphere.2015.06.016

Sojka, R.E., D.T. Westermann, and R.D. Lentz. 1998. Water and erosion management with multiple applications of polyacrylamide in furrow irrigation. Soil Sci. Soc. Am. J. 62:1672–1680. doi:10.2136/sssaj1998.03615995006200060027x

Sommer, S.G., and S. Husted. 1995. The chemical buffer system in raw and digested animal slurry. J. Agric. Sci. 124(1):45–53. doi:10.1017/S0021859600071239

Song, C., S. Shan, K. Müller, S. Wu, N.K. Niazi, S. Xu, Y. Shen, J. Rinklebe, D. Liu, and H. Wang. 2017. Characterization of pig manure-derived hydrochars for their potential application as fertilizer. Environ. Sci. Poll. Res. 25(26): 25772–25779. p. 1–8.

Stone, K.C., P.G. Hunt, A.A. Szögi, F.J. Humenik, and J.M. Rice. 2002. Constructed wetland design and performance for swine lagoon wastewater treatment. Trans. ASAE 45:723–730. doi:10.13031/2013.8828

Stingone, J.A., and S. Wing. 2011. Poultry litter incineration as a source of energy: Reviewing the potential for impacts on environmental health and justice. New Solut. 21(1):27–42. doi:10.2190/NS.21.1.g

Stone, K.C., M.E. Poach, P.G. Hunt, and G.B. Reddy. 2004. Marsh-pond-marsh constructed wetland design analysis for swine lagoon wastewater treatment. Ecol. Eng. 23:127–133. doi:10.1016/j.ecoleng.2004.07.008

Stratful, I., M. Scrimshaw, and J. Lester. 2001. Conditions influencing the precipitation of magnesium ammonium phosphate. Water Res. 35:4191–4199. doi:10.1016/S0043-1354(01)00143-9

Strous, M., J.J. Heijnen, J.G. Kuenen, and M.S.M. Jetten. 1998. The sequencing batch reactor as a powerful tool for the study of slowly growing anaerobic ammonium-oxidizing microorganisms. Appl. Microbiol. Biotechnol. 50:589–596. doi:10.1007/s002530051340

Sun, K., B. Gao, K.S. Ro, J.M. Novak, Z. Wang, S. Herbert, and B. Xing. 2012. Assessment of herbicide sorption by biochars and organic matter associated with soil and sediment. Environ. Pollut. 163:167–173. doi:10.1016/j.envpol.2011.12.015

Sun, K., K. Ro, M. Guo, J. Novak, H. Mashayekhi, and B. Xing. 2011. Sorption of bisphenol A, 17a-ethinyl estradiol and phenanthrene on thermally and hydrothermally produced biochars. Bioresour. Technol. 102:5757–5763. doi:10.1016/j.biortech.2011.03.038

Suzin, L., F.G. Antes, G.C. Bedendo, M. Bortoli, and A. Kunz. 2018. Chemical removal of phosphorus from swine effluent: The impact of previous effluent treatment technologies on process efficiency. Water Air Soil Pollut. (229):341 doi:10.1007/s11270-018-4018-4.

Szögi, A.A., P.J. Bauer, and M.B. Vanotti. 2010. Phosphorus recovery from poultry litter. Agron. J. 102:723–727. doi:10.2134/agronj2009.0355

Szögi, A.A., P.J. Bauer, and M.B. Vanotti. 2012. Vertical distribution of phosphorus in a sandy soil fertilized with recovered manure phosphates. J. Soils Sediments 12:334–340. doi:10.1007/s11368-011-0452-2

Szögi, A.A., P.G. Hunt, and F.J. Humenik. 2000. Treatment of swine wastewater using a saturated-soil-culture soybean and flooded rice system. Trans. ASAE 43:327–335. doi:10.13031/2013.2708

Szögi, A.A., P.G. Hunt, and F.J. Humenik. 2003. Nitrogen distribution in soils of constructed wetlands treating lagoon wastewater. Soil Sci. Soc. Am. J. 67:1943–1951. doi:10.2136/sssaj2003.1943

Szögi, A.A., P.G. Hunt, E.J. Sadler, and D.E. Evans. 2004a. Characterization of oxidation-reduction processes in constructed wetlands for swine wastewater treatment. Appl. Eng. Agric. 20:189–200. doi:10.13031/2013.15891

Szögi, A.A., and M.B. Vanotti. 2009. Removal of phosphorus from livestock effluents. J. Environ. Qual. 38:576–586. doi:10.2134/jeq2007.0641

Szögi, A.A., M.B. Vanotti, and P.G. Hunt. 2006. Dewatering of phosphorus extracted from liquid swine waste. Bioresour. Technol. 97(1):183–190. doi:10.1016/j.biortech.2005.02.001

Szögi, A.A., M.B. Vanotti, and P.G. Hunt. 2008. Phosphorus recovery from poultry litter. Trans. ASABE 51:1727–1734. doi:10.13031/2013.25306

Szögi, A.A., M.B. Vanotti, and P.G. Hunt. 2014a. Process for removing and recovering phosphorus from animal waste. U.S. Patent 8673,046 B1. U.S. Patent and Trademark Office, Washington, D.C.

Szögi, A.A., M.B. Vanotti, J.M. Rice, F.J. Humenik, and P.G. Hunt. 2004b. Nitrification options for pig wastewater treatment. N. Z. J. Agric. Res. 47:439–448. doi:10.1080/00288233.2004.9513612

Szögi, A.A., M.B. Vanotti, and M.J. Rothrock. 2014b. Gaseous ammonia removal system. U.S. Patent 8906,332 B2. U.S. Patent and Trademark Office, Washington, D.C.

Szögi, A.A., M.B. Vanotti, and P.G. Hunt. 2015. Phosphorus recovery from pig manure prior to land application. J. Environ. Manage. 157:1–7. doi:10.1016/j.jenvman.2015.04.010

Szögi, A.A., M.B. Vanotti, and P.D. Shumaker. 2018. Economic recovery of calcium phosphates from swine lagoon sludge using Quick Wash process and geotextile filtration. Frontiers in Sustainable Food Systems 2:37, doi:10.3389/fsufs.2018.00037

Tchobanoglous, G., F.L. Burton, and H.D. Stensen. 2003. Wastewater engineering treatment and reuse. 4th edition. Metcalf and Eddy, Inc. McGraw-Hill, New York.</bok>

Tilche, A., E. Bacilieri, G. Bortone, F. Malaspina, S. Piccinini, and L. Stante. 1999. Biological phosphorus and nitrogen removal in a full scale sequencing batch reactor treating piggery wastewater. Water Sci. Technol. 40:199–206. doi:10.2166/wst.1999.0043

USDA/ERS. 2014. Fertilizer use and prices. United States Department of Agriculture, Economic Research Service. http://www.ers.usda.gov/data-products/fertilizer-use-and-price.aspx#26727 (Accessed 28 Mar. 2019).

USDA/NRCS. 2008. Vegetative treatment area, practice Introduction. Practice Code 635. Natural Resources Conservation Service, Washington, D.C. https://www.nrcs.usda.gov/Internet/FSE_DOCUMENTS/nrcs143_026548.pdf (Accessed 28 Mar. 2019).

USDA/NRCS. 2015. Vegetated Treatment Area. Code 635. Conservation Practice Standard 635-CPS-1. Natural Resources Conservation Service, Washington, D.C. https://www.dec.ny.gov/docs/water_pdf/vta.pdf (Accessed 28 Mar. 2019).

Vaccari, D.A. 2011. Sustainability and the phosphorus cycle: Inputs, outputs, material flow, and engineering. Environmental Engineer and Scientist 12:29–38.

Van der Heyden, C., P. Demeyer, and E.I.P. Volcke. 2015. Mitigating emissions from pig and poultry housing facilities through air scrubbers and biofilters: State-of-the-art and perspectives. Biosystems Eng. 134:74–93. doi:10.1016/j.biosystemseng.2015.04.002

Van Groenestijn, J.W., and M.E. Lake. 1999. Elimination of alkanes from off-gases using biotrickling filters containing two liquid phases. Environ. Prog. 18(3):151–155. doi:10.1002/ep.670180310

Vaneeckhaute, C., V. Lebuf, E. Michels, E. Belia, P.A. Vanrolleghem, F.M.G. Tack, and E. Meers. 2017. Nutrient recovery from digestate: Systematic technology review and product classification. Waste Biomass Valorization 8:21–40. doi:10.1007/s12649-016-9642-x

Vanotti, M.B., P.J. Dube, A.A. Szögi, and M.C. Garcia-Gonzalez. 2017. Recovery of ammonia and phosphate minerals from swine wastewater using gas-permeable membranes. Water Res. 112:137–146. doi:10.1016/j.watres.2017.01.045

Vanotti, M.B., and P.G. Hunt. 1999. Solids and nutrient removal from flushed swine manure using polyacrylamides. Trans. ASAE 42(6):1833–1840. doi:10.13031/2013.13347

Vanotti, M.B., and P.G. Hunt. 2000. Nitrification treatment of swine wastewater with acclimated nitrifying sludge immobilized in polymer pellets. Trans. ASAE 43(2):405–413. doi:10.13031/2013.2719

Vanotti, M.B., J.H. Loughrin, P.G. Hunt, J.M. Rice, A. Kunz, and C.M. Williams. 2014. Generation 3 treatment technology for diluted swine wastewater using high-rate solid-liquid separation and nutrient removal processes. Proc. 2014 ASABE and CSBE/SCGAB Annual International Meeting, Montreal, QC. 13–16 July 2014. Paper No. 141901300, ASABE, St. Joseph, MI. p. 1-10. doi:10.13031/aim.20141901300.

Vanotti, M.B., J. Martinez, T. Fujii, A.A. Szögi, and D. Hira. 2012. Ammonia removal using nitrification and anammox in a single reactor. Proc. ASABE International Meeting, Dallas, TX. 29 July–1 Aug. 2012. Paper No. 1121337837. ASABE, St. Joseph, MI. doi:10.13031/2013.41817

Vanotti, M.B., P.D. Millner, P.G. Hunt, and A.Q. Ellison. 2005a. Removal of pathogen and indicator microorganisms from liquid swine manure in multi-step biological and chemical treatment. Bioresour. Technol. 96:209–214. doi:10.1016/j.biortech.2004.05.010

Vanotti, M.B., P.D. Millner, A.A. Szögi, C.R. Campbell, and L.M. Fetterman. 2006. Aerobic composting of swine manure solids mixed with cotton gin waste. Proc. 2006 ASABE Annual Int. Meeting, Portland, OR. 9–12 July 2006. Paper No. 06406 ASABE, St. Joseph, MI. doi: 10.13031/2013.21112.

Vanotti, M.B., D.M.C. Rashash, and P.G. Hunt. 2002. Solid-liquid separation of flushed swine manure with PAM: Effect of wastewater strength. Trans. ASAE 45:1959–1969. doi:10.13031/2013.11422

Vanotti, M.B., K.S. Ro, A.A. Szögi, J.H. Loughrin, and P.D. Millner. 2018a. High-rate solid-liquid separation coupled with nitrogen and phosphorus treatment of swine manure: Effect on water quality. Frontiers in Sustainable Food Systems 2:49, doi:10.3389/fsufs.2018.00049.

Vanotti, M.B., and A.A. Szögi. 2008. Water quality improvements of CAFO wastewater after advanced treatment. J. Environ. Qual. 37:S86–S96. doi:10.2134/jeq2007.0384

Vanotti, M.B., and A.A. Szögi. 2015. Systems and methods for reducing ammonia emissions from liquid effluents and for recovering ammonia. U.S. Patent 9005,333 B1. U.S. Patent and Trademark Office, Washington, D.C.

Vanotti, M.B., A.A. Szögi, M.P. Bernal, and J. Martinez. 2009a. Livestock waste treatment systems of the future: A challenge to environmental quality, food safety, and sustainability. OECD Workshop. Bioresour. Technol. 100(22):5371–5373. doi:10.1016/j.biortech.2009.07.038

Vanotti, M.B., A.A. Szögi, and P.J. Dube. 2018b. Systems and methods for recovering ammonium and phosphorus from liquid effluents. U.S. Patent 9926,213 B2. U.S. Patent and Trademark Office, Washington, D.C.

Vanotti, M.B., A.A. Szögi, and T.F. Ducey. 2013a. High performance nitrifying sludge for high ammonium concentration and low temperature wastewater treatment. U.S. Patent No. 8445,253 B2. U.S. Patent and Trademark Office, Washington, D.C.

Vanotti, M.B., A.A. Szögi, and P.G. Hunt. 2003. Extraction of soluble phosphorus from swine wastewater. Trans. ASAE 46(6):1665–1674. doi:10.13031/2013.15637

Vanotti, M.B., A.A. Szögi, and P.G. Hunt. 2005b. Wastewater treatment system. US Patent No. 6893,567. U.S. Patent & Trademark Office, Washington, D.C.

Vanotti, M.B., A.A. Szögi, P.G. Hunt, P.D. Millner, and F.J. Humenik. 2007. Development of an environmentally superior treatment system to replace anaerobic lagoons in the USA. Bioresour. Technol. 98:3184–3194. doi:10.1016/j.biortech.2006.07.009

Vanotti, M.B., A.A. Szögi, P.D. Millner, and J.H. Loughrin. 2009b. Development of a second-generation environmentally superior technology for treatment of swine manure in the USA. Bioresour. Technol. 100(22):5406–5416. doi:10.1016/j.biortech.2009.02.019

Vanotti, M.B., A.A. Szögi, and M.J. Rothrock. 2013b. Novel anammox bacterium isolate. U.S. No. Patent 8574,885 B2. U.S. Patent and Trademark Office, Washington, D.C.

Vidal, G., C.P. de Los Reyes, and O. Sáez. 2018. The performance of constructed wetlands for treating swine wastewater under different operating conditions. In: A.I. Stefanakis, editor, Constructed wetlands for industrial wastewater treatment. Wiley-Blackwell, Hoboken, NJ. doi:10.1002/9781119268376.ch10

Vohla, C., M. Koiv, H.J. Bavor, F. Chazarenc, and U. Mander. 2011. Filter materials for phosphorus removal from wastewater in treatment wetlands—A review. Ecol. Eng. 37:70–89. doi:10.1016/j.ecoleng.2009.08.003

Wang, L., Y. Li, P. Chen, M. Min, Y. Chen, J. Zhu, and R.R. Ruan. 2010. Anaerobic digested dairy manure as a nutrient supplement for cultivation of oil-rich green microalgae Chlorella sp. Bioresour. Technol. 101:2623–2628. doi:10.1016/j.biortech.2009.10.062

Wang, J.C., and J.K. Park. 2001. Effect of anaerobic-aerobic contact time on the change of internal storage energy in two different phosphorus-accumulating organisms. Water Environ. Res. 73:436–443. doi:10.2175/106143001X139489

Wang, Y., J. Wang, X. Zhao, X. Song, and J. Gong. 2016. The inhibition and adaptability of four wetland plant species to high concentration of ammonia wastewater and nitrogen removal efficiency in constructed wetlands. Bioresour. Technol. 202:198–205. doi:10.1016/j.biortech.2015.11.049

Westholm, L.J. 2006. Substrates for phosphorus removal—Potential benefits for on-site wastewater treatment? Water Res. 40:23–36. doi:10.1016/j.watres.2005.11.006

Wilkie, A.C., and W.W. Mulbry. 2002. Recovery of dairy manure nutrients by benthic freshwater algae. Bioresour. Technol. 84:81–91. doi:10.1016/S0960-8524(02)00003-2

Yanosek, K.A., M.L. Wolfe, and N.G. Love. 2003. Assessment of enhanced biological phosphorus removal for dairy manure treatment. In: Proceedings of the Ninth International Symposium of Animal, Agricultural and Food Processing Wastes, Raleigh, North Carolina. 12–15 Oct. 2003. ASAE, St. Joseph, MI. p. 212-220.

Yetilmezsoy, K., and Z.S. Zengin. 2009. Recovery of ammonium nitrogen from the effluent of UASB treating poultry manure wastewater by MAP precipitation as a slow release fertilizer. J. Hazard. Mater. 166:260–269. doi:10.1016/j.jhazmat.2008.11.025

Yilmazel, D., and G. Demirer. 2013. Nitrogen and phosphorus recovery from anaerobic co-digestion residues of poultry manure and maize silage via struvite precipitation. Waste Manag. Res. 31:792–804. doi:10.1177/0734242X13492005

Zarebska, A., D. Romero Nieto, K.V. Christensen, L. Fjerbæk Søtoft, and B. Norddahl. 2015. Ammonium fertilizers production from manure: A critical review. Crit. Rev. Environ. Sci. Technol. 45(14):1469–1521. doi:10.1080/10643389.2014.955630

Zeng, L., C. Mangan, and X. Li. 2006. Ammonia recovery from anaerobically digested cattle manure by steam stripping. Water Sci. Technol. 54:137–145. doi:10.2166/wst.2006.852

Zhang, L., and D. Jahng. 2010. Enhanced anaerobic digestion of piggery wastewater by ammonia stripping: Effects of alkali types. J. Hazard. Mater. 182:536–543. doi:10.1016/j.jhazmat.2010.06.065

Zhang, R.H., and F. Lei. 1998. Chemical treatment of animal manure for solid–liquid separation. Trans. ASAE 41:1103–1108. doi:10.13031/2013.17255

Zhao, Q., T. Zhang, C. Frear, K. Bowers, J. Harrison, and S. Chen. 2010. Phosphorous recovery technology in conjunction with dairy anaerobic digestion. CSANR Research Report 001, Washington State University, Pullman, WA.

Ziobro, J. 2017. The nutrient recycling challenge. Livestock and Poultry Environmental Learning Center, University of Nebraska, Lincoln, NE. https://articles.extension.org/pages/73942/us-epa-nutrient-recycling-challenge (Accessed 25 Mar. 2019).

Pelletizing Animal Manures for On- and Off-Farm Use

Xiying Hao* and Zhongqi He

Abstract

Pelleting is a promising option for managing animal manure; however, as this chapter will show, there is considerable need for more research on all aspects of this technology. Today, poultry litter is the most commonly pelleted animal manure, but interest is growing in applying similar processes to cattle, hog, and horse manure, or compost. Animal manure can be pelleted (or granularized) using various processes such as die and roller, extruder or granulator, alone or in combination with other organic or inorganic additives, binding agents, or synthetic fertilizer. On-farm, manure pellets are used (or are at least being tested) for field crops and vegetable production, greenhouse operations, and as animal feedstuff. Off-farm uses include horticulture plant production, fertilizing sport fields and parks, reclaiming contaminated soil, aquaculture, as a fuel source, and for making biochar. Compared with unpelleted animal manure and compost, pelleting offers ease of handling and uniformity of application, redistribution of manure nutrients away from areas with excess nutrient accumulation to areas with nutrient deficiency, reduced reliance on synthetic fertilizers, and the ability to modify the nutrient level and composition to meet specific crop needs. However, animal manure pellets are a slow-release fertilizer (SRF) so they need to be applied earlier than mineral fertilizers and raw animal manure in order to have sufficient time for mineralization and nutrient release prior to crop uptake. Other environmental impacts, particularly increased N_2O emission, should also be considered when making decisions on pellets for on-farm or off-farm use.

By definition, animal (or livestock) manure is animal excreta (urine and feces) and bedding materials. Animal manure may also contain dropped feed, scurf, water, and soil, depending upon specific animal management practices (He, 2012; He et al., 2016). Over one billion Mg of wet weight manure are produced annually in the United States of America, and 180 million Mg in Canada, the majority coming from confined feeding operations by the animal industry (USEPA, 2013; Statistics Canada, 2015). Animal manure has a dichotomous nature, a valuable resource if used judiciously as a soil amendment but an environmental pollutant if mismanaged (Larney et al., 2011; Zhang and Schroder, 2014). Efficient use of animal manure can make an important contribution toward meeting the challenges of increasing global food production to feed a population expected to grow from the current 7.6 to 9.7 billion by 2050 (United Nations, 2015). "Nutrients excreted in beef feedlots in the USA alone would cost over $461 million if purchased as fertilizer" (Eghball and Power, 1994).

Manure application for food and feed production provides immediate and delayed nutrient supply, modifies soil pH, improves soil structure, and enhances soil biological activity. Improved soil properties associated with manure application also lead to reductions in runoff and soil erosion (Gilley and Risse, 2000; He and Zhang, 2014). However, soil salinity problems can result when excessive rates of cattle

X. Hao, Agriculture and Agri-Food Canada, Lethbridge Research and Development Centre, 5403 1st Ave S., Lethbridge, AB Canada T1J 4B1; Z. He, USDA-ARS Southern Regional Research Center, 1100 Robert E. Lee Blvd., New Orleans, LA USA 70124.

* Corresponding author (xiying.hao@canada.ca)

doi:10.2134/asaspecpub67.c23

manure are applied in semi-arid regions (Hao and Chang, 2003), typically near animal operations to reduce transportation cost.

Traditionally, most animal manure, whether in solid or liquid form, has been applied to cropland as is without any processing, but now composting is increasingly being used to manage solid animal manure for easier handling, transportation, and application (Larney et al., 2006; Wichuk and McCartney, 2012). Separating dairy slurry into solid and liquid fractions allows the solid fraction to be applied to land further away from the dairy operation (Meyer et al., 1997) or as bedding material for animal operations (Spencer, 2016; Cornell Waste Management Institute, 2018). Others have used animal manure directly, without any processing, as feed in aquaculture fish farms (Wohlfarth and Schroeder, 1979).

Pelleting animal manure offers greater opportunities to alleviate challenges faced by both farmers and the general global population (He et al., 2012). This technology has been offered commercially for over 40 years by Mars Mineral (http://www.manureintomoney.com/) for poultry litter (PL), which remains the most common animal manure to be pelletized. Pellets have also been produced using manure from dairy and beef cattle, swine and horses, sometimes with added bulking agents or other supplements. Pelletizing animal manure removes water content, reducing both weight and volume, and odor, and creates a product that is preferred over fresh manure for its reduced cost of storage, transportation handling, and application (Kleinman et al., 2012; Zafari and Kianmehr, 2012b; Mieldazys et al., 2016). Thus, excess nutrients can be redistributed at less cost away from regions where large amounts of manure are being produced (Hamilton and Sims, 1995).

The pelleting process can also modify the product; for example, adding mineral fertilizer can change the N and P nutrient ratio in animal manure to meet specific crop needs (Mazeika et al., 2016; Sakurada et al., 2016). In general, pelletization combined with enrichment with inorganic fertilizer produces a higher (and customized) nutrient content along with better physical properties for easier transport and application by commercial fertilizer equipment. Pellet application also generates a tenth or less of the dust created by ordinary compost (Hara 2001).

Less than 20% of animal manure is composted in Canada (Larney et al., 2006) with a much smaller fraction currently being pelleted. Similar to managing digestate products from biogas production, the major challenges are marketing the pellets, which can have varying nutrient values, as well as the initial capital investment cost (Nagy et al., 2018). The marketability of animal manure pellets is limited by insufficient demand, low market value, and high transportation costs. The market prices need to be high enough to cover production, distribution, marketing and related business costs (USEPA, 1997).

Research on the use of pellets is ongoing and producing a variety of results. This is to be expected given the different manure sources (primarily poultry, swine and cattle), pellet production methods, possible supplements, application methods, and intended purposes. In addition to farm use, the off-farm market, including land reclamation, landscaping, home gardens, and greenhouses, may benefit from manure pellets. Some of the horticulture market will be attracted to organic products and, in some cases, be less sensitive to cost. The possible combinations of these factors are almost endless and provide substantial scope for ongoing research.

We will briefly outline the pelleting process, examine pellet quality as an organic fertilizer, and explore the potential of pellets for on-farm and off-farm use.

Pelleting Animal Manure
Pelleting Process

There are several approaches that can be used to process animal manure into pellets. The basic steps, illustrated in Fig. 1, include (i) drying the manure to ensure that its moisture content will be at the optimum level for the pelleting machine used; (ii) homogenizing and crushing the manure so it will be at the optimum size (2-5 mm diameter) for pelleting, while removing stones which can damage the machinery; (iii) pelleting the manure; (iv) cooling the pellets; and (v) storing the pellets. The last step may include bagging the pellets, particularly if they are intended for off-farm use, and then placement on wrapped pallets ready for delivery to retail stores.

Pelleting is the process of compressing or molding a material into the shape of a pellet (Fig. 2). Extruded pellets are created with pressure agglomeration. In contrast, granulation produces material that is not as dense, and the formed pellets are round rather than cylindrical (Prasai et al., 2018). The granulation process can be simpler because the granulator is a rotating disk where the ground manure material and fertilizer are placed (Sakurada et al., 2016). Some use the two terms interchangeably (e.g., Rao et al., 2007; Mieldazys et al., 2017). In this work, both manure pellets and granules are discussed without strict distinction.

The production capacity of animal manure pelletizers ranges from a few kilos (e.g., to be used in laboratories for experimentation purposes) to a few thousand kilograms per hour (commercial scale). Flat or ring die pellet mill extrusion is most commonly used in pelletizers (Whirlston Machinery, 2018a; 2018b). The advantage of using a flat over a ring die pellet mill is their simple design and structure, which also makes them easy to clean. On the other hand, ring die pellet mills are less prone to the uneven die and roller wear experienced with flat dies, thus reducing maintenance and repair costs for roller and ring die mills. The ring die pellet mill is also more energy efficient than flat die pelleting. Other pelletizing processes include non-pressure (tumble growth or mixer-drier approach), disc pelletizing, and disc or rotary drum granulation (Feeco International, 2018), and there are differences between hydraulic vs. mechanical processes as well. Each approach has its advantages and disadvantages, largely depending on the material to be pelleted, volume of material, end products desired, and both fixed (purchase/lease) and operational costs. While large animal operators could install pelleting operations on-farm, a centralized manure pelleting system, as employed in northeast Spain, might be more practical; this option will depend on the density and the intensity of farming in a given area (Flotats et al., 2009).

Pelleting Process Management and Pellet Quality

For pellets to be sold commercially, the production process must minimize variability in nitrogen form and content, phosphorus content, and pH and possible contamination during manufacturing (Lopez-Mosquera et al., 2008). The properties or quality

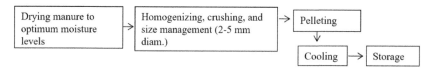

Fig. 1. Brief outline of the animal manure pelleting process.

Fig. 2. Pelletized and granulized manure products. Panels A and B are images of organomineral-granulated (OG) and organomineral-pelletized (OM) fertilizers used in Sakurada et al. (2016) (courtesy of Marcelo Batista). OG is synthetic fertilizer and poultry litter ground and pelletized into a single granule, and OM is pelletized litter and fertilizer granules separated into two products. Panel C is the image of pelletized separated solid from anaerobically digesting cattle manure used in Thomas et al. (2017). Panel D shows the three organic amendments used in Chiyoka et al. (2014a,b): cattle feedlot manure (top row), separated solid after cattle manure anaerobically digested (middle row) and pellets produced from the separated solid (bottom row) (courtesy of Wara Chiyoka).

of pellets is largely influenced by the pelleting process/machinery selected (disk pelletizer, extruder, granulator, etc.), raw material properties (including the moisture level at the time of production), and manufacturing conditions (such as the pelleting temperature and pressure used). Hara (2001) compared two types of pellet machines (disk pelletizer vs extruder) and found that for both types, the most important factor is the moisture content which influences the strength of the pellets and the processing speed. Hara (2001) reported that the best raw material moisture content is about 40% for an extruder and about 20 to 25% for a disk pelletizer.

Brambilla et al. (2017) investigated how physical and chemical properties of pellets relate to a bulking agent (sawdust) during separated swine manure solid composting and pelleting moisture levels (10% and 25%); they found that adding sawdust during composting better preserved important characteristics (such as total N and humification rate) when pelleting at 25% moisture levels. Increasing the moisture level from 15% to 25% also decreased the pellet surface roughness. However, increasing water content increases transportation cost.

Pocius et al. (2017) tested rheological and strength properties of granulated organic fertilizer pellets produced using beef cattle and poultry manure and organic waste compost. Their results show that there is limited strength in vertical and horizontal directions (4-mm and 6-mm pellets), and strength varied from 10% to 20% for poultry manure pellets. Using an open die approach, Zafari and Kianmehr (2012b) reported cattle manure with 50% moisture content, processed at a medium temperature about 40 °C and pressure at 6 MPa, resulted in maximum cattle manure pellet durability among the three moisture levels (50, 55 and 60%), three temperatures (40, 60 and 80 °C) and three pressure levels (3.5, 6 and 8 Mpa) tested. For the non-pressure granulation system, 20% moisture for poultry and 60% to 80% for hog and dairy manure in cake form instead of slurry is preferred (Feeco International, 2018). More information can be found at the Feeco International website.

Romano et al. (2014) produced four different pelleted organic fertilizer formulations made of swine manure solid fraction composted with or without sawdust, wood chips, and wheat straw, and tested them to determine differences in physicochemical and land distribution features. The four pellets were compared with two pelleted organic fertilizers ordinarily available at retailers. Their results show that, as far as physical and chemical features are concerned, the greatest difference from the commercial products used in their study is found in pellet size distribution after spreading, since the disintegrating action of the rotating vanes does not affect the tested formulation with the same intensity as the commercial products. Distribution tests showed that sawdust–swine compost pellets provided better longitudinal and transverse distribution, while swine compost pellets had good transverse but poor longitudinal distribution.

In addition to adding bulking agents during composting to improve animal manure suitability for pelleting (Rao et al., 2007; Brambilla et al., 2017; Romano et al., 2014), supplements or binders could also be added if desired during the pelleting process. Rao et al. (2007) investigated the possibility of producing organic fertilizer pellets using composted swine waste solids (20% w/w) blended with other locally available biodegradable wastes, such as poultry litter (26% w/w), spent mushroom compost (26% w/w), cocoa husks (18% w/w) and moistened shredded paper (10% w/w). They concluded that the resulting compost nutrient content (2.3% total N, 1.6% P and 3.1% K) was too low. However, they found that incorporating dried blood or feather meal amendments increased the organic N-content while reducing the moisture in mature compost mixtures to aid the granulation process. Inclusion of mineral supplements, for example, sulphate of ammonia, rock phosphate, and sulphate of potash, will also yield SRF with higher nutrient content pellets and could be used to produce N/P/K ratios that are suitable for fertilizing lawns and golf courses (Rao et al., 2007).

Mazeika et al. (2016) reported that the presence of hygroscopic KCl in combination with diammonium phosphate resulted in significant moisture absorption and loss of granulized poultry manure pellet structural integrity after 72 h at 30 °C and 80% relative humidity. In their study, poultry manure was dried and shredded to no more than 10% moisture content at 95 °C and granulation was performed via extrusion with and without mineral (diammonium phosphate and potassium chloride) additives. Mieldazys et al. (2017) investigated the granulation process parameters and factors affecting the pellets produced using a small capacity 7.5 kW granulator with a horizontal granulator matrix to produce 6 mm diameter pellets. The cattle manure compost was mixed with molasses or lime sludge to produce five types of pellets. They reported that pellets produced at the optimal moisture level of 28% had a diameter of 5.74 mm and average length of 13.2 mm. Pellet density ranged from 395 kg·m^{-3} to 789 kg·m^{-3}, depending on pelleting process conditions and manure compost factors. Molasses increased and lime sludge decreased pellet strength.

Toor et al. (2007) reported that granulation increased water-soluble P with PL, raising the concern that land application of the granulized litter product may increase P loss to ground and surface water. This concern could be alleviated by granulation of PL with biosolids (He et al., 2010). They found that granulating PL alone transformed moderately available phosphorus ($NaHCO_3$- and NaOH-extractable P) into labile water-soluble P and stable HCl–extractable P in the granulated PL, but did not substantially change the distribution of H_2O-, $NaHCO_3$-, and NaOH-extractable P in the granules produced using both PL and biosolids. The high concentration of metals (especially Al) in biosolids reduced the water-soluble P increase resulting from granulation. Phosphorus bioavailability in poultry compost pellets was also strongly affected by its moisture content (> 1 kg kg^{-1}), and indirectly by soil moisture conditions after application (Takahashi et al., 2016).

Pelletized composted swine manure has been shown to be an effective SRF for corn. Using composted swine solids separated from swine slurry, Purnomo et al. (2017) recommended that the best technical options for swine manure pellet production are to include a bulking agent, such as wood chips, before composting and create small diameter pellets (6 mm vs 8 mm).

Pellet Quality During Storage and Transportation

Preserving pellet quality during storage is an important aspect when promoting pellets as way to manage animal manure nutrients. As with any solid fertilizer, storage in a cool, dry, well-ventilated area is recommended. Hara (2001) reported that pellets made from animal manure compost could become moldy in humid regions if they are packed and stored without proper drying, and recommended that the moisture content of the swine compost pellets should be 20% or less in these locations. There is an inverse relationship between the moisture content of the pellets and the force required to rupture them (i.e., pellet durability) (McMullen et al., 2005; Alcaraz et al., 2018).

At 40 °C temperature conditions, after 40 hours of exposure to 80% relative humidity (RH) air, pellet moisture content increased to about 16% from the initial of 8.5%; the increase was much less when exposed to 60% RH air, but decreased to about 6.5% from initial 8.5% when exposed to 45% RH air (inferred from data in Fig. 6, McMullen et al., 2005). The same authors concluded that the rate of moisture absorption increases with the air temperature and RH. The values of the thermal properties of the pellets were not significantly affected by moisture content and were comparable to those for commercial fertilizer. Urea, the most common N fertilizer, has higher critical relative humidity (70-75% at 30°C) and can withstand hot and humid conditions with insignificant loss when stored at temperatures up to 35 °C over 3 months (Jewell, 1962).

Pellet Nutrient Availability

The pelleting production process should minimize variability in N form and content, P content and pH and reduce possible contamination during manufacture (Lopez-Mosquera et al., 2008). As discussed in the previous section, nutrient availability of animal manure pellets is directly linked to the source materials used to make them, the pelleting process and conditions, size and density of pellets produced, and pellet storage conditions. The rate of N release from pellets depends not only on the rate of pellet decomposition after application but also soil conditions (Souri et al., 2018) such as pH, soil moisture, microbial activity, and other nutrients and weather conditions (precipitation and temperature).

One of the most useful indicators of nutrient release after animal manure application, regardless of whether in pellet form or not, is the C/N ratio as demonstrated by Qian and Schoenau (2002). They used 12 kinds of solid organic amendments (fresh unpelleted and pelleted animal manure) with C/N ratios ranging from 7.6 to 21.7 in the two soil types investigated. The C/N ratio in the manure and N mineralization were negatively correlated (Qian and Schoenau, 2002).

The suitability of poultry manure for producing SRF in pellet form can be tested by immersing the pellets in water. Purnomo et al. (2017) reported that pellets made using the granulating method release nutrients faster than pellets produced using extrusion. Moreover, the energy and time to produce them is lower with granulation than with the extrusion process. Thus, a complete cost-benefit analysis must include the quality of the SLF matrix required (desired rate of nutrient release), processing time, cost, and energy when selecting the pelleting method.

In addition to moisture, temperature, and pressure used during pelleting, the density of the pellets produced also affects nutrient release. Souri et al. (2018) tested two different density (400 vs. 800 kg m^{-3}) pellets produced using 50% fully decomposed cattle manure and 50% (w/w) urea; they found the high-density pellets did not release sufficient nutrients over 13 to 17 weeks for coriander [*Corianderum sativum* L.], garden cress [*Lepidium sativum* L.] and parsley [*Petroselinium crispom* Mill.] compared with low density pellets or urea fertilizer.

Blair et al. (2014) compared the effects of pelleted poultry litter (PPL) with inorganic osmocote fertilizer on phosphorus nutrient availability and enzyme activities over one growing season, and found even at high soil P conditions, PPL had higher alkaline phosphatase activity and released greater phosphate than the inorganic osmocote fertilizer. Applying poultry compost pellets resulted in more liable P in soil than the unpelleted form based on a 56-day lab incubation study (Takahashi et al., 2016). Equal or lower mineral N and a higher NH_4 concentration in soil amended with pellets than solar-dried, mixed and ground manure prior to pelleting are also reported in a controlled 90-day laboratory incubation experiment (Hadas et al., 1983). In their study, temperature, ranging from 14 to 35 °C, had little impact on the rate of N mineralization with their pellets, which were produced using poultry or a poultry–dairy manure mixture by first solar drying, mixing and grinding, followed by compaction into cylindrical pellets 5 to 6 mm in length and 4 mm in diameter. The temperature increased to 75 °C for 5 to 10 min during the compression while making the pellets. In contrast, after incubating poultry pellets enriched with NPK fertilizer over 112 days (16 weeks), there was a 50% decrease in available N after six weeks, while soil-available N remained consistently low through the 16 weeks in the unpelleted PL. They also reported a net P immobilization for both PPL and PL in the two soils tested.

Dried pelletized chicken litter has more stable nutrient characteristics than the fresh product (Lopez-Mosquera et al., 2008). Pellets do not contain faecal bacteria and do not have significant noxious odour. From a practical point of view, the pelletizing process eases storage, transport and field application of the broiler litter, and can also facilitate incorporation of other components into the final product, including mineral elements, herbicides or nitrification inhibitors.

Pelleting poultry manure is also a way to produce SRF. Dried poultry manure mixed with rice husk was first milled to powder before mixing with a binder (starch) and nitrogen source (urea) to produce two types of pellets using a screw extruder and pan granulator. Purnomo et al. (2017) reported SRF made using an extruder has longer nutrient retention capability than granule SRF.

Poultry manure pellets, and two green waste–based composts (a straw-based compost and a vermicast), were incubated in a coarse-textured soil at 15 °C for 142 d, and gross N mineralization from poultry pellets was reported to be much higher than all the other amendments used. This suggests its application rates should be reduced, or alternative amendments used, to minimize leaching losses in regions where ground water quality is a concern (Flavel and Murphy, 2006). Using ryegrass [*Lolium multiflorum* L.] and sorghum-sudangrass [*Sorghum bicolor* L.] as the test plants in a greenhouse experiment, Hammac II et al. (2007) concluded that pelletized broiler litter can serve as N and P sources for plants, but N and P in broiler litter pellets may not be as available as from inorganic fertilizer.

How pellets are applied also affects nutrient release and availability. In a field study, PPL application methods (no pellets, pellets subsurface banded, broadcast, or incorporated with tillage) were investigated and compared with cover crop residue (Poffenbarger et al., 2015a). Incorporation of cover crop residues and pellets increased the rate of mass loss and N release for pure hairy vetch and hairy vetch (*Vicia villosa* subsp. *villosa*)–cereal rye (*Secale cereale* L.) mixtures. Although broadcast pellet application and incorporation affected decomposition patterns, subsurface banded pellet application did not. Subsurface banding is the recommended pellet application method to conserve surface residues.

In a 70-day laboratory incubation experiment using two types of soil to investigate the availability of N in a soil amended with undigested beef cattle feedlot manure, separated solid, or pelletized separated solid from anaerobically digested cattle manure, Chiyoka et al., (2014a) reported a net N immobilization in soils, that is, at the end of 70 days, the soil available N content in pellet-amended soil was lower than the unamended soil. The immobilization of N was further confirmed in a greenhouse pot experiment that showed there is less N uptake by barley biomass in pellet-amended than unamended soil (Chiyoka et al., 2014b).

In addition to the total amount of N mineralized, the form of N release is also important. A greater fraction of N in NH_4 than NO_3 form was reported with pelleted than unpelleted swine compost (Hara, 2001), suggesting there is less leaching potential with pelleted swine compost as the NO_3 form is more subject to leaching loss. An incubation study (Reiter et al., 2014) examined N availability from N–fortified PL granular fertilizers containing different additives and compared them with available data for commercially available inorganic and organic fertilizers applied to a silt loam soil under nonleached aerobic conditions. Averaged over the entire incubation period (112 days), granulated PL products had apparent net N mineralization of 71.5% while urea, biosolid, and PL averaged 80.5, 16.8, and 36.7%, respectively. Binding agents (lignosulfonate, urea formaldehyde, or water) and biosolid additions used in granulation had no statistical impact on NH_4–N or NO_3–N soil concentrations. Thus, Reiter et al. (2014) concluded that N-fortified PL granules may improve N efficiency over fresh PL and biosolids due to more N availability and potentially lower environmental N losses over a growing season.

Manure Pellets For On-farm Use as Organic Fertilizer

On-farm use is loosely defined as pellets being used to promote plant growth and yield increase in production of human food and animal feed. As with any organic fertilizer or amendment, information on the levels, availability, and rate of nutrient release from animal manure pellets (Table 1) is needed before they can be widely accepted for use.

Table 1. On-farm use of pelletized manure products from selected studies.

Manure or fraction	Additive	Pelleting process	Crop system	Application method	Outcome or impact	Reference
Composted swine manure	No	Extruder or roller and die disk (5 or 10 mm diameter, 20 mm long pellets)	Spring and summer spinach, pak choi, marrow vs. winter cabbage and spinach	Not specified	Substituting 100%, 50% and 20% N fertilizer as basal dressing with pellets increased summer vegetable yield, but lowered winter vegetable yield. Applying pellets earlier can solve the slow nitrification problem with winter vegetable production.	Hara, 2001
Cattle manure	Various rates of urea addition	Die and hydraulic compress force	Corn	Split application before seeding, at six-leaf stage, and anthesis stage to 3 cm depth between rows	A higher corn yield, 1000-kernel weight, and protein content when cattle manure in combination with various rates of urea (600 kg manure combined with 46, 92, 138 and 184 kg urea) was applied to soil in pelleted and unpelleted form. Better corn response attributed to the slow release of nutrients from pellets for plant uptake at different corn growth stages.	Bagheri et al., 2011
Poultry litter	No info	No info	Corn	Surface broadcast vs. subsurface banded (8–10-cm deep and 5 cm wide, 15 cm away from seed row)	Surface application led to higher corn grain yield, total N uptake, and apparent N use efficiency for pelletized than nonpelletized poultry litter, with no difference between pelletized and nonpelletized litter when subsurface banded.	Adeli et al., 2012
Poultry litter	No info	No info	Four year cotton	Four weeks after planting, PPL was subsurface banded	Four-year cotton lint yields were 5% higher with PPL than UAN fertilizer (1378 vs 1303 kg ha⁻¹). PPL enhanced soil fertility and soil aggregate stability, and minimized postharvest residual NO_3-N concentration in the soil profile.	Adeli et al., 2016
Poultry litter	No info	No info	Three year soybean	Band application for cotton production in previous three years	Residual effect of three annual applications of PPL were investigated using soybean in following three years. Soybean yields were higher in PPL treatment in 1 out of 3 years and PPL influenced soil fertility for several years beyond the year of application.	Adeli et al., 2015
Poultry manure	No info	No info provided, pellets were purchased as Nutriwave 4-1-2 from Environ Technologies, Fredericton, NB Canada		Broadcast and incorporated by disking 1 day prior to planting	When soil moisture levels were adequate, application of a pelletized poultry manure product increased potato yields (+ 5.8 Mg ha⁻¹ average) at low N rate of 300 kg ha⁻¹, but residual soil nitrate levels rose to 61 kg N ha⁻¹. Pellet yield benefit decreased at higher rates (600 kg N ha⁻¹). The effects of hog manure-sawdust compost were inconsistent.	Lynch et al., 2008
Separated solid of cattle manure digestate	No	Roller and flat die	Barley forage	Surface application followed by disk into top 10 cm soil, seed within one week.	Pelletized separated solid of cattle manure digestate may be an effective SRF, while also supplying C. But the pellet N utilization is low (2%) compared to digestate slurry (19%), separated solid without pelleting (9%) and undigested raw manure (10%).	Thomas et al., 2017
Composted solid poultry, solid dairy and liquid hog manure=	urea, $(NH_4)_2PO_4NH_4H_2PO_4$ and KCl to produce pelletized organo-mineral fertilizer 7-4-4	Not specified	Potato	The inorganic fertilizer band-applied (7.5 cm to each side, 5 cm below the potato seed). The pellets hand-applied as a band on the soil surface. Split N applications were made using a custom disc.	Application of equivalent rates of mineral or organo-mineral fertilizer at planting produced similar tuber yield, size distribution and quality parameters and soil NO_3-N content at tuber harvest.	Zebarth et al., 2005
Poultry litter	No info provided	Not specified	Rice	PPL and fresh PL broadcast by hand, incorporated to depth of 4 cm with rotator tiller before seeding and compared to urea application at preflooding	Similar net-N uptake and grain yields between FPL and PPL, but significantly lower than urea applied at preflood. Their data suggests only 25% of TN in FPL and PPL was used for rice production (equivalent to urea applied at preflood).	Golden et al., 2006

Table 1 cont.

Poultry litter	Feather meal as 6-3-3 organic fertilizer	Purchased from manufacturers	Rice	Broadcast before seeding and lightly incorporated into the soil using a roller.	Pelletized fertilizers were more effective than PL in supplying N to the crop when fields were continuously flooded. However, both PL and pelletized organic fertilizers were less effective (lower N recovery efficiency caused by denitrification) in supplying N when fields were drained for weed control. The organic fertilizer application may not be economically viable under such circumstances.	Wild et al., 2011
Swine solids	Wood chips (+/-) during composting	Mechanical pelletizer (CLM200E, La Meccanica Srl, Padua, Italy) 6- and 8-mm size	Corn	Surface and incorporation	Manure pellets with or without wood chip additive are effective SRF for corn. Increased nutrient availability in pig slurry solid fraction compost pellets by adding a bulking agent before composting, using small diameter pellets and incorporating pellets in soil.	Pampuro et al., 2017
Composted cattle manure	50% N provided by urea	Laboratory-scale pellet extruder	Basil	Mixed with soil in pot	The basil chlorophyll content, total weight, leaf weight, root weight and leaf area were higher with pellets than urea applications when basil was grown for 10 weeks.	Zafari and Kianmehr, 2012a
Pelleted poultry manure	No info	No info	Pepper	Mixed with soil in pot	Applying poultry manure pellets increased pepper fruit yield, leaf and fruit P concentrations, but not K, Mg, Si, Al, Ni, and Fe contents. Applying poultry pellets also increased leaf Zn, Cl, Cu and Br, and fruit Zn, Cl, Mg, Rb and Ce concentrations.	Sahin et al., 2014
Pelleted broiler litter	Broiler litter from 20 farms collected, mixed with sugarcane molasses (at 8%)	Using roller-and die pellet press (heat up to 90.0°C), pellets size were 6 mm in diameter and 1.5-2 cm in length.	Crossbred steers of Thai indigenous and Brahman cattle	Broiler litter pellets substitute level at 0, 10, 20, 30 40, and 50% of total cattle dietary level.	50% broiler litter pellets may be used effectively as a feedstuff for cattle; beef cattle may consume properly handled pelleted BL without increasing the likelihood of carcass/meat contamination with pathogenic bacteria. Should have 15-day withdrawal time for exclusion of BL from diets of cattle before slaughter to avoid drug residue in meat.	Suppadit and Poungsuk, 2010

Animal Manure Pellets for Field Crop Production

Conflicting results have been reported when comparing the agronomic value of animal manure pellets to mineral fertilizers. Most research conducted so far has focused on PPL with limited work using pelleted swine, dairy or cattle manure, and compost. Pelleted PL led to higher corn (*Zea mays* L.) grain yield, total N uptake, and apparent N use efficiency than PL when surface-applied, but no differences were found when subsurface banded over two growing seasons (Adeli et al., 2012). However, in a four-year field study when PPL was annually subsurface band applied (6.7 Mg ha^{-1} ~ 122 kg available N ha^{-1}) and compared with inorganic N fertilizer (urea ammonium nitrate, UAN solution) injected at the recommended rate of 128 kg N ha^{-1} for cotton production in Mississippi, (Adeli et al., 2016), the benefit of PPL in increased cotton (*Gossypium* spp.) lint yield did not show up until Years 3 and 4. Following cotton production, there were significant PPL residual effects on soybean (*Glycine max* L.) for one year and soil fertility was affected for several years beyond the last manure pellet application (Adeli et al., 2015).

When poultry pellets (3.5 Mg ha^{-1}) were applied using three methods (broadcast at planting, subsurface banded at the fifth-leaf stage, and broadcast and incorporated at planting) in combination with three cover crop residues (hairy vetch, cereal rye, and a hairy vetch–cereal rye mixture), Poffenbarger et al. (2015b) found corn N uptake tended to be greatest with hairy vetch residue and broadcast or subsurface band poultry litter pellet application.

Pelletized poultry manure was compared with commercial hog manure–sawdust compost, both applied at 300 and 600 kg total N ha^{-1} (plus an unamended control), for potato [*Solanum tuberosum* L.] production in Prince Edward Island, Canada (Lynch et al., 2008). When soil moisture was adequate, pellet application significantly increased total and marketable potato yields by an average of 5.8 and 7.0 Mg ha^{-1} compared with unamended or compost-amended treatments over the three-year field study. They concluded that given current premiums for certfied organic potatoes, improving yields through application of poultry litter pellets to supply moderate rates of N or organic matter appears feasible.

Again in Atlantic Canada, a pelletized organo-mineral fertilizer product with a nutrient analysis of approximately 7-4-4 produced from composted solid poultry, solid dairy and liquid hog manure and fortified with mineral fertilizer (urea, $(NH_4)_2HPO_4$, $NH_4H_2PO_4$, and KCl) as the sole amendment or in combination with mineral fertilizer as the N source were compared with mineral fertilizer for processing potato production, at N rates of 0, 50, 100 and 200 kg N ha^{-1} (Zebarth et al., 2005). The authors reported that applying equivalent rates of N as mineral or pellet fertilizer at planting generally resulted in comparable values of tuber yield, size distribution and quality parameters, and soil NO_3-N content at tuber harvest. They recommended application of 1.5 t ha^{-1} of pellets at planting, with additional mineral fertilizer applied as a split application if warranted.

Golden et al. (2006) compared pre-plant incorporated fresh poultry litter (FPL) and PPL with urea applied at preflood to determine the urea-nitrogen (N) equivalence of PL for rice (*Oryza sativa* L.) production in Arkansas. They found similar net N uptake and grain yield between FPL and PPL, but both were significantly lower when compared with urea applied preflood. Based on net grain yield, Golden et al. (2006) recommended that multiplying the total-N content in FPL and PPL by 0.25 would reasonably estimate its equivalence to urea applied preflood. Similarly, Wild et al. (2011) evaluated N availability from PL and pelletized organic amendments for organic rice production in California. Laboratory and field experiments were conducted to determine the effectiveness of the commonly used PL, pelletized organic fertilizers (blood/meat/feather meal, feather meal, PL plus feather meal), and $(NH4)_2SO_4$, in synchronizing the supply of mineralized N with the demand for N by rice. Unlike the work of Golden et al. (2006), Wild et al. (2011) found pelletized fertilizers significantly more effective than PL in supplying N to crops when fields were continuously flooded. However, they did report that both PL and pelletized organic fertilizers were less effective in supplying N when fields were drained for weed control due to lower N recovery efficiency (26%) and N loss through denitrification, indicating that organic fertilizer application may not be economically viable under such circumstances.

Investigating cabbage (*Brassica oleracea* L.) and spinach (*Spinacia oleracea* L.) response to pelleted swine compost, Hara (2001) reported that winter vegetable yields from field plots with pellets were lower than plots where mineral fertilizer was used. He attributed the lower yield with pellets to limited nitrification at low temperatures over the winter cropping season in upland soil in Japan. Thus, if pellets are to be used for winter vegetable production, an earlier pellet application time is recommended or application at a higher rate, providing residual nutrients for the crop that follows cabbage harvest. Due to their slow nutrient-release nature, pellets are not suitable as a basal dressing for vegetables grown in autumn and winter in Japan, but with adequate mineralization or nitrification are well suited to vegetables such as spinach, cabbage, pak choi (*Brassica rapa* L. subsp. chinensis [L.] Hanelt) and marrow (*Cucurbita pepo* L.) grown in spring and summer when temperatures are higher. Hara (2001) also recommended that 50% mineral fertilizer could be substituted with swine manure compost pellets for spring and summer crops in Japan.

Bagheri et al. (2011) reported a higher corn yield, 1000-kernel weight, and protein content when cattle manure in combination with various rates of urea (600 kg manure combined with 46, 92, 138, or 184 kg urea) was applied to soil in pellet form than cattle manure applied without pelleting. They attributed the better corn response to the slow release of nutrients from pellets for plant uptake at different corn growth stages. In contrast, a much lower barley (*Hordeum vulgare* L.) forage yield and N utilization resulted when applying pellets produced using separated solid of the cattle manure digestate compared with fresh cattle manure, digestate liquid or the separated solid from digestate without pelleting in a four-year field study conducted in central Alberta, Canada (Thomas et al., 2017). They found that only a small fraction of applied N was used by the barley forage crop with digestate at 19%, separated solid in loose form at 9%, and separated solid in pellet form at 2%, when compared with fresh (undigested) cattle manure at 10%. The lower yield and N utilization were attributed to the relatively low temperature in central Alberta Canada, and insufficient time for nutrients in pellets to mineralize or release to soil for barley use.

Animal Manure Pellets for High-End Crop or Greenhouse Production

In a research study under greenhouse conditions, 6-mm and 8-mm pellets produced from composted swine separated solid with or without wood chips as a bulking agent were applied at a 200 kg N ha^{-1} rate to the soil surface or mixed with soil (Pampuro et al., 2017). They reported similar corn yield and N and P uptake from the four combinations tested when compared with mineral fertilizer and both pellet types were effective as SRF for corn (as a test crop). To increase availability of the nutrients contained in solid swine compost, they suggested adding a bulking agent before composting, using small-diameter pellets (6 mm instead of 8 mm) and incorporating the pellets into soil instead of surface application.

Urea-enriched composted cattle manure pellets (urea and manure each providing 50% N) were compared to urea fertilizer with both applied at rates equivalent to 200 kg N ha^{-1} in greenhouse basil (*Ocimum basilicum* L.) production. The 6-mm diam. and 8- to 20-mm length pellets were produced using composted cattle manure and urea mixture at 35% moisture level (wet basis) with no external binding agent at room temperature. The chlorophyll content, total weight, leaf weight, root weight, and leaf area were significantly higher with pellets than urea applications when basil was harvested at end of 10 weeks (Zafari and Kianmehr, 2012a).

When growing pepper (*Capsicum annuum* L.) in a greenhouse with poultry pellet application rates of 0, 10, 20, and 40 g kg^{-1} DM (supplemented with 200 mg N kg^{-1} soil, 50 mg P kg^{-1} soil, and 62.5 mg K kg^{-1} soil), Sahin et al. (2014) reported pelleted poultry manure applications significantly improved pepper fruit yield, and increased leaf and fruit P concentrations but not N, K, Mg, Si, Al, Ni, and Fe. Applying PPL also increased leaf Zn, Cl, Cu and Br, and fruit Zn, Cl, Mg, Rb, and Ce concentrations.

Poultry Litter Pellets as Feedstuff to On-farm Animals

The presence of amino acids in PL is well-known (He and Olk, 2011; He et al., 2014). The potential reuse of these amino acids as feed ingredients for animals has been continuously studied for decades (Blair, 1974; Obeidat et al., 2016). However, direct use of fresh PL as feedstuff may lead to unacceptable problems associated with worms, insects, parasites, palatability and odor (Suppadit and Poungsuk, 2010).

Thus, research efforts have also been directed towards application of PPL as animal feedstuff (Suppadit et al., 2012). The PL pellets may be used effectively as a short-term feedstuff for meat goats (Jackson et al., 2005). Including broiler litter pellets for up to 50% of the cattle diets may be used effectively as a feedstuff for crossbred steers without increasing the likelihood of carcass or meat contamination with pathogenic bacteria, but a 15-day withdrawal time before slaughter is suggested in order to avoid drug residue in meat (Suppadit and Poungsuk. 2010). However, the market potential is still limited (Wolfe et al., 2002).

Animal Manure Pellets for Off-farm Use

Off–farm use covers a wide range of both commercial and noncommercial applications not involving food and feed production (Table 2), including (i) greenhouse or nursery nonfood plants; (ii) landscaping and golf courses; (iii) land reclamation; (iv) fuel production and as source materials for making biochar. Wolfe et al. (2002) assessed the feasibility of pelleting PL at a facility in South Georgia, and concluded PPL may not effectively compete in the fertilizer market as pellets are considered a weak fertilizer, but they appear to have potential in the homeowner market if the product is packaged and marketed appropriately.

Animal Manure Pellets for Horticulture Plant Production

There are readily available manure pellets for sale online, often labeled "organic" to appeal to customers who prefer this approach. Typically, the NPK is lower than commercial fertilizers and often slower releasing. Depending on region, the soil temperature can affect nutrient release and decomposition of the pellets. In addition to being a source of organic fertilizer, PL, whether pelleted or not, is a good control agent for carnation (*Dianthus caryophyllus* L.) Fusarium wilt, improves carnation yield and quality, and is a preferred alternative to methyl bromide (Melero-Vara et al., 2011). Applications should be made prior to each crop cycle to ensure continuous disease control, but the rate of application could be reduced to half for the third and fourth crop cycles, thereby reducing undesirable environmental effects.

Animal Manure Pellets as Fertilizer for Sport Fields and Parks

Rao et al. (2007) reported that pellets incorporating dried blood or feather meal amendments enriched the organic N-content, reduced moisture in mature compost mixtures, and aided the granulation process, while including mineral supplements such as ammonia sulphate, rock phosphate, and potassium potash

Table 2. Off-farm use of pelletized manure products from selected studies.

End use	Manure or fraction	Additive	Pelleting process	Application field	Application method	Outcome or impact	Reference
Horti-culture crop production	Commercial PPL	Not reported	Not reported	Field plastic house for carnation production	Surface applied, incorporate into top 15 cm soil	Both as source of organic fertilizer and control agent for carnation Fusarium wilt by improving carnation yield and quality. PPL is preferred over methyl bromide to control Fusarium wilt. Should apply PPL prior to each crop cycle to ensure continuous disease control, but reduce the application by half for the third and fourth crop cycles minimize undesirable environmental impact.	Melero-Vara et al., 2011
Turf-grass, sport fields	PPL obtained from the Perdue AgriRecycle Pelletizing Plant in Seaford, DE.	Not reported	Not reported	Perennial ryegrass (*Lolium perenne* L.) and tall fescue (*Festuca arundinacea*).	Valmar Airflow 1255PF granular applicator	No changes in measured soil parameters during the 2.5-year study compared to the synthetic fertilizer, but greater turf-grass quality when compared to the same N rate of synthetic fertilizer.	Sprinkle et al., 2011
Fuel source	Composted swine manure and wood bark or saw dust	Engine oil 0-6%, or hull meal at 5% or 10% weight ratios (w/w).	Roller and die	Not applicable	Not applicable	The pellet durability index (PDI) decreased when oil > 1%. Mixing with rice hull meal produced higher durability. An optimal fuel pellet when swine compost moisture content between 15% and 20%, oil $\leq$ 1%, and passing through a mesh 3 mm sieve. The overall and net efficiencies of a boiler were higher using pellets made from compost with wood bark than those with pellets made from swine compost with sawdust.	Oh et al., 2013
Biochar	PL and other animal manure not specified	Not reported	Die and roller (PMCL5 Lab pellet mill)	Not applicable	Not applicable	One-step charring process using PPL to make biochar	Ausmus, 2005; Lima and Marshall, 2005a, b; Lima et al. (2016)
Biochar	70% manure pellets mixed with 30% fir pellets	Not reported	Not reported	Not reported	Not reported	The Lucia Stove would not work with pure manure pellets; adding 30% fir tree pellets was sufficient to obtain complete pyrolysis. The element ratios of biochars produced using fir pellets alone (ABE) were 0.04 (H/C), 0.22 (O/C) and 35.0 (C/N), 0.05 (H/C) and using manure with 30% fir pellets (MAN) were 408 (C/N), 0.04 (H/C), 0.22 (O/C) and 35.0 (C/N), 0.05 (H/C) and 0.39 (O/C), respectively. The ash contents were: 1.4% for ABE and 36% for MAN.	Fellet et al., 2014

yielded SRF with nutrient N/P/K ratios of 10:3:6 and 3:5:10. The authors suggested these pellets should be well suited for spring or summer application and autumn dressing on golf courses and city parks, but they have yet to report results. Sprinkle et al. (2011) evaluated turfgrass response to the use of PL pellets as a nutrient source compared with inorganic fertilizer, under two aeration (core vs. vibrating) conditions for athletic turf fields. Application of PL pellets did not produce significant changes in measured soil parameters compared to the synthetic fertilizer during the 2.5-year study. However, application of PL pellets resulted in greater turfgrass quality when compared with the same N rate of synthetic fertilizer, regardless of aeration method.

Animal Manure Pellets as Fuel Source

Oh et al. (2013) used swine manure compost in combination with either wood bark or sawdust to produce fuel pellets used in stoves, furnaces, and power generation. They examined how compost material size (1-, 2-, 3-, and 5-mm size sieve), industry oil level (0%, 1%, 3%, or 6%), incorporation of rice hull meal (5% or 10%) and moisture content (10%, 15%, or 20%) affected fuel pellet durability, density, and calorie content. They reported that "optimal fuel pellets can be obtained by controlling the moisture content of the swine manure compost to between 15% and 20%, mixing with oil at a ratio of 1% or lower, and sieving through a mesh with a 3 mm or smaller diameter." The overall and net efficiencies of a boiler were higher using pellets made from compost with wood bark than with pellets made from compost with sawdust.

When making dairy manure pellets as a fuel source, Alcaraz et al. (2018) reported that removing sand from dairy manure not only increased the heating value of the pellets from 3.29 MJ kg^{-1} to 10.20 MJ kg^{-1}, but also increased pellet durability to withstand 1.12 kilogram-force (kg-f) to 1.80 kg-f (5-10% moisture levels).

Animal Manure Pellets for Making Biochar

Biochar is a form of charcoal that is produced by exposing organic waste matter (such as wood chips, crop residue, or manure) to heat in a low-oxygen environment and is used primarily as a soil amendment. Poultry pellets (Lima et al., 2008; Lima and Marshall, 2005a, b) and manure (type not specified) pellets (70%) mixed with fir (*Abies* sp.) tree pellets (30%) (Fellet et al., 2014) have been used as source materials for making active biochar. Biochar has been produced using animal manure directly (Cely et al., 2015), or from pelleted manure (Lima et al., 2008; 2016). The pelleted form should have the advantage of easier handling and transport, but the disadvantage of requiring two processes, particularly if they are in different locations. The activated biochar produced using PL has a powerful metallic attraction that steam-activated plant-based carbons lack.

Compared with biochar derived from woody materials, the manure-pellet-derived biochar has the advantages of a less complex structure and a higher O/C ratio (Fellet et al., 2014). The higher O/C ratio indicates a greater presence of hydroxyl, carboxylate, and carbonyl groups that are able to chelate metals (Lee et al., 2010), which was confirmed by greater reduction in metal bioavailability with manure pellet biochar than fir tree pellet biochar when applied to mine tailings (Fellet et al., 2014). The large surface area, high porosity and P levels of PL biochar are ideal for attracting hard-to-snag metals such as Cu, Cd, and Zn in tainted waters (Ausmus, 2005).

Using Manure Pellets and Manure Biochar for Land Reclamation

To alleviate subsoil compaction near well sites, pipeline corridors, and mining and construction sites, manure pellets (at rate of 20 Mg ha^{-1}) were injected along with deep subsoil tilling (40 cm depth) and compared with the same deep subsoil tilling without pellets in seven field sites (Leskiw et al., 2012). One year after treatment implementation, Leskiw et al. (2012) reported consistently lower soil bulk density (BD) with pellet applications, but BD in the deep tillage alone treatment returned to values similar to the unamended control. In another study, three types of biochar from different feedstocks (pruning residues, fir tree pellets and manure pellets) were tested on their ability to remediate mine tailings to promote plant growth for the phytostabilization of contaminants in a greenhouse pot experiment. Fellet et al. (2014) reported that biochar from manure pellets and pruning residues reduced shoot Cd and Pb accumulations for the three plant species tested: kidney-vetch [*Anthyllis vulneraria* L. subsp. *Polyphylla* (Dc.) Nyman), *Noccaea rotundifolium* (L.) *Moench* subsp. *cepaeifolium* and *Poa alpina* L. subsp. *alpina*). They found manure pellet–sourced biochar also led to higher biomass production, and concluded that this form of biochar has great potential as an amendment for phytoremediation but its effects depend on the feedstock source. There is limited research in this area.

Other Considerations When Using Animal Manure Pellets

Benefits of Pelleting Animal Manure

Pelletizing manure or compost provides a volume and weight reduction to 20 to 50% of the original manure (including moisture) and 50 to 90% of the compost (mostly due to compression), as indicated by increases in bulk density (Hara 2001; Alemi et al., 2010; Virk et al., 2013; McMullen et al., 2005). Pellets, produced using either fresh or composted animal manure, are generally more uniform as the pre-pelleting process of drying, mixing and size management ensures uniformity of materials to be pelleted. Since pellets take up smaller volume than in their initial form, they can be easily stored and transported (Sharara et al., 2016) with reduced transportation and application costs, but the net economic benefit will be reduced by the cost of pelleting.

Various other benefits have also been suggested. For example, adding particular nutrients to create a "designer fertilizer" can better meet the nutrient needs of specific crops and reduce runoff and/or leaching of nutrients from manure-rich farms into nearby water bodies (Hara, 2001). Also, when the pellets are applied to land, they generate only one-tenth or less of the dust generated by ordinary compost (Hara, 2001). Finally, pellets do not contain faecal bacteria and pathogens, do not have significantly noxious odor (Lopez-Mosquera et al., 2008), and do not contain possible contaminants in the final product (Wolfe et al., 2002).

An additional benefit in humid regions is that the low moisture in pellets (< 10%) means less water is applied to soil compared with fresh manure (50-80% water content) or compost (10-40% water content). Similar to fresh manure and compost, pellet application also increases organic matter, water retention, resistance to erosion, nutrient holding capacity, and nutrient uptake by plants while reducing air pollution when stored or applied to fields. Pellets are easy to handle compared with fresh manure during shipping and field application (Lopez-Mosquera et al., 2008). Poultry manure pellets can be easily applied on farms using conventional fertilizer spreaders and planting equipment (Hamilton and Sims, 1995).

Concerns with Adopting Pelleting Manure

Management Technology

Nitrogen can be lost during the initial drying stage needed to achieve the optimum moisture content for pelleting animal manure. According to Awiszus et al. (2018), drying the separated solid in digestate liquid from 24% to 91 to 93% dry matter content at 70 to 80 °C caused a total nitrogen content reduction of 47 to 48%, ammonium nitrogen by 82 to 87%, phosphorus by 6 to 8%, and sulphur content by 6 to 12% in comparison to the separated material before drying.

A higher runoff P concentration was observed during simulated rainfall events comparing surface-applied PPL to PL on a pasture field with a tall fescue (*Festuca arundincea* Shreb.), bermudagrass and clover mix on a silt loam soil (Haggard et al., 2005). This was attributed to the higher water-extractable P level in the source material used for the PPL; the PL came from a different source. When comparing PPL to commercial hog manure–sawdust compost for potato production in PEI, Canada, Lynch et al. (2008) reported greater residual nitrate in soil after potato harvest with pellets than with compost, and leaching losses of NO_3-N occurring in non-growing seasons, especially when pellets were applied at a higher rate (600 kg N ha⁻¹).

Using PPL and PL (120 kg N ha⁻¹) for the production of Komatsuna (*Brassica rapa* L.) in Japan, Hayakawa et al. (2009) reported cumulative N_2O emissions from pelleted poultry manure (2.72 ± 0.22 kg N ha⁻¹ y⁻¹) were 3.9 and 7.1 times higher than from poultry manure and mineral fertilizer, respectively. Incubation tests indicated that anaerobic conditions inside the pellets, caused by rainfall and heterotrophic microbial activities, led to denitrification and resulted in high N_2O fluxes. In a controlled 28-day laboratory study testing CO_2 and N_2O emissions from surface-applied PL (in fine particles or pellets) and soil moisture levels (58 or 90% water-filled pore space), Cabrera et al. (1994) found "PPL may produce similar CO_2 emission as fine-particle litter, and may cause equal or larger emissions of N_2O than fine-particle PL."

Higher N_2O emissions from soil amended with pellets than unpelleted separated solid from anaerobically digested cattle manure or un-digested cattle manure, likely due to the concentrated microsites of N in pellets, were also reported in a controlled laboratory study using two common Canadian soils (Chiyoka et al., 2011). Greater N_2O emission resulted when swine compost pellets were mixed with soil rather than surface-applied based on a controlled 57-day lab incubation study (Pampuro et al., 2017). The pH of the pellets affected the occurrence of a N_2O emission peak; lowering the pH of cattle manure compost pellets to pH 5.3 (the pH of N-enriched cattle manure compost pellets) resulted in the elimination of the N_2O emission peak at the start of 90 day incubation, but it did not decrease the overall cumulative emission (Yamane and Kubotera, 2016). Therefore, factors other than pellet pH, such as high nitrate contents of pellets, might also explain the mechanism underlying the suppressed N_2O emission in N-enriched cattle manure compost pellets.

The factor that has most impeded the adoption of manure pelleting technologies to manage animal manure is the cost of the pelleting process. Using dairy manure as an example, Sharara et al. (2016) concluded that "pelletization only marginally reduces the transportation cost (i.e., extremely large transportation distance is needed to offset the processing cost)" although the cost per unit of pellets produced decreases as the dairy farm size (number of cows) increases.

Summary and Conclusion

Pelleting is a promising option to manage animal manure. Compared with unpelleted animal manure, the benefits of pelleting include, but are not limited to, redistribution of manure nutrients further away from areas with excess nutrient accumulation to areas with nutrient deficiency, reduced reliance on synthetic fertilizers in those more distant areas, ability to modify the nutrient level and composition to meet specific crop needs, and reduced cost of storage, transportation, and field application. However, we should take into consideration that animal manure pellets are a SRF. When using pellets, application time needs to be earlier than for mineral fertilizers and unpelleted animal manure in order to have sufficient time for the mineralization and release of nutrient for current year crop uptake. The cost of producing manure pellets and other environmental impacts, particularly increased N_2O emission, should also be taken into consideration when deciding on the technology to produce pellets for on-farm or off-farm use.

References

Adeli, A., J.C. McCarty, J.J. Read, J.L. Willers, G. Feng, and J.N. Jenkins. 2016. Subsurface band placement of pelletized poultry litter in cotton. Agron. J. 108:1356–1366. doi:10.2134/agronj2015.0373

Adeli, A., J.J. Read, J. McCarty, J.N. Jenkins, and G. Feng. 2015. Soybean yield and nutrient utilization following long-term pelleted broiler litter application to cotton. Agron. J. 107:1128–1134. doi:10.2134/agronj14.0497

Adeli, A., H. Tewolde, and J.N. Jenkins. 2012. Broiler litter type and placement effects on corn growth, nitrogen utilization, and residual soil nitrate-nitrogen in a no-till field. Agron. J. 104:43–48. doi:10.2134/agronj2011.0093

Alcaraz, J., E. Baticados, S. Capareda, A. Maglinao, Jr., and H. Nam. 2018. Drying and pellet characterization of sand seperated dairy manure from Stephenville Texas, USA. Appl. Ecol. Environ. Res. 16(1):29–38. doi:10.15666/aeer/1601_029038

Alemi, H., M.H. Kianmehr, and A.M. Borghaee. 2010. Effect of pellet processing of fertilizer on slow-release nitrogen in soil. Asian J. Plant Sci. 9:74–80. doi:10.3923/ajps.2010.74.80

Ausmus, S. 2005. ARS research turns poultry waste into toxin-grabbing char. United States Department of Agriculture. AgResearch Magazine. Washington, D.C. https://agresearchmag.ars.usda.gov/2005/jul/char last (Accessed 25 May 2018).

Awiszus, S., K. Meissner, S. Reyer, and J. Müller. 2018. Ammonia and methane emissions during drying of dewatered biogas digestate in a two-belt conveyor dryer. Bioresour. Technol. 247:419–425. doi:10.1016/j.biortech.2017.09.099

Bagheri, R., G. Ali-Akbari, M. Hossein-Kianmehr, Z.A.-T. Sarvastani, and M.-Y. Hamzekhanlu. 2011. The effect of pellet fertilizer application on corn yield and its components. Afr. J. Agric. Res. 6(10):2364–2371.

Blair, B. 1974. Evaluation of dehydrated poultry waste as a feed ingredient for poultry. Fed. Proc. Fed. Am. Soc. Exp. Biol. 33:1934–1936.

Blair, R.M., M.C. Savin, and P. Chen. 2014. Phosphatase activities and available nutrients in soil receiving pelletized poultry litter. Soil Sci. 179:182–189. doi:10.1097/SS.0000000000000061

Brambilla, M., N. Pampuro, M. Cutini, E. Romano, P.E. Foppa, E. Cavallo, and C. Bisaglia. 2017. Effect of composted biomass moisture on pelleted fertilizers from swine manure solid fraction. In 25th European Biomass Conference and Exhibition, 12-15 June 2017. Stockholm, Sweden. p.1318–1323.

Cabrera, M.L., S.C. Chiang, W.C. Merka, O.C. Pancorbo, and S.A. Thompson. 1994. Pelletizing and soil water effects on gaseous emissions from surface-applied poultry litter. Soil Sci. Soc. Am. J. 58:807–811. doi:10.2136/sssaj1994.03615995005800030024x

Cely, P., G. Gascó, J. Paz-Ferreiro, and A. Méndez. 2015. Agronomic properties of biochars from different manure wastes. J. Anal. Appl. Pyrolysis 111:173–182. doi:10.1016/j.jaap.2014.11.014

Chiyoka, W.L., X. Hao, F. Zvumoya, and X. Li. 2011. Nitrous oxide emissions from Chernozemic soils amended with anaerobically digested beef cattle feedlot manure: A laboratory study. Anim. Feed Sci. Technol. 166-167:492–502. doi:10.1016/j.anifeedsci.2011.04.035

Chiyoka, W.L., F. Zvomuya, and X. Hao. 2014a. Changes in nitrogen availability during laboratory incubation of Chernozemic soils amended with solid digestate from anaerobically digested cattle manure. Soil Sci. Soc. Am. J. 78:843–851. doi:10.2136/sssaj2013.07.0297

Chiyoka, W.L., F. Zvomuya, and X. Hao. 2014b. A bioassay of nitrogen availability in soils amended with solid digestate from anaerobically digested beef cattle feedlot manure. Soil Sci. Soc. Am. J. 78:1291–1230. doi:10.2136/sssaj2013.01.0030

Cornell Waste Management Institute. 2018. Using manure solids as bedding. Cornell Waste Management Institute, Ithaca, NY. http://cwmi.css.cornell.edu/bedding.htm (Accessed 31 May 2018).

Eghball, B., and J.F. Power. 1994. Beef cattle feedlot manure management. J. Soil Water Conserv. 49(2):113–122.

Feeco International. 2018. The organic granulation hand book series: Manure Feeco International, Green Bay, WI. http://feeco.com/organic-systems/ (Accessed 31 May 2018).

Fellet, G., M. Marmiroli, and L. Marchiol. 2014. Elements uptake by metal accumulator species grown on mine tailings amended with three types of biochar. Sci. Total Environ. 468–469:598–608. doi:10.1016/j.scitotenv.2013.08.072

Flavel, T.C., and D.V. Murphy. 2006. Carbon and nitrogen mineralization rates after application of organic amendments to soil. J. Environ. Qual. 35:183–193. doi:10.2134/jeq2005.0022

Flotats, X., A. Bonmatí, B. Fernández, and A. Magrí. 2009. Manure treatment technologies: On-farm versus centralized strategies. NE Spain as case study. Bioresour. Technol. 100(22):5519–5526. doi:10.1016/j.biortech.2008.12.050

Gilley, J.E., and L.M. Risse. 2000. Runoff and soil loss as affected by the application of manure. Trans. ASAE 43(6):1583–1588. doi:10.13031/2013.3058

Golden, B.R., N.A. Slaton, R.J. Norman, E.E. Gbur, K.R. Brye, and R.E. DeLong. 2006. Recovery of nitrogen in fresh and pelletized poultry litter by rice. Soil Sci. Soc. Am. J. 70:1359–1369. doi:10.2136/sssaj2005.0298

Hadas, A., B. Bar-Yosef, S. Davidov, and M. Sofer. 1983. Effect of pelleting, temperature, and soil type on mineral nitrogen release from poultry and dairy manures. Soil Sci. Soc. Am. J. 47:1129–1133. doi:10.2136/sssaj1983.03615995004700060014x

Haggard, B.E., P.B. DeLaune, D.R. Smith, and P.A. Moore, Jr. 2005. Nutrient and b17-estradiol loss in runoff water from poultry litters. J. Am. Water Resour. Assoc. 41:245–256. doi:10.1111/j.1752-1688.2005.tb03732.x

Hamilton, C.M., and J.T. Sims. 1995. Nitrogen and phosphorus availability in enriched, pelletized poultry litters. J. Sustain. Agric. 5(3):115–132. doi:10.1300/J064v05n03_09

Hammac, W.A., II, C.W. Wood, B.H. Wood, O.O. Fasina, Y. Feng, and J.N. Shaw. 2007. Determination of bioavailable nitrogen and phosphorus from pelletized broiler litter. Sci. Res. Essays 2(4):89–94.

Hao, X., and C. Chang. 2003. Does long-term heavy cattle manure application increase salinity of a clay loam soil in semi-arid southern Alberta? Agric. Ecosyst. Environ. 94(1):89–103. doi:10.1016/S0167-8809(02)00008-7

Hara, M. 2001. Fertilizer pellets made from composted livestock manure. Extension Bulletin-Food and Fertilizer Technology Center, Central Asian Pacific Region, Taipei, Taiwan. p. 1–12. http://www.fftc.agnet.org/library.php?func=view&id=20110801154610 (Accessed 31 May 2018).

Hayakawa, A., H. Akiyama, S. Sudo, and K. Yagi. 2009. N2O and NO emissions from an Andisol field as influenced by pelleted poultry manure. Soil Biol. Biochem. 41:521–529. doi:10.1016/j.soilbio.2008.12.011

He, Z., editor. 2012. Applied research of animal manure: Challenges and opportunities beyond the adverse environmental concerns. Nova Science Publishers, New York.

He, Z., M. Guo, N. Lovanh, and K.A. Spokas. 2012. Applied manure research-Looking forward to the benign roles of animal manure in agriculture and the environment. In: Z. He, editor, Applied research of animal manure: Challenges and opportunities beyond the adverse environmental concerns. Nova Science Publishers, New York. p. 299–309.

He, Z., and D.C. Olk. 2011. Manure amino compounds and their bioavailability. In: Z. He, editor, Environmental chemistry of animal manure. Nova Science Publishers, Inc., New York. p. 179–199.

He, Z., P.H. Pagliari, and H.M. Waldrip. 2016. Applied and environmental chemistry of animal manure: A review. Pedosphere 26:779–816. doi:10.1016/S1002-0160(15)60087-X

He, Z., Z.N. Senwo, H. Zou, I.A. Tazisong, and D.A. Martens. 2014. Amino compounds in poultry litter, litter-amended pasture soils and grass shoots. Pedosphere 24:178–185. doi:10.1016/S1002-0160(14)60004-7

He, Z., and H. Zhang, editors. 2014. Applied manure and nutrient chemistry for sustainable agriculture and environment. Springer, Amsterdam, Netherlands. p. 1–379. doi:10.1007/978-94-017-8807-6

He, Z., H. Zhang, G.S. Toor, Z. Dou, C.W. Honeycutt, B.E. Haggard, and M.S. Reiter. 2010. Phosphorus distribution in sequentially-extracted fractions of biosolids, poultry litter and granulated products. Soil Sci. 175:154–161. doi:10.1097/SS.0b013e3181dae29e

Jackson, D.J., B.J. Rude, K.K. Karanja, and N.C. Whitley. 2005. Utilization of poultry litter pellets in meat goat diets. Small Ruminant Research 66(1-3):278–281. doi:10.1016/j.smallrumres.2005.09.005

Jewell, R.E. 1962. Drying and storage of NPK granular fertilisers based on superphosphate and containing urea as a plant nutrient. J. Sci. Food Agric. 13:414–422. doi:10.1002/jsfa.2740130803

Kleinman, P., K. Blunk, R. Bryant, L. Saporito, D. Beegle, K. Czymmek, Q. Ketterings, T. Sims, J. Shortle, J. McGrath, F. Coale, M. Dubin, D. Dostie, R. Maguire, R. Meinen, A. Allen, K. O'Neill, L. Garber, M. Davis, B. Clark, K. Sellner, and M. Smith. 2012. Managing manure for sustainable livestock production in the Chesapeake Bay Watershed. J. Soil Water Conserv. 67(2):54A–61A. doi:10.2489/jswc.67.2.54A

Larney, F.J., K. Buckley, X. Hao, and P. McCaughey. 2006. Fresh, stockpiled, and composted beef cattle feedlot manure. J. Environ. Qual. 35(5):1844–1854. doi:10.2134/jeq2005.0440

Larney, F.J., X. Hao, and E. Topp. 2011. Manure management. In: J.L. Hatfield and T. J. Sauer, editors, "Soil management: Building a stable base for agriculture. ASA, SSSA, Madison, WI. p. 246–263.

Lee, J.W., M. Kidder, B.R. Evans, S. Paik, A.C. Buchanan, C.T. Garten, and R.C. Brown. 2010. Characterization of biochars produced from cornstovers for soil amendment. Environ. Sci. Technol. 44:7970–7974. doi:10.1021/es101337x

Leskiw, L.A., C.M. Welsh, and T.B. Zeleke. 2012. Effect of subsoiling and injection of pelletized organic matter on soil quality and productivity. Can. J. Soil Sci. 92:269–276. doi:10.4141/cjss2011-003

Lima, I.M., K.T. Classon, and M. Uchimiya. 2016. Selective release of inorganic constituents in broiler manure biochars under different post-activation treatments. J. Residuals Sci. Technol. 13(1):37–48. doi:10.12783/issn.1544-8053/13/1/6

Lima, I.M., and W.E. Marshall. 2005a. Adsorption of selected environmentally important metals by poultry manure-based granular activated carbons. J. Chem. Technol. Biotechnol. 80:1054–1061. doi:10.1002/jctb.1283

Lima, I.M., and W.E. Marshall. 2005b. Utilization of turkey manure as granular activated carbon: Physical, chemical and adsorptive properties. Waste Manag. 25(7):726–732. doi:10.1016/j.wasman.2004.12.019

Lima, I.M., A. McAloon, and A.A. Boateng. 2008. Activated carbon from broiler litter: Process description and cost of production. Biomass Bioenergy 32:568–572. doi:10.1016/j.biombioe.2007.11.008

Lopez-Mosquera, M., F. Cabaleiro, M. Sainz, A. López-Fabal, and E. Carral. 2008. Fertilizing value of broiler litter: Effects of drying and pelletizing. Bioresour. Technol. 99:5626–5633. doi:10.1016/j.biortech.2007.10.034

Lynch, D., Z. Zheng, B. Zebarth, and R. Martin. 2008. Organic amendment effects on tuber yield, plant N uptake and soil mineral N under organic potato production. Renew. Agric. Food Syst. 23(03):250–259. doi:10.1017/S1742170508002330

Mazeika, R., G. Staugaitis, and J. Baltrusaitis. 2016. Engineered pelletized organo-mineral fertilizers (OMF) from poultry manure, diammonium phosphate and potassium chloride. ACS Sustain. Chem.& Eng. 4:2279–2285. doi:10.1021/acssuschemeng.5b01748

McMullen, J., O.O. Fasina, C.W. Wood, and Y. Feng. 2005. Storage and handling characteristics of pellets from poultry litter. Appl. Eng. Agric. 21(4):645–651. doi:10.13031/2013.18553

Melero-Vara, J.M., C.J. López-Herrera, A.M. Prados-Ligero, M.D. Vela-Delgado, J.A. Navas-Becerra, and M.J. Basallote-Ureba. 2011. Effects of soil amendment with poultry manure on carnation Fusarium wilt in greenhouses in southwest Spain. Crop Prot. 30:970–976. doi:10.1016/j.cropro.2011.03.022

Meyer, D.M., I. Garnett, and J. Guthrie. 1997. A Survey of dairy manure management practices in California. J. Dairy Sci. 80(8):1841–1845. doi:10.3168/jds.S0022-0302(97)76119-8

Mieldazy, R., E. Jotautiene, A. Jasinskas, and A. Aboltins. 2017. Valuation of physical mechanical properties of experimental granulated cattle manure compost fertilizer. 16th International Scientific Conference Engineering for Rural Development. May 24-26, 2017. Jelgava, Latvia. Page 575–580.

Mieldazys, R., E. Jotautiene, A. Pocius, and A. Jasinskas. 2016. Analysis of organic agricultural waste usage for fertilizer production. Agronomy Research 14:143–149.

Nagy, D., P. Balogh, Z. Gabnai, J. Popp, J. Oláh, and A. Bai. 2018. Economic analysis of pellet production in co-digestion biogas plants. Energies 11:1135. doi:10.3390/en11051135

Obeidat, B.S., A.Y. Abduallah, M.A. Mayyas, and M.S. Awawdeh. 2016. The potential use of layer litter in Awassi lambs' diet: It's effects on nutrient intake, digestibility, N balance, and growth performance. Small Rumin. Res. 137:24–27. doi:10.1016/j.smallrumres.2016.02.022

Oh, I.H., E.Y. Choi, and H. Park. 2013. Optimizing the production process of fuel pellets from swine dung compost. Appl. Eng. Agric. 29:539–545.

Pampuro, N., C. Bertora, D. Sacco, E. Dinuccio, C. Grignani, P. Balsari, E. Cavallo, and M.P. Berna. 2017. Fertilizer value and greenhouse gas emissions from solid fraction pig slurry compost pellets. J. Agric. Sci. 155:1646–1658. doi:10.1017/S002185961700079X

Pocius, A., E. Jotautiene, E. Zvicevicius, and S. Savickiene. 2017. Investigation of effects of organic fertilizer pellet rheological and geometric properties on mechanical strength. 16th International Scientific Conference Engineering for Rural Development. 24–26 May 2017. p. 1503–1508. doi:10.22616/ERDev2017.16.N339

Poffenbarger, H.J., S.B. Mirsky, M. Kramer, R.R. Weil, J.J. Meisinger, M.A. Cavigelli, and J.T. Spargo. 2015b. Cover crop and poultry litter management influence spatiotemporal availability of topsoil nitrogen. Soil Sci. Soc. Am. J. 79:1660–1673. doi:10.2136/sssaj2015.03.0134

Poffenbarger, H.J., S.B. Mirsky, R.R. Weil, M. Kramer, J.T. Spargo, and M.A. Cavigelli. 2015a. Legume proportion, poultry litter, and tillage effects on cover crop decomposition. Agron. J. 107:2083–2096. doi:10.2134/agronj15.0065

Prasai, T.P., K.B. Walsh, D.J. Midmore, B.E.H. Jones, and S.P. Bhattarai. 2018. Manure from biochar, bentonite and zeolite feed supplemented poultry: Moisture retention and granulation properties. J. Environ. Manage. 216:82–88. doi:10.1016/j.jenvman.2017.08.040

Purnomo, C.W., S. Indarti, C. Wulandari, H. Hinode, and K. Nakasaki. 2017. Slow release fertiliser production from poultry manure. Chem. Eng. Trans. 56:1531–1536.

Qian, P., and J. Schoenau. 2002. Availability of nitrogen in solid manure amendments with different C:N ratios. Can. J. Soil Sci. 82(2):219–225. doi:10.4141/S01-018

Rao, J.R., M. Watabe, T.A. Stewart, B.C. Millar, and J.E. Moore. 2007. Pelleted organo-mineral fertilisers from composted pig slurry solids, animal wastes and spent mushroom compost for amenity grasslands. Waste Manag. 27:1117–1128. doi:10.1016/j.wasman.2006.06.010

Reiter, M.S., T.C. Daniel, N.A. Slaton, and R.J. Norman. 2014. Nitrogen availability from granulated fortified poultry litter fertilizers. Soil Sci. Soc. Am. J. 78:861–867. doi:10.2136/sssaj2013.02.0065

Romano, E., M. Brambilla, C. Bisaglia, N. Pampuro, E.F. Pedretti, and E. Cavallo. 2014. Pelletization of composted swine manure solid fraction with different organic co-formulates: effect of pellet physical properties on rotating spreader distribution patterns. International Journal of Recycling of Organic Waste in Agriculture 3:101–111. doi:10.1007/s40093-014-0070-2

Sahin, O., M.B. Taskin, Y.K. Kadioglu, A. Inal, D.J. Pilbeam, and A. Gunes. 2014. Elemental composition of pepper plants fertilized with pelletized poultry manure. J. Plant Nutr. 37:458–468. doi:10.1080/01904167.2013.864307

Sakurada L., R., M.A. Batista, T.T. Inoue, A.S. Muniz, and P.H. Pagliari. 2016. Organomineral phosphate fertilizers: agronomic efficiency and residual effect on initial corn development. Agron. J. 108:2050–2059. doi:10.2134/agronj2015.0543

Sharara, M.A., Q. Yang, T.L. Cox, and T. Runge. 2016. Techno-economic assessment of dairy manure granulation. 2016 ASABE Annual International Meeting, Orlando, FL. 17–20 July 2016. ASABE, St. Jospeh, MI. doi:10.13031/aim.20162460305 (Accessed 11 July 2018).

Souri, M.K., M. Rashidi, and M.H. Kianmehr. 2018. Effects of manure-based urea pellets on growth, yield, and nitrate content in coriander, garden cress, and parsley plants. J. Plant Nutr. 41(11):1405–1413. doi:10.1080/01904167.2018.1454471

Spencer, R. 2016. Dairy beds with manure solids. BioCycle 57:62 https://www.biocycle.net/2016/09/15/dairy-beds-manure-solids/ (Accessed 22 Feb. 2018).

Sprinkle, A.L., D.J. Hansen, and S.E. White-Hansen. 2011. Pelletized poultry litter as a nutrient source for turfgrass sports fields. J. Agric. Sci. Tech. 5:399–411.

Statistics Canada. 2015. A geographical profile of livestock manure production in Canada. Ottawa, ON. http://www.statcan.gc.ca/pub/16-002-x/2008004/tbl/manure-fumier/tbl001-man-fum-eng.htm (Accessed 31 May 2018).

Suppadit, T., S. Pongpiachan, and S. Panomsri. 2012. Effects of moisture content in quail litter on the physical characteristics after pelleting using a Siriwan Model machine. Anim. Sci. J. 83:350–357. doi:10.1111/j.1740-0929.2011.00961.x

Suppadit, T., and P. Poungsuk. 2010. Utilization of broiler litter pellets to substitute mixed feed pellets in fattening steers. J. Int. Soc. Southeast Asian Agr. Sci. 16(1):55–67.

Takahashi, S., H. Ihara, and T. Karasaw. 2016. Compost in pellet form and compost moisture content affect phosphorus fractions of soil and compost. Soil Sci. Plant Nutr. 62(4):399–404. doi:10.1080/00380768.2016.1198680

Thomas, B.W., X. Li, V. Nelson, and X. Hao. 2017. Anaerobically digested beef cattle manure supplies more nitrogen with less phosphorus accumulation than undigested manure. Agron. J. 109:836–844. doi:10.2134/agronj2016.12.0719

Toor, G.S., B.E. Haggard, M.S. Reiter, T.C. Daniel, and A.M. Donoghue. 2007. Phosphorus solubility in poultry litters and granulates: Influence of litter treatments and extraction ratios. Trans. ASAE 50:533–542. doi:10.13031/2013.22641

United Nations. 2015. World population projected to reach 9.7 billion by 2050. United Nations, New York. http://www.un.org/en/development/desa/news/population/2015-report.html (Accessed 1 June 2018).

USEPA. 1997 Alternative technologies/uses for manure. USEPA, Washington, D.C. https://www3.epa.gov/npdes/pubs/cafo_report.pdf (Accessed 31 May 2018)

USEPA. 2013. Literature review of contaminants in livestock and poultry manure and implications for water quality. EPA 820-R-13–002. Office of Water, Washington, D.C.

Virk, S.S., J.P. Fulton, O.O. Fasina, and T.P. McDonald. 2013. Influence of broiler litter bulk density of metering and distributing for a spinner-disc spreader. Appl. Eng. Agric. 29:473–482.

Whirlston Machinery. 2018a. Organic fertilizer pellets extrusion line with a 30,000 tons annual output. Whirlston Machinery, Zhengzhou, China. https://organicfertilizermachine.com/product/production-line/manure-fertilizer-pellets-production-line (Accessed 31 May 2018).

Whirlston Machinery. 2018b. Small scale organic fertilizer plant – 800 kg/h. Whirlston Machinery, Zhengzhou, China. http://organicfertilizermachine.com/product/production-line/small-scale-organic-fertilizer-production-plant (Accessed 31 May 2018).

Wichuk, K.M., and D. McCartney. 2012. Animal manure composting: stability and maturity evaluation. In: Z. He, editor, Applied research of animal manure-Challenges and opportunities beyond the adverse environmental concerns. Nova Science Publishers, New York. p. 203–262.

Wild, P.L., C. van Kessel, J. Lundberg, and B.A. Linquist. 2011. Nitrogen availability from poultry litter and pelletized organic amendments for organic rice production. Agron. J. 103:1284–1291. doi:10.2134/agronj2011.0005

Wohlfarth, G., and G. Schroeder. 1979. Use of manure in fish farming - A review. Agric. Wastes 1(4):279–299. doi:10.1016/0141-4607(79)90012-X

Wolfe, K., C. Ferland, and J. McKissick. 2002. The feasibility of operating a poultry litter pelletizing facility in south Georgia. USA. University of Georgia, Center for Agribusiness and Economic Development. Athens, GA.

Yamane, T., and H. Kubotera. 2016. Impacts of pellet pH on nitrous oxide emission rates from cattle manure compost pellets. Soil Sci. Plant Nutr. 62(5–6):561–566. doi:10.1080/00380768.2016.1239510

Zafari, A., and M. Kianmehr. 2012a. Management and reduction of chemical nitrogen consumption in agriculture. Am. J. Plant Sci. 3(12):1827–1834. doi:10.4236/ajps.2012.312A224

Zafari, A., and M.H. Kianmehr. 2012b. Effect of temperature, pressure and moisture content on durability of cattle manure pellet in open-end die method. J. Agric. Sci. 4(5):203–208.

Zebarth, B., R. Chabot, J. Coulombe, R. Simard, J. Douheret, and N. Tremblay. 2005. Pelletized organo-mineral fertilizer product as a nitrogen source for potato production. Can. J. Soil Sci. 85(3):387–395. doi:10.4141/S04-071

Zhang, H., and J. Schroder. 2014. Animal manure production and utilization in the US, p. 1-21, In Z. He and H. Zhang, eds. Applied manure and nutrient chemistry for sustainable agriculture and environment. Springer, Amsterdam, the Netherlands.

Whole Farm Modeling: A Systems Approach to Understanding and Managing Livestock for Greenhouse Gas Mitigation, Economic Viability and Environmental Quality

Mukhtar Ahmed,* Shakeel Ahmad, Heidi M. Waldrip, Mohammad Ramin, Muhammad Ali Raza

Abstract

Climate change is a current major concern worldwide, largely due to increasing global emissions of greenhouse gases (GHG), such as carbon dioxide (CO_2), methane (CH_4), and nitrous oxide (N_2O). These gases are also derived from animal respiration (CO_2) and manure management (N_2O and CH_4). Increased population and rising food security concerns has promoted the practices of intensive use of resources such as CAFO (Concentrated Animal Feeding Operations). This resulted to the production of GHG emissions (11% of global) from agriculture sector (Bovine enteric CH_4 and N fertilization). The level of methane [100-yr Global Warming Potential (GWP)] is 28–36 times greater than CO_2 equivalent (CO_2e) mainly comes from anthropogenic activities which includes animal husbandry (27%), rice cultivation (26%), fossil fuel (28%), waste management (13%) and biomass production (9%). Livestock accounts for 14.5% of total GHG in which 44% are enteric CH_4 while 29% are from animal excreta as N_2O. Manure management is responsible for the production of 8% and 40% of CH_4 and N_2O emissions, respectively. To ensure environmental sustainability and minimization of GHG emission, adaptation and mitigation strategies are required. Accurate estimation/quantification is the first step to have a successful plan of action. Process-based modeling could be an option to quantify and evaluate emission in any system. Farm models that can efficiently simulate detailed biochemical process should be used to predict GHG emissions. We suggest here the use of Whole Farm Models as they help in testing the efficacy of various strategies to reduce GHG emissions.

Increased population and industrial activities have resulted in negative environmental impacts, globally. With population gains, reduced land availability for cropping, and improved standards of living in developing nations, food security (i.e., a condition related to food supply and accessibility) is a major concern. We are faced with the challenge of feeding 9 to 10 billion people by 2050, many of whom have become accustomed to an animal-based Western-style diet (Smith, 2013). This increase in consumption of meat, milk, and eggs has led to the construction of large numbers of concentrated animal feeding operations (CAFO) that may contain hundreds of thousands of animals at any one time, which produce manure

M. Ahmed and M. Ramin, Department of Agricultural Research for Northern Sweden, Swedish University of Agricultural Sciences, Umeå-90183, Sweden; M. Ahmed, Department of Agronomy, Pir Mehr Ali Shah Arid Agriculture University, Rawalpindi-46300, Pakistan; M. Ahmed, Biological Systems Engineering, Washington State University, Pullman, WA 99164-6120; S. Ahmad, Department of Agronomy, Bahauddin Zakariya University, Multan-60800, Pakistan; S. Ahmad, Department of Biological and Agricultural Engineering, The University of Georgia, Griffin, GA 30223, USA; H.M. Waldrip, USDA-ARS Conservation and Production Research Laboratory PO Drawer 10, 300 Simmons Rd Bushland, TX 79012, Heidi.waldrip@ars.usda.gov; M.A. Raza, College of Agronomy, Sichuan Agricultural University, Chengdu, 611130, PR China. *Corresponding author email: muhtar.ahmed@slu.se

doi:10.2134/asaspecpub67.c25

Animal Manure: Production, Characteristics, Environmental Concerns and Management. ASA Special Publication 67. Heidi M. Waldrip, Paulo H. Pagliari, and Zhongqi He, editors.
© 2019. ASA and SSSA, 5585 Guilford Rd., Madison, WI 53711, USA.

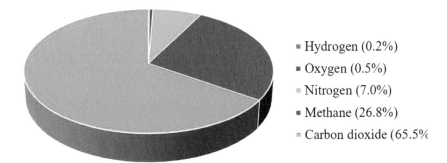

- Hydrogen (0.2%)
- Oxygen (0.5%)
- Nitrogen (7.0%)
- Methane (26.8%)
- Carbon dioxide (65.5%

Fig. 1. Average percentage of typical composition of rumen gases (Source: Sniffen and Herdt, 1991).

that can cause environmental risk to air, soil, and water if improperly managed. However, it has proven difficult to accurately predict the fate of airborne nitrogen (N) and carbon (C) from animals and their manure. Atmospheric concentrations of greenhouse gases (GHG) (i.e., CO_2, CH_4 and N_2O) have increased since preindustrial times due to human (i.e., anthropogenic) activities, including fossil fuel use, deforestation, industry, transportation, and crop and livestock production. The predominant issue with GHG release is the greenhouse effect, where in accumulated GHG blocks solar energy refraction from the earth and leads to a general warming trend, along with unpredictable weather patterns.

Smith et al. (2014) reported that a quarter of GHG comes from agriculture, forestry, and other land use change. Among them, 11% of global GHG emissions have been attributed to agriculture, primarily from bovine enteric CH_4 and N fertilization of cropland. Cropping systems are a significant source of both direct and indirect N_2O emissions following application of inorganic fertilizer or manure to soil. Indirect N_2O emissions occur when volatilized ammonia (NH_3) from fertilized land or livestock operations is deposited downward via either wet or dry deposition, and then undergoes nitrification in the soil to form N_2O (VanderZaag et al., 2011; Smith and Mukhtar, 2015).

The industry provides meat, milk, and eggs to feed the world's population, but CAFO must be managed carefully to ensure environmental sustainability and minimize GHG emissions (Fig. 1). There has been a push toward regulation, monitoring, and reduction of CAFO-derived GHG. Enteric fermentation, a complex anaerobic decomposition process facilitated by microorganisms, is largely responsible for livestock-derived CH_4 emissions (Capper and Bauman, 2013; Vergé et al., 2007), although manure storages can also be sources of CH_4 (Todd et al., 2011; Sun et al., 2015).

Lassey (2007) reported that CH_4, with a 100-yr global warming potential (GWP) of 28 to 36 times greater than CO_2 equivalent (CO_2e), was the second largest contributor to climate change after CO_2."Carbon dioxide equivalent" or "CO_2e" is a term for describing different greenhouse gases in a common unit. For any quantity and type of greenhouse gas, CO_2e signifies the amount of CO_2 which would have the equivalent global warming impact. Anthropogenic activities are responsible for 70% of CH_4 production, including animal husbandry (27%), rice cultivation (26%), fossil fuels (26%), waste management (13%) and biomass burning (9%) (Sejian et al., 2018). Despite the varied sources of anthropogenic CH_4, the livestock sector has received a good deal of attention due to their large contribution to GHG emissions, which can have global impacts on the environment (Steinfeld et al., 2006). Gerber et al. (2013) reported that livestock account for 14.5% of total GHG emissions. Among that, 44% are enteric CH_4, while 29% are derived from animal excreta as N_2O. Mosier et al. (1998a) reported that agriculture accounts for 50% anthropogenic emission of CH_4 and major sources are anaerobic turnover in rice paddies and enteric fermentation in ruminant animals. Manure-derived CH_4 accounts for 8% of the agricultural emissions. Similarly, 80% of global anthropogenic N_2O emissions comes from agriculture and 40% comes from manure management (Mosier et al., 1998b).

Globally, cattle produces 5335 M tons of CO_2e (11% of all human induced GHG emissions). The bovine rumen can store up to 40 to 60 gallons (151–227 L) of material, which contains about 150 billion microorganisms per teaspoon (4.9 mL). The rumen essentially functions as a large fermentation vat, where complex polymers are degraded via bacteria, fungi, protozoa and archea. Methanogens and denitrifiers exist within this diverse population, leading to the production of CH_4 and N_2O as byproducts, which are released by animal eructation or flatulence: these GHGs contribute to climate change (Sniffen and Herdt, 1991). Steinfeld et al. (2006) reported 7516 M tons per year of CO_2e (18% of anthropogenic gases).

Emission Factors

The fundamental tools used for GHG emissions inventory and control strategies are emission factors (EF). The EF approach has been applied to inventory local, regional, national and global GHG footprints of different livestock systems (EPA, 2018). An EF is a "representative" value, typically derived from literature review or limited data, that quantifies a pollutant released to the atmosphere and source of that pollutant. The amounts of ammonia (NH_3) or dinitrogen oxide (N_2O) released from N sources applied are determined by using EF. According to Dong et al. (2006) N_2O emission from soil receiving organic amendments are equal to 1%. The United Kingdom currently uses the Intergovernmental Panel on Climate Change (IPCC) tier 1 EF in its national N_2O inventory. Default EF by IPCC for NH_3 emission after manure application is 20% of the N applied. Emission factors for N_2O can be calculated from following equation:

$$EF = \left[\frac{\text{Cumulative } N_2O \text{ flux (Kg } N_2O-N) - \text{Cumulative } N_2O \text{ flux from control (Kg } N_2O-N)}{N \text{ applied (Kg N)}} \right]$$

Emission factors for NH_3 could be calculated by following equations where EFs for control plot were assumed to be zero.

$$EF = \left[\frac{\text{Cumulative } NH_3 \text{ flux (kg } NH_3-N)}{N \text{ applied (Kg N)}} \right] \times 100$$

The general equation for EF (kg of GHG) estimation is:

$$E = A \times EF \times (1 - \frac{ER}{100})$$

Where E = emissions; A = activity or production rate (varies for each specific GHG); EF = emission factor, and ER = overall emission reduction efficiency (%).

Grassland ecosystems can act as a source or sink for GHG such as CO_2, N_2O and CH_4 and this is possible by management of grazing and the fertilizer regime (Oates and Jackson, 2014). In grazing systems, CH_4 and N_2O are two main sources of GHG emissions (Beauchemin et al., 2011). The IPCC direct EF for N_2O is 2% of N deposited as urine and feces, or 0.02 g N_2O–N g^{-1} N deposited. However, large variations have been reported for N_2O EF, ranging from 0.1% to 4.0% (de Klein et al., 2003). This uncertainty could be because of different variable such as type of fertilizer, crop residue, atmospheric deposition, land use, soil type, and precipitation. The IPCC (Dong et al., 2006) has defined these as Tier 2 factors; an example is when enteric CH_4 is represented as a function of gross energy intake of the animal. A similar but usually more detailed model is an empirical relationship where a process is described as a function of multiple factors. This may be a purely statistical model based on measured data without much understanding of the underlying process or a relationship developed to represent the process using linear or nonlinear functions. The most detailed model is a more mechanistic process simulation that uses multiple relationships to represent the dynamics within the process. Examples of the various types of models will be discussed for each of the important sources of GHG emissions on a dairy farm. CH_4 yield (kJ MJ^{-1} or L kg^{-1} DM) were predicted by Ramin and Huhtanen, (2013) using regression based approach developed for dairy cows, growing cattle and sheep.

$$CH_4[kJ\ MJ^{-1}] = -0.6 - 0.70\,DMIBW + 0.076 \times OMD - 0.13 \times EE + 0.046 \times NDF + 0.044 \times NFC$$

$$CH_4[Lkg^{-1}DMI] = -5.0 - 0.35\,DMIBW + 0.031 \times OMD - 0.043 \times EE + 0.018 \times NDF + 0.016 \times NFC$$

Where DMIBW (gkg^{-1}) = Total dry matter intake per kg of body weight, OMD (gkg^{-1}) = Organic matter digestibility at maintenance level, EE ($gkg^{-1}DM$) = Ether extract, NDF (gkg^{-1} DM) = Neutral detergent fiber and NFC (gkg^{-1} DM) = Non-fibrous carbohydrates.

Total CH_4 (gd^{-1}) is calculated as CH_4 yield ($kJMJ^{-1}$ or gkg^{-1} DM) × Intake (DMI or GE, gross energy). Coefficient of 0.714 L can be used to convert CH_4 to grams.

The IPCC has given guidelines to be used for estimation of GHG inventories for reporting to the United Nations Framework Convention on Climate Change (UNFCCC) and the Kyoto protocol, which cover all economic sectors including land use, land-use change, and forestry (LULUCF). UNFCCC and the Kyoto Protocol have well established systems for monitoring inventories of developed countries, and this is basis for monitoring progress toward emission reduction targets. The United Nations Collaborative Program on Reducing Emissions from Deforestation and Forest Degradation in Developing Countries (REDD+) activities, inventory estimates are prerequisite for participation in results-based incentive schemes. The IPCC introduced the good practice guidance (GPG) in 2000 and added LULUCF sector in 2003 (GPG2003) for GHG emission (Penman et al., 2003). However, in 2006 the IPCC combined agriculture and LULUCF into a single sector, AFOLU (Agriculture, Forestry, and Other Land Uses) called as 2006 GL (IPCC Guidelines for National Greenhouse Gas Inventories) (Dong et al., 2006). The GPG2003 provides estimation methodologies for GHG emissions from five carbon pools (soil organic matter (OM), biomass above and below ground biomass, litter, and dead wood), and non-CO_2 GHG emissions for six categories of land use (settlements, cropland, grassland, wtland, forest land, and other land). Carbon dioxide emission and removal estimations were calculated by two generic methods provided by IPCC while for gases other than CO_2 product of EF and A (activity) data have been used. Three different approaches have been described in GPG2003 and 2006GL to have activity data. Furthermore, IPCC given tier concept called as the IPCC Tier concept which consist of three tiers. Tier 1 is default EF, Tier 2 is detailed country specific EF and Tier 3 is higher order models used. These models consider seasonal variation in animal population or feed quality and availability, diet composition in detail, concentration of products arising from ruminant fermentation and possible mitigation strategies. Moving from Tier 1 to 3 minimizes the uncertainty in GHG estimates but increased complexity in measurement systems (Processes and analyses). If large portion of country total emissions is coming from enteric fermentation, Tier 2 should be used. The IPCC CH_4 EF is estimated as:

$$EF = \left[GE \times \left(\frac{Y_m}{100} \right) \times 365 \text{ days} \right] \Big/ 55.65$$

Where *EF* (kg CH_4 animal^{-1} yr^{-1}) = emission factor, *GE* (MJ animal^{-1} d^{-1}) = gross energy intake, *Ym* (per cent of gross energy in feed converted to CH_4) = methane conversion factor. The factor 55.65 (MJkg^{-1} CH_4) is the energy content of methane. Cattle and buffaloes CH_4 conversion factors for feedlot fed is 3.0%. For dairy cows (Cattle and Buffalo) and their young, and cattle and buffaloes fed primarily on low quality crop residues, byproducts, or via grazing the *Ym* is 6.5%. However, for lamb and mature sheep *Ym* = 4.5 and 6.5%, respectively. If *Ym* is not reported for specific livestock types *Ym* from reported livestock that resembles most closely can be used (Dong et al., 2006). The average enteric CH_4 EF (kg animal^{-1} yr^{-1}) calculated by using the above equation was 29.5 kg animal^{-1} yr^{-1} for several African breeds of cattle ranged from 15.0 to 43.6, 16.9 to 46.3, and 24.7 to 64.9 kg CH_4 animal^{-1} yr^{-1} for Lagune (average body weight 152 Kg), Somba (stocky animals with good conformation for meat production, average body weight 149 Kg) and Borgou (average body weight 130 Kg) cattle, respectively. These breeds are multipurpose and have been used for milk, meat, and hide production as well as for draft power. They are grazed in open pastures and crop lands have tropical forages and crops (Kouazounde et al., 2014).

For milk production, Thoma et al. (2013) found that 72% of the CH_4 emissions from dairy occurred prior to milk leaving the farm with the remainder being associated with processing and transportation. To understand the magnitude and impact of agriculture on global GHG production, it is essential to know specific on-farm emission sources (e.g., animal housing, slurry and liquid manure storage lagoons/pits/retention ponds, dry manure stacks, composts, pastures, cropland, etc.) and to understand the processes linked with those emissions. This will help to design targeted and cost-effective mitigation strategies that are specific for each livestock facility. As an example, the primary sources of GHG emissions from dairy farms have been elaborated in Fig. 2.

Quantification of GHG Emissions from a Production System or Farm

Monitoring and quantification of GHG emissions from a production system or farm is impractical and expensive due to the requirements of advanced instrumentation and technical expertise (Waldrip et al., 2016). In addition, differences in management, animal diet and characteristics, available facilities, climate, and topography among livestock systems render erroneous the assumption that emissions from one farm

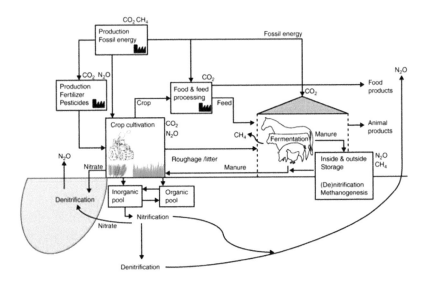

Fig. 2. GHG (CO_2, CH_4, and N_2O) emissions pathways during animal production (Source: de Boer et al., 2011).

would be the same as another, even within the same geographic region. Accurate measurement of dynamics of GHG exchanges between different systems is very important. Generally, for CO_2 fluxes eddy-covariance is widely used method (Baldocchi, 2014). However, nitrous oxide (N_2O) which accounts for 6% of the global greenhouse effect (Ciais et al., 2014) continues to be scarcely measured. Different manual and automated chambers methods have been used in the past to measure N_2O fluxes (Eugster and Merbold, 2015). Pattey et al. (2007) reported micrometeorological greenhouse gas measurements methods as an appropriate one and good alternative to traditional chamber methods. Eddy covariance methodology was reported promising to estimate N_2O budgets as compared with three other methods (automated chambers, manual chambers, and relaxed eddy accumulation) in NitroCOSMES project (Tallec et al., 2019). Similarly, different stationary and non-stationary methods have been recommended by Environmental Protection Agency (EPA) to monitor GHGs (https://www.epa.gov/). Methods for stationary point sources includes: FID (flame ionization detector), FTIS (Fourier transform infrared spectroscopy), GC (gas chromatography), NID (non-dispersive infrared detector) and TCD (thermal conductivity detector) while non-EPA stationary methods includes: AED (atomic emission detector), ECCA (electro-chemical and colorimetric analysis), ECD (electron capture device), MS (mass spectrometer) and PAS (photoacoustic absorption spectroscopy). However, nonpoints methods include: RPM (radial plume mapping) (Hashmonay et al., 2008), MTC (mobile tracer correlation), AAP (Aerostat aloft platforms), AL (airplane LIDAR), OGI (optical gas imaging) and EC (Eddy covariance) methods. Furthermore, now a day Gasmet is providing portable and powerful gas analyzers for greenhouse gas flux measurements (https://www.gasmet.com/applications/environment/greenhouse-gases-from-soil/). Mosquera et al. (2005) reported different methods for measurement of NH_3 emissions from animal houses. Zimmerman, (2011) developed new method to measure real-time CO_2 and CH_4 mass fluxes using GreenFeed (C-Lock Inc, Rapid City, SD). The system is equipped with sensors to measure concentration of CH_4 and carbon dioxide (CO_2) gases and eventually the flux of gases produced from the animal. The system is also equipped with a head sensor allowing reliable estimates of CH_4 production. GreenFeed can be used for 25 to 30 cows at the same time by attracting the cows with drops of concentrate to the manifold (Fig. 3a). In vitro gas production system is also another method measuring total gas and CH_4 production from small amount of feeds that are previously incubated in fermentation bottles (Fig. 3b). The application of the method was recently developed (Ramin and Huhtanen, 2012) in such a way that in vivo CH_4 production could be successfully estimated by further modeling approaches from data taken from the in vitro after the

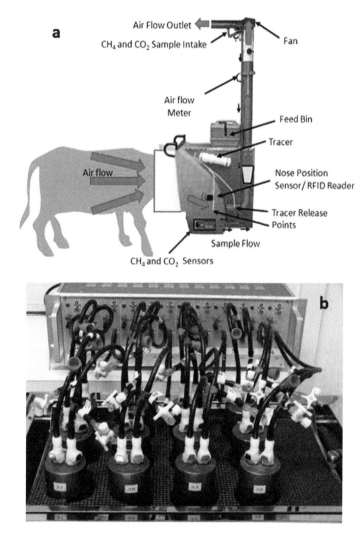

Fig. 3. General layout (a) of the GreenFeed system in the stand-alone feeder (Source: GreenFeed stand-alone feeder instruction manual, C-Lock; Zimmerman, 2011). Figure 3b shows the fully automated gas in vitro system used for recording total gas and CH_4 production (Source: Mohammad Ramin).

incubation is over (48 h). Thus, modeling could be an option to quantify and evaluate emissions among systems.

Models are defined by their complexity and include simple EF, process-driven EF (i.e., some information is available on factors that drive the specific gas emission), simple empirical models (i.e., mathematical equations that incorporate known influencing variables), and process-based models (i.e., mechanistic biochemical models). Hamilton et al. (2010) used statistical model to investigate the effects of feed additive and rumen microbial modifier on enteric CH_4 production in lactating dairy cows. Enteric CH_4 production for dairy farms can be represented by using IPCC approach where it was 121 kg CH_4 Cow^{-1} yr^{-1}, while for animals in a grazing operation in New Zealand it may be 81 kg CH_4 Cow^{-1} yr^{-1}. Process-based EF has been recommended by IPCC to represent enteric CH_4 production (Dong et al., 2006). However, many of regression-based models assume linear relationships among variables (e.g., DMI; daily dry matter intake (kg d^{-1}), N (g kg^{-1} of DM), NDF (g kg^{-1} of DM), ADF (g kg^{-1} of DM), Starch (g kg^{-1} of DM), WSC; water-soluble carbohydrate (g kg^{-1} of DM),CH_4 (MJ d^{-1}), forage proportion and ME; metabolized energy (MJ kg^{-1} of DM) which are not correct for enteric CH_4 production. Mills et al. (2003) developed a nonlinear model to represent enteric CH_4. Enteric fermentation is a complex process and requires a dynamic process simulation. Hence, models that can effectively simulate detailed biochemical processes (e.g., digestion, absorption and outflow of nutrients) should be used to predict enteric fermentation of cattle from different breeds, levels of production, and with diets that differ in digestibility and fat content. Molly is one example of a process-based model which uses the concept of a whole-farm system (Baldwin, 1995). Another example of a mechanistic model that predicts CH_4 production in dairy cows is the Karoline model (Huhtanen et al., 2015). Nordic dairy cow model Karoline in predicting CH_4 emissions was further evaluated by Ramin and Huhtanen, (2015) and conclude that Karoline is a useful tool in predicting CH_4 emissions and understanding the system behavior. Cattle produce enteric N_2O in small amounts, compared with enteric CH_4, but N_2O has 10 times greater global warming potential than CH_4 (Hamilton et al., 2010): estimation of small amounts of N_2O is important at the farm scale. Rotz et al. (2016) reported an enteric EF of 0.4 g N_2O cow^{-1} d^{-1}, or 0.8 g N_2O kg^{-1} N intake. Different sources of GHGs emissions are detailed in Fig. 4 and 5.

The overall CH_4 emissions from manure in housing and storage in the United States have EF values ranging from 48 to 78 kg CH_4 animal^{-1} yr^{-1}, while for New Zealand it was 27 kg CH_4 animal^{-1} yr^{-1}. These gross estimations can be made more specific by considering individual components of production systems (i.e., housing, long term storage and field applications). Different researchers reported different EF depending on type of housing (Chianese et al., 2009; Adviento-Borbe et al., 2010; Groenestein et al., 2012). Similarly, different process-based models, for example, Dairy CropSyst (http://modeling.bsyse.wsu.edu/rnelson/Dairy-CropSyst/index.html) were used to simulate beef cattle feed yard emissions by considering detailed biogeochemical processes (decomposition, hydrolysis, nitrification, denitrification and fermentation). Examples of this approach were the research by Li et al. (2012) and Bonifacio et al. (2015) with Manure DNDC and the Integrated Farm Systems Model (IFSM), respectively. These works showed that GHG emissions during manure handling and storage were minimal, but NH_3 emission from manure was large (Waldrip et al., 2013, 2014) which would indirectly contribute to N_2O emissions. Process-level dynamic simulations could also be used for manure storage as suggested by Li et al. (2012) and Bonifacio et al. (2015). Sándor et al. (2018) reported that process based biogeochemical models can

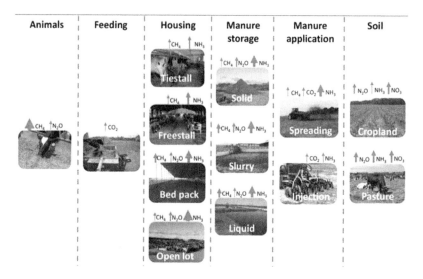

Fig. 4. Greenhouse gas sources (direct or indirect) and relative amounts (differently sized arrows) emitted from dairy farms (Source: Rotz, 2018).

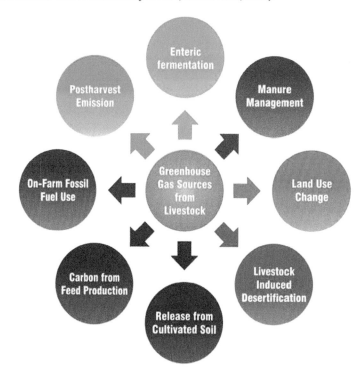

Fig. 5. Sources of GHGs from livestock farms (Source: Sejian et al., 2018).

be used to quantify C and N cycling in grassland ecosystems in response to changes in managements. Furthermore, they confirmed that grasslands can act as potential C and N sink compared to croplands if managed properly.

The processes of on-farm GHG emissions continue after incorporation of manure and inorganic fertilizer into cropland or pasture. Oxidation, fermentation, hydrolysis, nitrification, denitrification and other biochemical processes affect GHG emissions from pasture and cropland (Li et al., 2012). Methane is oxidized to CO_2 under moist soil conditions by methanotrophs, thus creating a net CH_4 sink (Li et al., 2005; Liu et al., 2017). Generally, this sink is ignored when modeling whole-farm GHG emissions (Chianese et al., 2009). Nitrification is a stepwise aerobic process performed by a consortia of microorganisms, where NH_4^+ is oxidized to NO_3^- with production/release of NO and N_2O as intermediates. Denitrification is an anaerobic microbial process that involves the stepwise reduction of nitrate (NO_3^-) to N_2O and then N_2. The primary end product of denitrification is N_2O if pH or some other factor limits the final step to N_2 in the pathway. For cropland emissions, different process-based models, such as DayCent (DayCent, 2008) and DNDC (Giltrap et al., 2010), which were developed for soils, could be used to monitor these processes as a function of thermodynamics, enzyme kinetics, weather, manure nutrient content, and cropping practices. However, modeling of individual emission sources has limited the use to a producer who is attempting to balance economic and environmental trade-offs for a system with many interacting components. Therefore, in whole-farm process-based models the various farm components are integrated to enable evaluation from soil operation, crop management, animal management, and economic component at the farm scale called as whole farm modeling (WFM). These WFM efforts could help to quantify total farm emissions via a comprehensive assessment of all GHG-producing components of a particular system. In addition, WFM can be a useful tool to evaluate overall environmental effects per unit of production.

Whole Farm Modeling

Whole farm modeling is a holistic tool that can be used to pinpoint "leaks" in a production system and assess new technologies without risks to animal health, environmental quality, or economics. The use of WFM can make it possible to evaluate how on-farm resources are allocated prior to physically designing and implementing changes in management or available facilities (Torkamani, 2005). Livestock producers operate in an uncertain decision-making environment, particularly in developing countries. Fluctuation in input and output prices, climate change impacts, animal health and productivity, and variation in crop yields are important risks in agriculture. Whole farm modeling provides a framework for assessing the potential impacts of perspective technology on sustainability. It employs numerous mathematical equations to form a comprehensive approach to simulate multiple interactions among different farm management components under diverse climatic conditions. Using the WFM approach allows for virtual evaluation of market benefits and trade-offs without experiencing actual risk to the enterprise under study. A WFM uses the concept of a farm as a complete working system. A farm or a whole farm system is not simply a collection of crops and animals to which one can apply any input and expect immediate results, rather it is a complex interwoven mesh of soil, plants, animals, implements, workers, other inputs and environmental influence. Individual system components in the WFM can be manipulated in attempt to cost-effectively and sustainably produce livestock products using available inputs and technology.

Why Whole Farm Modeling?

1. Account for the complexities livestock producers face when managing complicated farm businesses under risk and uncertainty.

2. Provide information at the farm business level– the level at which farmers think and make decisions.

3. Increase rate of productivity and build adaptive capacity.

4. Estimate or predict GHG emissions from various farm sources and the total farm.

5. Develop and evaluate novel farm systems. For example, modeling different levels of supplementation and different levels of crop N application, to predict how these changes could affect environmental integrity and profitability.

6. Mathematical representation of planning, property design and management of a farm based on natural resources and economic factors.

7. Simulate short- and long-term (i.e., > 20 yr) changes in all farm assets (physical and non-physical).

8. Life cycle assessment (LCA), as component of WFM, provides an environmental accounting tool for evaluating emissions over the full life cycle of the produced animal products (i.e., milk and dairy products, eggs, meat, etc.).

9. WFM provides important tools for quantifying emissions, identifying opportunities for reduction, and evaluating mitigation strategies.

Whole farm modeling includes the following components for integration:

- Soil
- Crops
- Pasture
- Forages, Silage
- Environment
- Vehicles
- Livestock– dynamic management
- Climate change
- Economics

Various methods and technologies are being adapted when they fit within a farm manager's resources, objectives and goals (Torkamani, 200). These new methods may include different preferences, such as new crops or crop varieties, improved soil conditions and crop and water management. Potential technologies may be adapted if they are easy to adapt from the farmers' view point. Whole farm modeling approaches are recommended as a means to evaluate new methods, particularly when risk is a possibility. Whole farm modeling has potential to deliver a representative assessment of the appropriateness and suitability of new skills.

Many whole farm modeling strategies have been established which explore climate change mitigation approaches for livestock systems (Salcedo, 2015; Sándor et al., 2018). Although whole farm decision making is very complex, it is important for producers to look at the entire system holistically for sustainability, improved production, and environmental stewardship. Whole farm modeling integrates crops, livestock, and other farm components as a whole system and deals with management of manure for economic viability while reducing nutrient losses to soil, air and water. Comprehensive whole-farm models provide necessary tools for integrating the effects and interactions of all sources on the farm. Farm models are available where processes may be represented using relationships ranging from EFs to detailed mechanistic simulations and various combinations of model types. Models provide important tools for quantifying emissions, identifying opportunities for reduction, and evaluating mitigation strategies.

A Typical Farm Representation

Producers are continuously losing income due to land degradation, suboptimal crop and livestock production, and climate change. Meanwhile, short-term nutrient dynamics could benefit significantly from enhanced management of manures and other organic residues in combination with inorganic fertilizer applications (Fonte et al., 2012). Significant research effort has been conducted in the advancement of agricultural models. Application of WFM to individual facilities has been low and random, despite the potential advantages that could be elucidated. Technologies supporting smart farms offer an open door for empowering the development and use of farm models by management, consulting agents or extension personnel. This objective is attainable and offers a radical innovation for demonstrating the overall effects of different management practices (O'Grady and O'Hare, 2017).

The whole-farm model MELODIE was primarily developed for farm system simulations that include crop and/or animal production. The current MELODIE version can be used for simulating common or innovative farm strategies. For example, MELODIE could be used to study the dynamics of nutrient flow and the dynamic behavior of an operation under diverse feeding systems on dairies. It illustrates that small differences in nitrogen (N) losses between different systems can be reflected by an average N farm gate budget indicator (Chardon et al., 2012).

'DairyCant' is another whole farm model based on research and statistical analysis that simulates management aspects related to milk production and environmental health on dairy farms. Manure production is calculated from the dry matter and nutrient intake (Salcedo, 2015).

The DairyWise model is an experimental model that simulates environmental, technical, and financial procedures on a dairy farm. The output of the feed supply model was utilized as contribution for a few specialized ecological and economic sub-models. The sub-models simulated a range of farm aspects, including N and phosphorus (P) cycling, NO_3^- leaching and runoff, ammonia (NH_3) emissions, GHG emissions, energy utilization and a financial farm budget plan (Schils et al., 2007).

NAMASTE is a dynamic computer-based model for management of water resources at the farm level. It was developed to create collaborations between decisions and methods under different situations, including climate-change, fiscal sustainability and water-administration. The model was first created to address basic issues of groundwater and farming practices within a watershed (Robert et al., 2018).

Models of various scales have been developed and used for approximation of GHG emissions from different farm sources. These include constant EF, variable process-related EF, empirical or statistical models, detailed process-level simulation, and LCA. Important direct emissions from farms include CH_4 and N_2O from enteric, manure, and soil sources, and CO_2 from the combustion of fossil fuels and the decomposition of field smeared lime and urea. Indirect emissions should also

Table 1. Whole farm models for the estimation of the emissions of GHGs from dairy farms.

Model	Description	Application	Developer
AgRE Calc	Emission factor-based carbon calculator that determines a carbon footprint of various types of farms, including dairy. (http://www.agrecalc.com)	Decision support and education	SAC Consulting, United Kingdom
COMET-Farm	Emission factor and process model primarily for estimating carbon sequestration and emissions of various types of farms, including dairy. (http://cometfarm.nrel.colostate.edu/)	Decision support	USDA/Natural Resource Conservation Service, Colorado State University, Fort Collins
Cool Farm Tool	Emission factor-based carbon accounting tool for a wide range of cropping systems and includes a dairy livestock component. (https://coolfarmtool.org/)	Decision support and education	Cool Farm Alliance, England
DairyGEM	Emission factor and process simulation tool that estimates GHG, NH_3, and other gaseous emissions and the carbon footprint of dairy production systems. (https://www.ars.usda.gov/northeast-area/up-pa/pswmru/docs/dairy-gas-emissions-model/)	Education and decision support	USDA-Agricultural Research Service, University Park, PA
DairyMod	Biophysical process simulation of pastoral dairy systems predicting GHG dynamics including direct and indirect emissions and soil carbon balance. (http://imj.com.au/dairymod/)	Research and education	IMJ Consultants, Dairy Australia, University of Melbourne, Australia
DairyWise	An empirical model that simulates the technical, environmental, and financial processes on a dairy farm that includes nitrogen and phosphorus cycling and losses, GHG emissions, and energy use. (Schils et al., 2007)	Research and education	Wageningen UR, the Netherlands
FarmAC	Process-related emission factors represent carbon and nitrogen flows on arable and livestock farms quantifying GHG, soil C sequestration, and N losses to the environment. (http://www.farmac.dk/)	Education and decision support	Aarhus University, Denmark
FASSET	Process simulation used to evaluate consequences of changes in regulations, management, prices and subsidies on farm production, profitability, nitrogen losses, energy consumption and GHG emissions. (http://www.fasset.dk/)	Research	Aarhus University, Denmark
Holos	Process-based emission factors estimate all important direct and indirect sources of GHG emissions of livestock operations. (http://www.agr.gc.ca/eng/science-and-innovation/results-of-agricultural-research/holos/?id=1349181297838)	Education, decision support, research	Agriculture and Agri-Food Canada
IFSM	Process simulation of all important farm components representing the performance, economics, and environmental impacts including direct and indirect GHG emissions and carbon footprint. (https://www.ars.usda.gov/northeast-area/up-pa/pswmru/docs/integrated-farm-system-model/)	Research and education	USDA-Agricultural Research Service, University Park, PA
Manure DNDC	Simulation of soil and manure biogeochemical processes producing GHG and NH_3 emissions. (http://www.dndc.sr.unh.edu/)	Research	University of New Hampshire, Durham
MELODIE	Dynamic simulation of the flows of carbon, nitrogen, phosphorus, copper, zinc and water within animal, pasture, crop and manure components. (Chardon et al., 2012)	Education and decision support	French National Institute for Agricultural Research, INRA, France
SIMS(Dairy)	Process simulation of the effects of management, climate and soil properties on nitrogen, phosphorus, and carbon losses along with profitability, biodiversity, soil quality, and animal welfare. (Del Prado et al., 2013)	Education and research	BC3-Basque Centre for Climate Change, Spain

be considered, which include NH_3 and NO_3–losses that potentially transform to N_2O beyond the farm boundaries. Models provide important tools for quantifying emissions, identifying opportunities for reduction, and evaluating mitigation strategies. Greenhouse and related gases are emitted from manure from the point of excretion by the animal until they are incorporated into soil to produce feed

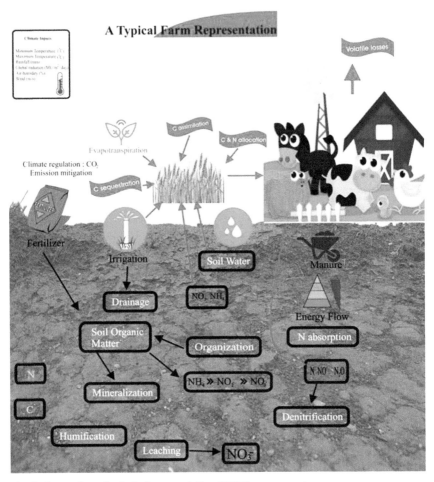

Fig. 6. Illustration of whole farm modeling (WFM) components.

crops and pasture (Rotz, 2018). These emissions could be predicted using models by considering different variables. Models assist in the development and evaluation of mitigation strategies for sustainability of dairy farms through reduction of emissions. The details of WFM components are presented in Fig. 6.

Farm-scale process simulation models include DairyMod, MELODIE, FASSET, SIMS (Dairy), IFSM, and Manure DNDC. These models were primarily developed by scientists to better understand the processes involved and predict how they interact; thus, they often provide the best understanding of the effects of farm management changes. The addition of new processes requires detailed process models, which often require much time and effort to develop or adapt to the existing model structure.

Available process models were developed in different countries and often for different types of dairy production systems. DairyMod was developed in Australia solely for pasture-based dairy production, with the major components being pasture production and animal utilization. Models such as MELODIE and SIMS (dairy) were primarily developed around pasture-based dairy production in

Europe. Both FASSET and IFSM were developed around confinement production systems but also include pasture and grazing components; IFSM provides one of the most comprehensive tools in that it includes components for tillage, planting, harvest, feed storage, feeding, manure handling, and economics, along with crop and pasture growth and animal consumption and performance (Rotz et al., 2012). Manure DNDC provides the most detailed representation of the biogeochemical processes producing emissions from the soil and manure, but has little or no detail in representing some of the other farm components (Li et al., 2012).

Livestock effluent "surplus" is an extremely delicate issue for farmers who have numerous management challenges to guarantee environmentally safe disposal or utilization. Manure composting provides a method to reduce the majority of environmental impacts related with manure management. The production of on-farm compost can be an answer for surplus livestock wastes (Pergola et al., 2018).

A whole-farm emissions model, IFSM, predicts emissions from livestock housing and manure storages based on the assumption that the temperature is equivalent to the normal surrounding over the past 10 d. Another whole-farm model is manure-DNDC, which is based on the DeNitrification and DeComposition Model that is frequently used for cropping systems. Emissions of CH_4, N_2O, and NH_3 from liquid manure lagoons/storages are an imperative issue with respect to GHG alleviation, nutrient management, and scent reduction. These emissions are strongly influenced by the manure temperature: generally considered equivalent to air temperature in process-based models designed to track the biochemistry of manure components (Rennie et al., 2017).

Efforts to recycle nutrients in livestock manure for crop production can reduce some pollution risks. It is important to consider pollution control strategies for a farm in the framework of local and regional pollution control planning (Petersen et al., 2007). The livestock bio-economic farm model 'Orfee' was intended to optimize and simulate the production, economic results, and contribution to climate change of beef and dairy systems in conjunction with forage and cash crop production. The key features of Orfee are an ability to simulate over a large range of levels of intensification and integration (Mosnier et al., 2017).

Whole farm models are potentially relevant tools for policy impact assessment. Governments and international organizations use impact assessment (IA) as an exante policy process and procedure to evaluate policy options prior to new policy introduction. Reidsma et al. (2018) performed a systematic review based on 202 studies from the period 2007 to 2015 and results were discussed in a science-policy workshop and they emphasized on the development and use of farm models.

A development version of the coupled Dairy and CropSyst model, with a simple user interface and example project is another example of WFM (Fig. 7). This WFM is a decision support tool for researchers and extension personnel. It could be used to evaluate the effects of manure treatment options, including anaerobic digestion, nutrient recovery and solid separation, on net GHG emission and manure nutrient fate from a whole-farm perspective. This model integrates previously established models, performance parameters of different manure treatments, and a cropping systems model, CropSyst, to predict the soil nutrient budget and carbon (C) footprint. At various stages during the manure life cycle, from excretion in housing to manure storage to land application, the nutrients and organic matter in manure undergo different chemical reactions driven by environmental factors, such as temperature and water content. The mass balance of these manure components in organic and

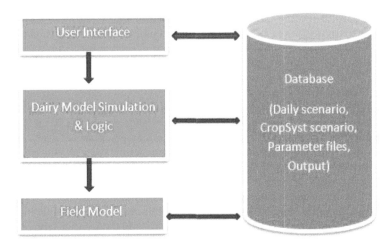

Fig. 7. Relationshps among major components of Dairy-CropSyst is shown above. The user interface writes scenario-specific information to the database and executes the dairy models and simulation logic, which in turn executes the field model.

inorganic forms is modeled as they pass from one stage to another on a daily time step. For dairies, an animal submodel uses input variables (i.e., feed intake, milk production, and body weight) and simulates CO_2 and CH_4 emissions during respiration and enteric fermentation, manure production, and manure nutrient composition. Dairy-CropSyst uses an empirical model to estimate CO_2 and CH_4, while N_2O emission is estimated using an EF. CropSyst is a well-known, tested model that has been used to study the effect of cropping systems management on productivity and the environment by simulating the soil water budget, soil–plant nitrogen budget, biomass production, crop yield, and C footprint (Stöckle et al., 2003).

Inputs for the dairy component include daily weather and specific parameters for unit operations. A fertigation schedule may be defined if field emissions are to be simulated based on weather, soil, and crop management. The model output is an Excel file with detailed daily assessment of nutrient production, transformation, volatilization and emission, leaching, flow between component facilities, and residual soil nutrient balance in soil. Manure flow, GHG emissions, and value-added product recovery for whole farm cycle are elaborated in Fig. 8.

The DNDC model is a computer simulation model of C and N biogeochemistry in agroecosystems. The DNDC model can be used for predicting crop growth, soil temperature and moisture regimes, soil C dynamics, N leaching, and emissions of trace gases, including (N_2O), nitric oxide (NO), dinitrogen (N_2), NH_3, CH_4 and CO_2. Developed for quantifying C sequestration and emissions of GHG, DNDC has been extensively evaluated against datasets of trace gas fluxes measured worldwide. Over the past 25 yr, improved scientific understanding of biogeochemical reactions has been incorporated to refine DNDC. The DNDC model has been intensively and independently tested by a wide range of researches worldwide with encouraging results. The model consists of two components. The first module consists of submodules for soil climate, crop growth, and decomposition to predict soil temperature, moisture, pH, Eh, and substrate concentration profiles (e.g., ammonium [NH_4^+], NO_3^-, and dissolved organic C [DOC]) based on ecological drivers (e.g., climate, soil, vegetation and anthropogenic activity). The second component consists of nitrification, denitrification, and fermentation submodels, and predicts C and N fluxes, such as NO, N_2O, CH_4 and NH_3, based on the soil environmental variables (Fig. 9).

The entire DNDC model bridges the C and N biogeochemical cycles and their primary ecological drivers. In DNDC, daily weather data (i.e., air temperature, precipitation and wind speed), soil properties [i.e., bulk density, texture, soil organic C (SOC) content and pH], vegetation characteristics, and field management activities are inputs to determine the nutrient and C dynamics in the soil profile. Farming management practices (FMP), such as tillage, fertilization, manure application, flooding, irrigation, and cultivation of cover crops, have been parameterized to regulate soil environmental conditions. When manure is applied and incorporated into soil the C and N bound in the organic matter is released through decomposition and then distributed into the relevant soil C and N pools where the principles of enzyme kinetics and thermodynamics are used to model nitrification and other

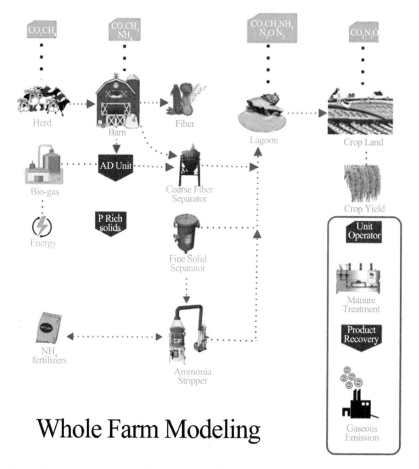

Whole Farm Modeling

Fig. 8. Schematic representation of manure flow, gaseous emissions, and value-added product recovery for a whole-farm cycle from barn to cropland application.

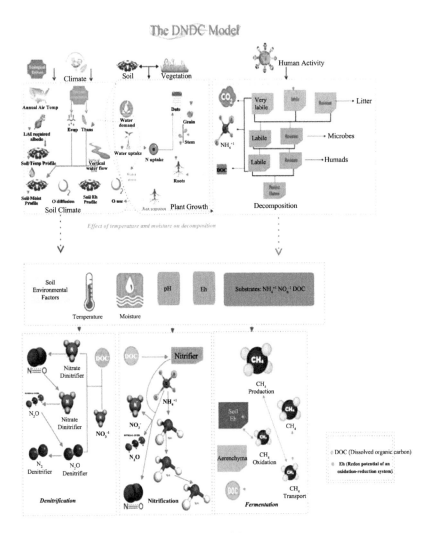

Fig. 9. Schematic representation of DNDC Model parameters.

processes relevant to losses to air and water. Whole farm models for the estimation of the emissions of GHGs from dairy farms are further presented in Table 1.

Case Studies of Application of Whole Farm Modeling Approaches

Farm-level N cycling was evaluated by Alvarez et al. (2014) under four contrasting crop–livestock farm systems in the highlands of Madagascar. The farm types were (i) Large cattle farms, (ii) Farms with fewer dairy cows and significant diversification with swine), (iii) Upland farms with forage-fed dairy cows and (iv) very small herds of Zebu-cross cattle that had low milk production. These scenarios were assessed for effects of (i) increased supplemental dietary N on milk production, (ii) addition of mineral fertilizer N on crop production, (iii) N conservation during manure storage and field application, and (iv) the economic profitability of combined dietary N supplementation and improved manure management (scenarios i and iii). The results showed that improving manure management (scenario iii) had a positive impact on gross profit margin, while a combination with Scenario "IV" led to increased whole farm N use efficiencies from 2% to 50%, N cycling from 9 to 68% and food self-sufficiency from 12 to 37% across farm types. Scenario IV had the highest impact on farm productivity, gross profit margin, food self-sufficiency, and environmental sustainability.

A WFM model, APS Farm (Rodriguez and Sadras, 2011), was used to examine the sensitivity of four case farm studies to climate change scenario (Rodriguez et al., 2014). The information generated was used to develop profitable and sustainable farming systems. The results showed that participating farmers operates close to the efficiency frontier (i.e., in the relationship between profits and risks). Investments in time, labor, new crop varieties, alternative cropping practices, equipment, etc. can help to develop innovative cropping and grazing systems or transform existing farming systems to adapt to climate change. Ryschawy et al. (2012) reported through sensitivity analysis that nutrient cycling could be improved with mixed crop-livestock systems. This mixed system reduced chemical inputs and generated more income at farm level. The study revealed that N pollution was minimal in mixed crop–livestock systems (livestock and cash crops at farm level) and beef production systems as compared to dairy and crop production systems. Therefore, mixed crop–livestock systems appeared to be environmentally and economically sustainable. The Livestock Simulator (LIVSIM) by Rufino (2008) simulates livestock production on a monthly time step by considering genetic potential and feed intake. The LIVSIM evaluates different feeding regimes and other management practices on livestock production. The MITERRA which is integrated framework of mitigation and adaptation options for sustainable livestock production under climate change is a good example of such models which can provide GHGs emissions under different scenarios. FarmGHG by Olesen et al. (2006) is a model dealing with farm-scale dairy C and N flows. It can model on-farm GHG emissions and eutrophication impacts. The model simulates import, exports and flow of all products and was practically used to simulate N flows, N surpluses and GHG emissions for European-type dairy farms (Fig. 10) (Olesen et al., 2006).

Manure management is an important component to consider for intensification and specialization of livestock production. Petersen et al. (2013) discussed the diversity of livestock production systems and manure management in four regions: Sub-Saharan Africa, Southeast Asia, China and Europe. These researchers used modeling approach for estimation of GHG emissions under different mitigation practices. Livestock sector in developing countries is less productive as they produce large amounts of GHG emissions due to incompetent rearing system, feed production and manure management. Emissions intensity in developing countries is high therefore, application of appropriate tools such as whole farm modeling is recommended to reduce emission intensity and increase livestock productivity (Forabosco et al., 2017). Management and having balanced diets that fulfills the requirement of animals is the first and most promising scenario in reducing GHG emissions in developing countries rather than using sophisticated/expensive feed additives to reduce GHG.

Quantification of GHG emissions from dairy farms were evaluated using two standard methodologies, LCA and the IPCC method. GHG emissions were modeled by both methods and results indicated that estimated GHG emissions were greater when calculated using the LCA approach. Similarly, if target is net reduction in global GHG emissions for projected increase in milk and meat production, LCA and holistic

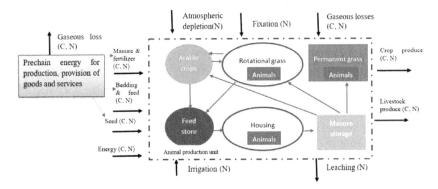

Fig. 10. Farm GHG model flows of C and N (Modified from Olesen et al., 2006).

approaches should be used to assess emissions (O'Brien et al., 2011). Similarly, carbon footprint per unit of milk could be assessed by LCA as reported by O'Brien et al. (2014).

An LCA of broiler production was evaluated from a farm to fork perspective. The five subsystems (broiler houses, slaughterhouse, meat processing plant, retail sales, and household consumption) were considered, while five environmental impacts were calculated: (i) GWP, (ii) acidification potential (AP), (iii) eutrophication potential (EP), (iv) ozone layer depletion (OLD), (v) and cumulative energy demand (CED). The GWP, AP, and EP are considered three comprehensive crucial impact categories for a LCA (Perrin et al., 2014). The results from the LCA by Skunca et al. (2018) showed that largest overall contributor to environmental risk from broiler production are feed production (i.e., cropping and processing) and energy usage during production process. Therefore, mitigation strategies, such as use of grain legumes, biogas digestion of chicken litter, recycling of household waste, and use of energy efficient equipment should be opted.

Chardon et al. (2012) developed whole-farm model called MELODIE which can simulate C, N, P, Cu, Zn, and H_2O flow within the whole farm over short and long-term scenarios. Whole farm models are valuable tools to study feedback and forward interactions between mitigation of GHG emissions and climate change adaptation for ruminant based production systems. The GHG emissions are governed by cycling of C and N within the farm and it is affected by climate change as reported by Del Prado et al. (2013). Smart farming could be another option to generate information which could be used effectively to design economically sustainable business models as suggested by Wolfert et al. (2017). Smart farming helps to have efficient, productive, and profitable farming enterprises. Similarly, incorporation of modeling tools such as GPFARM, APSIM, and GRAZPLAN could help to have sustainable intensification, reducing methane emissions, and sustainable decision-making on long term basis (O'Grady and O'Hare, 2017).

Integrated farm system model (IFSM) was evaluated by Sejian et al. (2018) to see how management strategies affect the GHG emissions. Carbon foot print and GHG emissions were monitored for four commercial farms using IFSM (DariryGHG model). The results depicted that dairy farm GHG assessments could be a good step to design uniform methodology for the quantification of dairy farm emissions.

Whole farm models could be used to study impact assessments (IA) which is a process and procedure to evaluate impacts of policy options on range of impact areas. Reidsma et al. (2018) used IA and concluded that farmer decision making could be improved by having good interactions between farmers and other related sectors. Similarly, model evaluation and comparison are needed with more attention toward its sensitivity and uncertainty. Reidsma et al. (2018) also suggested the organization of network of farm modelers same as AgMIP, MACSUR, and GYGA and stronger science-policy interaction and quality data collection.

FarmAC (semi-dynamic whole farm model) was used to evaluate the potential of agronomic management for mitigating GHG emissions and enhancing N fluxes for on-farm forage resources by Doltra et al. (2018). Results depicted that increasing forage productivity could reduce the external dependence for feeding animals and mitigating GHG emissions. Nitrogen and carbon flow of FarmAC model is presented in Fig. 11.

Life cycle assessment (LCA) is a well-known methodology used to assess the impact of a product or a process or a service during its life cycle on environment (Finnveden et al., 2009). Winkler et al. (2016) used LCA to study environmental impacts occurred during the production of pork. Global warming potential (GWP) (CO_2–equivalents), soil acidification (SO_2–equivalents) and eutrophication (NO_3–equivalents) were used to express the results. The results indicated that main environmental burden is during farming stage i.e., GWP (92.3%), soil acidification (98.4%) and eutrophication (95.4%). Therefore, authors suggested that more focus should be given to farming stage improvement in future as it is giving more damage. Furthermore, they also suggested use of protection measures such as improving the choice of commodities and feed production. Environmental effects of nutrition's have been also evaluated in earlier studies in the context of foot print called as nutritional foot print but LCA is a more holistic approach that can detect environmental hotspots. It can analyze the environmental impacts of the various commodities during its production cycle so it should be integrated with farm models.

McClelland et al. (2018) conducted LCA by considering Jolliet et al. (2004) categorization approach which considers twelve impact points: (i) acidification (ii) biodiversity (iii) climate change (or global warming potential) (iv) ecotoxicity (v) eutrophication (vi) human toxicity (vii) ionizing radiation (viii) land use or land occupation (ix) ozone depletion (x) Particulate matter (xi) Photochemical ozone formation or photooxidant formation and (xii) Resource depletion (including biotic and abiotic resources; e.g., fossil fuel, electricity, water, etc.) for livestock LCA. The review results indicated that usage of multi-category livestock LCA is increased over time but simplified LCA is the most prevalent approach used in literature.

Environmental impacts of GHG emissions and emissions intensity (Ei) for the small ruminants were evaluated using LCA method with a system boundary from Cradle to farm gate. The aim was to check the weakness and robustness of the

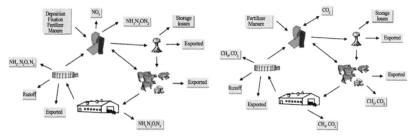

Fig. 11. The FarmAC nitrogen and carbon flow (Source: https://www.farmac.dk/)

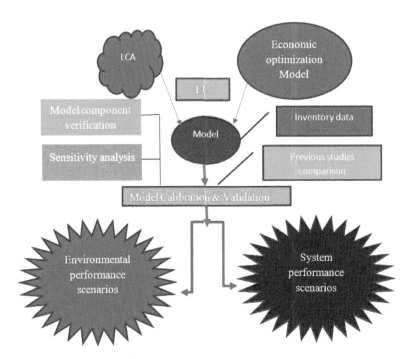

Fig. 12. Schematic view of LCA analysis.

livestock production system. The emissions were estimated using whole farm GHG model. LCA estimated the GHG intensity of 5.7 gigatonnes CO_2e per annual, which represents 80% of livestock emissions from ruminant supply chain. Enteric methane was the largest contributor of GHG which accounted for 47% while N_2O contributed 24%. Similarly, land use change contributed 9% of overall GHG emissions. However, LCA should be applied on large number of practical farms to avoid wrong conclusions. LCA wraps the entire farming system changes in GHG emissions due to possible mitigation strategies. Whole farm system approach in combination with LCA could be considered beneficial as reducing GHG emissions in one part of the system might increase in another part. It can be categorized as system analysis model as it will help to design alternative mitigation strategies to minimize GHG emissions. HOLOS and MDSM are examples of GHG emissions models which could be used to quantify the internal flow of C and N on a farm scale (Olesen et al., 2006). Functional Unit (FU) defined as mass of product (e.g., L of milk, kg of protein, kg of fat, kg of protein or fat corrected milk, or kg of energy corrected milk) leaving the farm gate was designed in LCA to study environmental impacts (Fig. 12). Life cycle assessment is an effective tool to evaluate the environmental impact of a product throughout its life cycle with the consideration of whole farm modeling approach. Similarly, different climatic and system performance scenario could be considered as an important indicator for environmental sustainability of livestock systems.

Conclusion

The WFM approach has great potential to quantify and predict GHG emissions from different sources in farm production. They can integrate the effects and interactions of all sources on the farm. Whole farm modeling includes different emission factors, empirical or statistical models, detailed simulation, and LCA. They can act as vital research tools to simulate direct and indirect emissions. Since livestock sector is a big contributor of GHG emissions however, estimation of GHG emission through experimentation is impossible. Simulation models offer huge scope to quantify emission of GHG under different set of experiments. Generated information from these mechanistic models could help to design effective mitigation strategies to reduce emissions and improve the sustainability of livestock production. The above case studies suggested that process-based models accurately predicted GHG emissions from farms. These models can consider entire livestock farm operations to evaluate GHG emissions. Therefore, alternative solutions for livestock production related to climate change and under other risks could be easily designed by these processes based on mechanistic models. Furthermore, crop–livestock systems simulation under changing climate and design of adaptation strategies could be modeled by these models. LCA acts as an environmental accounting tool for estimating emissions over the full life cycle of the farm produce. Finally, models are tools which can quantify emissions, identify opportunities and design adaptation and mitigation strategies.

References

Adviento-Borbe, M., E.F. Wheeler, N.E. Brown, P.A. Topper, R. Graves, V. Ishler, and G.A. Varga. 2010. Ammonia and greenhouse gas flux from manure in freestall barn with dairy cows on precision fed rations. Trans. ASABE 53:1251–1266. doi:10.13031/2013.32590

Alvarez, S., M.C. Rufino, J. Vayssières, P. Salgado, P. Tittonell, E. Tillard, and F. Bocquier. 2014. Whole-farm nitrogen cycling and intensification of crop-livestock systems in the highlands of Madagascar: An application of network analysis. Agr Syst 126: 25–37. doi:10.1016/j.agsy.2013.03.005.

Baldocchi, D. 2014. Measuring fluxes of trace gases and energy between ecosystems and the atmosphere–The state and future of the eddy covariance method. Glob. Change Biol. 20:3600–3609. doi:10.1111/gcb.12649

Baldwin, R.L. 1995. Modeling ruminant digestion and metabolism. Chapman and Hall, London.

Beauchemin, K.A., H.H. Janzen, S.M. Little, T.A. McAllister, and S.M. McGinn. 2011. Mitigation of greenhouse gas emissions from beef production in Western Canada– Evaluation using farm-based life cycle assessment. Anim Feed Sci Tech 166-167: 663–677. doi:10.1016/j.anifeedsci.2011.04.047.

Bonifacio, H.F., C.A. Rotz, A.B. Leytem, H.M. Waldrip, and R.W. Todd. 2015. Process-based modeling of ammonia and nitrous oxide from open lot beef and dairy facilities. Trans. ASABE 58:827–846.

Capper, J.L., and D.E. Bauman. 2013. The role of productivity in improving the environmental sustainability of ruminant production systems. Annu. Rev. Anim. Biosci. 1:469–489. doi:10.1146/annurev-animal-031412-103727

Chianese, D., C. Rotz, and T. Richard. 2009. Simulation of methane emissions from dairy farms to assess greenhouse gas reduction strategies. Trans. ASABE 52:1313–1323. doi:10.13031/2013.27781

Chardon, X., C. Rigolot, C. Baratte, S. Espagnol, C. Raison, R. Martin-Clouaire, J.-P. Rellier, et al. 2012. MELODIE: A whole-farm model to study the dynamics of nutrients in dairy and pig farms with crops. Animal 6:1711–1721. doi:10.1017/S1751731112000687

Ciais, P., C. Sabine, G. Bala, L. Bopp, V. Brovkin, J. Canadell, A. Chhabra, et al. 2014. Carbon and other biogeochemical Cycles. In: IPCC, editor, Climate Change 2013: The physical science basis. Contribution of Working Group I to the Fifth Assessment Report of the Intergovernmental Panel on Climate Change. Cambridge Univ. Press, Cambridge, U.K. p. 465-570.

DayCent. 2008. DayCent version 4.5. Natural Resource Ecology Laboratory. Colorado State University, Fort Collins, CO. https://www2.nrel.colostate.edu/projects/daycent/ (Accessed 5 Feb. 2019).

de Boer, I.J.M., C. Cederberg, S. Eady, S. Gollnow, T. Kristensen, and M. Macleod. 2011. Greenhouse gas mitigation in animal production: Towards an integrated life cycle sustainability assessment. Curr. Opin. Env. Sust. 3: 423–431. doi:10.1016/j.cosust.2011.08.007.

de Klein, C.A., L. Barton, R.R. Sherlock, Z. Li and R.P. Littlejohn. 2003. Estimating a nitrous oxide emission factor for animal urine from some New Zealand pastoral soils. Soil Research 41: 381-399.

Del Prado, A., P. Crosson, J.E. Olesen, and C.A. Rotz. 2013. Whole-farm models to quantify greenhouse gas emissions and their potential use for linking climate change mitigation and adaptation in temperate grassland ruminant-based farming systems. Animal 7:373–385. doi:10.1017/S1751731113000748

Doltra, J., A. Villar, R. Moros, G. Salcedo, N.J. Hutchings, and I.S. Kristensen. 2018. Forage management to improve on-farm feed production, nitrogen fluxes and greenhouse gas emissions from dairy systems in a wet temperate region. Agr Syst. 160: 70-78. doi:10.1016/j.agsy.2017.11.004.

Dong, H., J. Mangino, T.A. McAllister, J.L. Hatfield, D.E. Johnson, K.R. Lassey, M. Aparecida de Lima, et al. 2006. Chapter 10: Emissions from Livestock and Manure Management. In: IPCC, editor, 2006 IPCC Guidelines for National Greenhouse Gas Inventories. Intergovernmental Panel on Climate Change, National Greenhouse Gas Inventories Programme (IPCC-NGGIP). United Nations Framework Convention on Climate Change (UNFCCC), Geneva, Switzerland.

EPA. 2018. Inventory of U.S. greenhouse gas emissions and sinks: 1990-2016. United States Environmental Protection Agency, Washington, D.C. https://www.epa.gov/ghgemissions/inventory-us-greenhouse-gas-emissions-and-sinks-1990-2016 (Accessed 3 Oct. 2018).

Eugster, W., and L. Merbold. 2015. Eddy covariance for quantifying trace gas fluxes from soils. SOIL 1:187–205. doi:10.5194/soil-1-187-2015

Finnveden, G., M.Z. Hauschild, T. Ekvall, J. Guinée, R. Heijungs, S. Hellweg, A. Kohler, D. Pennington, and S. Sangwon. 2009. Recent developments in life cycle assessment. J. Environ. Manage. 91: 1–21. doi:10.1016/j.jenvman.2009.06.018.

Fonte, S.J., S.J. Vanek, P. Oyarzun, S. Parsa, D.C. Quintero, I.M. Rao, and P. Lavelle. 2012. Chapter four- Pathways to agroecological intensification of soil fertility management by smallholder farmers in the Andean Highlands. In: D.L. Sparks, editor, Advances in agronomy. Vol. 116. Academic Press, p. 125–184.

Forabosco, F., Z. Chitchyan, and R. Mantovani. 2017. Methane, nitrous oxide emissions and mitigation strategies for livestock in developing countries: A review. S. Afr. J. Anim. Sci. 47:268–280. doi:10.4314/sajas.v47i3.3

Gerber, P.J., D.H. Steinfel, B. Henderson, A. Mottet, C. Opio, J. Dijkman, A. Falcucci, and G. Tempio. 2013. Tackling climate change through livestock: A global assessment of emissions and mitigation opportunities. Food and Agriculture Organization of the United Nations, Rome.

Giltrap, D.L., C. Li, and S. Saggar. 2010. DNDC: A process-based model of greenhouse gas fluxes from agricultural soils. Agric., Ecosyst. Environ. 136: 292-300. doi:10.1016/j.agee.2009.06.014

Groenestein, K., J. Mosquera, and S. Van der Sluis. 2012. Emission factors for methane and nitrous oxide from manure management and mitigation options AU-Groenestein, Karin. J. Integr. Environ. Sci. 9:139–146. doi:10.1080/1943815X.2012.698990

Hamilton, S.W., E.J. DePeters, J.A. McGarvey, J. Lathrop, and F.M. Mitloehner. 2010. Greenhouse gas, animal performance, and bacterial population structure responses to dietary monensin fed to dairy cows. J. Environ. Qual. 39:106–114. doi:10.2134/jeq2009.0035

Hashmonay, R.A., R.M. Varma, M.T. Modrak, R.H. Kagann, R.R. Segall and P.D. Sullivan. 2008. Radial Plume Mapping: A US EPA test method for area and fugitive source emission monitoring using optical remote sensing. In: Y. J. Kim and U. Platt, editors, Advanced environmental monitoring. Springer Netherlands, Dordrecht. p. 21-36.

Huhtanen, P., M. Ramin, and P. Udén. 2015. Nordic dairy cow model Karoline in predicting methane emissions: 1. Model description and sensitivity analysis. Livest. Sci. 178: 71-80. doi:10.1016/j.livsci.2015.05.009.

Jolliet, O., R. Müller-Wenk, J. Bare, A. Brent, M. Goedkoop, R. Heijungs, N. Itsubo, et al. 2004. The LCIA midpoint-damage framework of the UNEP/SETAC life cycle initiative. Int. J. Life Cycle Assess. 9:394. doi:10.1007/BF02979083

Kouazounde, J.B., J.D. Gbenou, S. Babatounde, N. Srivastava, S.H. Eggleston, C. Antwi, et al. 2014. Development of methane emission factors for enteric fermentation in cattle from Benin using IPCC Tier 2 methodology. Animal 9:526–533. doi:10.1017/S1751731114002626

Lassey, K.R. 2007. Livestock methane emission: From the individual grazing animal through national inventories to the global methane cycle. Agri. Forest Meteorol. 142: 120–132. doi:10.1016/j.agrformet.2006.03.028.

Li, C., W. Salas, R. Zhang, C. Krauter, A. Rotz, and F. Mitloehner. 2012. Manure-DNDC: A biogeochemical process model for quantifying greenhouse gas and ammonia emissions from livestock manure systems. Nutr. Cycling Agroecosyst. 93:163–200. doi:10.1007/s10705-012-9507-z

Li, J., X. Tong, and Q. Yu. 2005. Methane uptake and oxidation by unsaturated soil. Acta Ecol. Sin. 25:141–147.

Liu, H., X. Wu, Z. Li, Q. Wang, D. Liu, and G. Liu. 2017. Responses of soil methanogens, methanotrophs, and methane fluxes to land-use conversion and fertilization in a hilly red soil region of Southern China. Environ. Sci. Pollut Res. 24:8731–8743. doi:10.1007/s11356-017-8628-y

McClelland, S.C., C. Arndt, D.R. Gordon, and G. Thoma. 2018. Type and number of environmental impact categories used in livestock life cycle assessment: A systematic review. Livest. Sci. 209:39–45. doi:10.1016/j.livsci.2018.01.008

Mills, J.A.N., E. Kebreab, C.M. Yates, L.A. Crompton, S.B. Cammell, M.S. Dhanoa, R.E. Agnew, and J. France. 2003. Alternative approaches to predicting methane emissions from dairy cows. J. Anim. Sci. 81:3141–3150. doi:10.2527/2003.81123141x

Mosier, A.R., J.M. Duxbury, J.R. Freney, O. Heinemeyer, K. Minami, and D.E. Johnson. 1998a. Mitigating agricultural emissions of methane. Clim. Change 40:39–80. doi:10.1023/A:1005338731269

Mosier, A., C. Kroeze, C. Nevison, O. Oenema, S. Seitzinger, and O. van Cleemput. 1998b. Closing the global N2O budget: Nitrous oxide emissions through the agricultural nitrogen cycle. Nutr. Cycling Agroecosyst. 52:225–248. doi:10.1023/A:1009740530221

Mosnier, C., A. Duclos, J. Agabriel, and A. Gac. 2017. Orfee: A bio-economic model to simulate integrated and intensive management of mixed crop-livestock farms and their greenhouse gas emissions. Agr. Syst. 157: 202–215. doi:10.1016/j.agsy.2017.07.005

Mosquera, J., G.J. Monteny and J.W. Erisman. 2005. Overview and assessment of techniques to measure ammonia emissions from animal houses: the case of the Netherlands. Environmental Pollution 135: 381-388. doi:10.1016/j.envpol.2004.11.011.

Oates, L.G., and R.D. Jackson. 2014. Livestock management strategy affects net ecosystem carbon balance of subhumid pasture. Rangeland Ecol. Manag. 67: 19–29. doi:10.2111/REM-D-12-00151.1

O'Brien, D., J.L. Capper, P.C. Garnsworthy, C. Grainger, and L. Shalloo. 2014. A case study of the carbon footprint of milk from high-performing confinement and grass-based dairy farms. J. Dairy Sci. 97:1835–1851. doi:10.3168/jds.2013-7174

O'Brien, D., L. Shalloo, F. Buckley, B. Horan, C. Grainger, and M. Wallace. 2011. The effect of methodology on estimates of greenhouse gas emissions from grass-based dairy systems. Agr. Ecosyst. Environ. 141: 39–48. doi:10.1016/j.agee.2011.02.008

O'Grady, M.J., and G.M.P. O'Hare. 2017. Modelling the smart farm. Inform. Proc. Agric. 4: 179–187. doi:10.1016/j.inpa.2017.05.001

Olesen, J.E., K. Schelde, A. Weiske, M.R. Weisbjerg, W.A.H. Asman, and J. Djurhuus. 2006. Modelling greenhouse gas emissions from European conventional and organic dairy farms. Agr. Ecosyst. Environ. 112: 207–220. doi:10.1016/j.agee.2005.08.022

Pattey, E., G.C. Edwards, R.L. Desjardins, D.J. Pennock, W. Smith, B. Grant, and J.I. MacPherson. 2007. Tools for quantifying N2O emissions from agroecosystems. Agr. Forest Meteorol. 142: 103–119. doi:10.1016/j.agrformet.2006.05.013.

Penman, J., M. Gytarsky, T. Hiraishi, T. Krug, D. Kruger, R. Pipatti, L. Buendia, K. Miwa, T. Ngara, K. Tanabe, and F. Wagner. 2003. Good practice guidance for land use, land-use change and forestry. Intergovernmental Panel on Clima

Pergola, M., A. Piccolo, A.M. Palese, C. Ingrao, V. Di Meo, and G. Celano. 2018. A combined assessment of the energy, economic and environmental issues associated with on-farm manure composting processes: Two case studies in South of Italy. J. Clean. Prod. 172: 3969–3981. doi:10.1016/j.jclepro.2017.04.111.

Perrin, A., C. Basset-Mens, and B. Gabrielle. 2014. Life cycle assessment of vegetable products: A review focusing on cropping systems diversity and the estimation of field emissions. Int. J. Life Cycle Assess. 19:1247–1263. doi:10.1007/s11367-014-0724-3

Petersen, S.O., M. Blanchard, D. Chadwick, A. Del Prado, N. Edouard, J. Mosquera, and S.G. Sommerl. 2013. Manure management for greenhouse gas mitigation. Animal 7:266–282. doi:10.1017/S1751731113000736

Petersen, S.O., S.G. Sommer, F. Béline, C. Burton, J. Dach, J.Y. Dourmad, and R. Mihelic. 2007. Recycling of livestock manure in a whole-farm perspective. Livest. Sci. 112(3):180–191. doi:10.1016/j.livsci.2007.09.001.

Ramin, M., and P. Huhtanen. 2015. Nordic dairy cow model Karoline in predicting methane emissions: 2. Model evaluation. Livestock Sci. 178: 81-93. doi:10.1016/j.livsci.2015.05.008.

Ramin, M., and P. Huhtanen. 2013. Development of equations for predicting methane emissions from ruminants. J. Dairy Sci. 96:2476–2493. doi:10.3168/jds.2012-6095

Ramin, M., and P. Huhtanen. 2012. Development of an in vitro method for determination of methane production kinetics using a fully automated in vitro gas system—A modelling approach. Animal Feed Science and Technology 174: 190-200. doi:10.1016/j.anifeedsci.2012.03.008.

Reidsma, P., S. Janssen, J. Jansen, and M.K. van Ittersum. 2018. On the development and use of farm models for policy impact assessment in the European Union– A review. Agr. Syst. 159: 111-125. doi:10.1016/j.agsy.2017.10.012.

Rennie, T.J., H. Baldé, R.J. Gordon, W.N. Smith, and A.C. VanderZaag. 2017. A 3-D model to predict the temperature of liquid manure within storage tanks. Biosyst. Eng. 163: 50-65. doi:10.1016/j.biosystemseng.2017.08.014.

Robert, M., A. Thomas, M. Sekhar, H. Raynal, É. Casellas, P. Casel, P. Chabrier, et al.. 2018. A dynamic model for water management at the farm level integrating strategic, tactical and operational decisions. Environ. Modell. Softw. 100: 123-135. doi:10.1016/j.envsoft.2017.11.013.

Rodriguez, D., H. Cox, P. deVoil, and B. Power. 2014. A participatory whole farm modelling approach to understand impacts and increase preparedness to climate change in Australia. Agr. Syst. 126: 50–61. doi:10.1016/j.agsy.2013.04.003

Rodriguez, D., and V. Sadras. 2011. Opportunities from integrative approaches in farming systems design. Field Crops Res. 124:137–141. doi:10.1016/j.fcr.2011.05.022

Rotz, C.A. 2018. Modeling greenhouse gas emissions from dairy farms. J. Dairy Sci. 101: 6675-6690. doi:10.3168/jds.2017-13272.

Rotz, C.A., M.S. Corson, D.S. Chianese, F. Montes, S.D. Hafner, H.F. Bonifacio, and C.U. Coiner. 2016. The Integrated farm system model: Reference manual, Version 4.3. USDA-ARS Pasture Systems and Watershed Management Research Unit. https://www.ars.usda.gov/ARSUserFiles/80700500/Reference%20Manual.pdf (Accessed Oct. 31, 2018).

Rotz, C.A., M.S. Corson, D.S. Chianese, F. Montes, and S.D. Hafner. 2012. The integrated farm system model. NRAES-176. Natural Resource, Agriculture and Engineering Service, Ithaca, NY.

Rufino, M. 2008. Quantifying the contribution of crop-livestock integration to African farming. PhD thesis. Wageningen UR, The Netherlands.

Ryschawy, J., N. Choisis, J.P. Choisis, A. Joannon, and A. Gibon. 2012. Mixed crop-livestock systems: An economic and environmental-friendly way of farming? Animal 6:1722–1730. doi:10.1017/S1751731112000675

Salcedo, G. 2015. DairyCant: A model for the reduction of dairy farm greenhouse gas emissions. Adv. Anim. Biosci. 6:26–28. doi:10.1017/S2040470014000466

Sándor, R., F. Ehrhardt, L. Brilli, M. Carozzi, S. Recous, P. Smith, V. Snow, et al. 2018. The use of biogeochemical models to evaluate mitigation of greenhouse gas emissions from managed grasslands. Sci. Total Environ. 642, 292–306. doi:10.1016/j.scitotenv.2018.06.020

Schils, R.L.M., M.H.A. de Haan, J.G.A. Hemmer, A. van den Pol-van Dasselaar, J.A. de Boer, A.G. Evers, G. Holshof, et al. 2007. DairyWise, A whole-farm dairy model. J. Dairy Sci. 90:5334–5346. doi:10.3168/jds.2006-842

Sejian, V., R.S. Prasadh, A.M. Lees, J.C. Lees, Y.A.S. Al-Hosni, M.L. Sullivan, and J.B. Gaughan. 2018. Assessment of the carbon footprint of four commercial dairy production systems in Australia using an integrated farm system model. Carbon Manage. 9:57–70. doi:10.1080/17583004.2017.1418595

Skunca, D., I. Tomasevic, I. Nastasijevic, V. Tomovic, and I. Djekic. 2018. Life cycle assessment of the chicken meat chain. J. Clean. Prod. 184: 440-450. doi:10.1016/j.jclepro.2018.02.274.

Smith, D.W., and S. Mukhtar. 2015. Estimation and attribution of nitrous oxide emissions following subsurface application of animal manure: A review. Trans. ASABE. 58:429–438.

Smith, P. 2013. Delivering food security without increasing pressure on land. Global Food Secur. 2: 18-23. doi:10.1016/j.gfs.2012.11.008

Smith, P., M. Bustamante, H. Ahammad, H. Clark, H. Dong, E.A. Elsiddig, H. Haberl, et al. 2014. Agriculture, forestry and other land use (AFOLU). In: O. Edenhofer, R. Pichs-Madruga, Y. Sokona, E. Farahani, S. Kadner, K. Seyboth, A. Adler, I. Baum, S. Brunner, P. Eickemeier, B. Kriemann, J. Savolainen, J. Schlomer, C. van Stechow, T. Zwickel, and J. Minx, editors, Climate Change 2014: Mitigation of climate change. Contribution of Working Group III to the fifth assessment report of the Intergovernmental Panel on Climate Change. Chapter 11. Cambridge Univ. Press, Cambridge, U.K.

Sniffen, C.J., and H.H. Herdt. 1991. Group management and physical facilities. Vet. Clin. North Am. Food Anim. Pract. 7(2):465–471. doi:10.1016/S0749-0720(15)30793-3

Steinfeld, H., P. Gerber, T. Wassenaar, V. Castel, M. Rosales, and C. de Haan. 2006. Livestock's long shadow: Environmental issues and options. FAO, Rome. http://www.europarl.europa.eu/climatechange/doc/FAO%20report%20executive%20summary.pdf (Accessed 9 Feb. 2019).

Stöckle, C.O., M. Donatelli, and R. Nelson. 2003. CropSyst, A cropping systems simulation model. Eur. J. Agron. 18:289–307. doi:10.1016/S1161-0301(02)00109-0

Sun, C., W. Cao, and R. Liu. 2015. Kinetics of methane production from swine manure and buffalo manure. Appl. Biochem. Biotechnol. 177:985–995. doi:10.1007/s12010-015-1792-y

Tallec, T., A. Brut, L. Joly, N. Dumelié, D. Serça, P. Mordelet, N. Claverie. 2019. N2O flux measurements over an irrigated maize crop: A comparison of three methods. Agr. Forest Meteorol. 264: 56–72. doi:10.1016/j.agrformet.2018.09.017.

Thoma, G., J. Popp, D. Nutter, D. Shonnard, R. Ulrich, M. Matlock, D.S. Kim, et al. 2013. Greenhouse gas emissions from milk production and consumption in the United States: A cradle-to-grave life cycle assessment circa 2008. Int. Dairy J. 31: S3–S14. doi:10.1016/j.idairyj.2012.08.013.

Todd, R.W., N.A. Cole, K.D. Casey, R. Hagevoort, and B.W. Auvermann. 2011. Methane emissions from southern High Plains dairy wastewater lagoons in the summer. Anim. Feed Sci. Tech. 166-167: 575-580. doi:10.1016/j.anifeedsci.2011.04.040

Torkamani, J. 2005. Using a whole-farm modelling approach to assess prospective technologies under uncertainty. Agr. Syst. 85: 138–154. doi:10.1016/j.agsy.2004.07.016

VanderZaag, A.C., S. Jayasundara, and C. Wagner-Riddle. 2011. Strategies to mitigate nitrous oxide emissions from land applied manure. Anim. Feed Sci. Tech. 166-167: 464-479. doi:10.1016/j.anifeedsci.2011.04.034

Vergé, X.P.C., C. De Kimpe, and R.L. Desjardins. 2007. Agricultural production, greenhouse gas emissions and mitigation potential. Agr. Forest Meteorol. 142: 255-269. doi:10.1016/j.agrformet.2006.06.011.

Waldrip, H.M., R.W. Todd, D.B. Parker, N.A. Cole, C.A. Rotz, and K.D. Casey. 2016. Nitrous oxide emissions from open-lot cattle feedyards: A review. J. Environ. Qual. 45:1797–1811. doi:10.2134/jeq2016.04.0140

Waldrip, H.M., C.A. Rotz, S.D. Hafner, R.W. Todd, and N.A. Cole. 2014. Process-based modeling of ammonia emission from beef cattle feedyards with the integrated farm systems model. J. Environ. Qual. 43:1159–1168. doi:10.2134/jeq2013.09.0354

Waldrip, H.M., R.W. Todd, C. Li, N. Andy Cole, and W.H. Salas. 2013. Estimation of ammonia emissions from beef cattle feedyards using the process-based model manure-DNDC. Trans. ASABE 56: 1103–1114. doi:10.13031/trans.56.9906.

Winkler, T., K. Schopf, R. Aschemann, and W. Winiwarter. 2016. From farm to fork– A life cycle assessment of fresh Austrian pork. J. Clea. Prod. 116: 80–89. doi:10.1016/j.jclepro.2016.01.005

Wolfert, S., L. Ge, C. Verdouw, and M.-J. Bogaardt. 2017. Big data in smart farming– A review. Agr. Syst. 153: 69-80. doi:10.1016/j.agsy.2017.01.023

Zimmerman, P. 2011. Method and System for Monitoring and Reducing Ruminant Methane Production. U.S. Patent 7966971. Date issued: 28 June.

Advances and Outlook of Manure Production and Management

Zhongqi He*, Paulo Pagliari, and Heidi M. Waldrip

Abstract

With the global awareness on sustainable utilization of limited natural resources, manure researchers are facing both challenges and opportunities to promote more effective utilization of animal manure as a beneficial resource with fewer environmental burdens. This book is a scientifically sound and up-to-date manure reference. It provides a comprehensive coverage of major manure-producing species (beef, dairy, swine, poultry, goats/sheep, horse, and deer) and manure management systems. This chapter, as the last chapter, highlights the activities and accomplishments synthesized in the previous 20 chapters in this volume, while also presenting and discussing several topics and emerging issues that are not covered in those chapters, but are worthy of mention and deserve attention from manure researchers in future exploration. The insights and visions conceived in the individual chapters, and as a whole, provides a valuable resource for manure researchers to stimulate new ideas and directions for applied and environmental manure research.

Animal manure resulting from animal farming demonstrates large variability in form and composition. Careful management of animal manure, especially from concentrated animal feeding operations (CAFOs) in beef cattle feedyards, broiler and layer houses and large swine production facilities, is crucial for environmentally sound and sustainable agriculture. Manure accumulation in and around CAFO structures may lead to contamination of air, soil, and water if improperly managed. With the global awareness on sustainable utilization of limited natural resources, manure researchers are facing both challenges and opportunities to promote more effective utilization of animal manure as a valuable resource with less environmental burden than traditional manure management practices (He et al., 2016). Hatfield and Stewart (1998) compiled a collection on the effective use of manure as a soil resource. He (2011) organized a book reviewing the state-of-the-art manure environmental chemistry. He (2012) further edited a monograph focusing on the applied manure research with challenges and opportunities beyond the adverse environmental concerns. Through a special publication by John Wiley & Sons, Sommer et al. (2013) presented efforts toward managing animal manure more efficiently and sustainably, primarily in European production systems. In the work of He and Zhang (2014), several case studies were presented where modern techniques were used to characterize manure and soil. The present book contains 21 chapters that are divided into five parts, reporting on the latest basic and applied manure research activities. These chapters include issues related to adverse environmental concerns; however, the

Abbreviations: CAFO, concentrated animal feeding operation; GHG, greenhouse gas.

Z. He, USDA-ARS Southern Regional Research Center, 1100 Robert E. Lee Blvd. New Orleans, LA 70124; P. Pagliari, Department of Soil, Water, and Climate, University of Minnesota. Southwest Research and Outreach Center. 23669 130th St. Lamberton, MN 56152.; H.M. Waldrip, USDA-ARS Conservation and Production Research Laboratory, Bushland, TX 79012. * Corresponding author (Zhongqi.he@usda.gov)

doi:10.2134/asaspecpub67.c28

scope expands much further by providing comprehensive and current information on manure characteristics and how management practices potentially influence air, soil, and water quality. The first two chapters (Pagliari et al., 2019b; Zhang et al., 2019) are introductory information. The chapter of Pagliari et al. (2019b) provides an overview of the impact of modern CAFOs on animal manure production and utilization in the United States. Six pairs of terms frequently used in animal manure research and management (i.e., Animal versus livestock, Animal head versus animal unit, Animal manure versus animal waste, As-excreted versus as-stored manure, Manure liquid versus slurry, semi-solid, and solid, and Poultry manure versus poultry litter) are clarified in meaning and application scope. The chapter of Zhang et al. (2019) presents case studies to analyze the temporal changes of manure chemical compositions and environmental awareness in the Southern Great Plains using data from two service laboratories in Kansas and Oklahoma over the last two decades. The data presented in this chapter showed that dry matter contents, pH, and macro- and micronutrient contents of the manures had few changes over time in the last 5 to 20 years; however, there was a trend toward phosphorus decrease over time in swine effluent, reflecting various environmental regulations related to animal manure management that have been established and implemented. The next 18 chapters, in three parts, covered the four aspects of animal manure production, characteristics, environmental concerns and management. The present chapter, as the last chapter, highlights the activities and accomplishments in the preceding 18 chapters presented in this volume. In addition, this final chapter also discusses several topics and emerging issues that are not covered in those chapters yet are worthy of mention and deserve future exploration by manure researchers.

Diversity of Animal Farming Practices and Impacts on Manure Characteristics

Most animals raised indoors are housed in buildings specifically designed for the animals being reared, to provide environments suitable for their health, welfare, and optimal production. Janni and Cortus (2019) reviewed several common systems for production of swine, poultry, dairy, and beef in the United States, along with relevant manure storage and management methods. The images, illustrations and/or schemes in this chapter clearly illustrate the relevant production systems and storage methods, providing useful visual information to assist with understanding these management practices. The following five chapters covered the generation and nutrient characteristics of beef and dairy manure (Pagliari et al., 2019a), poultry manure and litter (Ashworth et al., 2019), swine manure and wastewater (Wilson et al., 2019), small ruminant and cervid livestock (Norris and Smith, 2019), and horses and other equids (Westendorf et al., 2019). The last chapter in this part (He, 2019) reviewed the general organic animal farming practices and comparative studies of conventional and organic manures.

Altogether, those chapters provided helpful references of the production and characteristics of different types of animal manure. However, the features of organic matter and heavy metals in livestock manure are not covered in-depth, due to the fact that these components of animal manure are not as critical as nitrogen (N) and phosphorus (P) as nutrients for crop growth or issues of environmental concern, such as gaseous emissions and eutrophication of surface waters. In addition, the contents of heavy elements in animal manure are generally not high enough to impact, directly or after application to soil, environmental quality in an imminent way (He et al., 2009; Kingery et al., 1994). However, these two factors (organic matter and heavy metals) are still worthy of research for recycling and/or environmental monitoring in the long run. The research of manure organic matter characterization could be found in He and Waldrip (2015), He et al. (2017), Miller et al. (2018), Park et al. (2019), and Waldrip et al. (2014). Schroder et al. (2011) synthesized the information on sources and contents of heavy metal elements in manure. Church et al. (2011) and Tazisong et al. (2011) focused specifically on manure-related arsenic (As) and mercury (Hg) topics: their occurrence in manures, environmental fate and transport, and ecosystem impacts. Updated information could be seen in Azeez et al. (2019), Leclerc and Laurent (2017), Lv et al. (2016), Qian et al. (2018), and Wang et al. (2018).

Environmental Aspects of Animal Manure and Mitigation Technologies

There are five chapters in the present book covering the issues of environmental effects of animal manure and potential mitigation measures. Given the expansion of eutrophication in water bodies around the world, the improved management of manure to mitigate P losses to water has become a global concern. The chapter of Kleinman et al. (2019) framed and illustrates the diversity of strategies and practices for manure management that constitute the state of knowledge and the state of the art of managing manures to minimize P losses to water. It focuses on manure properties, land application practices, farmstead infrastructure, and specific farming systems. Greenhouse gas (GHG) emissions derived from animal respiration (CO_2 and CH_4) and manure management (N_2O and CH_4) are also an environmental concern worldwide, due to GHG effects on global climate change. While adaptation and mitigation strategies are required, accurate estimation and/or quantification is the first step to have a successful plan of action. The chapter of Ahmed et al. (2019) reviews available simulation models for quantifying emissions, identifying opportunities, and designing adaptation and mitigation strategies. Whole Farm Modeling (WFM) was recommended as a powerful tool in testing the efficacy of various strategies to reduce GHG emissions. This holistic whole-farm approach integrates the effects and interactions of all variables that influence manure emissions, nutrient contents, animal production, and animal diet within a specific system. The chapter reviews different emission factors, empirical or statistical models, and detailed process-level simulations, such as WFM) and the life cycle assessment (LCA) approach. The authors concluded that WFM had great potential to quantify and predict GHG emissions from different sources in livestock production. While GHG emissions are a global concern, manure odor negatively impacts local and regional environmental air quality around the CAFO facilities. Readers may refer to Miller and Varel (2011) for origins and identities of key manure odor components. Recent advances in deodorizing or reducing odor in animal manure can be found in Guo et al. (2019a), Yan et al. (2017), and Zang et al. (2016). Large CAFOs also pose significant risks to human health from the emitted organic and inorganic compounds (Ribaudo et al., 2003; Hooiveld et al., 2016). Studies conducted to assess the health of communities located around CAFOs have shown that those communities are more susceptible to respiratory and gastrointestinal problems compared with communities located further away (Wing and Wolf, 2000; Radon et al., 2006; Hooiveld et al., 2016). Other issues of concern with CAFO management are emissions of hydrogen sulfide (H_2S) and dust, which can cause both long- and short-term effects on human and animal health. Under anaerobic conditions, excreted sulfur (S)-containing amino acids (i.e., cysteine, cystine and methionine) in manure are broken down into over 27 different forms of organic S compounds that contribute to odor. Although not as extensively studied as N transformations, organic S in manure can be oxidized to form H_2S and then volatilize into the atmosphere (Clanton and Schmidt, 2000). Emissions of H_2S within livestock housing can produce symptoms in both production animals and human CAFO personnel, with effects ranging from mild eye irritation to death via respiratory failure. Fugitive dust from animal housing and manure storage reduces the quality of life for regional residents, can cause ocular and respiratory issues, and provides a means of transport for potential pathogens and pharmaceuticals from CAFOs. In addition, dust within livestock housing can affect animal performance and long-term respiratory problems in CAFO personnel (Al Homindan and Robertson, 2003; Andersen et al., 2004). Further studies are needed to characterize H_2S and dust emissions from CAFOs, as well as to develop effective mitigation measures to reduce their impacts both at the farm- and regional-levels. Trabue et al. (2019) recently reported a case study on the sources of odorous compounds and their transport from a swine deep-pit finishing operation. They reported that odorous compounds generated during agitation and pumping of the deep pits was mainly H_2S, and were mainly transported in the gas phase with less than 0.1% being associated with dust particle matter (PM10). Per the case study, they recommended that odor mitigation efforts should focus on gaseous compounds emitted from deep-pits and especially during manure agitation and deep-pit pumping in swine production management.

A long history of use of veterinary pharmaceuticals in CAFOs has resulted in annual environmental discharges of 3000 to 27000 Mg of pharmaceuticals and their residues as components animal manure (Song and Guo, 2014). The chapter by Bai (2019) reviews the excretion and types of estrogens and estrogen conjugates in animal manure, the ecological risks, the persistence and mobility in manure-amended soils and adjacent water systems, and fate and transport mechanisms in soil-plant-water ecosystems. It provides critical information to better understand the environmental risks and behaviors of pharmaceutical contaminants with manure land application practices. The chapter by Zheng et al. (2019) addresses the current use of veterinary antibiotics in animal agriculture, potential risks from disposing of antibiotic residue-containing animal manure, and the environmental fate and transport of manure-associated antibiotics. Veterinary antibiotics may migrate from manure-receiving fields to surrounding water bodies via surface runoff and leaching; thus, this chapter highlights three mitigation strategies to reduce their load into the environment and minimize their negative effects on agro-ecosystems. The chapter by Chen et al. (2019) further reviews and discusses the relationships among veterinary pharmaceuticals, pathogens, and antibiotic resistance in animal manure and manure-affected environments. Prolonged exposure to non-lethal doses of antibiotics in the environment can lead to the development of resistance in microorganisms, which can transfer to pathogenic bacteria. This chapter also synthesizes information on the potential impact of pathogens and antibiotic resistance on human and environmental health. Lastly, the chapter provides insights on the strategies for minimizing the spread and proliferation of antibiotic resistance to the environment with the ultimate purpose to protect human health.

Beneficial Manure Utilization

Beneficial utilization of animal manure is to recycle nutrients, energy, and other valuable components of animal manure while minimizing the adverse environmental impacts (He et al., 2012). Six chapters are included in this part. Composting manure is a process in which animal feces, urine, and bedding are stacked, turned, and managed in a way that promotes decomposition. Composting has many benefits including destruction of harmful pathogens and weed seeds, and reduction in overall volume and odor. A previous work (Wichuk and McCartney, 2012) reviewed the process, and provided insight into stability and maturity of manure composting. In this volume, Modderman (2019) updated the information on composting with a focus on "composting with or without additives". The chapter of Harrison and Ndegwa (2019) reviews and analyzes the anaerobic digestion of dairy and swine manures. Anaerobic digestion technologies are most commonly associated with dairy production, although they are also utilized in the management of manure and waste water at swine and poultry operations. While the biogas (active ingredient CH_4) produced in anaerobic digestion is used as a fuel, the liquid after anaerobic digestion (digestate) can be used as a nutrient source for crop production. Furthermore, the solid fraction of digestate can be used as a bedding material or as a soil amendment. The high organic matter and mineral nutrient contents make animal manure a readily available feedstock for bioenergy and biochar production. For example, the higher heating values of dry solid cattle manure, swine solid manure, and poultry litter were 18.0, 17.6 and 15.8 MJ kg^{-1} on dry weight basis, respectively (Uchimiya and He, 2012). In this volume, Guo et al. (2019b) summarized and synthesized the thermochemical techniques for processing animal manure into biochar and biofuels. Pyrolysis and gasification of animal manure generate three products (i.e., biochar, biooil, and syngas). Hydrothermal liquefaction of animal manure yields hydrochar and biocrude oil. Per the review, they concluded that, currently, thermochemical conversion of animal manure to bioenergy and biochar is technically feasible, but needs to be improved in economic viability. As manure is bulky and expensive to transport long distances for treatment, other researchers have found that solid–liquid separation of raw or digested manure improves the range of opportunities for removal and recovery of N and P as plant nutrients. Vanotti et al. (2019) contributed the chapter on removing and recovering N and P from animal manure. Phosphorus can be harvested from liquid dairy and swine manure by precipitation with magnesium ammonium phosphate crystals (struvite). Methods have been also developed to efficiently precipitate and recover P from manures as calcium phosphates. Other alternatives are iron (Fe) and aluminum (Al) salts, which can be used to remove P from wastewater; however, there are limitations due to high dosage rates of Fe and Al along with low plant availability of the products. Alum $[Al_2(SO_4)_3–14H_2O]$ has been used successfully to treat poultry litter and was found to be a very sustainable practice: approximately one billion broiler chickens are being grown in the United States with alum-amended bedding (Moore Jr., 2011). Comprehensive coverage on improving the sustainability of animal agriculture by treating manure with alum is provided by Moore (2011). The long-term (up to 20-years) effects of alum-treated litter on P distribution and availability in soils are further investigated by Abdala et al. (2018), Anderson et al. (2018), and Huang et al. (2016).

Regarding the environmental and human health issues of rogue N emissions, the chapter of Vanotti et al. (2019) points out that ammonia (NH_3) can be captured from the air in the animal barns using chemical and biological scrubbers that trap NH_3 before it enters the atmosphere. Stripping processes are also available for removing the NH_3 from liquid manures by using strong acids favored by high temperatures and pH. Gas-permeable membrane processes have been used to remove and recover N from liquid manure by submerging membrane manifolds in the liquid manure and capturing the NH_3 before it escapes into the air. In addition, they also reviewed available biological N and P removal processes. While multiple technologies are available, the authors pointed out that the agronomic efficacy of recovered nutrient materials for use as plant fertilizers is an important consideration in the selection of the best approach for N and P removal and recovery from manure.

While direct land application is common for manure as fertilizers (Tewolde et al., 2018; He et al., 2019), pelletization or granulation offers a greater opportunity to alleviate challenges faced in both on- and off-farm manure utilization by creating a product that is easier to handle, transport and apply. The chapter of Hao and He (2019) reviews these manure processing techniques. Pelletizing animal manure removes its moisture, reduces its weight, volume and odor, thus, creates a product that is preferred over fresh manure for its reduced cost of storage, and its convenience for transportation, handling, and application. While poultry litter is the most common animal manure to be pelletized, pellets have also been produced using manure from dairy and beef cattle, swine and horses, sometimes with added bulking agents or other supplements. This chapter also evaluates the quality of manure pellets as an organic fertilizer, and explores the potential of manure pellets for on-farm and off-farm use. Latest examples of such studies are the use of pelletized manure products for tomato (*Solanum lycopersicum* var. *cerasiforme*) growth in soilless media (Liu et al., 2019) and for organic spring wheat production (Alam et al., 2018). The last chapter in this part discusses organomineral fertilizers and their field application (Smith et al., 2019). Organomineral fertilizers belong to the category of pellets and granules. More specifically, they are combined products of mineral fertilizers and manure or other organic wastes through industrial processes. Compared with manure, these products are significantly higher in nutrient content and nutrient balance as well as material stability and uniformity and, therefore, lower application rates and general application methods can be used. These organomineral fertilizers can be packaged and sold commercially, with labels that clearly articulate the nutrient values for the amounts of nutrients applied to specific crops. The literature and data presented in the chapter have shown that organomineral fertilizers improve plant growth in parameters, such as yield and nutrient uptake, to a significantly greater degree than when manure or fertilizers are used alone. Future research is needed to determine the best fertilizer management practices for organomineral fertilizers in different soil types. For example, Frazão et al. (2019) recently compared a granular poultry litter-derived organomineral fertilizer with triple superphosphate in two contrasting soil types (Oxisol and Entisol). Their data indicated that soil-P adsorption capacity affected the effectiveness of the poultry litter-derived organomineral fertilizer. In oxisols and other highly weathered soils, organomineral fertilizer is as effective as conventional water-soluble P fertilizers (e.g., triple superphosphate). On the other hand, water-soluble P fertilizer may be more effective than organomineral fertilizer on sandy soils. Rodrigo Sakurada et al. (2019) characterized the organomineral fertilizer residues recovered from an Oxisol by chemical, thermal, and spectroscopic analysis. They compared the characteristics of two phosphate organominerals and one inorganic fertilizer before and after their introduction into the soil for four 35-d cropping cycles, confirming that organomineral and inorganic fertilizers have different solubilities and capacities to make P available in the soil. In the meantime, Mumbach et al. (2019) noted that the organomineral fertilizer can be used as a substitute for mineral fertilization, but it does not present any additional improvements in relation to the other sources when it comes to wheat growth.

Conclusion

The chapters in this book volume addressed the environmental concerns, while looking at numerous options for beneficial utilization of manure as a valuable commodity. Topics covered include characteristics of manure from different animals and under differing management systems, environmental aspects of animal manure, existing and evolving mitigation technologies to reduce environmental impact, and beneficial utilization practices for maximal fertilizer value with minimal environmental degradation. This volume provides a timely, comprehensive reference that is applicable and useful for the scientific community, regulatory agencies, environmental advocacy groups, and animal farm managers. The insights and visions conceived in the individual chapters and as a whole book would be also helpful for manure researchers to develop new ideas and directions for future exploration of applied and environmental manure research.

References

Abdala, D., P. Moore, M. Rodrigues, W. Herrera, and P.S. Pavinato. 2018. Long-term effects of alum-treated litter, untreated litter and NH_4NO_3 application on phosphorus speciation, distribution and reactivity in soils using K-edge XANES and chemical fractionation. J. Environ. Manage. 213:206–216. doi:10.1016/j.jenvman.2018.02.007

Ahmed, M., S. Ahmad, H.M. Waldrip, M. Ramin, and M.A. Raza. 2019. Whole farm modeling: A systems approach to understanding and managing livestock for greenhouse gas mitigation, economic viability and environmental quality. In: H.M. Waldrip, P.H. Pagliari, and Z. He editors, Animal manure: Production, characteristics, environmental concerns and management. Vol. ASA Spec. Publ. 67. ASA and SSSA, Madison, WI. doi:10.2134/asaspecpub67.c25

Al Homindan, A., and J.F. Robertson. 2003. Effect of litter type and stocking density on ammonia, dust concentrations, and broiler performance. Br. Poult. Sci. 44(supp. 1):7–8. doi:10.1080/00071669708417953

Alam, M.Z., D.H. Lynch, G. Tremblay, R. Gillis-Madden, and A. Vanasse. 2018. Optimizing combining green manures and pelletized manure for organic spring wheat production. Can. J. Soil Sci. 98:638–649. doi:10.1139/cjss-2018-0049

Anderson, K.R., P.A. Moore, D.M. Miller, P.B. DeLaune, D.R. Edwards, P.J. Kleinman, and B.J. Cade-Menun. 2018. Phosphorus leaching from soil cores from a twenty-year study evaluating alum treatment of poultry litter. J. Environ. Qual. 47:530–537. doi:10.2134/jeq2017.11.0447

Andersen, C.I., S.G. Von Essen, L.M. Smith, J. Spencer, R. Jolie, and K.J. Donham. 2004. Respiratory symptoms and airway obstruction in swine veterinarians: A persistent problem. Am. J. Ind. Med. 46(4):386–392. doi:10.1002/ajim.20080

Ashworth, A.J., J.P. Chastain, and J.P.A. Moore. 2019. Nutrient characteristics of poultry manure and litter. In: H.M. Waldrip, P.H. Pagliari, and Z. He, editors, Animal manure: Production, characteristics, environmental concerns and management. Vol. Spec. Publ. 67. ASA and SSSA, Madison, WI.

Azeez, J.O., T.B. Olowoboko, M.D. Ajenifuja, N. Ilebor, and E. Adekoya. 2019. Speciation of some heavy metals as influenced by poultry manure application in dumpsite soils. J. Appl. Sci. 19:487–494. doi:10.3923/jas.2019.487.494

Bai, X. 2019. Fate and transport of estrogens and estrogen conjugates in manure-amended soils, p. In: press. In: H.M. Waldrip, P.H. Pagliari, and Z. He, editors, Animal manure: Production, characteristics, environmental concerns and management, Vol. ASA Spec. Publ. 67. ASA and SSSA, Madison, WI.

Chen, C., S. Hilaire, and K. Xia. 2019. Veterinary pharmaceuticals, pathogens and antibiotic resistance, p. In: press. In: H.M. Waldrip, P. Pagliari, and Z. He, editors, Animal manure: Production, characteristics, environmental concerns and management, Vol. ASA Spec. Publ. 67. ASA and SSSA, Madison, WI.

Church, C.D., J.E. Hill, and A.L. Allen. 2011. Fate and transport of arsenic from organoarsenicals fed to poultry. In: Z. He, editor, Environmental chemistry of animal manure. Nova Science Publishers, New York. p. 415–426.

Clanton, C.J., and D.R. Schmidt. 2000. Sulfur compounds in gases emitted from stored manure. Trans. ASABE 43:1229–1239. doi:10.13031/2013.3016

Frazão, J.J., V. de Melo Benites, J.V.S. Ribeiro, V.M. Pierobon, and J. Lavres. 2019. Agronomic effectiveness of a granular poultry litter-derived organomineral phosphate fertilizer in tropical soils: Soil phosphorus fractionation and plant responses. Geoderma 337:582–593. doi:10.1016/j.geoderma.2018.10.003

Guo, H., H. Hao, Q. Zhang, J. Wang, and J. Liu. 2019a. Components and dispersion characteristics of organic and inorganic odorous gases in a large-scale dairy farm. J. Air Waste Manage. Assoc. doi:10.1080/10962247.2018.1562389

Guo, M., H. Li, B. Baldwin, and J. Morrison. 2019b. Thermochemical processing of animal manure for bioenergy and biochar. In: H.M. Waldrip, P. Pagliari, and Z. He, editors, Animal manure: Production, characteristics, environmental concerns and management, Vol. ASA Spec. Publ. 67. ASA and SSSA, Madison, WI.

Hao, X., and Z. He. 2019. Pelletizing animal manures for on- and off-farm use. In: H.M. Waldrip, P.H. Pagliari, and Z. He, editors, Animal manure: Production, characteristics, environmental concerns and management. ASA Spec. Publ. 67. ASA and SSSA, Madison, WI.

Harrison, J.H., and P.M. Ndegwa. 2019. Anaerobic digestion of dairy and swine waste. In: H.M. Waldrip, P.H. Pagliari, and Z. He, editors, Animal manure: Production, characteristics, environmental concerns and management, Vol. ASA Spec. Publ. 67. ASA and SSSA, Madison, WI.

Hatfield, J.L., and B.A. Stewart, editors. 1998. Animal waste utilization: Effective use of manure as a soil resource. Ann Arbor Press, Chelsea, MI.

He, Z., editor. 2011. Environmental chemistry of animal manure. Nova Science Publishers, NY. p. 1–459.

He, Z., editor. 2012. Applied research of animal manure: Challenges and opportunities beyond the adverse environmental concerns. Nova Science Publishers, New York. p. 1–325.

He, Z. 2019. Organic animal farming and comparative studies of conventional and organic manures, p. In: press. In: H.M. Waldrip, P.H. Pagliari, and Z. He, editors, Animal manure: Production, characteristics, environmental concerns and management. ASA Spec. Publ. 67. ASA and SSSA, Madison, WI.

He, Z., and H.W. Waldrip. 2015. Composition of whole and water-extractable organic matter of cattle manure affected by management practices. In: Z. He and F. Wu, editors, Labile organic matter–chemical compositions, function, and significance in soil and the environment. SSSA Spec. Publ. 62. Soil Science Society of America, Madison, WI. p. 41–60. doi:10.2136/sssaspecpub62.2014.0034

He, Z., and H. Zhang, editors. 2014. Applied manure and nutrient chemistry for sustainable agriculture and environment. Springer, Amsterdam, the Netherlands. p. 1–379. doi:10.1007/978-94-017-8807-6

He, Z., P.H. Pagliari, and H.M. Waldrip. 2016. Applied and environmental chemistry of animal manure: A review. Pedosphere 26:779–816. doi:10.1016/S1002-0160(15)60087-X

He, Z., D.M. Endale, H.H. Schomberg, and M.B. Jenkins. 2009. Total phosphorus, zinc, copper, and manganese concentrations in Cecil soil through ten years of poultry litter application. Soil Sci. 174:687–695. doi:10.1097/SS.0b013e3181c30821

He, Z., M. Guo, N. Lovanh, and K.A. Spokas. 2012. Applied manure research-Looking forward to the benign roles of animal manure in agriculture and the environment. In: Z. He, editor, Applied research of animal manure: Challenges and opportunities beyond the adverse environmental concerns. Nova Science Publishers, New York. p. 299–309.

He, Z., I.A. Tazisong, X. Yin, D.B. Watts, Z.N. Senwo, and H.A. Torbert. 2019. Long-term cropping system, tillage, and poultry litter application affect the chemical properties of an Alabama Ultisol. Pedosphere 29:180–194. doi:10.1016/S1002-0160(19)60797-6

He, Z., M. Zhang, A. Zhao, H.M. Waldrip, P.H. Pagliari, and R.D. Harmel. 2017. Impact of management practices on water extractable organic carbon and nitrogen from 12-year poultry litter amended soils. Open J. Soil Sci. 7:259–277. doi:10.4236/ojss.2017.710019

Hooiveld, M., L.A. Smit, F. van der Sman-de Beer, I.M. Wouters, C.E. van Dijk, P. Spreeuwenberg, et al. 2016. Doctor-diagnosed health problems in a region with a high density of concentrated animal feeding operations: A cross-sectional study. Environ. Health 15:24. doi:10.1186/s12940-016-0123-2

Huang, L., P.A. Moore, P.J. Kleinman, K.R. Elkin, M.C. Savin, D.H. Pote, and D.R. Edwards. 2016. Reducing phosphorus runoff and leaching from poultry litter with alum: Twenty-year small plot and paired-watershed studies. J. Environ. Qual. 45:1413–1420. doi:10.2134/jeq2015.09.0482

Janni, K., and E. Cortus. 2019. Common animal production systems and manure storage methods. In: H.M. Waldrip, P.H. Pagliari, Z. He, editors, Animal manure: Production, characteristics, environmental concerns and management, Vol. Spec. Publ. 67. ASA and SSSA, Madison, WI.

Kingery, W.L., C.W. Wood, D.P. Delaney, J.C. Williams, and G.L. Mullins. 1994. Impact of long-term application of broiler litter on environmentally related soil properties. J. Environ. Qual. 23:139–147. doi:10.2134/jeq1994.00472425002300010022x

Kleinman, P.J.A., S. Spiegal, J. Liu, M. Holly, C. Church, and J. Ramirez-Avila. 2019. Managing animal manure to minimize phosphorus losses from land to water. In: H.M. Waldrip, P.H. Pagliari, and Z. He, editors, Animal manure: Production, characteristics, environmental concerns and management. Vol. ASA Specl. Pub. 67. ASA and SSSA, Madison, WI. doi:10.2134/asaspecpub67.c13

Leclerc, A., and A. Laurent. 2017. Framework for estimating toxic releases from the application of manure on agricultural soil: National release inventories for heavy metals in 2000–2014. Sci. Total Environ. 590-591:452–460. doi:10.1016/j.scitotenv.2017.01.117

Liu, Z., J. Howe, X. Wang, X. Liang, and T. Runge. 2019. Use of dry dairy manure pellets as nutrient source for tomato (*Solanum lycopersicum* var. cerasiforme) growth in soilless media. Sustainability 11:811. doi:10.3390/su11030811

Lv, B., M. Xing, and J. Yang. 2016. Speciation and transformation of heavy metals during vermicomposting of animal manure. Bioresour. Technol. 209:397–401. doi:10.1016/j.biortech.2016.03.015

Miller, D.N., and V.H. Varel. 2011. Origins and identities of key manure odor components. In: Z. He, editor, Environmental chemistry of animal manure. Nova Science Publishers, New York. p. 153–177.

Miller, J., P. Hazendonk, and C. Drury. 2018. Influence of manure type and bedding material on carbon content of particulate organic matter in feedlot amendments using 13C NMR-DPMAS. Compost Sci. Util. 26:27–39. doi:10.1080/1065657X.2017.1342106

Modderman, C. 2019. Composting with or without additives. In: H.M. Waldrip, P.H. Pagliari, and Z. He, editors, Animal manure: Production, characteristics, environmental concerns and management. ASA Spec. Publ. 67. ASA and SSSA, Madison, WI. doi:10.2134/asaspecpub67.c19

Moore Jr., P.A. 2011. Improving the sustainability of animal agriculture by treating manure with alum. In: Z. He, editor, Environmental chemistry of animal manure. Nova Science Publishers, N.Y. p. 349–381.

Mumbach, G.L., L.C. Gatiboni, F.D. de Bona, D.E. Schmitt, D.J. Dall'Orsoletta, C.A. Gabriel, and É.B. Bonfada. 2019. Organic, mineral and organomineral fertilizer in the growth of wheat and chemical changes of the soil. Revista Brasileira de Ciências Agrárias 14:e5618.

Norris, A.B., and W.B. Smith. 2019. Farming characteristics and manure management of small ruminant and cervid livestock. In: H.M. Waldrip, P.H. Pagliari, and Z. He, editors, Animal manure: Production, characteristics, environmental concerns and management. Vol. Spec. Publ. 67. ASA and SSSA, Madison, WI. doi:10.2134/asaspecpub67.c7

Pagliari, P.H., S. Niraula, M. Wilson, C. Modderman, H.M. Waldrip, and Z. He. 2019a. Nitrogen and phosphorus characteristics of beef and dairy manure. In: H.M. Waldrip, P.H. Pagliari, and Z. He, editors, Animal manure: production, characteristics, environmental concerns and management, Vol. Spec. Publ. 67. ASA and SSSA, Madison, WI. p. 45–62. doi:10.2134/asaspecpub67.c22

Pagliari, P.H., M. Wilson, and Z. He. 2019b. Animal manure production and utilization: Impact of modern concentrated animalfeeding. In: H.M. Waldrip, P.H. Pagliari, and Z. He, editors, Animal manure: Production, characteristics, environmental concerns and management, Vol. Spec. Publ. 67. ASA and SSSA, Madison, WI. 1–14.

Park, J., K.H. Cho, M. Ligaray, and M.-J. Choi. 2019. Organic matter composition of manure and its potential impact on plant growth. Sustainability 11:2346. doi:10.3390/su11082346

Qian, X., Z. Wang, G. Shen, X. Chen, Z. Tang, C. Guo, H. Gu, and K. Fu. 2018. Heavy metals accumulation in soil after 4 years of continuous land application of swine manure: A field-scale monitoring and modeling estimation. Chemosphere 210:1029–1034. doi:10.1016/j.chemosphere.2018.07.107

Radon, K., A. Schulze, and D. Nowak. 2006. Inverse association between farm animal contact and respiratory allergies in adulthood: Protection, underreporting or selection? Allergy 61:443–446. doi:10.1111/j.1398-9995.2006.00995.x

Ribaudo, M., J.D. Kaplan, L.A. Christensen, N. Gollehon, R. Johansson, V.E. Breneman, et al. 2003. Manure management for water quality costs to animal feeding operations of applying manure nutrients to land. USDA-ERS Agricultural Economic Report No. 8. USDA-ERS, Washington, D.C. 97 pages. doi:10.2139/ssrn.757884.

Rodrigo Sakurada, L., A.S. Muniz, F. Sato, T.T. Inoue, A.M. Neto, and M.A. Batista. 2019. Chemical, thermal, and spectroscopic analysis of organomineral fertilizer residue recovered from an Oxisol. Soil Sci. Soc. Am. J. 83:409–418. doi:10.2136/sssaj2018.08.0294

Schroder, J.L., H. Zhang, J.R. Richards, and Z. He. 2011. Sources and contents of heavy metals and other trace elements in animal manures. In: Z. He, editor, Environmental chemistry of animal manure. Nova Science Publishers, NY. p. 385–414.

Smith, W.B., M. Wilson, and P.H. Pagliari. 2019. Organomineral fertilizers and their application to field. In: H.M. Waldrip, P.H. Pagliari, and Z. He, editors, Animal manure: Production, characteristics, environmental concerns and management. Vol. ASA Spec. Publ. 67. ASA and SSSA, Madison, WI. p. 203–218.

Sommer, S.G., M.L. Christensen, T. Schmidt, and L.S. Jensen, editors. 2013. Animal manure recycling: Treatment and management. John Wiley & Sons, West Sussex, U.K. p. 1–384. doi:10.1002/9781118676677

Song, W., and M. Guo. 2014. Residual veterinary pharmaceuticals in animal manures and their environmental behaviors in soils. In: Z. He and H. Zhang, editors, Applied manure and nutrient chemistry for sustainable agriculture and environment. Springer, Amsterdam, the Netherlands. p. 23–52. doi:10.1007/978-94-017-8807-6_2

Tazisong, I.A., Z.N. Senwo, R.W. Taylor, and Z. He. 2011. Mercury in animal manure and impacts on environmental health. In: Z. He, editor, Environmental chemistry of animal manure. Nova Science Publishers, NY. p. 427–442.

Tewolde, H., M.W. Shankle, T.R. Way, D.H. Pote, K.R. Sistani, and Z. He. 2018. Poultry litter band placement affects accessibility and conservation of nutrients and cotton yield. Agron. J. 110:675–684. doi:10.2134/agronj2017.07.0387

Trabue, S., K. Scoggin, J. Tyndall, T. Sauer, G. Hernandez-Ramirez, R. Pfeiffer, and J. Hatfield. 2019. Odorous compounds sources and transport from a swine deep-pit finishing operation: A case study. J. Environ. Manage. 233:12–23. doi:10.1016/j.jenvman.2018.10.110

Uchimiya, M., and Z. He. 2012. Calorific values and combustion chemistry of animal manure. In: Z. He, editor, Applied research of animal manure: Challenges and opportunities beyond the adverse environmental concerns. Nova Science Publishers, New York. p. 45–62.

Vanotti, M.B., M.C. Garcia-Gonzalez, A.A. Szogi, J.H. Harrison, W.B. Smith, and R. Moral. 2019. Removing and recovering nitrogen and phosphorus from animal manure. In: H.M. Waldrip, P.H. Pagliari, and Z. He, editors, Animal manure: Production, characteristics, environmental concerns and management, Vol. ASA Spec. Publ. 67. American Society of Agronomy and Soil Science Society of America, Madison, WI. p. 249–296.

Waldrip, H.M., Z. He, R.W. Todd, J.F. Hunt, M.B. Rhoades, and N.A. Cole. 2014. Characterization of organic matter in beef feedyard manure by ultraviolet-visible and Fourier transform infrared spectroscopies. J. Environ. Qual. 43:690–700. doi:10.2134/jeq2013.09.0358

Wang, R., Y. Wei, Y. Wu, J. Zheng, and B. Zheng. 2018. Changes of arsenic speciation during swine manure windrow composting at full scale. Environ. Eng. Manag. J. 17:1563–1573. doi:10.30638/eemj.2018.155

Westendorf, M.L., C.A. Williams, S. Murphy, and L. Kenny. 2019. Generation and management of manure from horses and other equids. In: H.M. Waldrip, P.H. Pagliari, and Z. He, editors. Animal manure: Production, characteristics, environmental concerns and management, Vol. Spec. Vol. 67. ASA and SSSA, Madison, WI.

Wichuk, K.M., and D. McCartney. 2012. Animal manure composting: Stability and maturity evaluation. In: Z. He, editor, Applied research of animal manure-Challenges and opportunities beyond the adverse environmental concerns. Nova Science Publishers, New York. p. 203–262.

Wing, S., and S. Wolf. 2000. Intensive livestock operations, health, and quality of life among eastern North Carolina residents. Environ. Health Perspect. 108:233–238. doi:10.1289/ehp.00108233

Wilson, M.L., S. Niraula, and E.L. Cortus. 2019. Nutrient characteristics of swine manure and wastewater. In: H.M. Waldrip, P.H. Pagliari, and Z. He, editors, Animal manure: Production, characteristics, environmental concerns and management. Vol. Spec. Publ. 67. ASA and SSSA, Madison, WI.

Yan, Z., J. Li, X. Liu, Y. Yuan, Y. Liao, and X. Li. 2017. Deodorization of swine manure using a Lactobacillus strain. Environ. Eng. Manag. J. 16:2191–2198. doi:10.30638/eemj.2017.226

Zang, B., S. Li, F. Michel, Jr., G. Li, Y. Luo, D. Zhang, and Y. Li. 2016. Effects of mix ratio, moisture content and aeration rate on sulfur odor emissions during pig manure composting. Waste Manag. 56:498–505. doi:10.1016/j.wasman.2016.06.026

Zhang, H., F. Vocase, J. Antonangelo, and C. Gillespie. 2019. Temporal changes of manure chemical compositions and environmental awareness in the Southern Great Plains. In: H.M. Waldrip, P.H. Pagliari, and Z. He, editors, Animal manure: Production, characteristics, environmental concerns and management, Vol. Spec. Publi. 67. ASA and SSSA, Madison, WI.

Zheng, W., M. Guo, and G. Czapar. 2019. Environmental fate and transport of veterinary antibiotics derived from animal manure, p. In: press. In: Z. He, P. Pagliari, and H.M. Waldrip, editors, Animal manure: Production, characteristics, environmental concerns and management. Vol. ASA Spec. Publ. 67. American Society of Agronomy and Soil Science Society of America, Madison, WI.

Veterinary Pharmaceuticals, Pathogens and Antibiotic Resistance

Chaoqi Chen, Sheldon Hilaire, and Kang Xia*

Abstract

The emergence of human infections caused by resistant pathogens is becoming a global concern; antibiotic resistance is one of the most critical human health challenges of the 21st century as declared by the World Health Organization. In recent years, many countries have begun to take actions to tackle the challenge of antibiotic resistance. Although overuse and prescription of antibiotics for humans has contributed largely to the development of antibiotic resistance, the intensive use of antibiotics in pet care, and more so in animal production for therapeutic and subtherapeutic purposes has been linked to the rapid rising of antibiotic resistance in nonclinical environments. This review is, therefore, focused on veterinary antibiotics, pathogens, and antibiotic resistance, with particular emphasis on: i) history and current state of veterinary antibiotic usage and regulations; ii) occurrence of veterinary antibiotics, pathogens, and antibiotic resistance in animal manure and the manure-affected environment; iii) potential impact of pathogens and antibiotic resistance on human and environmental health; iv) strategies for reducing their input to the environment; and v) what the future holds based on the past lessons and existing information.

Antibiotic resistance is one of the most critical human health challenges of the 21st century, as declared by the World Health Organization (World Health Organization, 2014). The emergence of human infections caused by resistant pathogens is becoming a global concern. Antibiotic resistance compromises the effectiveness of antibiotics, leading to longer hospital stays, higher medical costs, and increased mortality. In the European Union and United States, an estimated 48,000 deaths and more than $3.5 billion in health care costs every year are believed to be caused by antibiotic resistance (Centers for Disease Control and Prevention, 2013; Boonyasiri et al., 2014; Thorpe et al., 2018). Some important resistant bacteria, known as superbugs, including methicillin-resistant *Staphylococcus aureus*, penicillin-resistant *Streptococcus pneumonia*, and vancomycin-resistant *Enterococcus spp.* have caused untreatable and fatal infections to a large number of people around the world (Ventola, 2015).

In recent years, many countries have begun to take actions to tackle the challenge of antibiotic resistance. In 2014, the World Health Organization (WHO) reported that antibiotic resistant bacteria have reached alarming levels in many parts of the world (World Health Organization, 2014). Later, a draft global action plan was endorsed to "ensure, for as long as possible, continuity of successful treatment and prevention of infectious diseases with effective and safe medicines that are quality-assured, used in a responsible way, and accessible to all who need them" (World Health Organization, 2015).

Although overuse and prescription of antibiotics for humans have contributed largely to the development of antibiotic resistance, the intensive use of antibiotics

---

School of Plant and Environmental Sciences, Virginia Tech, Blacksburg, VA 24061.

*Corresponding author (kxia@vt.edu)

doi:10.2134/asaspecpub67.c26

Animal Manure: Production, Characteristics, Environmental Concerns and Management. ASA Special Publication 67. Heidi M. Waldrip, Paulo H. Pagliari, and Zhongqi He, editors.

in pet care, and more so in animal production for therapeutic and subtherapeutic purposes, has been linked to the rapid rising of antibiotic resistance in nonclinical environments (Lathers, 2001; Landers et al., 2012; Kumar et al., 2018). This review is, therefore, focused on veterinary antibiotics, pathogens, and antibiotic resistance, with particular emphasis on: i) history and current state of veterinary antibiotic usage and regulations; ii) occurrence of veterinary antibiotics, pathogens, and antibiotic resistance in animal manure and the manure-affected environments; iii) potential impact of pathogens and antibiotic resistance on human and environmental health; iv) strategies for reducing their input to the environment; and v) what the future holds based on the past lessons and existing information.

Veterinary Antibiotics
Veterinary Antibiotic Use

The majority of antibiotics used in the United States are for livestock or animal husbandry purpose. In 2013, it was estimated that 14,800 tons of antibiotics were sold for use in livestock in the United States, which accounted for 80% of total US antibiotic consumption (Food and Drug Administration, 2015). Similarly, it was estimated that 84,000 tons of antibiotics were used in livestock in China in 2013, accounting for 52% of the total antibiotic use in China (Zhang et al., 2015).

Due to the growing number of animals raised for food production to fulfill the increasing worldwide meat consumption, the global veterinary antibiotic use was predicted to increase from an average of 131,109 tons in 2013 to 200,235 tons by 2030 (Van Boeckel et al., 2015). It is estimated that the human population is rising at roughly 1.2% a year, while livestock population is rising at around 2.4% a year (Herrero, 2010). There is no doubt that the total worldwide consumption of veterinary antibiotics will continue to increase in the future due to increasing world population if no action is taken to curtail their widespread uses.

As detailed in later sections, the consequences of widespread veterinary antibiotic use are elevated levels of antibiotic resistance in the environment, increased likelihood of human pathogens to acquire antibiotic resistance from the environment, and reduced effectiveness of antibiotic therapy for infectious disease due to drug resistance. Veterinary antibiotics used in food animal production are for two major purposes: subtherapeutic use for growth promotion and therapeutic use for treating animal diseases.

Regulations on Subtherapeutic Use of Veterinary Antibiotics

Since the first scientific article supporting use of antibiotics for the purpose of growth promotion and weight gain in food animal production was published in 1955 (Libby and Schaible, 1955), antibiotics have been often used as animal feed additives. For example, in year 1978, subtherapeutic use of veterinary antibiotics in the United State reached to 48% of total veterinary antibiotics (National Research Council, 1980). However, since 1960s, this practice has been questioned due to the evidence of increasing emergency of resistant bacteria and the concern for its impact on human health (Manten, 1963; Smith, 1971; Walton, 1971). In 1986, Sweden was the first in the world to institute a national ban on the use of antibiotics for animal growth promotion (Wierup, 2001). Since then, many other countries

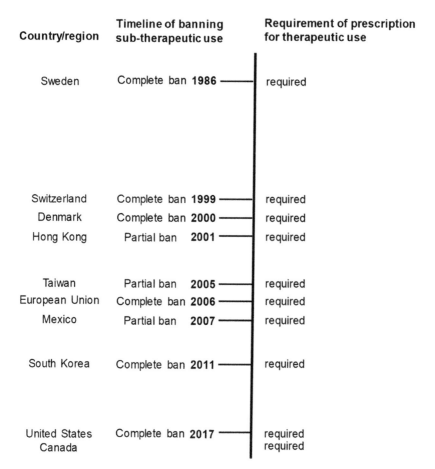

Fig. 1. Regulation on veterinary antibiotic use for sub-therapeutic and therapeutic purpose around the world (U.S. Government Accountability Office, 2011; Rehder, 2012; Maron et al., 2013).

have imposed their own regulations on usage of veterinary antibiotics as animal growth promoters (Fig. 1).

As early as in 1977, the U.S. FDA's first proposed ban on using penicillin and tetracycline in animal feed sold in the United States failed due to opposition of farmers, antibiotic producers, and members in the Congress (Cima, 2012). In 2013, the FDA introduced a final guidance for industry to phase out the use of human medically-important antibiotics in animal production (U.S. Department of Health and Human Services, 2013). In 2015, the U.S. Department of Public Health and Human Services published the Veterinary Feed Directive Final Rule (Guidance #213) prohibiting the use of antibiotics that are important for human health as animal growth promoters; this was finally implemented on January 1, 2017. As a consequence of this rule, the amount of antibiotic used for livestock in the United States dropped, for the first time, by 10% from 2015 to 2016 (Food and

Drug Administration, 2017), implying that the livestock and poultry industries were preparing the change before this rule was fully implemented.

In the European Union, limitations on antibiotic use began in 1997, when avoparcin was first banned for growth promotion (van den Bogaard et al., 2000). In 1999, more antibiotics, including tylosin, spiramycin, bacitracin, virginiamycin, carbadox, and olaquindox were also banned for growth promotion use. Since 2006, the growth promotion use of all antibiotics has been completely banned in the European Union (U.S. Government Accountability Office, 2011). In 2005, South Korea began cutting down the use of 44 antibiotics in animal feed. Finally, in 2011, South Korea became the first Asian country to ban the addition of all antibiotics in animal feeds (Johnson, 2011).

However, many other countries currently still do not have any restrictions on antibiotic use for animal growth promotion (Maron et al., 2013). Due to the urgent human health concerns of antibiotic resistance (World Health Organization, 2014), it is expected that the regulations on antibiotic usage for animal growth promotion would soon be developed and implemented in more countries. For example, China has proposed plans to reduce the use of antibiotics for poultry and livestock by 2020 and to gradually ban antibiotics for growth promotion (Wang, 2017).

Regulations of Therapeutic Use of Veterinary Antibiotics

While the use of antibiotics for animal growth promotion is becoming more and more restricted and regulated, their therapeutic use for treating and preventing animal diseases is, and will still be a common and important practice due to increasing worldwide food animal production and demand for concentrated animal production operations. Many countries require veterinary prescriptions for antibiotic use in food animal productions (Fig. 1), to avoid their excessive and unnecessary use. For example, the FDA announced in 2012 a voluntary program to require a prescription and veterinarian supervision of therapeutic antibiotic use for food animal production. Later, in 2017 the US FDA fully implemented the Veterinary Feed Directive Rule, which requires producers to obtain a legal form or prescription signed by a licensed veterinarian before administration of antibiotics to animals for therapeutic purpose.

Occurrence of Veterinary Antibiotics in Animal Manure and the Manure-Affected Environment

Most antibiotics are weakly absorbed and incompletely metabolized in animal guts once administrated and eventually are excreted via feces or urine as unchanged forms and metabolites (Sarmah et al., 2006). For instance, the daily excretion masses of 50 antibiotics from swine and cattle were estimated to be 18.2 mg d^{-1} swine^{-1} and 4.24 mg day cattle^{-1}. It was estimated that antibiotics used as growth promoters contributed more than 80% of the excreted mass in swine farms in China (Zhou et al., 2013). Concentrations of three major antibiotic classes, including sulfonamides, tetracyclines, and quinolones in manure and manure-affected environments are summarized in Table 1.

Antibiotics have been widely detected in manure, with a wide range from several mg kg^{-1} to hundreds of mg kg^{-1}, depending on the source of manure and animal production (Xie et al., 2018). The intensive use of antibiotics in animal husbandry and livestock would unavoidably result in the release of veterinary antibiotics and metabolites into the environment via manure land application. The concentrations of antibiotics in the surface soils of manure-amended fields vary from below detection limits to several mg kg^{-1} (Xie et al., 2018). The detected concentrations of antibiotics in manure-applied soils are generally lower than those in raw manures (Xie et al., 2018). It is because antibiotics can degrade before manure application, particularly with proper manure management approaches (e.g., composting) or they can also be diluted by the soil matrix. Among the environmental compartment, soil is the main receiver of veterinary antibiotics via grazing livestock and manure application. After being released into soils, antibiotics may enter the aquatic environment, mainly thorough surface runoff and leaching (Chee-Sanford et al., 2009). The concentrations of antibiotics in aquatic environment range from below detection limits to several mg L^{-1} (Kummerer, 2009).

Pathogens

Animal guts are recognized as a source and reservoir for human pathogens (Wolfe et al., 2007). About 61% of the 1415 pathogens known to infect humans are zoonotic (Taylor et al., 2001). It is estimated that zoonotic pathogens have caused 56% of over 10,000 human infectious disease outbreak events from 1980 to 2013 globally (Smith et al., 2014).

Animal manure is a source of several significant zoonotic pathogens, including protozoan parasites, manure-borne pathogenic bacteria, viruses, and fungi (Milinovich and Klieve, 2011). Numerous manure-borne pathogens, including several important pathogenic bacteria such as *Salmonella spp., Escherichia coli* O157:H7, *Salmonella spp., Listeria spp., Streptococcus spp., Campylobacter spp.,* and *Clostridium spp.* (Islam et al., 2004; Van Donkersgoed et al., 2009; Rodriguez et al., 2017) and protozoan parasites such as *Giardia* and *Cryptosporidium* (Berry et al., 2007; Dreelin et al., 2014; Chuah et al., 2016) have been detected in animal manure and manure affected environments. According to an investigation of over 5000

Table 1. Occurrence of antibiotics in manure, compost, soils, and aquatic environment

Compartment	Antibiotic	Concentrations (µg kg^{-1} or µg L^{-1})	References
Manure	Tetracyclines	80–184,000 (oxytetracylcine)	
	Sulfonamides	20–18,000 (sulfamethazine)	
	Quinolones	100–94,000	
Composted manure/compost	Tetracyclines	MQL-851*	Xie et al., 2018
	Sulfonamides	MQL-93	
	Quinolones	0.1–5,447	
Soil (0–20 cm)	Tetracyclines	MQL-3,064	
	Sulfonamides	3.6–321	
	Quinolones	0.2–1,527	
Surface water	Tetracyclines	MQL-0.69 (chlortetracycline)	
	Sulfonamides	MQL-1.9 (sulfamethoxazole)	Kummerer, 2009
	Quinolones	MQL-0.02 (fluoroquinolone)	
Groundwater	Tetracyclines	MQL	
	Sulfonamides	MQL-0.47	
	Quinolones	MQL	

recorded zoonotic disease outbreaks, *Salmonella* caused the most outbreaks (855), followed by *Escherichia coli* which caused 460 outbreaks (Smith et al., 2014).

Salmonella spp. and *Escherichia coli* O157:H7 have been detected in environmental samples, including manure, compost, soil, and water (Johannessen et al., 2004; Van Donkersgoed et al., 2009; Wilkes et al., 2011; Micallef et al., 2012). For example, *Salmonella* was detected from 6 out of 380 manure samples from pig farms (Holzel and Bauer, 2008) and 6 out of 96 water samples from irrigation and non-irrigation sources on leafy green produce farms (Benjamin et al., 2013). In market-ready composts from 94 facilities, one compost contained salmonella and 6% had detectable *Escherichia coli* O157:H7 (Brinton et al., 2009). However, the two pathogens were rarely detected in fresh produce (Mukherjee et al., 2004; Johnston et al., 2006; Tango et al., 2018). For example, *Salmonella* was isolated from one lettuce and one green pepper out of 476 fresh produce samples from 32 organic farms (Mukherjee et al., 2004). *Escherichia coli* O157:H7 was found in 1 out of 360 fresh produce samples (Tango et al., 2018). A trackback investigation identified 22 outbreaks caused by vegetable-associated *Escherichia coli* O157:H7 between the year 1995 to 2006 in the United States. Nine of these outbreaks have been traced to a region producing lettuce and spinach. Consistently, *Escherichia coli* O157:H7 was frequently detected from environmental samples, including soil, sediment, feces, and plants collected from this region (Cooley et al., 2007).

Manure-borne pathogens can survive during livestock manure storage and following land application (Guan and Holley, 2003; Nicholson et al., 2005; Berry et al., 2007; Oliveira et al., 2012; Nyberg et al., 2014). For example, *Salmonella spp.* and *Escherichia coli* O157:H7 can survive in livestock manure for up 21 mo and in soils up to 405 d following land application (Ongeng et al., 2011). They can be transported further through surface runoff and leaching from manure-amended fields to surface water (Tanaro et al., 2014), groundwater (Forslund et al., 2011), and even into drinking water (Ogden et al., 2001). They can enter the human food chain via vegetables grown in contaminated soils or irrigated with contaminated water (Wang et al., 2015a; Beneduce et al., 2017; Alegbeleye et al., 2018).

Antibiotic Resistance

Antibiotic resistance, the major consequence of overuse of antibiotics in both animals and humans, is the ability of bacteria and other microorganisms to resist the effects of an antibiotic to which they were once sensitive to. On average, bacteria can develop antibiotic resistance within 10 yr after an antibiotic is introduced to the market (Ventola, 2015). In 2014, the WHO released the first global report on antibiotic resistance based on the surveillance data from 114 countries (World Health Organization, 2014), revealing that antibiotic resistance is widespread across the world and has reached to alarmingly high levels in some countries. The prevalence of isolations of antibiotic resistant bacteria in every region of the world has indicated that it is not just a prediction of the future but in fact already a serious threat to human health.

The development of antibiotic resistance in humans is largely due to the overuse of antibiotics in clinical situations (Ventola, 2015); however, contribution of veterinary antibiotic use to the development of antibiotic resistance in the nonclinical environment should not be neglected (Walsh, 2013). Administering antibiotics to animals may result in selection of resistant bacteria and resistance genes in animal guts (Looft et al., 2012), in manure (Pu et al., 2018), and in the subsequent manure affected environment (Zhang et al., 2017b). Antibiotics released into soils with manure can remain active and exert selection pressure on soil bacteria (Heuer et al., 2008). Upon cell replication, antibiotic resistance genes carried by the bacteria can amplify and pass to the next generation through vertical gene transfer (Martinez et al., 2007). Further, susceptible bacteria in soils can become resistant by acquiring antibiotic resistance genes from resistant bacteria present in land-applied manure via horizontal gene transfer (Heuer et al., 2011). Both vertical gene and horizontal gene transfer can cause dissemination of antibiotic resistant bacteria and antibiotic resistance genes in the environment. Additionally, manure provides nutrients, which can further enhance the proliferation of resident antibiotic-resistant bacteria in soils (Heuer and Smalla, 2007; Udikovic-Kolic et al., 2014).

As listed in Tables 2 and 3, numerous studies have found antibiotic-resistant bacteria and antibiotic resistance genes in manure (Quirk, 2001; Allen, 2014; Boonyasiri et al., 2014) and manure affected soils (Marti et al., 2013; Liu et al., 2016; Zhao et al., 2017) and water (Oluyege et al., 2009; Moore et al., 2010).

The elevated level of antibiotic resistance has been associated with the use of antibiotics in livestock (Heuer and Smalla, 2007; Knapp et al., 2010; Huttner et al., 2013; Liu et al., 2016). For instance, the levels of antibiotic resistance genes in anthropogenic soils near a swine raising facility was found to be enriched 192-fold compared with the control soils, corresponding to the intensive use of antibiotics in China (Zhu et al., 2013). Comparably, the levels of antibiotic resistance genes in soils applied with manure from animals administered with low level of antibiotics increased for two to seven-fold compared with control soils in a study in Finland (Ruuskanen et al., 2016). Finland has banned antibiotic use for growth promotion. It has the fourth lowest use of veterinary antibiotics per production animal among the European Union countries. It exemplifies that although the reduction of antibiotic use can result in a lower enrichment of antibiotic resistance genes in soils, development of antibiotic resistance cannot be completely avoided because of the need for therapeutic uses of antibiotics.

Strategies for Minimizing the Flow of Antibiotics, Pathogens, and Antibiotic Resistance Genes from Animal Production to Human

As discussed in previous sections, antibiotic resistance has been frequently detected at elevated levels in manure of animal production facilities using

Table 2. Occurrence of antibiotic resistant bacteria in manure, compost, and soils.

Compartment	Antibiotic resistant bacteria	Absolute abundance (CFU g^{-1} or CFU L^{-1})	References
Manure/feces	Tetracycline-resistant	1.3×10^6 to 3.0×10^9	Kobashi et al., 2005
	Sulfonamide-resistant	9.0×10^7	Wang et al., 2014b
	β-lactam-resistant	3.8×10^6 to 7.5×10^{11}	Kobashi et al., 2005
Composted manure/ compost	Tetracycline-resistant	2.8×10^6	Arikan et al., 2007
	Antibiotic resistant	10^3 to 3.2×10^8	Andrews et al., 1994
Topsoil (0–20 cm)	Tetracycline-resistant	1.2×10^8 to 4.6×10^8	Yang et al., 2016
	Sulfonamide-resistant	4.5×10^5 to 9.0×10^7	Wang et al., 2014b

Table 3. Occurrence of antibiotic resistance genes in manure, compost, soils, and aquatic environment.

Compartment	Antibiotic resistance gene	Absolute abundance (Copies g⁻¹ or copies L⁻¹)	Relative Abundance (ARG/16s rRNA genes)	Reference
Manure/slurry	Tetracycline-resistance	Up to 10^{11}	Up to 10^2	Chen and Xia, 2017
	Sulfonamide-resistance	Up to 10^{11}	Up to 10^2	
	Quinolone-resistance	Up to 10^9	Up to 10^{-4}	
Composted manure/compost	Tetracycline-resistance	Up to 10^9	Up to 10^{-2}	
	Sulfonamide-resistance	Up to 10^9	Up to 10^{-2}	
	Quinolone-resistance	Up to 10^6	Up to 10^{-5}	
Topsoil (0–20 cm)	Tetracycline-resistance	Up to 10^9	Up to 10^{-2}	Xie et al., 2018
	Sulfonamide-resistance	Up to 10^8	Up to 10^{-3}	
Surface water	Tetracycline-resistance		Up to 10^{-2}	Wu et al., 2015
	Sulfonamide-resistance		Up to 10^{-1}	

antibiotics. Animal manure is also a source of various pathogens. Land application provides a means to dispose of animal manure while also serving as an abundant, low-cost source of fertilizer in replacement of chemical fertilizers for crop production. However, the use of animal manure can pose risks in terms of introducing pathogens and antibiotic resistance to the receiving environment (Chee-Sanford et al., 2009) and potentially to humans. While restricting the antibiotic use in animal production can reduce the development of antibiotic resistant microorganisms and resistance genes, proper animal manure management, manure land application approach, and post-harvest treatment of crops would provide additional means for reducing the spread of antibiotic resistance and pathogens.

Animal Manure Management

Manure-indigenous organisms can suppress the survival of manure-borne pathogens because of predation, substrate competition and antagonism (van Elsas et al., 2011). Therefore, enforcing a wait period for manure application prior to crop harvest may provide benefits in reducing risk of exposure to manure-borne pathogens. USDA National Organic Program recommends a 90-d and 120-d waiting period between application of raw animal manure and harvest of crops that are not in contact or in contact with soil, respectively (USDA, 2012). However, survival of manure-borne pathogens in soils ranged from a few days to several months (Lemunier et al., 2005; Nicholson et al., 2005; Wang et al., 2014a), indicating that a waiting period of 120 d may not be long enough to completely avoid the contamination of pathogens in fresh produce.

To further minimize levels of pathogens in manure, pretreatment of raw animal manure is currently recommended before land application (Food and Drug Administration, 2014). The FDA states that manure management is acceptable before manure land application if the numbers of indicator pathogens in manure are reduced to below standardized levels (Table 4) (Food and Drug Administration, 2014). However, the threshold levels of manure-borne pathogens vary among different countries. For example, the threshold level of *Salmonella* in manure is absence per 25 g fresh masses (wet weight basis) in France and United Kingdom (Manyi-Loh et al., 2016). The threshold level for *Escherichia coli* is 100 and < 1000 per g fresh mass (wet weight basis) in France and United Kingdom, respectively. Common methods for pretreatment of animal manure include chemical, physical, and biological methods or a combination of any of these. Many of these methods are found to be effective for reducing the levels of pathogens (Table 5).

Studies examining the effectiveness of manure management approaches for antibiotics and antibiotic resistance genes have been focused on biological processes, particularly including composting and digestion (Massé et al., 2014; Youngquist et al., 2016). These biological treatments have shown mixed effects on the removal of antibiotics and antibiotic resistance genes (Tables 6 and 7).

Of the methods for pretreating animal manure, composting is frequently recommended for pathogen reduction; recent research has also shown its effectiveness for removal of antibiotics and reduction of pathogens and certain antibiotic resistance genes (Tables 5, 6, and 7). However, many countries do not require composting of animal manure before land application because of concerns of significant nitrogen loss during composting (Hoornweg et al., 2000), resulting in insufficient nutrient input to the amended soils. In the United States, raw animal manure is not required to be composted if it is applied to land used for crops not intended for human consumption, or applied to soils 90 or 120 d before harvest of a product for human consumption depending on whether the edible portion is in direct contact with the soil or not (USDA, 2012). In the Netherlands, raw manure is allowed to be used as organic fertilizer if the levels of manure nutrients (nitrogen and phosphorus) do not exceed certain levels (Schröder and Neeteson, 2008). In

Table 4. Microbial standards for scientifically valid controlled treatment processes (Food and Drug Administration, 2014).

Treatment Process	Microbes	Standards
Physical processes (for example, thermal), chemical processes (for example, high alkaline pH), or combinations	*L. monocytogenes*	Not detected using a method that can detect one CFU per õve gram analytical portion
	Salmonella species	Less than three (3) MPN per four (4) grams of total solids (dry weight basis)
	E. coli O157:H7	Less than 0.3 MPN per one gram analytical portion
Physical processes (for example, thermal), chemical processes (for example, high alkaline pH), or combinations	*Salmonella* species	Less than three (3) MPN per four (4) grams of total solids (dry weight basis)
	Fecal Coliforms	Less than 1000 MPN fecal coliforms per gram of total solids (dry weight basis)
Composting processes	*Salmonella* species	Less than three (3) MPN per four (4) grams of total solids (dry weight basis)
	Fecal Coliforms	Less than 1000 MPN fecal coliforms per gram of total solids (dry weight basis).

Table 5. Manure management approaches and their effectiveness in reducing pathogens or fecal coliforms.

Method types	Method description	Effectiveness	References
Chemical	Addition of lime substance, pig manure mixed with alkaline coal fly ash and lime, 8 d	No detection of Salmonella, fecal coliforms, *E. coli* and *Streptococcus*	Wong and Selvam, 2009
	Addition of disinfectant compounds, didecyldimethylammonium chloride, glutaraldehyde, hydrogen peroxide, phenol, and sodium hypochlorite, several min to several hours	3 log CFU mL^{-1} reduction for *E.coli* O157:H7	Skinner et al., 2018
Physical	UV radiation, *Salmonella* inoculated manure dust were exposed to a wavelength of 365 nm, 7 d	1.5 log CFU reduction of *Salmonella*	Oni et al., 2013
Biological	Composting, liquid pig manure, 56 d	No detection of *Salmonella*, *E.coli*, and *Enterococcus*	Mc Carthy et al., 2011
	Vermicomposting, cow manure, 8 wk	3 log CFU reduction of fecal coliforms	Karimi et al., 2017
	Aerobic Digestion, cow manure, 12 d	5 log CFU reduction of *E.coli*	Pandey and Soupir, 2013
	Anaerobic digestion, pig manure, 2 yr	No detection of *Salmonella*, *Campylobacter* spp., and *Y. enterocolitica*	Massé et al., 2011

Table 6. Manure management approaches and their effectiveness in reducing antibiotics.

Method	Method description	Effectiveness	References
Composting	Aerobic composting, cattle or dairy manure, 42 d	71 to 84% for chlortetracycline; 97 to 98% for sulfamethazine; 62 to 86% for tylosin; > 99% for pirlimycin;	Ray et al., 2017
	Aerobic composting, turkey manure, 35 d	No removal for sulfamethazine; 99% for chlortetracycline; 54 to 76% for monensin and tylosin	Dolliver et al., 2008
	Anaerobic composting with biochar addition, swine manure, 15 d	100% for tetracycline and oxytetracycline	Shan et al., 2018
	Vermicomposting, larvae manure, 6 d	34% to 72% for tetracyclines, sulfonamides, and fluoroquinolones	Zhang et al., 2014
Anaerobic digestion	Anaerobic digestion, pig manure, 40 d	No removal for sulfadiazine and sulfamethizole; > 99% for sulfamethoxazole, erythromycin, and trimethoprim	Feng et al., 2017
	Anaerobic digestion, swine or cattle manure, 25 or 28 d	7% to 98% for chlortetracycline 3 to 27% for monensin	Varel et al., 2012

contrast, Finland requires composting animal manure used for growing products intended for direct human consumption (Heinonen, 2006). In China, manure without composting is not considered as organic fertilizer, although farmers do use raw manure to improve soil quality and provide nutrients for crop production (Li, 2017).

Animal Manure Land Application

Land application strategy is another option to control the spread of antibiotic and antibiotic resistance. A proper manure land application approach can provide benefits in reducing the loss of nutrients including nitrogen and phosphorus and sediments (Sharpley et al., 2004; Maguire et al., 2011). However, only a few studies examined the impact of manure application approaches on the fate of antibiotics, antibiotic-resistant bacteria, and antibiotic resistance genes. For example, the concentrations of pirlimycin in the surface runoff from soils applied with dairy manure using sub-surface injection were six times lower compared with the fields when manure was amended using surface application (Kulesza et al., 2016). A similar conclusion was also drawn for tylosin, chlortetracycline, and sulfamerazine in a recent field study (Le et al., 2018). In the same study, it was also demonstrated that regardless of the manure application method used, the antibiotic surface runoff could be reduced by 9 to 45 times if manure was applied more than 3 d before a subsequent rain event com-pared to manure land application right before a rain event (Le et al., 2018). An early rainfall simulation study showed that land application of manure using the incorpo-ration and injection methods resulted in a significantly lower abundance of antibiotic resistance genes in the surface runoff compared to using broadcast methods (Joy et al., 2013). These results indicated that manure injection or incorporation can be a bet-ter manure land management practice to prevent loss of antibiotics and antibiotic resistance with surface runoff. However, little is known about the impact of manure land application methods on leaching of antibiotics, pathogens, and resistant micro-organisms and genes and their fate in the manure-affected soils.

Post-harvest Treatment of Crops

Crop post-harvest treatment is an essential procedure to reduce the risk of food-borne illness due to pathogen contamination on fresh produces (Mahajan et al., 2014; Ali et al., 2017). Numerous studies have examined the effectiveness of various post-harvest treatments, including physical, chemical, and gaseous treat-ments, on reducing the levels of pathogens and resistant bacteria (Table 8). These studies usually rely on bacteria inoculated crops. The effectiveness of treatment

Table 7. Manure management approaches and their effectiveness in reducing antibiotic resistance genes (ARGs).

Method	Method description	Effectiveness	References
Composting	No chemical addition, cattle manure, 30 d	Decrease by up to > 90% or increase by up to 1000 times	Xu et al., 2018
	Biochar, pig and duck manure, 42 d	Averaged decrease by > 99%	Cui et al., 2017
	Zeolite, superphosphate, zeolite and ferrous sulfate, pig and duck manure, 42 d	Averaged decreased by 86.5%, 68.6%,72.2%, respectively	Peng et al., 2018
	Superabsorbent polymer, swine manure, 35 d	Decrease by 8.1 to 96.7%	Guo et al., 2017
	Surfactants (rhamnolipid or Tween 80), chicken manure, 30 d	Decrease by > 90% and > 84.2%, respectively	Zhang et al., 2016
	Vermicomposting, swine manure, 6 d	Decrease by up to > 90%, or increase by up to 100 times	Wang et al., 2015b
Aerobic biofiltration	Swine manure	No change except decrease of erm(X) by 96%	Chen et al., 2010
	No chemical addition dairy manure, 60 d	Decrease by up to > 90%, or increase by up to ~10 times	Sun et al., 2016
Anaerobic digestion	Biochar, cattle manure, 60 d	Decrease by up to 87% (compared to without biochar addition)	Sun et al., 2018
	Zinc, swine manure, 52 d	Increase by up to 21.5 times (compared to without zinc addition)	Zhang et al., 2017a

Table 8. Post-harvest treatments and their effectiveness in reducing pathogens or resistant bacteria.

Treatment	Treatment description	Effectiveness (Log reduction CFU g-1 or CFU cm-2 leaf)	References
Acidified sodium benzoate	Cherry tomatoes, 3000 ppm, pH 2.0, 21 °C, washing 3 min	*E. coli* O157:H7 (5.0) *S. enterica* (4.1) *L. monocytogenes* (4.9)	Chen and Zhong, 2018
Ultrasound	Lettuce leaf, washing 10 min	*E. coli* O157:H7 (0.5) *L. innocua* (0.5)	Huang et al., 2018
Chlorine	Tomatoes and Spinach, 200 ppm, 20 °C, washing	*L. monocytogenes* (< 3)	Ijabadeniyi et al., 2011
potassium acetate, potassium citrate, and calcium lactate	Blueberries, washing 5 min	*L. monocytogenes* (4) *E. coli* O157:H7 (3.5)	Liato et al., 2017
High pressure processing	Tomatoes, pressures up to 550 MPa, 120 s	*S. Braenderup* (0.46 to 3.7)	Maitland et al., 2011
Ozone	Alfalfa sprouts, 21 ppm, 4 °C, 64 min	*E. coli* O157:H7 (0.85)	Sharma et al., 2003
Low temperature blanching and calcium treatment	Spinach, 64.9°C,0.52% Ca(OH)2, 42.4 s	*E. coli* O157:H7 (6.6)	Kim et al., 2015
Cold plasma	Corn salad leaf, power, gas flow of 5 L min-1, up to 2 min	*E. coli* O157:H7 (3.3) *E. coli* O104:H4 (3.2)	Baier et al., 2015
Pseudomonas fluorescens	Spinach, stored at 20 °C, 24 or 48 h	*E. coli* O157:H7 (1.5–2.4)	Olanya et al., 2013
Geraniol loaded nanoparticles	Spinach, stored for 24 h	*E. coli* O157:H7 and *S. Typhimurium* (0.3 to 4.2)	Yegin et al., 2016
Tap water	Carrot, washing 2 to 2.5 min	Antibiotic resistant- E.coli O157:H7 (0.6) and *P. aeruginosa* (0.5)	Pulido, 2016
Sodium hypochlorite		Antibiotic resistant- *E.coli* O157:H7 (1.9) and *P. aeruginosa* (1.6)	
Peroxyacetic acid/hydrogen peroxide		Antibiotic resistant- *E.coli* O157:H7 (2.6) and *P. aeruginosa* (2.2)	

methods varied largely depending on treatment conditions (e.g., temperature, treatment duration) as well as types of crops. Besides, there is a lack of assessment of the potential regrowth of bacteria after a treatment. In addition, less is known about the effect of these post-harvest treatments on the removal of antibiotic resistance genes which can be released after cell death.

Topics Need to Be Further Explored

Up to date, the severity of antibiotic resistance in human health has been well acknowledged, many causes to this threat have been identified, and strategies to minimize and control this urgent issue have been proposed. As shown in Fig. 2, combating the rising threat of pathogen contamination and antibiotic resistance requires "one health" approach for which multi-level critical control points within the people-animals-environment system need to be first identified and understood and then timely and effectively addressed and balanced (Robinson et al., 2016; Monnier et al., 2018). Built on the existing information, three important topics need to be further explored are: i) long term impact assessment of regulations on veterinary antibiotic use; ii) other contaminants (e.g., pesticides)-introduced antibiotic resistance; and iii) novel remediation technology for the pathogen- and antibiotic resistance-contaminated environments.

Long-term Impact Assessment of Regulations on Veterinary Antibiotic Use

The aim of the regulations on veterinary antibiotics is to reduce the total use of antibiotics in animal production to minimize the environmental input of antibiotics, lower the levels of antibiotics, and reduce the occurrence of antibiotic resistance in the affected environment. A recent study showed that the overall antibiotic concentrations in manure amended soils from countries with regulations are considerably lower than those in countries without regulations (Xie et al., 2018). However, other factors may also explain such differences. The quantity and patterns of antibiotic use differ largely among countries. It is possible that the amount of historical antibiotic use in the countries with regulations has been already less than in the countries without regulations. Employment of antibiotic regulations does not necessarily result in a reduction of overall antibiotic use and release if total animal populations are growing. Lower antibiotic concentrations in manure amended soils may also be a result of a smaller livestock size or lower livestock intensity. In addition, transformation rates of antibiotics in the manure-amended soils vary widely depending on their physicochemical characteristics and on environmental factors (Chen et al., 2018).

Currently, there is a lack of data comparing the long term change of environmental levels of antibiotics within the same regions before and after the employment of veterinary antibiotic regulations. Because most of the countries only began to implement veterinary antibiotic regulations in recent years,

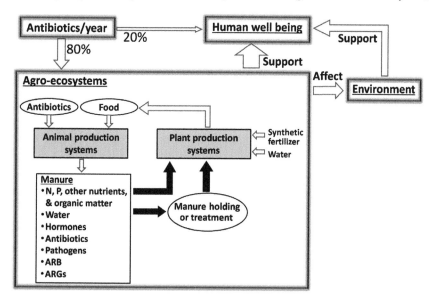

Fig. 2. The ultimate goal of antibiotic use in animal production is to support human well being by enhancing the overall productivity of agro-ecosystems. However, unregulated antibiotic usage, improper manure management, and land application practices may result in short term productivity gain but long term adverse impact on the world's sustainable development (Jasovský et al., 2016). The percentages shown reflect the yearly antibiotic usage distribution in the United States in 2013 (FDA, 2015).

long-term comprehensive monitoring and modeling are needed to better assess the effectiveness of these regulations in minimizing antibiotic environmental input and the occurrence of antibiotic resistance.

Other Contaminant-introduced Antibiotic Resistance

In recent years, concerns have been raised regarding the development of antibiotic resistance in microorganisms exposed to different classes of chemicals, such as heavy metals and pesticides (Chen et al., 2015; Wales and Davies, 2015).

Heavy metals are important contributors for antibiotic resistance (Seiler and Berendonk, 2012). Arsenate, copper, and zinc were found to induce tetracycline-resistance in selected bacterial strains (Chen et al., 2015). Heavy metals can drive the co-selection of antibiotic resistance in both soil and water (Seiler and Berendonk, 2012). Therefore, a better understanding of the impact of heavy metals on antibiotic resistance is important because some heavy metals, such as copper and zinc, have been widely used in livestock to promote growth and for therapeutic purpose.

A recent study suggested that three most commonly used herbicides, including 2,4-dichlorophenoxyacetic acid, dicamba, or glyphosate could induce adaptive responses by *Escherichia coli* and *Salmonella* to multiple antibiotics (Kurenbach et al., 2017). It was suggested that efflux pump genes are major contributors to the observed pesticides-induced antibiotic resistance. Similarly, 25 bacterial isolates from agricultural soils were found to be resistant to both insecticides and antibiotics, suggesting that the persistence of insecticides in the soils can lead to increasing development of multidrug resistance among soil bacteria (Rangasamy et al., 2017). More studies are needed to find out if other widely-used pesticides in agro-ecosystems and their environmental persistence can also cause antibiotic resistance in the environment and the significance of this impact.

Remediation Technologies for Antibiotic Resistance in a Contaminated Environment

Antibiotic resistance is prevalent in the environment; however, there is no doubt that human activities have further increased the levels of antibiotic resistance (Knapp et al., 2010). Several remediation technologies, including physicochemical treatment, bioremediation, phytoremediation have been developed for antibiotics (Awad et al., 2010). However, the effectiveness of these remediation technologies on reducing antibiotic resistance, including antibiotic bacteria and antibiotic resistance gene is less understood.

In a recent study, the performance of the electrokinetic remediation process in the removal of antibiotic resistant bacteria and antibiotic resistance genes was examined (Li et al., 2018). At a voltage of 1.2 V cm^{-1}, a removal of antibiotic resistant bacteria (26.0 to 31.5%) and antibiotic resistance genes (37.9% to 83.1%) in soil was obtained. The efficiency of antibiotic resistance removal was increased under several conditions, including low pH, high current density, and even voltage, as well as increased voltage and prolonged reaction time. It was found that tetracycline-resistance genes are more vulnerable in an electric field compared with sulfonamides-resistance gene (Li et al., 2018).

A mesocosm-scale wetland was constructed to examine its removal efficiencies for antibiotics and antibiotic resistance genes (Chen et al., 2016). Three flow types including surface flow, horizontal subsurface flow, and vertical subsurface flow were tested. The removal efficiencies of antibiotics and antibiotic resistance genes ranged from 75.8 to 98.6% and 63.9 and 84.0%, respectively, while the subsurface flow resulted in a higher removal efficiency of antibiotics compared with the surface flow. The presence of plants in the wetland was beneficial to the removal of antibiotics and antibiotic resistance genes (Chen et al., 2016). However, the biological mechanisms for plant-assisted removal of antibiotics and resistant genes are still not fully understood.

There are many other existing remediation technologies for cleaning up contaminated sites (https://www.epa.gov/remedytech). The effectiveness of these technologies in cleaning up antibiotic resistance in contaminated sites has yet to be evaluated.

Conclusion

Antibiotics have given us the ability to treat and cure diseases that are otherwise life endangering or fatal for humans. They have also enabled us to operate large-scale intensified animal productions while maintaining animal health. Intensified animal productions worldwide are a direct result of increasing demand of animal products by a rapidly-growing human population. However, numerous studies have demonstrated that our over reliance on antibiotics has ultimately resulted in the ramping up of a natural phenomenon (resistance) among bacteria to levels negatively impact human health. It is becoming apparent that due to over-usage of antibiotics in human and for animal productions, antibiotic resistance has become a key issue and a major contributor to global human death tolls from diseases previously treatable with antibiotics. To minimize the spread and proliferation of antibiotic resistance and to ultimately protect human health, there is an urgent need to understand the critical points along the pathways of antibiotics (Fig. 2) and to take corrective actions accordingly.

Policy changes in some countries have stopped subtherapeutic use of antibiotics, while others continue to do so. Although therapeutic veterinary antibiotic use continues to increase due to high demand for animal products, there is an increasing worldwide effort to make this use more regulated with the introduction of laws that require veterinary prescriptions for antibiotic classes used in human medicine. Even so, antibiotics continue to enter the environment through land application of manure from antibiotic-treated animals. Prolonged exposure to non-lethal doses of antibiotics in the environment can lead to resistance development in microorganisms, and this resistance can transfer to pathogenic bacteria. Strategies surrounding pretreatment of animal manure such as composting and anaerobic digestion are useful for reducing the input of antibiotics to the environment. Another useful consideration is soil subsurface injection of manure, as opposed to surface broadcasting, because this manure land application technique can significantly reduce levels of antibiotics in surface runoff after rainfall events. Steps can also be taken at post-harvest of crops to ensure that animal manure–related resistance elements are not unknowingly consumed by humans. While actions are taken to reduce the overall input of antibiotics to the environment, remediation strategies downstream of these pathways need to be investigated further.

References

Alegbeleye, O.O., I. Singleton, and A.S. Sant'Ana. 2018. Sources and contamination routes of microbial pathogens to fresh produce during field cultivation: A review. Food Microbiol. 73:177–208. doi:10.1016/j.fm.2018.01.003

Ali, A., W.K. Yeoh, C. Forney, and M.W. Siddiqui. 2017. Advances in postharvest technologies to extend the storage life of minimally processed fruits and vegetables. Crit. Rev. Food Sci. Nutr. doi:10.1080/10408398.2017.1339180

Allen, H.K. 2014. Antibiotic resistance gene discovery in food-producing animals. Curr. Opin. Microbiol. 19:25–29. doi:10.1016/j.mib.2014.06.001

Andrews, S.A., H. Lee, and J.T. Trevors. 1994. Bacterial species in raw and cured compost from a large-scale urban composter. J. Ind. Microbiol. 13:177–182. doi:10.1007/BF01584004

Arikan, O.A., L.J. Sikora, W. Mulbry, S.U. Khan, and G.D. Foster. 2007. Composting rapidly reduces levels of extractable oxytetracycline in manure from therapeutically treated beef calves. Bioresour. Technol. 98:169–176. doi:10.1016/j.biortech.2005.10.041

Awad, Y.M., S.S. Lee, S.-C. Kim, J.E. Yang, and Y.S. Ok. 2010. Novel approaches to monitoring and remediation of veterinary antibiotics in soil and water: A review. Korean Journal of Environmental Agriculture 29:315–327. doi:10.5338/KJEA.2010.29.4.315

Baier, M., T. Janssen, L.H. Wieler, J. Ehlbeck, D. Knorr, and O. Schluter. 2015. Inactivation of Shiga toxin-producing Escherichia coli O104:H4 using cold atmospheric pressure plasma. J. Biosci. Bioeng. 120:275–279. doi:10.1016/j.jbiosc.2015.01.003

Beneduce, L., G. Gatta, A. Bevilacqua, A. Libutti, E. Tarantino, M. Bellucci, M. Troiano, and G. Spano. 2017. Impact of the reusing of food manufacturing wastewater for irrigation in a closed system on the microbiological quality of the food crops. Int. J. Food Microbiol. 260:51–58. doi:10.1016/j.ijfoodmicro.2017.08.009

Benjamin, L., E.R. Atwill, M. Jay-Russell, M. Cooley, D. Carychao, L. Gorski, and R.E. Mandrell. 2013. Occurrence of generic Escherichia coli, E. coli O157 and Salmonella spp. In: water and sediment from leafy green produce farms and streams on the Central California coast. Int. J. Food Microbiol. 165:65–76. doi:10.1016/j.ijfoodmicro.2013.04.003

Berry, E.D., B.L. Woodbury, J.A. Nienaber, R.A. Eigenberg, J.A. Thurston, and J.E. Wells. 2007. Incidence and persistence of zoonotic bacterial and protozoan pathogens in a beef cattle feedlot runoff control vegetative treatment system. J. Environ. Qual. 36:1873–1882. doi:10.2134/jeq2007.0100

Boonyasiri, A., T. Tangkoskul, C. Seenama, J. Saiyarin, S. Tiengrim, and V. Thamlikitkul. 2014. Prevalence of antibiotic resistant bacteria in healthy adults, foods, food animals, and the environment in selected areas in Thailand. Pathog. Glob. Health 108:235–245. doi:10.1179/20 47773214Y.0000000148

Brinton, W.F., Jr., P. Storms, and T.C. Blewett. 2009. Occurrence and levels of fecal indicators and pathogenic bacteria in market-ready recycled organic matter composts. J. Food Prot. 72:332–339.

Centers for Disease Control and Prevention. 2013. Antibiotic resistance threats in the United States. U.S. Department of Health and Human Services, Washington, D.C.

Chee-Sanford, J.C., R.I. Mackie, S. Koike, I.G. Krapac, Y.F. Lin, A.C. Yannarell, S. Maxwell, and R.I. Aminov. 2009. Fate and transport of antibiotic residues and antibiotic resistance genes following land application of manure waste. J. Environ. Qual. 38:1086–1108. doi:10.2134/jeq2008.0128

Chen, C., P. Ray, K.F. Knowlton, A. Pruden, and K. Xia. 2018. Effect of composting and soil type on dissipation of veterinary antibiotics in land-applied manures. Chemosphere 196:270–279. doi:10.1016/j.chemosphere.2017.12.161

Chen, C.Q., and K. Xia. 2017. Fate of Land Applied Emerging Organic Contaminants in Waste Materials. Curr. Pollut. Rep. 3:38–54. doi:10.1007/s40726-017-0048-6

Chen, H.Q., and Q.X. Zhong. 2018. Antibacterial activity of acidified sodium benzoate against Escherichia coli O157:H7, Salmonella enterica, and Listeria monocytogenes in tryptic soy broth and on cherry tomatoes. Int. J. Food Microbiol. 274:38–44. doi:10.1016/j.ijfoodmicro.2018.03.017

Chen, J., F.C. Michel, S. Sreevatsan, M. Morrison, and Z.T. Yu. 2010. Occurrence and persistence of erythromycin resistance genes (erm) and tetracycline resistance genes (tet) in waste treatment systems on swine farms. Microb. Ecol. 60:479–486. doi:10.1007/s00248-010-9634-5

Chen, J., G.G. Ying, X.D. Wei, Y.S. Liu, S.S. Liu, L.X. Hu, L.Y. He, Z.F. Chen, F.R. Chen, and Y.Q. Yang. 2016. Removal of antibiotics and antibiotic resistance genes from domestic sewage by constructed wetlands: Effect of flow configuration and plant species. Sci. Total Environ. 571:974–982. doi:10.1016/j.scitotenv.2016.07.085

Chen, S., X. Li, G. Sun, Y. Zhang, J. Su, and J. Ye. 2015. Heavy metal induced antibiotic resistance in bacterium LSJC7. Int. J. Mol. Sci. 16:23390–23404. doi:10.3390/ijms161023390

Chuah, C.J., N. Mukhaidin, S.H. Choy, G.J.D. Smith, I.H. Mendenhall, Y.A.L. Lim, and A.D. Ziegler. 2016. Prevalence of Cryptosporidium and Giardia in the water resources of the Kuang River catchment, Northern Thailand. Sci. Total Environ. 562:701–713. doi:10.1016/j. scitotenv.2016.03.247

Cima, G. 2012. FDA cancels 1977 drug withdrawal bids. J. Am. Vet. Med. Assoc. 240:360–361.

Cooley, M., D. Carychao, L. Crawford-Miksza, M.T. Jay, C. Myers, C. Rose, C. Keys, J. Farrar, and R.E. Mandrell. 2007. Incidence and tracking of Escherichia coli O157:H7 in a major produce production region in California. PLoS One 2:E1159. doi:10.1371/journal.pone.0001159

Cui, E., Y. Wu, Y. Jiao, Y. Zuo, C. Rensing, and H. Chen. 2017. The behavior of antibiotic resistance genes and arsenic influenced by biochar during different manure composting. Environ. Sci. Pollut. Res. Int. 24:14484–14490. doi:10.1007/s11356-017-9028-z

Dolliver, H., S. Gupta, and S. Noll. 2008. Antibiotic degradation during manure composting. J. Environ. Qual. 37:1245–1253. doi:10.2134/jeq2007.0399

Dreelin, E.A., R.L. Ives, S. Molloy, and J.B. Rose. 2014. Cryptosporidium and Giardia in surface water: A case study from Michigan, USA to inform management of rural water systems. Int. J. Environ. Res. Public Health 11:10480–10503. doi:10.3390/ijerph111010480

Food and Drug Administration. 2015. Summary report on antimicrobials sold or distributed for use in food-producing animals. Food and Drug Administration, Washington, D.C. https://www.fda.gov/downloads/ForIndustry/UserFees/AnimalDrugUserFeeActADUFA/UCM534243.pdf (Accessed 15 Feb. 2019).

Feng, L., M.E. Casas, L.D.M. Ottosen, H.B. Moller, and K. Bester. 2017. Removal of antibiotics during the anaerobic digestion of pig manure. Sci. Total Environ. 603-604:219–225. doi:10.1016/j.scitotenv.2017.05.280

Food and Drug Administration. 2014. Food Safety Modernization Act Facts: Biological soil amendments: Subpart F. U.S. Department of Health and Human Services, Washington, D.C. http://www.fda.gov/downloads/Food/GuidanceRegulation/FSMA/UCM359281.pdf (Accessed 15 Feb. 2019).

Food and Drug Administration. 2017. Summary Report on Antimicrobials Sold or Distributed for Use in Food-Producing Animals. U.S. Food and Drug Administration, Washington, D.C. https://www.fda.gov/downloads/ForIndustry/UserFees/AnimalDrugUserFeeActADUFA/UCM588085.pdf. (Accessed 15 Feb. 2019).

Forslund, A., F. Plauborg, M.N. Andersen, B. Markussen, and A. Dalsgaard. 2011. Leaching of human pathogens in repacked soil lysimeters and contamination of potato tubers under subsurface drip irrigation in Denmark. Water Res. 45:4367–4380. doi:10.1016/j.watres.2011.05.009

Guan, T.Y., and R.A. Holley. 2003. Pathogen survival in swine manure environments and transmission of human enteric illness–a review. J. Environ. Qual. 32:383–392. doi:10.2134/jeq2003.3830

Guo, A., J. Gu, X. Wang, R. Zhang, Y. Yin, W. Sun, X. Tuo, and L. Zhang. 2017. Effects of superabsorbent polymers on the abundances of antibiotic resistance genes, mobile genetic elements, and the bacterial community during swine manure composting. Bioresour. Technol. 244:658–663. doi:10.1016/j.biortech.2017.08.016

Heinonen, S.T. 2006. Manure fertilizers, application- FI Luomuliitto standards for "Leppäkerttu" quality label 2004. CertCost, University of Southampton, Southampton, U.K. http://organicrules.org/1203/ (Accessed 18 Feb. 2019).

Herrero, P.K.T.M. 2010. The inter-linkages between rapid growth in livestock production, climate change, and the impacts on water resources, land use, and deforestation. Policy Research Working Paper. World Bank, Geneva, Switzerland. https://openknowledge.worldbank.org/bitstream/handle/10986/19942/WPS5178.pdf (Accessed 15 Feb. 2019).

Heuer, H., A. Focks, M. Lamshöft, K. Smalla, M. Matthies, and M. Spiteller. 2008. Fate of sulfadiazine administered to pigs and its quantitative effect on the dynamics of bacterial resistance genes in manure and manured soil. Soil Biol. Biochem. 40:1892–1900. doi:10.1016/j.soilbio.2008.03.014

Heuer, H., H. Schmitt, and K. Smalla. 2011. Antibiotic resistance gene spread due to manure application on agricultural fields. Curr. Opin. Microbiol. 14:236–243. doi:10.1016/j.mib.2011.04.009

Heuer, H., and K. Smalla. 2007. Manure and sulfadiazine synergistically increased bacterial antibiotic resistance in soil over at least two months. Environ. Microbiol. 9:657–666. doi:10.1111/j.1462-2920.2006.01185.x

Hölzel, C., and J. Bauer. 2008. Salmonella spp. In: Bavarian liquid pig manure: Occurrence and relevance for the distribution of antibiotic resistance. Zoonoses Public Health 55:133–138. doi:10.1111/j.1863-2378.2007.01102.x

Hoornweg, D., L. Thomas, and L. Otten. 2000. Composting and its applicability in developing countries (English). Urban waste management working paper series; no. 8. The World Bank, Washington, D.C. http://documents.worldbank.org/curated/en/483421468740129529/Composting-and-its-applicability-in-developing-countries (Accessed 15 Feb. 2019).

Huang, K., S. Wrenn, R. Tikekar, and N. Nitin. 2018. Efficacy of decontamination and a reduced risk of cross-contamination during ultrasound-assisted washing of fresh produce. J. Food Eng. 224:95–104. doi:10.1016/j.jfoodeng.2017.11.043

Huttner, A., S. Harbarth, J. Carlet, S. Cosgrove, H. Goossens, A. Holmes, V. Jarlier, A. Voss, D. Pittet, and the World Healthcare-Associated Infections Forum participants. 2013. Antimicrobial resistance: A global view from the 2013 World Healthcare-Associated Infections Forum. Antimicrob. Resist. Infect. Control 2:31. doi:10.1186/2047-2994-2-31

Ijabadeniyi, O.A., A. Minnaar, and E.M. Buys. 2011. Effect of attachment time Followed by chlorine washing on the survival of inoculated listeria monocytogenes on tomatoes and spinach. J. Food Qual. 34:133–141. doi:10.1111/j.1745-4557.2011.00375.x

Islam, M., M.P. Doyle, S.C. Phatak, P. Millner, and X. Jiang. 2004. Persistence of enterohemorrhagic Escherichia coli O157:H7 in soil and on leaf lettuce and parsley grown in fields treated with contaminated manure composts or irrigation water. J. Food Prot. 67:1365–1370. doi:10.4315/0362-028X-67.7.1365

Jasovský, D., J. Littmann, A. Zorzet, and O. Cars. 2016. Antimicrobial resistance—a threat to the world's sustainable development. Ups. J. Med. Sci. 121:159–164. doi:10.1080/03009734.2016.1195900

Johannessen, G.S., R.B. Froseth, L. Solemdal, J. Jarp, Y. Wasteson and L.M. Rørvik. 2004. Influence of bovine manure as fertilizer on the bacteriological quality of organic Iceberg lettuce. J. Appl. Microbiol. 96:787–794.

Johnson, R. 2011. Potential trade implications of restrictions on antimicrobial use in animal production. CRS Report for Congress. Congressional Research Service, Washington, D.C.

Johnston, L.M., L.A. Jaykus, D. Moll, J. Anciso, B. Mora, and C.L. Moe. 2006. A field study of the microbiological quality of fresh produce of domestic and Mexican origin. Int. J. Food Microbiol. 112:83–95. doi:10.1016/j.ijfoodmicro.2006.05.002

Joy, S.R., S.L. Bartelt-Hunt, D.D. Snow, J.E. Gilley, B.L. Woodbury, D.B. Parker, D.B. Marx, and X. Lu. 2013. Fate and transport of antimicrobials and antimicrobial resistance genes in soil and runoff following land application of swine manure slurry. Environ. Sci. Technol. 47:12081–12088. doi:10.1021/es4026358

Karimi, H., M. Mokhtari, F. Salehi, S. Sojoudi, and A. Ebrahimi. 2017. Changes in microbial pathogen dynamics during vermicomposting mixture of cow manure–organic solid waste and cow manure–sewage sludge. International Journal of Recycling of Organic Waste in Agriculture 6:57–61. doi:10.1007/s40093-016-0152-4

Kim, N.H., N.Y. Lee, S.H. Kim, H.J. Lee, Y. Kim, J.H. Ryu, and M.S. Rhee. 2015. Optimization of low-temperature blanching combined with calcium treatment to inactivate Escherichia coli O157:H7 on fresh-cut spinach. J. Appl. Microbiol. 119:139–148. doi:10.1111/jam.12815

Knapp, C.W., J. Dolfing, P.A. Ehlert, and D.W. Graham. 2010. Evidence of increasing antibiotic resistance gene abundances in archived soils since 1940. Environ. Sci. Technol. 44:580–587. doi:10.1021/es901221x

Kobashi, Y., A. Hasebe, and M. Nishio. 2005. Antibiotic-resistant bacteria from feces of livestock, farmyard manure, and farmland in Japan. Microbes Environ. 20:53–60. doi:10.1264/jsme2.20.53

Kulesza, S.B., R.O. Maguire, K. Xia, J. Cushman, K. Knowlton, and P. Ray. 2016. Manure injection affects the fate of pirlimycin in surface runoff and soil. J. Environ. Qual. 45:511–518. doi:10.2134/jeq2015.06.0266

Kumar, S., P. Pornsukarom, G.K. Sivaraman, and S. Thakur. 2018. Environmental dissemination of multidrug methicillin-resistant staphylococcus sciuri after application of manure from commercial swine production systems. Foodborne Pathog. Dis. 15:210–217. doi:10.1089/fpd.2017.2354

Kümmerer, K. 2009. Antibiotics in the aquatic environment–a review–part I. Chemosphere 75:417–434. doi:10.1016/j.chemosphere.2008.11.086

Kurenbach, B., P.S. Gibson, A.M. Hill, A.S. Bitzer, M.W. Silby, W. Godsoe, and J.A. Heinemann. 2017. Herbicide ingredients change Salmonella enterica sv. Typhimurium and Escherichia coli antibiotic responses. Microbiology. doi:10.1099/mic.0.000573

Landers, T.F., B. Cohen, T.E. Wittum, and E.L. Larson. 2012. A review of antibiotic use in food animals: Perspective, policy, and potential. Public Health Rep. 127:4–22. doi:10.1177/003335491212700103

Lathers, C.M. 2001. Role of veterinary medicine in public health: Antibiotic use in food animals and humans and the effect on evolution of antibacterial resistance. J. Clin. Pharmacol. 41:595–599. doi:10.1177/00912700122010474

Le, H.T.V., R.O. Maguire, and K. Xia. 2018. Method of dairy manure application and time before rainfall affect antibiotics in surface runoff. J. Environ. Qual. doi:10.2134/jeq2018.02.0086

Lemunier, M., C. Francou, S. Rousseaux, S. Houot, P. Dantigny, P. Piveteau, and J. Guzzo. 2005. Long-term survival of pathogenic and sanitation indicator bacteria in experimental biowaste composts. Appl. Environ. Microbiol. 71:5779–5786. doi:10.1128/AEM.71.10.5779-5786.2005

Li, C., Q. Wei, R.W. Melse, L. Lujun, F.E. de Buiisonjé, W. Yajing, and D. Renjie. 2017. Patterns of dairy manure management in China. Int. J. Agric. Biol. Eng. 10(3):227–236.

Li, H., B. Li, Z. Zhang, Y. Tian, J. Ye, X. Lv, and C. Zhu. 2018. Factors influencing the removal of antibiotic-resistant bacteria and antibiotic resistance genes by the electrokinetic treatment. Ecotoxicol. Environ. Saf. 160:207–215. doi:10.1016/j.ecoenv.2018.05.028

Liato, V., R. Hammami, and M. Aider. 2017. Influence of electro-activated solutions of weak organic acid salts on microbial quality and overall appearance of blueberries during storage. Food Microbiol. 64:56–64. doi:10.1016/j.fm.2016.12.010

Libby, D.A., and P.J. Schaible. 1955. Observations on growth responses to antibiotics and arsonic acids in poultry feeds. Science 121:733–734. doi:10.1126/science.121.3151.733

Liu, J., Z. Zhao, L. Orfe, M. Subbiah, and D.R. Call. 2016. Soil-borne reservoirs of antibiotic-resistant bacteria are established following therapeutic treatment of dairy calves. Environ. Microbiol. 18:557–564. doi:10.1111/1462-2920.13097

Looft, T., T.A. Johnson, H.K. Allen, D.O. Bayles, D.P. Alt, R.D. Stedtfeld, B. Chai, J.R. Cole, S.A. Hashsham, J.M. Tiedje, and T.B. Stanton. 2012. In-feed antibiotic effects on the swine intestinal microbiome. Proc. Natl. Acad. Sci. USA 109:1691–1696. doi:10.1073/pnas.1120238109

Maguire, R.O., P.J.A. Kleinman, and D.B. Beegle. 2011. Novel manure management technologies in no-till and forage systems: Introduction to the special series. J. Environ. Qual. 40:287–291. doi:10.2134/jeq2010.0396

Mahajan, P.V., O.J. Caleb, Z. Singh, C.B. Watkins, and M. Geyer. 2014. Postharvest treatments of fresh produce. Philos Trans A Math Phys. Eng. Sci. 372:20130309. doi:10.1098/rsta.2013.0309

Maitland, J.E., R.R. Boyer, J.D. Eifert, and R.C. Williams. 2011. High hydrostatic pressure processing reduces Salmonella enterica serovars in diced and whole tomatoes. Int. J. Food Microbiol. 149:113–117. doi:10.1016/j.ijfoodmicro.2011.05.024

Manten, A. 1963. The non-medical use of antibiotics and the risk of causing microbial drug-resistance. Bull. World Health Organ. 29:387–400.

Manyi-Loh, C.E., S.N. Mamphweli, E.L. Meyer, G. Makaka, M. Simon, and A.I. Okoh. 2016. An overview of the control of bacterial pathogens in cattle manure. Int. J. Environ. Res. Public Health 13. doi:10.3390/ijerph13090843

Maron, D.F., T.J.S. Smith, and K.E. Nachman. 2013. Restrictions on antimicrobial use in food animal production: An international regulatory and economic survey. Global. Health 9:48. doi:10.1186/1744-8603-9-48

Marti, R., A. Scott, Y.C. Tien, R. Murray, L. Sabourin, Y. Zhang, and E. Topp. 2013. Impact of manure fertilization on the abundance of antibiotic-resistant bacteria and frequency of detection of antibiotic resistance genes in soil and on vegetables at harvest. Appl. Environ. Microbiol. 79:5701–5709. doi:10.1128/AEM.01682-13

Martínez, J.L., F. Baquero, and D.I. Andersson. 2007. Predicting antibiotic resistance. Nat. Rev. Microbiol. 5:958–965. doi:10.1038/nrmicro1796

Massé, D., Y. Gilbert, and E. Topp. 2011. Pathogen removal in farm-scale psychrophilic anaerobic digesters processing swine manure. Bioresour. Technol. 102:641–646. doi:10.1016/j.biortech.2010.08.020

Massé, D.I., N.M. Saady, and Y. Gilbert. 2014. Potential of biological processes to eliminate antibiotics in livestock manure: An overview. Animals (Basel) 4:146–163. doi:10.3390/ani4020146

McCarthy, G., P.G. Lawlor, L. Coffey, T. Nolan, M. Gutierrez, and G.E. Gardiner. 2011. An assessment of pathogen removal during composting of the separated solid fraction of pig manure. Bioresour. Technol. 102:9059–9067. doi:10.1016/j.biortech.2011.07.021

Micallef, S.A., R.E. Rosenberg Goldstein, A. George, L. Kleinfelter, M.S. Boyer, C.R. McLaughlin, A. Estrin, L. Ewing, J. Jean-Gilles Beaubrun. 2012. Occurrence and antibiotic resistance of multiple Salmonella serotypes recovered from water, sediment and soil on mid-Atlantic tomato farms. Environ. Res. 114:31–39. doi:10.1016/j.envres.2012.02.005

Milinovich, G.J. and A.V. Klieve. 2011. Manure as a source of zoonotic pathogens. In: D. Krause and S. Hendrick, editors, Zoonotic Pathogens in the Food Chain. CABI, Oxfordshire, U.K. p. 59-83.

Monnier, A.A., B.I. Eisenstein, M.E. Hulscher, I.C. Gyssens and DRIVE-AB WPI group. 2018. Towards a global definition of responsible antibiotic use: Results of an international multidisciplinary consensus procedure. J. Antimicrob. Chemother. 73:Vi3–vi16. doi:10.1093/jac/dky114

Moore, J.E., M. Watabe, B.C. Millar, A. Loughrey, M. McCalmont, C.E. Goldsmith, et al. 2010. Screening of clinical, food, water and animal isolates of Escherichia coli for the presence of blaCTX-M extended spectrum beta-lactamase (ESBL) antibiotic resistance gene loci. Ulster Med. J. 79:85–88.

Mukherjee, A., D. Speh, E. Dyck, and F. Diez-Gonzalez. 2004. Preharvest evaluation of coliforms, Escherichia coli, Salmonella, and Escherichia coli O157:H7 in organic and conventional produce grown by Minnesota farmers. J. Food Prot. 67:894–900. doi:10.4315/0362-028X-67.5.894

National Research Council. 1980. The use of antimicrobial agents. In: National Research Council, editors, The effects on human health of subtherapeutic use of antimicrobials in animal feeds. National Research Council (US) Committee to study the human health effects of subtherapeutic antibiotic use in animal feeds. National Academies Press, Washington, D.C. https://www.ncbi.nlm.nih.gov/books/NBK216522/.

Nicholson, F.A., S.J. Groves, and B.J. Chambers. 2005. Pathogen survival during livestock manure storage and following land application. Bioresour. Technol. 96:135–143. doi:10.1016/j.biortech.2004.02.030

Nyberg, K.A., J.R. Ottoson, B. Vinneras, and A. Albihn. 2014. Fate and survival of Salmonella Typhimurium and Escherichia coli O157:H7 in repacked soil lysimeters after application of cattle slurry and human urine. J. Sci. Food Agric. 94:2541–2546. doi:10.1002/jsfa.6593

Ogden, L.D., D.R. Fenlon, A.J. Vinten, and D. Lewis. 2001. The fate of Escherichia coli O157 in soil and its potential to contaminate drinking water. Int. J. Food Microbiol. 66:111–117. doi:10.1016/S0168-1605(00)00508-0

Olanya, M.O., D.O. Ukuku, B.A. Annous, B.A. Niemira, and C.H. Sommers. 2013. Efficacy of Pseudomonas fluorescens for biocontrol of Escherichia coli O157:H7 on spinach. J. Food Agric. Environ. 11:86–91.

Oliveira, M., I. Vinas, J. Usall, M. Anguera, and M. Abadias. 2012. Presence and survival of Escherichia coli O157:H7 on lettuce leaves and in soil treated with contaminated compost and irrigation water. Int. J. Food Microbiol. 156:133–140. doi:10.1016/j.ijfoodmicro.2012.03.014

Oluyege, J.O., A.C. Dada, and A.T. Odeyemi. 2009. Incidence of multiple antibiotic resistant Gram-negative bacteria isolated from surface and underground water sources in south western region of Nigeria. Water Sci. Technol. 59:1929–1936. doi:10.2166/wst.2009.219

Ongeng, D., C. Muyanja, A.H. Geeraerd, D. Springael, and J. Ryckeboer. 2011. Survival of Escherichia coli O157:H7 and Salmonella enterica serovar Typhimurium in manure and manure-amended soil under tropical climatic conditions in Sub-Saharan Africa. J. Appl. Microbiol. 110:1007–1022. doi:10.1111/j.1365-2672.2011.04956.x

Oni, R.A., M. Sharma, S.A. Micallef, and R.L. Buchana. 2013. The effect of UV radiation on survival of Salmonella enterica in dried manure dust. http://cfs3.umd.edu/docs/IAFP_2013_Abstract_II.pdf (Accessed 9 Sept. 2018).

Pandey, P.K., and M.L. Soupir. 2013. Assessing dairy manure pathogen indicator inactivation under anaerobic and aerobic digestions in mesophilic temperature. 2013 IEEE International Conference on Bioinformatics and Biomedicine, Shanghai, China. 18–21 Dec. 2013. IEEE, New York. doi:10.1109/BIBM.2013.6732762

Peng, S., H.J. Li, D. Song, X.G. Lin, and Y.M. Wang. 2018. Influence of zeolite and superphosphate as additives on antibiotic resistance genes and bacterial communities during factory-scale chicken manure composting. Bioresour. Technol. 263:393–401. doi:10.1016/j.biortech.2018.04.107

Pu, C., L. Liu, M. Yao, H. Liu, and Y. Sun. 2018. Responses and successions of sulfonamides, tetracyclines and fluoroquinolones resistance genes and bacterial community during the short-term storage of biogas residue and organic manure under the incubator and natural conditions. Environ. Pollut. 242:749–759. doi:10.1016/j.envpol.2018.07.063

Pulido, N.A. 2016. Effect of standard post-harvest interventions on the survival and regrowth of antibiotic-resistant bacteria on fresh produce. MS thesis, Virginia Polytechnic Institute, Blacksburg, VA.

Quirk, M. 2001. Antibiotic-resistant bacteria in food animals on the rise. Lancet Infect. Dis. 1:293. doi:10.1016/S1473-3099(01)00137-2

Rangasamy, K., M. Athiappan, N. Devarajan, and J.A. Parray. 2017. Emergence of multi drug resistance among soil bacteria exposing to insecticides. Microb Pathogenesis 105: 153-165. doi:10.1016/j.micpath.2017.02.011

Ray, P., C. Chen, K.F. Knowlton, A. Pruden, and K. Xia. 2017. Fate and effect of antibiotics in beef and dairy manure during static and turned composting. J. Environ. Qual. 46:45–54. doi:10.2134/jeq2016.07.0269

Rehder, L.E. 2012. Germany plans to reduce farm use of antibiotics. USDA Foreign Agricultural Service, Washington, D.C.

Robinson, T.P., D.P. Bu, J. Carrique-Mas, E.M. Fèvre, M. Gilbert, D. Grace, S.I. Hay, J. Jiwakanon, M. Kakkar, S. Kariuki, R. Laxminarayan, J. Lubroth, U. Magnusson, P. Thi Ngoc, T.P. Van Boeckel, and M.E.J. Woolhouse. 2016. Antibiotic resistance is the quintessential One Health issue. Trans. R. Soc. Trop. Med. Hyg. 110:377–380. doi:10.1093/trstmh/trw048

Rodriguez, C., D.E. Hakimi, R. Vanleyssem, B. Taminiau, J. Van Broeck, M. Delmée, N. Korsak, and G. Daube. 2017. Clostridium difficile in beef cattle farms, farmers and their environment: Assessing the spread of the bacterium. Vet. Microbiol. 210:183–187. doi:10.1016/j.vetmic.2017.09.010

Ruuskanen, M., J. Muurinen, A. Meierjohan, K. Parnanen, M. Tamminen, C. Lyra, L. Kronberg, and M. Virta. 2016. Fertilizing with animal manure disseminates antibiotic resistance genes to the farm environment. J. Environ. Qual. 45:488–493. doi:10.2134/jeq2015.05.0250

Sarmah, A.K., M.T. Meyer, and A.B. Boxall. 2006. A global perspective on the use, sales, exposure pathways, occurrence, fate and effects of veterinary antibiotics (VAs) in the environment. Chemosphere 65: 725-759. doi:10.1016/j.chemosphere.2006.03.026

Schröder, J.J., and J.J. Neeteson. 2008. Nutrient management regulations in The Netherlands. Geoderma 144: 418-425. doi:10.1016/j.geoderma.2007.12.012

Seiler, C., and T.U. Berendonk. 2012. Heavy metal driven co-selection of antibiotic resistance in soil and water bodies impacted by agriculture and aquaculture. Front. Microbiol. 3:399. doi:10.3389/fmicb.2012.00399

Shan, S.D., H. Wang, C.R. Fang, Y.X. Chu, and L.H. Jiang. 2018. Effects of adding biochar on tetracycline removal during anaerobic composting of swine manure. Chem. Ecol. 34:86–97. do i:10.1080/02757540.2017.1400022

Sharma, R.R., A. Demirci, L.R. Beuchat, and W.F. Fett. 2003. Application of ozone for inactivation of Escherichia coli O157: H7 on inoculated alfalfa sprouts. J. Food Process. Preserv. 27:51–64. doi:10.1111/j.1745-4549.2003.tb00500.x

Sharpley, A., P. Kleinman, and J. Weld. 2004. Assessment of best management practices to minimise the runoff of manure-borne phosphorus in the United States. N. Z. J. Agric. Res. 47:461–477. doi:10.1080/00288233.2004.9513614

Skinner, B.M., A.T. Rogers, and M.E. Jacob. 2018. Susceptibility of Escherichia coli O157:H7 to disinfectants in vitro and in simulated footbaths amended with manure. Foodborne Pathog. Dis. doi:10.1089/fpd.2018.2457

Smith, H.W. 1971. The effect of the use of antibacterial drugs on the emergence of drug-resistant bacteria in animals. Adv. Vet. Sci. Comp. Med. 15:67–100.

Smith, K.F., M. Goldberg, S. Rosenthal, L. Carlson, J. Chen, C. Chen, and S. Ramachandran. 2014. Global rise in human infectious disease outbreaks. J. R. Soc. Interface 11:20140950. doi:10.1098/rsif.2014.0950

Sun, W., J. Gu, X. Wang, X. Qian, and X. Tuo. 2018. Impacts of biochar on the environmental risk of antibiotic resistance genes and mobile genetic elements during anaerobic digestion of cattle farm wastewater. Bioresour. Technol. 256:342–349. doi:10.1016/j.biortech.2018.02.052

Sun, W., X. Qian, J. Gu, X.J. Wang, and M.L. Duan. 2016. Mechanism and effect of temperature on variations in antibiotic resistance genes during anaerobic digestion of dairy manure. Sci. Rep. 6:30237. doi:10.1038/srep30237

Tanaro, J.D., M.C. Piaggio, L. Galli, A.M. Gasparovic, F. Procura, D.A. Molina, M. Vitón, G. Zolezzi, and M. Rivas. 2014. Prevalence of Escherichia coli O157:H7 in surface water near cattle feedlots. Foodborne Pathog. Dis. 11:960–965. doi:10.1089/fpd.2014.1770

Tango, C.N., S. Wei, I. Khan, M.S. Hussain, P.N. Kounkeu, J.H. Park, S. Kim, and D.H. Oh. 2018. Microbiological quality and safety of fresh fruits and vegetables at retail levels in Korea. J. Food Sci. 83:386–392. doi:10.1111/1750-3841.13992

Taylor, L.H., S.M. Latham, and M.E. Woolhouse. 2001. Risk factors for human disease emergence. Philos. Trans. R. Soc. Lond. B. Biol. Sci. 356: 983-989. doi:org/10.1098/rstb.2001.0888

Thorpe, K.E., P. Joski, and K.J. Johnston. 2018. Antibiotic-resistant infection treatment costs have doubled since 2002, now exceeding $2 billion annually. Health Aff. 37:662–669. doi:10.1377/hlthaff.2017.1153

U.S. Government Accountability Office. 2011. Antibiotic resistance: Agencies have made limited progress addressing antibiotic use in animals. U.S. Government Accountability Office, Washington, D.C.

U.S. Department of Health and Human Services. 2013. Guidance for Industry #213: New animal drugs and new animal drug combination products administered in or on medicated feed or drinking water of food producing animals: Recommendations for drug sponsors for voluntarily aligning product use conditions with GF. U.S. Department of Health and Human Services, Washington, D.C. http://www.fda.gov/downloads/AnimalVeterinary/GuidanceComplianceEnforcement/GuidanceforIndustry/UCM299624.pdf (Accessed 15 Feb. 2019).

Udikovic-Kolic, N., F. Wichmann, N.A. Broderick, and J. Handelsman. 2014. Bloom of resident antibiotic-resistant bacteria in soil following manure fertilization. Proc. Natl. Acad. Sci. USA 111:15202–15207. doi:10.1073/pnas.1409836111

USDA. 2012. Guide for organic crop producers. United States Department of Agriculture, Washington, D.C. https://www.ams.usda.gov/sites/default/files/media/GuideForOrganicCropProducers.pdf (Accessed 13 Feb. 2019).

Van Boeckel, T.P., C. Brower, M. Gilbert, B.T. Grenfell, S.A. Levin, T.P. Robinson, A. Teillant, and R. Laxminarayan. 2015. Global trends in antimicrobial use in food animals. Proc. Natl. Acad. Sci. USA 112:5649–5654. doi:10.1073/pnas.1503141112

van den Bogaard, A.E., N. Bruinsma, and E.E. Stobberingh. 2000. The effect of banning avoparcin on VRE carriage in The Netherlands. J. Antimicrob. Chemother. 46:146–148. doi:10.1093/jac/46.1.146

Van Donkersgoed, J., V. Bohaychuk, T. Besser, X.M. Song, B. Wagner, D. Hancock, et al. 2009. Occurrence of foodborne bacteria in Alberta feedlots. Can. Vet. J. 50:166–172.

van Elsas, J.D., A.V. Semenov, R. Costa, and J.T. Trevors. 2011. Survival of *Escherichia coli* in the environment: Fundamental and public health aspects. ISME J. 5:173–183. doi:10.1038/ismej.2010.80 [erratum: 5:367].

Varel, V.H., J.E. Wells, W.L. Shelver, C.P. Rice, D.L. Armstrong, and D.B. Parker. 2012. Effect of anaerobic digestion temperature on odour, coliforms and chlortetracycline in swine manure or monensin in cattle manure. J. Appl. Microbiol. 112:705–715. doi:10.1111/j.1365-2672.2012.05250.x

Ventola, C.L. 2015. The antibiotic resistance crisis: Part 1: Causes and threats. P&T 40:277–283.

Wales, A.D., and R.H. Davies. 2015. Co-selection of resistance to antibiotics, biocides and heavy metals, and its relevance to foodborne pathogens. Antibiotics (Basel) 4:567–604. doi:10.3390/antibiotics4040567

Walsh, F. 2013. Investigating antibiotic resistance in non-clinical environments. Front. Microbiol. 4:19. doi:10.3389/fmicb.2013.00019

Walton, J.R. 1971. The public health implications of drug-resistant bacteria in farm animals. Ann. N. Y. Acad. Sci. 182:358–361. doi:10.1111/j.1749-6632.1971.tb30671.x

Wang, F.H., M. Qiao, Z. Chen, J.Q. Su, and Y.G. Zhu. 2015a. Antibiotic resistance genes in manure-amended soil and vegetables at harvest. J. Hazard. Mater. 299:215–221. doi:10.1016/j.jhazmat.2015.05.028

Wang, H., H. Li, J.A. Gilbert, H. Li, L. Wu, M. Liu, L. Wang, Q. Zhou, J. Yuan, and Z. Zhang. 2015b. Housefly larva vermicomposting efficiently attenuates antibiotic resistance genes in swine manure, with concomitant bacterial population changes. Appl. Environ. Microbiol. 81:7668–7679. doi:10.1128/AEM.01367-15

Wang, H., T. Zhang, G. Wei, L. Wu, J. Wu, and J. Xu. 2014a. Survival of *Escherichia coli* O157:H7 in soils under different land use types. Environ. Sci. Pollut. Res. Int. 21:518–524. doi:10.1007/s11356-013-1938-9

Wang, N., X. Yang, S. Jiao, J. Zhang, B. Ye, and S. Gao. 2014b. Sulfonamide-resistant bacteria and their resistance genes in soils fertilized with manures from Jiangsu Province, Southeastern China. PLoS One 9:E112626. doi:10.1371/journal.pone.0112626

Wang, X. 2017. Use of antibiotics in poultry and livestock to be reduced. China Daily, 24 June. http://www.chinadaily.com.cn/china/2017-06/24/content_29869315.htm (Accessed 24 June 2018).

Wierup, M. 2001. The Swedish experience of the 1986 year ban of antimicrobial growth promoters, with special reference to animal health, disease prevention, productivity, and usage of antimicrobials. Microb. Drug Resist. 7:183–190. doi:10.1089/10766290152045066

Wilkes, G., T.A. Edge, V.P. Gannon, C. Jokinen, E. Lyautey, N.F. Neumann, N. Ruecker, A. Scott, M. Sunchara, E. Topp, and D.R. Lapen. 2011. Associations among pathogenic bacteria, parasites, and environmental and land use factors in multiple mixed-use watersheds. Water Res. 45:5807–5825. doi:10.1016/j.watres.2011.06.021

Wolfe, N.D., C.P. Dunavan, and J. Diamond. 2007. Origins of major human infectious diseases. Nature 447:279–283. doi:10.1038/nature05775

Wong, J.W., and A. Selvam. 2009. Reduction of indicator and pathogenic microorganisms in pig manure through fly ash and lime addition during alkaline stabilization. J. Hazard. Mater. 169:882–889. doi:10.1016/j.jhazmat.2009.04.033

World Health Organization. 2014. Antimicrobial resistance global report on surveillance. World Health Organization, Geneva, Switzerland. http://www.who.int/drugresistance/documents/ surveillancereport/en/ (Accessed 13 Feb. 2019).

World Health Organization. 2015. Global action plan for antimicrobial resistance. World Health Organization, Geneva, Switzerland. http://www.who.int/antimicrobial-resistance/global-action-plan/en/ (accessed 13 Feb. 2019).

Wu, D., Z.T. Huang, K. Yang, D. Graham, and B. Xie. 2015. Relationships between antibiotics and antibiotic resistance gene levels in municipal solid waste leachates in Shanghai, China. Environ. Sci. Technol. 49:4122–4128. doi:10.1021/es506081z

Xie, W.Y., Q. Shen, and F.J. Zhao. 2018. Antibiotics and antibiotic resistance from animal manures to soil: A review. Eur. J. Soil Sci. 69:181–195. doi:10.1111/ejss.12494

Xu, S., I.D. Amarakoon, R. Zaheer, A. Smith, S. Sura, G. Wang, T. Reuter, F. Zvomuya, A.J. Cessna, F.J. Larney, and T.A. McAllister. 2018. Dissipation of antimicrobial resistance genes in compost originating from cattle manure after direct oral administration or post-excretion fortification of antimicrobials. J. Environ. Sci. Health, Part A: Toxic/Hazard. Subst. Environ. Eng. 53: 373-384. doi:10.1080/10934529.2017.1404337.

Yang, Q., H. Zhang, Y. Guo, and T. Tian. 2016. Influence of chicken manure fertilization on antibiotic-resistant bacteria in soil and the endophytic bacteria of Pakchoi. Int. J. Environ. Res. Public Health 13. doi:10.3390/ijerph13070662

Yegin, Y., K.L. Perez-Lewis, M. Zhang, M. Akbulut, and T.M. Taylor. 2016. Development and characterization of geraniol-loaded polymeric nanoparticles with antimicrobial activity against foodborne bacterial pathogens. J. Food Eng. 170:64–71. doi:10.1016/j.jfoodeng.2015.09.017

Youngquist, C.P., S.M. Mitchell, and C.G. Cogger. 2016. Fate of antibiotics and antibiotic resistance during digestion and composting: A review. J. Environ. Qual. 45:537–545. doi:10.2134/ jeq2015.05.0256

Zhang, Q.Q., G.G. Ying, C.G. Pan, Y.S. Liu, and J.L. Zhao. 2015. Comprehensive evaluation of antibiotics emission and fate in the river basins of China: Source analysis, multimedia modeling, and linkage to bacterial resistance. Environ. Sci. Technol. 49:6772–6782. doi:10.1021/acs.est.5b00729

Zhang, R., X. Wang, J. Gu, and Y. Zhang. 2017a. Influence of zinc on biogas production and antibiotic resistance gene profiles during anaerobic digestion of swine manure. Bioresour. Technol. 244:63–70. doi:10.1016/j.biortech.2017.07.032

Zhang, Y., H. Li, J. Gu, X. Qian, Y. Yin, Y. Li, R. Zhang, and X. Wang. 2016. Effects of adding different surfactants on antibiotic resistance genes and intl1 during chicken manure composting. Bioresour. Technol. 219:545–551. doi:10.1016/j.biortech.2016.06.117

Zhang, Y.J., H.W. Hu, M. Gou, J.T. Wang, D. Chen, and J.Z. He. 2017b. Temporal succession of soil antibiotic resistance genes following application of swine, cattle and poultry manures spiked with or without antibiotics. Environ. Pollut. 231:1621–1632. doi:10.1016/j.envpol.2017.09.074

Zhang, Z., J. Shen, H. Wang, M. Liu, L. Wu, F. Ping, Q. He, H. Li, C. Zeng, and X. Xu. 2014. Attenuation of veterinary antibiotics in full-scale vermicomposting of swine manure via the housefly larvae (*Musca domestica*). Sci. Rep. 4:6844. doi:10.1038/srep06844

Zhao, X., J. Wang, L. Zhu, W. Ge, and J. Wang. 2017. Environmental analysis of typical antibiotic-resistant bacteria and ARGs in farmland soil chronically fertilized with chicken manure. Sci. Total Environ. 593-594:10–17. doi:10.1016/j.scitotenv.2017.03.062

Zhou, L.J., G.G. Ying, S. Liu, R.Q. Zhang, H.J. Lai, Z.F. Chen, and C.G. Panl. 2013. Excretion masses and environmental occurrence of antibiotics in typical swine and dairy cattle farms in China. Sci. Total Environ. 444:183–195. doi:10.1016/j.scitotenv.2012.11.087

Zhu, Y.G., T.A. Johnson, J.Q. Su, M. Qiao, G.X. Guo, R.D. Stedtfeld, S.A. Hashsham, and J.M. Tiedje. 2013. Diverse and abundant antibiotic resistance genes in Chinese swine farms. Proc. Natl. Acad. Sci. USA 110:3435–3440. doi:10.1073/pnas.1222743110

Environmental Fate and Transport of Veterinary Antibiotics Derived from Animal Manure

Wei Zheng,* Mingxin Guo, and George Czapar

Abstract

Therapeutic and non-therapeutic uses of veterinary antibiotics are common in concentrated animal feeding operations (CAFOs), resulting in a vast volume of manure with significant presence of antibiotic residues. This review addresses the current usage of veterinary antibiotics in animal agriculture, potential risks from disposing of antibiotic residue-containing animal manure, and the environmental fate and transport of manure-associated antibiotics. To date, more than 150 veterinary antibiotics have been used in animal farms, of which 80% are for non-therapeutic purposes. Most veterinary antibiotics are poorly absorbed in animal bodies, with 30 to 90% of the administered dosage being excreted via urine and feces either as parent compounds or metabolites. In the United States alone, this would lead annually to 18.9 to 56.7 million kg of antibiotic residues inherent in animal manure. Veterinary antibiotic residues are introduced into the environment through land application of animal manure as a nutrient source for crop production. To assess the persistence of antibiotic contaminants in the environment, we summarize the latest literature information to illustrate biotic and abiotic transformation of commonly used antibiotics in manure-containing water and manure-amended soils. The antibiotics, especially those hydrophobic species can interact with soil organic matter (SOM) and soil minerals, thereby limiting their availability for degradation and transport. Combined with our research finding, we discussed concurrent sorption and colloids-facilitated transport of manure-associated antibiotics in field soils. In addition, the uptake and accumulation of veterinary antibiotics in crop plants are also articulated. Considering that veterinary antibiotics may migrate from manure-receiving fields to surrounding water bodies via surface runoff and leaching, we highlighted three mitigation strategies to reduce their load into the environment and minimize their negative effects on agroecosystems.

Veterinary pharmaceuticals have been widely used to treat or prevent diseases or infections in livestock and companion animals for decades. Veterinary pharmaceuticals include antibiotics, antiparasitics, anti-inflammatory medications, anesthetics, pain medications, and specialized products for managing animal reproductive, cardiovascular, and metabolic conditions (Song and Guo, 2014). Veterinary antibiotics are a type of antimicrobial chemicals used to treat and prevent bacterial infections or as a feed supplement for promoting animal growth. Currently, the use of antibiotics accounts for more than 70% of the consumed veterinary pharmaceuticals in the United States. Over the past decades, the rapid development of concentrated animal feeding operations (CAFOs) has resulted in an unprecedented increase in the usage of veterinary antibiotics (Van Boeckel et al., 2015).

Most veterinary antibiotics administered to animals are poorly absorbed and not completely metabolized in animal bodies, resulting in approximately 30 to 90% of

Abbreviations: ARB, antimicrobial resistant bacteria; CAFO, concentrated animal feeding operations; DOM, dissolved organic matter; SOM, soil organic matter.

W. Zheng, Illinois Sustainable Technology Center, University of Illinois at Urbana-Champaign, 1 Hazelwood Drive, Champaign, IL 61820; M. Guo, Department of Agriculture and Natural Resources, Delaware State University, 1200 North DuPont Highway, Dover, DE 19901; G. Czapar, University of Illinois, 170 National Soybean Research Center, 1101 W. Peabody Drive, Urbana, IL 61801. *Corresponding author (weizheng@illinois.edu)

doi:10.2134/asaspecpub67.c15

Animal Manure: Production, Characteristics, Environmental Concerns and Management. ASA Special Publication 67. Heidi M. Waldrip, Paulo H. Pagliari, and Zhongqi He, editors.
© 2019. ASA and SSSA, 5585 Guilford Rd., Madison, WI 53711, USA.

the dosed antibiotics being excreted in urine and feces either as parent compounds or metabolites (Kumar et al., 2005a; Sarmah et al., 2006; Lin et al., 2017). Veterinary antibiotics and their active metabolites may be introduced into the environment through direct disposal of unused or expired medication and land application of animal wastes (i.e., manure or manure-containing wastewater). Manure-derived antibiotics may be adsorbed in agricultural soils and decomposed via various abiotic and biotic processes (Song and Guo, 2014). Meanwhile, antibiotic contaminants can be transported to the surrounding watershed from manure-amended soils via leaching and/or runoff (Dolliver and Gupta, 2008; Du and Liu, 2012) or may be taken up by crop plants and subsequently enter into the food chain (Dolliver et al., 2007; Chitescu et al., 2013; Sallach et al., 2018). Currently, antibiotic contaminants are frequently detected in the environment, which is being recognized as an emerging global issue (Kuppusamy et al., 2018).

In general, antibiotic contaminants occur in the environment at levels as low as parts-per-trillion (i.e., ng L^{-1} in water and $\mu g\ kg^{-1}$ in soil) (Kolpin et al., 2002; Gothwal and Shashidhar, 2015), greatly below therapeutic doses employed for medical purposes. However, some antibiotic residues may cause serious allergies or toxicity to public health (Kim et al., 2011; Kuppusamy et al., 2018). Moreover, chronic exposure to antibiotic contaminants through water and food supplies may also imperil human health and agroecosystems (Kuppusamy et al., 2018). The presence of antibiotics may directly induce multiple adverse effects on soil microbial populations and diversity (Qin et al., 2015). These antibiotics, even at very low concentrations, may increase antimicrobial resistance not only in pathogenic bacteria, but also in the indigenous microbial populations of animals and humans (Vilca et al., 2018; Xie et al., 2018). A study showed that multi-antimicrobial resistant bacteria (ARB) have been detected frequently in the urban waterways of Milwaukee, exposing recreational users to risks of drug-resistant infection (Kappell et al., 2015). Some studies also revealed that drinking water supplies were contaminated with ARB because of antibiotic residues in groundwater and surface water (Koike et al., 2007; O'Flaherty et al., 2018). As a result, antimicrobial resistance is a looming public health crisis because it jeopardizes the continued therapeutic efficacy of antibiotics in animal and human medication (Xie et al., 2018). More than 2 million Americans reportedly become ill with antimicrobial-resistant infections each year, resulting in more than 23,000 deaths (Paulson et al., 2015).

To assess the potential risks of veterinary antibiotics derived from land application of animal wastes, the occurrence of manure-associated antibiotics in the environment and their fate and transport in agroecosystems need to be addressed. The objective of this paper is to summarize the frontier studies on the environmental behaviors of veterinary antibiotics derived from animal wastes. Armed with the present compilation of knowledge of manure-associated antibiotics and our current studies, this chapter presents six themes: (i) usage of veterinary antibiotics in animal agriculture; (ii) residual antibiotics in animal wastes; (iii) transformation of manure-associated antibiotics; (iv) transport of manure-associated antibiotics; (v) uptake and accumulation of veterinary antibiotics in crop plants; and (vi) mitigation strategies for reducing antibiotic contamination.

Usage of Veterinary Antibiotics in Animal Agriculture

Veterinary antibiotics are administrated to various animal species for either therapeutic or non-therapeutic use. In the 1940s, antibiotics were initially adopted in U.S. animal agriculture for treatment, prevention, and control of specific diseases as "therapeutic use" (Dolliver, 2007). Subsequent experiments found that continuously feeding low levels of antibiotics (i.e., non-therapeutic use) improved feed efficiency and promoted animal growth (Moore et al., 1946). In 1949, the United States officially approved the nontherapeutic use of antibiotics as a feed additive in rearing domestic livestock for human consumption (Witte, 2000). Since then, the use of antibiotics in animal feed has significantly increased, especially with the modern development of large-scale animal farming (i.e., CAFOs). Today, more than 150 veterinary antibiotics have been approved for use in animal agriculture, of which 80% are used for nontherapeutic purposes (Ventola, 2015). Supplementing animal feed with antibiotics has been practiced in nearly all livestock production and aquaculture in many countries.

The main groups of veterinary antibiotics used in animal agriculture include aminoglycosides, ampicillin, cephalosporines, fluorochinolones, ionophores, β-lactams, lincosamides, macrolides, peptides, peptidomimetics, sulfonamides, tetracyclines, and trimethoprim (Kemper, 2008; Tasho and Cho, 2016; Kuppusamy et al., 2018). Tetracyclines are the most heavily used veterinary antibiotics, followed by sulfonamides and macrolides. These veterinary antibiotics are organic compounds that are produced by either bacteria and fungi (through secondary metabolism) or semi-synthetically and completely synthetically using chemicals. An estimated over 63 million kg of veterinary antibiotics are consumed annually in animal agriculture across the globe (Tasho and Cho, 2016). In the United States, between 16 and 23 million kg of antibiotics are produced annually, with up to 84% being used for animal agriculture purposes (Dolliver, 2007). In 2015, China exceeded the United States and became the largest producer and consumer of veterinary antibiotics (Wei et al., 2016; Kuppusamy et al., 2018). The global consumption of veterinary antibiotics is estimated to continuously rise because of an increasing requirement of the growing global population for meat foods. A study estimated that global animal agriculture may need more than 105 million kg of veterinary antibiotics by 2030, an increase of 67% compared with the consumption in 2010 (Van Boeckel et al., 2015). The dosage of veterinary antibiotics administrated varies from 3 to 220 mg kg^{-1}, depending on the antibiotic and type and growth stage of animal (McEwen and Fedorka-Cray, 2002). For example, the average annual consumption of antibiotics per kg of livestock produced is about 45, 148, and 172 mg kg^{-1} for cattle, chicken, and pigs, respectively (Van Boeckel et al., 2015; Kuppusamy et al., 2018). In addition, some antibiotics are added for a specific growth stage of animals, but some could be fed continuously up to the point of slaughter (Kumar et al., 2005a).

No doubt, the use of veterinary antibiotics benefits livestock welfare and the animal agricultural economy (Moore et al., 1946; Cromwell, 2002; Tedeschi et al., 2003). Particularly in large-scale animal farms, prevention of disease is critical because disease transmission can occur rapidly. A previous report summarized that the use of veterinary antibiotics decreased swine mortality rate up to 50 to 80% (Cromwell, 2002). In addition to preventing disease outbreaks, use of antibiotics in CAFOs can increase the ability of animals to absorb nutrients and reach the market weight with fewer feeding days, thereby reducing the cost of animal products and benefiting the economy. For example, adding veterinary antibiotics to animal feed has increased swine growth by 4.2 to 16.4% and the feed efficiency by 2.2 to 6.9% (Cromwell, 2002). In addition, there are several environmental benefits of antibiotic use in animal agriculture (Dolliver, 2007). Some veterinary antibiotics, such as monensin, have been found to reduce methane emissions from ruminant animals because of changes of the microbial composition in their stomach (Tedeschi et al., 2003). Antibiotics usage not only lowers the number of days needed to raise animals, but also reduces manure production and the nitrogen content of manure, which reduces ammonia emissions and excess nutrient contamination in the environment (Tedeschi et al., 2003). However, the extensive use of veterinary antibiotics is posing significant health and environmental risks, including antibiotic residues in food products, toxicity to nontargeted animals, and development and spread of ARB (Kuppusamy et al., 2018; Vilca et al., 2018; Xie et al., 2018). Because of the development of ARB, animal producers often add veterinary antibiotics into animal feeds at higher than the recommended levels to achieve the medication purposes. However, the overdosing of antibiotics may result in more residues in food products. In addition, it is commonly accepted that ARB developed in the animals may contaminate food products and be transmitted through the human food chain (Christou et al., 2017). Therefore, the nontherapeutic use of veterinary antibiotics as a supplement for growth promotion of livestock has been banned in European Union (EU) countries (European Commission, 2005), resulting in a significant reduction of veterinary antibiotics use in the EU since 2006.

Residual Antibiotics in Manure Wastes

Veterinary antibiotics are administrated to animals via injection, by external application, or as a feed additive. However, most of the administered antibiotics are not metabolized fully in animal digestive systems and are excreted shortly after medication, resulting in antibiotic residues in manure wastes (Campagnolo et al., 2002; Bradford et al., 2008). Depending on the chemical and animal species, antibiotic contaminants in manure wastes are excreted as parent compounds, conjugated metabolites, or oxidation or hydrolysis products (Tolls, 2001). Although many antibiotics undergo conjugation in animal bodies and are subsequently inactivated, the conjugated metabolites can be transformed back to their bioactive forms after excretion in animal urine and feces (Sarmah et al., 2006). Therefore, antibiotic contaminants in fresh animal wastes can be present either as metabolites or as parent compounds.

The excretion amounts of antibiotics vary with type and growth stage of animals as well as the dosage level. A previous study showed that swine manure had the highest levels of tetracycline residues, followed by poultry manure and cattle manure (Zhang et al., 2008). The same report showed that generally higher amounts of tetracycline were detected in manure from large-scale CAFOs compared with small household farms because the former used more antibiotics than the latter (Zhang et al., 2008). The excretion of veterinary antibiotics is also related to their types and properties. For example, more than 72% of tetracyclines was excreted within 2 d after their oral application (Winckler and Grafe, 2001), and more than 90% of fluoroquinolone antibiotics was excreted by swine after oral application (Sukul et al., 2009). To date, fluoroquinolones are the only type of veterinary antibiotics that were banned by the U.S. Food and Drug Administration because of high excretion rates and public health concerns (Dolliver, 2007).

Although veterinary antibiotics can be degraded by microorganisms during the storage of animal wastes (e.g., in storage shed and lagoons) and handling process (e.g., compost), their residues are still commonly detected in manure compost and manure-containing lagoon water. Bradford et al. (2008) listed illustrative concentrations of several common veterinary antibiotics in swine, poultry, dairy, and beef lagoon water, in which antibiotic concentrations ranged from less than 0.01 to 1340 µg L^{-1}, depending on the type of antibiotics and lagoon sources (Bradford et al., 2008). Veterinary antibiotics were detected in manure slurry, ranging from trace levels to more than 200 mg L^{-1} (Kumar et al., 2005a). Song and Guo (2014) also summarized that antibiotic residues in different manure wastes derived from various CAFOs, and the highest concentration of tetracycline antibiotics could be up to 765 mg kg^{-1} in swine manure. Given that the 63 million kg of veterinary antibiotics are consumed annually in the world and as much as 30 to 90% of them are excreted in animal waste, each year there would be about 18.9 to 56.7 million kg of antibiotic residues remaining in manure and manure-containing wastewater. The majority of manure and manure-containing wastewater is applied to agricultural land as a nutrient source for crop production. In the United States, about 1.3 billion metric tons (wet weight) of manure is produced annually from 2.2 billion head of beef and dairy cattle, swine, and poultry (USEPA, 2013), which is applied to approximately 9.2 million hectares of land. Once manure is applied to agricultural land, antibiotic residues and their metabolites in animal wastes enter into the environment.

Transformation of Manure-Associated Antibiotics

Antibiotics remaining in animal wastes may undergo a series of biotic and abiotic transformations/degradation during storage, transportation, and land application of solid manure or manure-containing wastewater (Fig. 1). Song and Guo (2014) summarized the degradation half-lives of a variety of veterinary antibiotics, suggesting that most antibiotics are degradable in soil, especially in the presence of manure, with a half-life time of less than 30 d. Some antibiotics such as roxithromycin and sarafloxacin are recalcitrant to degradation and remain largely unchanged in soil over 120 d (Song and Guo, 2014).

Biotic Transformation

Biodegradation plays a key role in determining the fate of antibiotic contaminants in the environment. A wealth of studies have been conducted to investigate

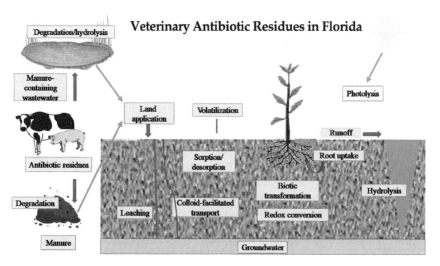

Fig. 1. Diagram of environmental fate and transport for veterinary antibiotics derived from animal manure and manure-containing lagoon water.

the biodegradation of a variety of veterinary antibiotics in manure, water, and soil (Ingerslev et al., 2001; Teeter and Meyerhoff, 2003; Kolz et al., 2005; Wang et al., 2006; Islas-Espinoza et al., 2012; Tasho and Cho, 2016). Because animal manure possesses high microbial activity, antibiotic residues can be readily decomposed by manure-associated bacteria. For example, a fermentation-derived macrolide antibiotic, tylosin, degraded rapidly in cattle, chicken, and swine excreta, suggesting its residues should not persist in the environment after manure land application (Teeter and Meyerhoff, 2003). Many veterinary antibiotics, however, cannot completely degrade in animal excreta, resulting in the spreading of their residues on agricultural fields with manure land application (Feng et al., 2017). Agricultural soils usually contain a diversity of bacteria that have been shown to break down veterinary antibiotics (Dantas et al., 2008). Moreover, manure applied to soil could stimulate microbial respiration and enhance the microorganism activity, which would facilitate the biodegradation of antibiotic residues in soils (Islas-Espinoza et al., 2012; Zhang et al., 2017). However, some studies also showed that the addition of manure-contained antibiotics is likely to impact microbial diversity, increase the spread of antibiotic resistance genes, and may inhibit the function of certain microorganisms participating in important biogeochemical cycles (Dantas et al., 2008; Kotzerke et al., 2008; Islas-Espinoza et al., 2012).

In general, temperature and incubation conditions are the main environmental factors determining biodegradation rates of manure-associated antibiotics because they significantly impact the activity of microorganisms (Wang et al., 2006). The optimum temperature for biodegradation of antibiotic ceftiofur was reported to be between 35 and 45 °C, which is the most suitable temperature for growth of active microorganisms in various environmental media (Li et al., 2011). A few laboratory incubation studies assessed the aerobic and anaerobic biodegradability of veterinary antibiotics and revealed that their biodegradation under anaerobic conditions was much slower compared with that under aerobic conditions (Ingerslev et al., 2001; Yang et al., 2009). Some studies also showed that anaerobic conditions could inhibit the biotransformation of some veterinary antibiotics (Sun et al., 2014; Feng et al., 2017). In contrast, anaerobic dissipation of the antibiotics sulfamethoxazole and trimethoprim through biodegradation was observed more rapid than aerobic dissipation (Wu et al., 2012c).

Abiotic Transformations

The main abiotic transformation mechanisms for veterinary antibiotics include hydrolysis, photolysis, and/or redox conversion. For most antibiotics, the processes of abiotic transformation are driven by their physicochemical properties (e.g., molecular structure, solubility, speciation, and hydrophobicity) and environmental conditions (e.g., pH and temperature) (Jeon et al., 2014). Hydrolysis is one of the most common reactions controlling abiotic degradation, and hydrolysis rates of organic compounds are often pH dependent (Zhang et al., 2015). Generally, most veterinary antibiotics are hydrolytically stable under typical environmental conditions. Some antibiotics such as β-lactams and sulphonamides are susceptible to hydrolysis under certain conditions (Li et al., 2011; Bialk-Bielińska et al., 2012; Braschi et al., 2013). For example, a study revealed that an acidic solution was most favorable for hydrolysis of sulphonamide antibiotics, followed by neutral and alkaline solutions (Bialk-Bielińska et al., 2012). Furthermore, manure or manure-containing wastewater carries dissolved organic matter and metals, which can catalyze hydrolytic reactions of veterinary antibiotics in water and soils. For instance, a study revealed that the hydrolytic transformation of β-lactam antibiotics were promoted by ferric ions [Fe(III)] (Chen et al., 2017). The Fe(III)-catalyzed hydrolysis likely occurred via complexation of β-lactam antibiotics with carboxyl group and tertiary nitrogen, and then enhancing the hydrolytic cleavage of the β-lactam ring (Chen et al., 2017).

Veterinary antibiotics may undergo direct photolysis in sunlight-exposed water or soils (Werner et al., 2009; Bonvin et al., 2013; Niu et al., 2013; Conde-Cid et al., 2018). Photodegradation of antibiotics may be minor compared with other degradations such as biodegradation and hydrolysis. Especially in soils, sunlight attenuation with depth in soils and antibiotic sorption to soils may significantly impede photolysis (Balmer et al., 2000). However, some veterinary antibiotics such as chlortetracycline and oxytetracycline were found to be vulnerable to photolysis via sunlight (Werner et al., 2009; Xuan et al., 2010; Conde-Cid et al., 2018), suggesting that the photochemical loss of these antibiotics is a potentially important process in their environmental fate. In addition to direct photolysis, photocatalytic degradation often plays an important role in manure-associated antibiotics because of the presence of photosensitizers in manure (Sukul et al., 2008). The photosensitizers can markedly accelerate the photo-degradation of antibiotics, indicating this indirect photolytic process is an important driver in reducing antibiotic persistence in soils or aqueous environments (Sukul et al., 2008; Niu et al., 2013).

In addition to hydrolysis and photolysis, other abiotic transformations may occur for veterinary antibiotics in soils amended with animal waste because manure contains large amounts of organic matters that possess a variety of functional groups such as amines, carboxyls, carbonyls, ethers, hydroxyls, nitriles, nitrosos, sulfides, and sulfonyls. These functional groups can react with veterinary antibiotics via redox conversion and/or radical reactions and thereby bind chemical contaminants in soils (Song and Guo, 2014).

Comprehensive Degradations

Degradation of veterinary antibiotics in agricultural soils or aqueous environments is a comprehensive result of microbial degradation, hydrolysis, photolysis, oxidation–reduction, and other chemical reactions. Many factors influence the rates and extent of antibiotic degradation, including structure and type of chemicals, physicochemical properties of manure, microbial community, nutrients, electronic acceptors, oxidation–reductive potential, and various environmental conditions (e.g., pH, temperature, or light) (Song and Guo, 2014). All these biotic and abiotic factors may contribute simultaneously to the antibiotic degradation via synergistic or antagonistic effects. Furthermore, the major degradation processes and mechanisms of veterinary antibiotics may vary under differing environmental conditions. Hydrolysis and direct photolysis were reported as the primary processes for degradation of four cephalosporin antibiotics (cefradine, cefuroxime, cefepime, and ceftriaxone) in the surface water of a lake, whereas biodegradation was responsible for their degradation in the sediment (Jiang et al., 2010). The variation in predominant degradation processes and mechanisms may result in different metabolites. For example, our previous study showed that the primary biodegradation of antibiotic ceftiofur in animal wastewater was the cleavage of the β-lactam ring, with cef-aldehyde as a main degradation product (Li et al., 2011). By contrast, the hydrolysis of ceftiofur generated two products (desfuroylceftiofur and furoic acid) and exhibited a different degradation pathway as described below (Li et al., 2011):

Unlike the parent compound, ceftiofur, and the hydrolysis product, desfuroylceftiofur, the biodegradation product cef-aldehyde does not contain a β-lactam ring and has less antimicrobial activity. This study indicated that biodegradation of ceftiofurin in animal wastewater was a detoxification process, but not for its hydrolysis (Li et al., 2011). In addition, the degradation kinetics of ceftiofur in the manure-containing wastewater was much more rapid compared with the corresponding sterile solutions, indicating that biodegradation could dominate the dissipation of this antibiotic in the water and soils containing CAFO waste (Li et al., 2011).

In most cases, the degradation of veterinary antibiotics can convert parent compounds to more water-soluble metabolites (Bradford et al., 2008). These degradation products may be less toxic than their parent compounds, but certain metabolites raise an equal or even greater health concern. Transformation products of antibiotic cephalosporins have been demonstrated to be more toxic and more persistent than the parent compounds (Ribeiro et al., 2018). Most current studies focus on the toxicity and environmental fate of parent antibiotics; relatively few investigations consider their degradation products. Consequently, greater efforts on ecotoxicological data generation and verification of the biological inactivation of metabolites are needed to comprehensively evaluate the environmental fates of antibiotics. In addition, knowledge about the mixture effects of multiple antibiotics and their degradation products on toxicity is also scarce.

Transport of Manure-Associated Antibiotics

Occurrence of Antibiotics in the Environment Derived from Manure Application

Applications of manure or manure-containing wastewater to agricultural fields serve as an important pathway to disseminate veterinary antibiotics and their metabolites into the soils. In addition to biotic and abiotic transformation/degradation, these manure-associated antibiotic contaminants may migrate from manure storage and application sites to surface water and groundwater through runoff and leaching (Fig. 1), and ultimately result in adverse impacts on terrestrial and aquatic agro-ecosystems (Kuchta et al., 2009; Popova et al., 2013; Sura et al., 2015). Numerous studies have reported that residual antibiotics can enter and contaminate surface water and groundwater after manure land application (Burkhardt et al., 2005; Kuchta et al., 2009; Spielmeyer et al., 2017). For example, a study investigated several large-scale swine and poultry feeding facilities and found that antibiotics were detected in 67% of surface and groundwater sources within close proximity to the farms (Campagnolo et al., 2002). Moreover, a long-term monitoring study revealed that soils amended with sulfonamides-fortified liquid manure were a long-time source for the transfer of the antibiotics into groundwater (Spielmeyer et al., 2017). Although veterinary antibiotics have been widely detected in the aquatic environment after land application of animal wastes, currently they are neither regulated nor monitored using maximum contaminant levels in the Safe Drinking Water Act (Bradford et al., 2008).

A recent report showed that soil subsurface injection of manure could reduce antibiotic surface runoff from manure-applied fields (Le et al., 2018). Accordingly, they recommended that using subsurface injection and avoiding manure application less than 3 d before rain would be a best management practice for manure land application (Le et al., 2018). However, it is uncertain if this subsurface injection practice can result in loss of antibiotics via leaching. In addition to manure application methods, many other factors influence the transport of manure-associated antibiotics from fields to nearby watersheds (e.g., environmental conditions, soil physicochemical properties, soil microorganisms, and antibiotic sorption capacities on soils) (Hu and Coats, 2009; Strauss et al., 2011; Spielmeyer et al., 2017).

Sorption of Veterinary Antibiotics

Sorption and desorption are crucial processes affecting the environmental fate and transport of veterinary antibiotics in soils, including their entry into aquatic systems and their availability for degradation. In general, chemical contaminants that are strongly adsorbed to soil can be retained in upper soil zones, whereas highly mobile chemicals have the potential to leach into groundwater. Studies of sorption–desorption for most common antibiotics in soils have been well documented (Wehrhan et al., 2010; Zhang et al., 2014; Pollard and Morra, 2018). In soils, antibiotic contaminants interact with clay minerals and soil organic matter (SOM), resulting in their sorption, binding, and fixation in the soil matrix. The strength of the interaction is primarily determined by the chemical structure and soil properties, as well as soil conditions (e.g., temperature and moisture) (Song and Guo, 2014). In general, sorption kinetics of antibiotics on soils are a rapid process; and they can approach their maximum sorption capacities in a short time. For example, more than 95% of the chlortetracycline sorption to soils occurred within 10 min and 95% of the typsoin sorption occurred within 3 h (Allaire et al., 2006).

The mobility and transport of antibiotic contaminants in the environment depend on their sorption capacities, which are typically predicted by extrapolating from an experimentally determined soil–water distribution coefficient (K_d). The K_d value represents the distribution of a compound between water and soil compartments. A low value of K_d means that a chemical is highly mobile in soils. For example, veterinary antibiotic sulfonamides exhibited much lower K_d values (0.03–0.47) than trimethoprim antibiotics (6.73–9.21) (Zhang et al., 2014), suggesting the former has a high migration potential to the groundwater, especially in the soil with low organic carbon and high pH (Wang et al., 2015). For most hydrophobic antibiotics, adsorption occurs predominantly by partitioning into SOM (Tolls, 2001). Thus, the soil organic carbon-water partitioning coefficient (Koc) often serves as a more useful and accurate parameter to estimate the sorption capacities of hydrophobic antibiotics and their environmental distribution. The

Koc is the ratio of the mass of a chemical that is adsorbed in the soil per unit mass of organic carbon in the soil per the equilibrium chemical concentration in solution. Song and Guo (2014) summarized the Koc values for several common veterinary antibiotics. Antibiotics with high values of Koc can strongly adsorb onto soil and are less likely to leach or become surface runoff. However, the potential for adverse effects on terrestrial organisms should be considered for antibiotics with very high Koc (e.g., log Koc > 4.5). For example, the most heavily used tetracycline antibiotics with high Koc values (27,800–93,300 L/kg) can strongly adsorb in soil, resulting in their long-term persistence in the environment (Tolls, 2001; Kong et al., 2012; Popova et al., 2017; Pollard and Morra, 2018).

For hydrophilic or ionizable antibiotics, partitioning into SOM may not be a dominant adsorption mechanism. Instead, soil minerals play a vital role in the transport and retention of these veterinary antibiotics in soil. Sulfonamides are ionizable and polar veterinary antibiotics. Clay minerals could strongly adsorb sulfathiazole cations, followed by neutral species; while ferrihydrite as a specific cation mineral could interact with an anion antibiotic species (Kahle and Stamm, 2007). A number of interaction mechanisms exist between antibiotics and soil minerals, including van der Walls interaction, electrostatic attraction, cation exchange, cation bridging at clay surfaces, surface complexation, and hydrogen bonding (Song and Guo, 2014). Cation exchange was reported as a major contributor to the sorption of cationic antibiotics (e.g., enrofloxacin) on clay minerals (Yan et al., 2012). Furthermore, the sorption capacity of antibiotics to soil minerals is affected by soil pH, soil solution ionic strength, and types of exchangeable cations. For example, the exchangeable cations in soils that significantly influence the sorption of ciprofloxacin and its sorption capacities decrease in the order of Na→Ca→Al-montmorillonite (Wu et al., 2012a). These sorption capacities are accounted for by organic carbon normalization, suggesting that Koc values are conceptually inappropriate to describe the sorption behaviors for hydrophilic or ionizable antibiotics.

Currently, most of the sorption processes of antibiotic contaminants have been demonstrated by simple simulated studies to investigate their environmental fate and transport. When in actuality, sorption and desorption processes of veterinary antibiotics in the environment are often coupled with other processes (e.g., biotic and abiotic degradation) and thus are difficult to distinguish. In soils, the sorption of antibiotics may influence their biodegradation since it reduces their bioavailability for soil microorganisms (Wegst-Uhrich et al., 2014; Gothwal and Shashidhar, 2015). In addition, little research attention has been given to the sorption behaviors of degradation products compared with those parent antibiotics in soils (Sittig et al., 2014).

Colloid-Facilitated Transport of Manure-Associated Antibiotics
In general, the occurrence of veterinary antibiotics in the environment is often associated with colloids. Colloids are defined as mineral and organic particles with a size between 1 nm and 10 μm in diameter. Dissolved organic matter (DOM), the major component of colloids, can affect the sorption of organic compounds in soils (Kretzschmar et al., 1999; Chefetz et al., 2008; Zou and Zheng, 2013; Dodgen and Zheng, 2016). In some cases, DOM can enhance the sorption of chemical contaminants in soils because of cosorption and cumulative sorption to the soil solid phases (Kretzschmar et al., 1999; Chefetz et al., 2008). However, some studies showed that DOM decreased the sorption of organic contaminants in soils and therefore enhanced their mobility (Kan and Tomson, 1990;

Martinez-Hernandez et al., 2014). Manure and manure-containing wastewater contain large volumes of colloids, including DOM, and the land application of animal wastes can load colloids and antibiotic residues simultaneously into the environment. Colloid-facilitated transport has been illustrated in numerous laboratory and field experiments for organic chemicals such as polycyclic aromatic hydrocarbons (PAHs), pesticides, and other hydrophobic organic contaminants (Kan and Tomson, 1990; Nelson et al., 2000; Ter Laak et al., 2009). However, few studies have been conducted on colloid interactions with antibiotic contaminants. The interaction of effluent-borne colloids and veterinary antibiotics may facilitate

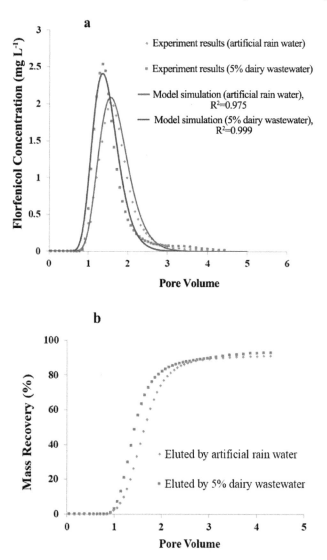

Fig. 2. (a) Florfenicol breakthrough curves from the saturated soil column. (b) Total florfenicol concentration eluting by 5% dairy manure-containing wastewater.

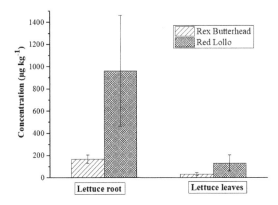

Fig. 3. Concentrations of sulfamethoxazole in lettuce roots and leaves (dry weight) after 3 wk of exposure to nutrient solution containing antibiotic at 0.5 µg L-1. Standard derivation of triplicate samples is shown as error bars.

their transport in soils (Zou and Zheng, 2013). Failure to account for this colloid-facilitated transport may severely underestimate the downward transport of manure-associated contaminants in soils amended with animal wastes.

To evaluate the effect of manure-borne colloids on the transport of veterinary antibiotics, we conducted a column experiment to investigate the transport of the antibiotic florfenicol in soils irrigated with manure-containing wastewater. The detailed experimental conditions and procedure can be found in our previous soil column study (Zou and Zheng, 2013). Flofenicol is a weakly hydrophobic antibiotic with low log Kow (-0.04) and Koc (24–52 L kg⁻¹) values that tends to be less bound to soils (McCarthy and Zachara, 1989). The soil column experiment showed that the breakthrough of florfenicol occurred in about one pore volume (Fig. 2a). This result further confirmed that florfenicol sorption on the soil was not very strong and was readily mobile (Zou and Zheng, 2013). As shown in Fig. 2a, the emergence peak of florfenicol breakthrough was facilitated when 5% dairy wastewater was used as an elution solution compared to artificial rainwater. Moreover, the peak concentration of florfenicol was 23% higher when eluting with dairy wastewater instead of simulated rainwater. These results clearly indicate that the transport of florfenicol can be facilitated by manure-borne colloids. Total florfenicol recovery from both elution solutions was high (> 90%) (Fig. 2b), suggesting this antibiotic is recalcitrant to degradation in soils and has a high potential for leaching or runoff. In previous studies, significantly facilitated transport was reported for hydrophobic organic compounds (Kan and Tomson, 1990; Grolimund and Borkovec, 2005; Ter Laak et al., 2009). By contrast, the facilitated transports were believed negligible for weakly hydrophobic organic compounds as they were less retarded due to weak adsorption in soils (Kan and Tomson, 1990). In this study, the manure wastewater contains an abundance of colloidal particles (e.g., DOM), and florfenicol might be preferentially bound to DOM. Moreover, Pagliari et al. (2011) found that dairy manure caused water soil repellency, which could be coating on soil particles and making more difficult for soils to hold colloidal particles. Thus, the transport of colloids in soils could carry over florfenicol and thereby facilitate the antibiotic transport. A two-site nonequilibrium adsorption model (Zou and Zheng, 2013) was used to simulate the florfenicol transport processes in the soil columns eluted by dairy wastewater and simulated rainwater. The model fits the breakthrough curves fairly well ($R^2 > 0.975$) (Fig. 2a).

Uptake and Accumulation of Veterinary Antibiotics in Crop Plants

Uptake and Accumulation of Antibiotics under a Hydroponic System

Since antibiotic residues have been frequently detected in the aquatic environment, we conducted a greenhouse experiment to investigate the possible contamination of food plants grown in antibiotic-containing water. Two lettuce (*Lactuca sativa* L.) cultivars (Green Rex Butterhead and Red Lollo) were grown in a continuously aerated nutrient solution containing the antibiotic sulfamethoxazole at a low level (0.5 µg L^{-1}). The detailed experimental procedure and greenhouse conditions can be found in our previous report (Zheng et al., 2014). After 3 wk of exposure, the accumulated concentration of sulfamethoxazole in lettuce roots was as high as 960 µg kg^{-1} (dry weight) for Red Lollo (Fig. 3), suggesting that this antibiotic has a strong potential to be taken up and accumulated in plant roots, even when exposed to low levels of contamination. Meanwhile, sulfamethoxazole was detected in the lettuce leaves at concentrations ranging from 32.7 to 132.6 µg kg^{-1} (Fig. 3). This result suggests that this antibiotic can be readily translocated from plant roots to leaves via water transpiration, thereby accumulating in plant leaves. Accumulation of antibiotics into lettuce leaves is an issue of food security as humans may be exposed to these contaminants through dietary consumption of edible plant parts. A previous study showed that not all organic contaminants could be detected in plant leaves because they could not be easily taken up by roots or readily transferred from roots to leaves (Zheng et al., 2014). Sulfamethoxazole is predominantly an unionized compound in neutral pH solutions and has low hydrophobicity with a relatively low log Kow (0.89), implicating its high potential of passing-through plant cell membranes and migrating to lettuce leaves via water transpiration (Miller et al., 2016). It is consistent with the previous study, which reveals that polar organic compounds have a great potential for root uptake and translocation (Zhang et al., 2013; Zheng et al., 2014). Considering the relatively high accumulation of sulfamethoxazole in both lettuce roots and leaves, this compound might be investigated as a potential marker for antibiotic uptake and accumulation in leafy vegetables.

Plant Uptake of Veterinary Antibiotics from Manure-Amended Soil

Agricultural application of manure and manure-containing wastewater can provide nutrients and organic matter, which improve plant growth and reduce chemical fertilizer needs. Meanwhile, manure-associated antibiotics could be introduced into field soils through this common agricultural practice. When crop plants are cultivated in manure-amended soils, antibiotic residues could be taken up by roots from the soil and accumulate in plants (Fig. 1) (Kumar et al., 2005b; Kang et al., 2013; Tasho and Cho, 2016; Wang et al., 2016; Chung et al., 2017; Pan and Chu, 2017). The antibiotics accumulated in crops exhibited negative effects on plant growth (Ahmed et al., 2015). Moreover, the antibiotic residues accumulated in food plants growing in manure-amended soils pose food safety and human health concerns. Previous studies have clearly shown that food plants were able to take up antibiotics and accumulate the contaminants in edible parts, indicating that these contaminants may enter food chains (Zheng et al., 2014; Tasho and Cho, 2016; Pan and Chu, 2017). The extents of uptake, internal transfer, and accumulation of antibiotic contaminants in plants are likely to be associated with compound properties and concentrations, plant species and cultivars, soil characteristics, growth conditions, and plant development stages (Tasho and Cho, 2016; Pan and Chu, 2017). The processes and mechanisms of root uptake of emerging organic contaminants including antibiotics have been well illustrated, especially in crops irrigated with reclaimed wastewater or grown in manure-amended soils (Miller et al., 2016).

However, some studies indicated that the potential acute risk of veterinary antibiotics to public health through dietary uptake was negligible, since detected concentrations of antibiotics in the edible parts of food plants were much lower than the acceptable daily intake (Dodgen et al., 2013; Kang et al., 2013; Ahmed et al., 2015). Concentrations of veterinary antibiotic residues in soil receiving animal manure are usually lower than the levels tested in most greenhouse or field trial studies (Song and Guo, 2014). Moreover, sorption of manure-associated antibiotics to the soil solid matrix makes the contaminants much less available for plant uptake (Kang et al., 2013). Thus, it is not clear if the accumulation of veterinary antibiotics in food crops receiving manure application indeed poses a health hazard to consumers. Unlike for pesticides, regulation criteria do not exist for monitoring the levels of veterinary antibiotic residues in crops. Thus, it is difficult to quantify their hazardous nature. Previous simple estimations of acute toxicity based on a few compounds and plant types do not address chronic toxicity endpoints and may not encompass all possible human health effects (Wu et al., 2012b). The potential accumulation of veterinary antibiotics in plants over time always raises public concern about animal waste application. Therefore, safe and effective mitigation strategies need to be developed to reduce the loading of these contaminants into agricultural fields.

Mitigation Strategies for Reducing Antibiotic Contamination

Mitigation Strategies

Control strategies are needed to minimize the adverse effects of veterinary antibiotics on ecological and human health resulting from animal manure land application. A series of mitigation strategies were proposed to limit the occurrence of antibiotic contaminants in agroecosystems (Kuppusamy et al., 2018). One strategy is to optimize or control the usage of antibiotics as feed additives to reduce their excretion from animal farms, for example, developing improved nutritional programs to maintain good animal health and thereby reduce antibiotic usage (Kuppusamy et al., 2018).

A second strategy is to eliminate antibiotic residues in animal wastes prior to their land disposal or agricultural applications. Previous studies found that antibiotic residues derived in animal farms could be degraded during manure anaerobic digestion or composting (Arikan et al., 2009; Selvam et al., 2012). Composting is an economical and feasible practice to stabilize nutrients and reduce pathogens and odors (Selvam et al., 2012). The composting processes may help to decompose certain amounts of antibiotic residues, but cannot completely avoid their release into the environment (Selvam et al., 2012; Tasho and Cho, 2016). Some advanced treatment technologies such as physical adsorption, membrane separation, and advanced chemical oxidation have been developed to remove many emerging contamaints including antibiotics from municipal sewage effluents (Ternes et al., 2003; Pronk et al., 2006). However, these advanced treatment technologies are too costly to be implemented at animal farms, even at large-scale operations such as CAFOs.

A third strategy is to develop best management practices for mitigating manure-associated antibiotics in the environment after animal wastes are applied to crop fields. Several practices that were suggested to reduce pharmaceuticals and personal care products in agricultural soils irrigated with reclaimed water (Qin et al., 2015) can be used to reduce the negative impact of veterinary antibiotics derived from animal wastes. These practices include (i) irrigating with manure-containing wastewater on soils with high organic matters; (ii) applying animal manure along with inorganic fertilizers to promote microbial degradation of antibiotic residues or to reduce their bioavailability; (iii) avoiding the use of manure or manure-containing wastewater during harvest seasons, especially for food plants; and (iv) avoiding direct contact of animal wastes with edible parts of crops, for example, minimizing spray irrigation manure-containing wastewater for aerial vegetables.

Removal of Manure-Associated Antibiotics by Oil Capture

Since many antibiotics are hydrophobic, an innovative treatment technique has been developed to remove antibiotic contaminants from CAFO's wastewater using oil extraction. Recently, we successfully captured estrogenic hormones by vegetable oils from manure-containing wastewater before it was used for field irrigation (Dodgen et al., 2018). Compared with activated carbon adsorption, the most common and effective contaminant treatment technique, the oil extraction approach has demonstrated at least three advantages: (i) after treatment, the activated carbon is usually shipped to a landfill. However, the treated oil can be recycled as a biofuel to harvest bioenergy, a process also eliminating the adsorbed contaminants; (ii) activated carbon simultaneously sorbs organic contaminants and desirable inorganic nutrients, while oils extract only the hydrophobic contaminants. Thus, after manure-containing wastewater is treated with oil, the treated water can still provide nutrients (e.g.,

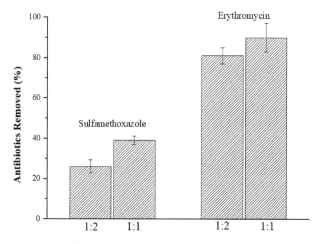

Fig. 4. Removal capacity of sulfamethoxazole and erythromycin from manure-containing wastewater by corn oil. Standard derivation of triplicate samples is shown as error bars.

ammonium ions) to crop plants; and (iii) activated carbon is fairly expensive, ranging from $1,500 to $3,000 per tonne, while soybean or corn oil is typically < $1,000 per tonne and used oil is even cheaper (i.e., < $500 per tonne). Thus, oil extraction is an economically feasible treatment to remove hydrophobic contaminants.

We have conducted a laboratory experiment to capture veterinary antibiotics from manure-containing wastewater using commercial corn oil. The detailed experimental procedure can be our previous study (Dodgen et al., 2018). The capacities of the corn oil to capture sulfamethoxazole and erythromycin from antibiotic-spiked water are shown in Fig. 4. The removal efficiencies of sulfamethoxazole were 26% and 39% when ratios of the oil to water (v/v) were 1:2 and 1:1, respectively (Fig. 4). For the antibiotic erythromycin, the removal capacities were much higher than those of sulfamethoxazole, and the removal efficiency was up to more than 90% at the 1:1 oil/water ratio (v/v) (Fig. 4). This high efficiency could be attributable to the higher log Kow value of erythromycin (3.06) than that of sulfamethoxazole (0.89). Our previous study reported that vegetable oils, including corn oil, could efficiently remove hydrophobic estrogenic compounds from manure-containing water, even when the ratio of oil to water was as low as 1:100 (v/v) (Dodgen et al., 2018). The present study confirmed that oil extraction could be used as a feasible and effective treatment approach to remove hydrophobic veterinary antibiotics from manure-containing wastewater. For low hydrophobic or hydrophilic antibiotics, multiple oil-capture steps may be needed to reach satisfactory removal efficiency, yet the involved treatment costs will increase correspondingly.

Conclusions

Veterinary antibiotics have been used worldwide to protect the health of farm animals by treating and preventing diseases. They have also been supplemented into animal diets to improve growth rate and feed efficiency. The extensive use of veterinary antibiotics in CAFOs has, however, resulted in large amounts of parent compounds and their metabolites remaining in animal manure and manure-containing wastewater. The concentrations of common veterinary antibiotics in swine, poultry, dairy, and beef lagoon water were reported in the range of less than 0.01 to 1340 $\mu g\ L^{-1}$, depending on the type of antibiotics and the lagoon sources. Antibiotic residues in most animal manure varied from trace levels to as high as 765 mg kg^{-1}. These manure-associated antibiotics and active metabolites can enter into the environment by land application of animal wastes. Although it has been recognized that the worldwide occurrence of veterinary antibiotics in the environment elevates the crisis of antimicrobial resistance and poses a critical risk on public health, there are few surveillance programs, legal regulations, and treatment practices to control or limit the loading of these contaminants into agroecosystems.

In animal wastes and manure-amended soils, antibiotic contaminants undergo biotic degradation and abiotic transformation such as hydrolysis, photolysis, redox conversion, and radical reaction. In soils, manure-associated antibiotics primarily interact with SOM and soil minerals, thereby limiting their availability for degradation and transport. Nevertheless, our soil column study showed that the presence of colloids derived from animal manure facilitated the antibiotic florfenicol transports. In addition, antibiotic residues can be readily taken up by the plant roots and translocated to edible plant parts. However, some studies suggested that the potential acute risk to public health through dietary uptake was negligible because accumulated concentrations of antibiotics in food plants were much lower than the acceptable daily intake. Therefore, it is still unclear if the accumulation of veterinary antibiotics in food crops receiving manure application indeed poses a health hazard to consumers. To minimize the adverse effects of veterinary antibiotics on the agroecosystem and public health resulting from animal manure land application, effective and feasible mitigation strategies are needed. The innovative oil capture technique might be an effective method to remove hydrophobic veterinary antibiotics from manure-containing wastewater.

References

Ahmed, M.B.M., A.U. Rajapaksha, J.E. Lim, N.T. Vu, I.S. Kim, H.M. King, S.S. Lee, and Y.S. Ok. 2015. Distribution and accumulative pattern of tetracyclines and sulfonamides in edible vegetables of cucumber, tomato, and lettuce. J. Agric. Food Chem. 63:398–405. doi:10.1021/jf5034637

Allaire, S.E., J. Del Castillo, and V. Juneau. 2006. Sorption kinetics of chlortetracycline and tylosin on sandy loam and heavy clay soisl. J. Environ. Qual. 35:969–972. doi:10.2134/jeq2005.0355

Arikan, O.A., W. Mulbry, and C. Rice. 2009. Management of antibiotic residues from agricultural sources: Use of composting to reduce chlortetracycline residues in beef manure from treated animals. J. Hazard. Mater. 164: 483–489. doi:10.1016/j.jhazmat.2008.08.019.

Balmer, M.E., K.-U. Goss, and R.P. Schwarzenbach. 2000. Photolytic transformation of organic pollutants on soil surfaces: An experimental approach. Environ. Sci. Technol. 34:1240–1245. doi:10.1021/es990910k

Bialk-Bielińska, A., S. Stolte, M. Matzke, A. Fabianska, J. Maszkowska, M. Kolodziejska, B. Liberek, P. Stepnowski, and J. Kumirska. 2012. Hydrolysis of sulphonamides in aqueous solutions. J. Hazard. Mater. 221–222:264–274. doi:10.1016/j.jhazmat.2012.04.044

Bonvin, F., J. Omlin, R. Rutler, W.B. Schweizer, P.J. Alaimo, T.J. Strathmann, K. McNeill, and T. Kohn. 2013. Direct photolysis of human metabolites of the antibiotic sulfamethoxazole: Evidence for abiotic back-transformation. Environ. Sci. Technol. 47:6746–6755. doi:10.1021/es303777k

Bradford, S.A., E. Segal, W. Zheng, Q.Q. Wang, and S.R. Hutchins. 2008. Reuse of concentrated animal feeding operation wastewater on agricultural lands. J. Environ. Qual. 37:S97–S115. doi:10.2134/jeq2007.0393

Braschi, I., S. Blasioli, C. Fellet, R. Lorenzini, A. Garelli, M. Pori, and D. Giacomini. 2013. Persistence and degradation of new beta-lactam antibiotics in the soil and water environment. Chemosphere 93:152–159. doi:10.1016/j.chemosphere.2013.05.016

Burkhardt, M., C. Stamm, C. Waul, H. Singer, and S. Muller. 2005. Surface runoff and transport of sulfonamide antibiotics and tracers on manured grassland. J. Environ. Qual. 34:1363–1371. doi:10.2134/jeq2004.0261

Campagnolo, E.R., K.R. Johnson, A. Karpati, C.S. Rubin, D.W. Kolpin, M.T. Meyer, J.E. Esteban, R.W. Currier, K. Smith, K.M. Thu, and M. McGeehin. 2002. Antimicrobial residues in animal waste and water resources proximal to large-scale swine and poultry feeding operations. Sci. Total Environ. 299:89–95. doi:10.1016/S0048-9697(02)00233-4

Chefetz, B., T. Mualem, and J. Ben-Ari. 2008. Sorption and mobility of pharmaceutical compounds in soil irrigated with reclaimed wastewater. Chemosphere 73:1335–1343. doi:10.1016/j.chemosphere.2008.06.070

Chen, J.B., Y. Wang, Y.J. Qian, and T.Y. Huang. 2017. Fe(III)-promoted transformation of beta-lactam antibiotics: Hydrolysis vs oxidation. J. Hazard. Mater. 335:117–124. doi:10.1016/j.jhazmat.2017.03.067

Chitescu, C.L., A.I. Nicolau, and A.A.M. Stolker. 2013. Uptake of oxytetracycline, sulfamethoxazole and ketoconazole from fertilised soils by plants. Food Addit. Contam., Part A. 30:1138–1146. doi:10.1080/19440049.2012.725479

Christou, A., A. Aguera, J.M. Bayona, E. Cytryn, V. Fotopoulos, D. Lambropoulou, C.M. Manaia, C. Michael, M. Revitt, P. Schroder, and D. Fatta-Kassinos. 2017. The potential implications of reclaimed wastewater reuse for irrigation on the agricultural environment: The knowns and unknowns of the fate of antibiotics and antibiotic resistant bacteria and resistance genes: A review. Water Res. 123:448–467. doi:10.1016/j.watres.2017.07.004

Chung, H.S., Y.J. Lee, M.M. Rahman, A.M. Abd El-Aty, H.S. Lee, M.H. Kabir, S. Kim, B.J. Park, J.E. Kim, F. Hacimuftuoglu, N. Nahar, H.C. Shin, and J.H. Shim. 2017. Uptake of the veterinary antibiotics chlortetracycline, enrofloxacin, and sulphathiazole from soil by radish. Sci. Total Environ. 605–606:322–331. doi:10.1016/j.scitotenv.2017.06.231

Conde-Cid, M., D. Fernandez-Calvino, J.C. Novoa-Munoz, M. Arias-Estevez, M. Diaz-Ravina, M.J. Fernandez-Sanjurjo, A. Nunez-Delgado, and E. Alvarez-Rodriguez. 2018. Biotic and abiotic dissipation of tetracyclines using simulated sunlight and in the dark. Sci. Total Environ. 635:1520–1529. doi:10.1016/j.scitotenv.2018.04.233

Cromwell, G.L. 2002. Why and how antibiotics are used in swine production. Anim. Biotechnol. 13:7–27. doi:10.1081/ABIO-120005767

Dantas, G., M.O.A. Sommer, R.D. Oluwasegun, and G.M. Church. 2008. Bacteria subsisting on antibiotics. Science 320:100–103. doi:10.1126/science.1155157

Dodgen, L.K., J. Li, D. Parker, and J.J. Gan. 2013. Uptake and accumulation of four PPCP/EDCs in two leafy vegetables. Environ. Pollut. 182:150–156. doi:10.1016/j.envpol.2013.06.038

Dodgen, L.K., and W. Zheng. 2016. Effects of reclaimed water matrix on fate of pharmaceuticals and personal care products in soil. Chemosphere 156:286–293. doi:10.1016/j.chemosphere.2016.04.019

Dodgen, L.K., K.N. Wiles, J. Deluhery, N. Rajagopalan, N. Holm, and W. Zheng. 2018. Removal of estrogenic hormones from manure-containing water by vegetable oil capture. J. Hazard. Mater. 343:125–131. doi:10.1016/j.jhazmat.2017.08.074

Dolliver, H.A.S. 2007. Fate and transport of veterinary antibiotics in the environment. Ph.D. thesis. Univ. of Minnesota, St. Paul, MN.

Dolliver, H., and S. Gupta. 2008. Antibiotic losses in leaching and surface runoff from manure-amended agricultural land. J. Environ. Qual. 37:1227–1237. doi:10.2134/jeq2007.0392

Dolliver, H., K. Kumar, and S. Gupta. 2007. Sulfamethazine uptake by plants from manure-amended soil. J. Environ. Qual. 36:1224–1230. doi:10.2134/jeq2006.0266

Du, L.F., and W.K. Liu. 2012. Occurrence, fate, and ecotoxicity of antibiotics in agro-ecosystems. A review. Agron. Sustain. Dev. 32:309–327. doi:10.1007/s13593-011-0062-9

European Commission. 2005. Ban on antibiotics as growth promoters in animal feed enters into effect. Europa press release, 22 December. europa.eu/rapid/press-release_IP-05-1687_en.htm (Accessed 1 Apr. 2019).

Feng, L., M.E. Casas, L.D.M. Ottosen, H.B. Moller, and K. Bester. 2017. Removal of antibiotics during the anaerobic digestion of pig manure. Sci. Total Environ. 603–604:219–225. doi:10.1016/j.scitotenv.2017.05.280

Gothwal, R., and T. Shashidhar. 2015. Antibiotic pollution in the environment: A review. Clean: Soil, Air, Water 43:479–489. doi:10.1002/clen.201300989

Grolimund, D., and M. Borkovec. 2005. Colloid-facilitated transport of strongly sorbing contaminants in natural porous media: Mathematical modeling and laboratory column experiments. Environ. Sci. Technol. 39:6378–6386. doi:10.1021/es050207y

Hu, D.F., and J.R. Coats. 2009. Laboratory evaluation of mobility and sorption for the veterinary antibiotic, tylosin, in agricultural soils. J. Environ. Monit. 11:1634–1638. doi:10.1039/b900973f

Ingerslev, F., L. Torang, M.L. Loke, B. Halling-Sorensen, and N. Nyholm. 2001. Primary biodegradation of veterinary antibiotics in aerobic and anaerobic surface water simulation systems. Chemosphere 44:865–872. doi:10.1016/S0045-6535(00)00479-3

Islas-Espinoza, M., B.J. Reid, M. Wexler, and P.L. Bond. 2012. Soil bacterial consortia and previous exposure enhance the biodegradation of sulfonamides from pig manure. Microb. Ecol. 64:140–151. doi:10.1007/s00248-012-0010-5

Jeon, D.S., T.K. Oh, M. Park, D.S. Lee, Y.J. Lim, J.S. Shin, S.G. Song, S.C. Kim, Y. Shinogi, and D.Y. Chung. 2014. Reactions and behavior relevant to chemical and physical properties of various veterinary antibiotics in soil. J. Fac. Agric. Kyushu Univ. 59:391–397.

Jiang, M.X., L.H. Wang, and R. Ji. 2010. Biotic and abiotic degradation of four cephalosporin antibiotics in a lake surface water and sediment. Chemosphere 80:1399–1405. doi:10.1016/j.chemosphere.2010.05.048

Kahle, M., and C. Stamm. 2007. Time and pH-dependent sorption of the veterinary antimicrobial sulfathiazole to clay minerals and ferrihydrite. Chemosphere 68:1224–1231. doi:10.1016/j.chemosphere.2007.01.061

Kan, A.T., and M.B. Tomson. 1990. Gound-water transport of hydrophobic organic-compounds in the presence of dissolved organic-matter. Environ. Toxicol. Chem. 9:253–263. doi:10.1002/etc.5620090302

Kang, D.H., S. Gupta, C. Rosen, V. Fritz, A. Singh, Y. Chander, H. Murray, and C. Rohwer. 2013. Antibiotic uptake by vegetable crops from manure-applied soils. J. Agric. Food Chem. 61:9992–10001. doi:10.1021/jf404045m

Kappell, A.D., M.S. DeNies, N.H. Ahuja, N.A. Ledeboer, R.J. Newton, and K.R. Hristova. 2015. Detection of multi-drug resistant Escherichia coli in the urban waterways of Milwaukee, WI. Front. Microbiol. 6:336. doi:10.3389/fmicb.2015.00336

Kemper, N. 2008. Veterinary antibiotics in the aquatic and terrestrial environment. Ecol. Indic. 8:1–13. doi:10.1016/j.ecolind.2007.06.002

Kim, K.R., G. Owens, S.I. Kwon, K.H. So, D.B. Lee, and Y.S. Ok. 2011. Occurrence and environmental fate of veterinary antibiotics in the terrestrial environment. Water Air Soil Pollut. 214:163–174. doi:10.1007/s11270-010-0412-2

Koike, S., I.G. Krapac, H.D. Oliver, A.C. Yannarell, J.C. Chee-Sanford, R.I. Aminov, and R.I. Mackie. 2007. Monitoring and source tracking of tetracycline resistance genes in lagoons and groundwater adjacent to swine production facilities over a 3-year period. Appl. Environ. Microbiol. 73:4813–4823. doi:10.1128/AEM.00665-07

Kolpin, D.W., E.T. Furlong, M.T. Meyer, E.M. Thurman, S.D. Zaugg, L.B. Barber, and H.T. Buxton. 2002. Pharmaceuticals, hormones, and other organic wastewater contaminants in US streams, 1999-2000: A national reconnaissance. Environ. Sci. Technol. 36:1202–1211. doi:10.1021/es011055j

Kolz, A.C., T.B. Moorman, S.K. Ong, K.D. Scoggin, and E.A. Douglass. 2005. Degradation and metabolite production of tylosin in anaerobic and aerobic swine-manure lagoons. Water Environ. Res. 77:49–56. doi:10.2175/106143005x41618

Kong, W.D., C.G. Li, J.M. Dolhi, S.Y. Li, J.Z. He, and M. Qiao. 2012. Characteristics of oxytetracycline sorption and potential bioavailability in soils with various physical-chemical properties. Chemosphere 87:542–548. doi:10.1016/j.chemosphere.2011.12.062

Kotzerke, A., S. Sharma, K. Schauss, H. Heuer, S. Thiele-Bruhn, K. Smalla, B.M. Wilke, and M. Schloter. 2008. Alterations in soil microbial activity and N-transformation processes due to sulfadiazine loads in pig-manure. Environ. Pollut. 153:315–322. doi:10.1016/j.envpol.2007.08.020

Kretzschmar, R., M. Borkovec, D. Grolimund, and M. Elimelech. 1999. Mobile subsurface colloids and their role in contaminant transport. Adv. Agron. 66:121–193. doi:10.1016/S0065-2113(08)60427-7

Kuchta, S.L., A.J. Cessna, J.A. Elliott, K.M. Peru, and J.V. Headley. 2009. Transport of lincomycin to surface and ground water from manure-amended cropland. J. Environ. Qual. 38:1719–1727. doi:10.2134/jeq2008.0365

Kumar, K., S.C. Gupta, Y. Chander, and A.K. Singh. 2005a. Antibiotic use in agriculture and its impact on the terrestrial environment. Adv. Agron. 87:1–54. doi:10.1016/S0065-2113(05)87001-4

Kumar, K., S.C. Gupta, S.K. Baidoo, Y. Chander, and C.J. Rosen. 2005b. Antibiotic uptake by plants from soil fertilized with animal manure. J. Environ. Qual. 34:2082–2085. doi:10.2134/jeq2005.0026

Kuppusamy, S., D. Kakarla, K. Venkateswarlu, M. Megharaj, Y.E. Yoon, and Y.B. Lee. 2018. Veterinary antibiotics (VAs) contamination as a global agro-ecological issue: A critical view. Agric. Ecosyst. Environ. 257:47–59. doi:10.1016/j.agee.2018.01.026

Le, H.T.V., R.O. Maguire, and K. Xia. 2018. Method of dairy manure application and time before rainfall affect antibiotics in surface runoff. J. Environ. Qual. 47:1310–1317. doi:10.2134/jeq2018.02.0086

Li, X.L., W. Zheng, M.L. Machesky, S.R. Yates, and M. Katterhenry. 2011. Degradation kinetics and mechanism of antibiotic ceftiofur in recycled water derived from a beef farm. J. Agric. Food Chem. 59:10176–10181. doi:10.1021/jf202325c

Lin, W., J. Flarakos, Y. Du, W.Y. Hu, H.D. He, J. Mangold, S.K. Tanaka, and S. Villano. 2017. Pharmacokinetics, distribution, metabolism, and excretion of omadacycline following a single intravenous or oral dose of C-14-omadacycline in rats. Antimicrob. Agents Chemother. 61:e01784-16. doi:10.1128/AAC.01784-16

Martinez-Hernández, V., R. Meffe, S. Herrera, E. Arranz, and I. de Bustamante. 2014. Sorption/desorption of non-hydrophobic and ionisable pharmaceutical and personal care products from reclaimed water onto/from a natural sediment. Sci. Total Environ. 472:273–281. doi:10.1016/j.scitotenv.2013.11.036

McCarthy, J.F., and J.M. Zachara. 1989. Subsurface transport of contaminants: Mobile colloids in the subsurface environment may alter the transport of contaminants. Environ. Sci. Technol. 23:496–502. doi:10.1021/es00063a602

McEwen, S.A., and P.J. Fedorka-Cray. 2002. Antimicrobial use and resistance in animals. Clin. Infect. Dis. 34:S93–S106. doi:10.1086/340246

Miller, E.L., S.L. Nason, K.G. Karthikeyan, and J.A. Pedersen. 2016. Root uptake of pharmaceuticals and personal care product ingredients. Environ. Sci. Technol. 50:525–541. doi:10.1021/acs.est.5b01546

Moore, P.R., A. Evenson, T.D. Luckey, E. McCoy, C.A. Elvehjem, and E.B. Hart. 1946. Use of sulfasuxidine, streptothricin, and streptomycin in nutritional studies with the chick. J. Biol. Chem. 165:437–441.

Nelson, S.D., J. Letey, W.J. Farmer, C.F. Williams, and M. Ben-Hur. 2000. Herbicide application method effects on napropamide complexation with dissolved organic matter. J. Environ. Qual. 29:987–994. doi:10.2134/jeq2000.00472425002900030038x

Niu, J.F., L.L. Zhang, Y. Li, J.B. Zhao, S.D. Lv, and K.Q. Xiao. 2013. Effects of environmental factors on sulfamethoxazole photodegradation under simulated sunlight irradiation: Kinetics and mechanism. J. Environ. Sci. (China) 25:1098–1106. doi:10.1016/S1001-0742(12)60167-3

O'Flaherty, E., C.M. Borrego, J.L. Balcazar, and E. Cummins. 2018. Human exposure assessment to antibiotic-resistant Escherichia coli through drinking water. Sci. Total Environ. 616–617:1356–1364. doi:10.1016/j.scitotenv.2017.10.180

Pagliari, P., M. Flores-Mangual, B. Lowery, D. Weisenberger, and C. Laboski. 2011. Manure-induced soil-water repellency. Soil Sci. 176(11):576–581. doi:10.1097/SS.0b013e3182316c7e

Pan, M., and L.M. Chu. 2017. Transfer of antibiotics from wastewater or animal manure to soil and edible crops. Environ. Pollut. 231:829–836. doi:10.1016/j.envpol.2017.08.051

Paulson, J.A., T.E. Zaoutis, The Council on Environmental Health, and The Committee on Infectious Diseases. 2015. Nontherapeutic use of antimicrobial agents in animal agriculture: Implications for pediatrics. Pediatrics 136:E1670–E1677. doi:10.1542/peds.2015-3630

Pollard, A.T., and M.J. Morra. 2018. Fate of tetracycline antibiotics in dairy manure-amended soils. Environ. Rev. 26:102–112. doi:10.1139/er-2017-0041

Popova, I.E., D.A. Bair, K.W. Tate, and S.J. Parikh. 2013. Sorption, leaching, and surface runoff of beef cattle veterinary pharmaceuticals under simulated irrigated pasture conditions. J. Environ. Qual. 42:1167–1175. doi:10.2134/jeq2013.01.0012

Popova, I.E., R.D.R. Josue, S.P. Deng, and J.A. Hattey. 2017. Tetracycline resistance in semi-arid agricultural soils under long-term swine effluent application. J. Environ. Sci. Health B. 52:298–305. doi:10.1080/03601234.2017.1281639

Pronk, W., H. Palmquist, M. Biebow, and M. Boller. 2006. Nanofiltration for the separation of pharmaceuticals from nutrients in source-separated urine. Water Res. 40:1405–1412. doi:10.1016/j.watres.2006.01.038

Qin, Q., X.J. Chen, and J. Zhuang. 2015. The fate and impact of pharmaceuticals and personal care products in agricultural soils irrigated with reclaimed water. Crit. Rev. Environ. Sci. Technol. 45:1379–1408. doi:10.1080/10643389.2014.955628

Ribeiro, A.R., B. Sures, and T.C. Schmidt. 2018. Cephalosporin antibiotics in the aquatic environment: A critical review of occurrence, fate, ecotoxicity and removal technologies. Environ. Pollut. 241:1153–1166. doi:10.1016/j.envpol.2018.06.040

Sallach, J.B., S.L. Bartelt-Hunt, D.D. Snow, X. Li, and L. Hodges. 2018. Uptake of antibiotics and their toxicity following routine irrigation with contaminated water in different soil types. Environ. Eng. Sci. 35:887–896. doi:10.1089/ees.2017.0376

Sarmah, A.K., M.T. Meyer, and A.B.A. Boxall. 2006. A global perspective on the use, sales, exposure pathways, occurrence, fate and effects of veterinary antibiotics (VAs) in the environment. Chemosphere 65:725–759. doi:10.1016/j.chemosphere.2006.03.026

Selvam, A., Z.Y. Zhao, and J.W.C. Wong. 2012. Composting of swine manure spiked with sulfadiazine, chlortetracycline and ciprofloxacin. Bioresour. Technol. 126:412–417. doi:10.1016/j.biortech.2011.12.073

Sittig, S., R. Kasteel, J. Groeneweg, D. Hofmann, B. Thiele, S. Koppchen, and H. Vereecken. 2014. Dynamics of transformation of the veterinary antibiotic sulfadiazine in two soils. Chemosphere 95:470–477. doi:10.1016/j.chemosphere.2013.09.100

Song, W., and M. Guo. 2014. Residual veterinary pharmaceuticals in animal manures and their environmental behaviors in soils. In: Z. He and H. Zhang, editors, Applied manure and nutrient chemistry for sustainable agriculture and environment. Springer, New York, NY. p. 23–52. doi:10.1007/978-94-017-8807-6_2

Spielmeyer, A., H. Hoper, and G. Hamscher. 2017. Long-term monitoring of sulfonamide leaching from manure amended soil into groundwater. Chemosphere 177:232–238. doi:10.1016/j.chemosphere.2017.03.020

Strauss, C., T. Harter, and M. Radke. 2011. Effects of pH and manure on transport of sulfonamide antibiotics in soil. J. Environ. Qual. 40:1652–1660. doi:10.2134/jeq2010.0535

Sukul, P., M. Lamshoft, S. Kusari, S. Zuhlke, and M. Spiteller. 2009. Metabolism and excretion kinetics of C-14-labeled and non-labeled difloxacin in pigs after oral administration, and antimicrobial activity of manure containing difloxacin and its metabolites. Environ. Res. 109:225–231. doi:10.1016/j.envres.2008.12.007

Sukul, P., M. Lamshoft, S. Zuhlke, and M. Spiteller. 2008. Photolysis of C-14-sulfadiazine in water and manure. Chemosphere 71:717–725. doi:10.1016/j.chemosphere.2007.10.045

Sun, P.Z., C.H. Huang, and S.G. Pavlostathis. 2014. Inhibition and biotransformation potential of veterinary ionophore antibiotics under different redox conditions. Environ. Sci. Technol. 48:13146–13154. doi:10.1021/es503005m

Sura, S., D. Degenhardt, A.J. Cessna, F.J. Larney, A.F. Olson, and T.A. McAllister. 2015. Transport of three veterinary antimicrobials from feedlot pens via simulated rainfall runoff. Sci. Total Environ. 521–522:191–199. doi:10.1016/j.scitotenv.2015.03.080

Tasho, R.P., and J.Y. Cho. 2016. Veterinary antibiotics in animal waste, its distribution in soil and uptake by plants: A review. Sci. Total Environ. 563–564:366–376. doi:10.1016/j.scitotenv.2016.04.140

Tedeschi, L.O., D.G. Fox, and T.P. Tylutki. 2003. Potential environmental benefits of ionophores in ruminant diets. J. Environ. Qual. 32:1591–1602. doi:10.2134/jeq2003.1591

Teeter, J.S., and R.D. Meyerhoff. 2003. Aerobic degradation of tylosin in cattle, chicken, and swine excreta. Environ. Res. 93:45–51. doi:10.1016/S0013-9351(02)00086-5

Ter Laak, T.L., J.C.H. Van Eijkeren, F.J.M. Busser, H.P. Van Leeuwen, and J.L.M. Hermens. 2009. Facilitated transport of polychlorinated biphenyls and polybrominated diphenyl ethers by dissolved organic matter. Environ. Sci. Technol. 43:1379–1385. doi:10.1021/es802403v

Ternes, T.A., J. Stuber, N. Herrmann, D. McDowell, A. Ried, M. Kampmann, and B. Teiser. 2003. Ozonation: A tool for removal of pharmaceuticals, contrast media and musk fragrances from wastewater? Water Res. 37:1976–1982. doi:10.1016/S0043-1354(02)00570-5

Tolls, J. 2001. Sorption of veterinary pharmaceuticals in soils: A review. Environ. Sci. Technol. 35:3397–3406. doi:10.1021/es0003021

USEPA. 2013. Literature review of contaminants in livestock and poultry manure and implications for water quality. EPA 820-R-13-002. Office of Water, Washington, D.C.

Van Boeckel, T.P., C. Brower, M. Gilbert, B.T. Grenfell, S.A. Levin, T.P. Robinson, A. Teillant, and R. Laxminarayan. 2015. Global trends in antimicrobial use in food animals. Proc. Natl. Acad. Sci. USA 112:5649–5654. doi:10.1073/pnas.1503141112

Ventola, C.L. 2015. The antibiotic resistance crisis: Part 1: Causes and threats. P&T 40:277–283.

Vilca, F.Z., W.G. Angeles, E.T. Palma, C.N.C. Quiroz, T.D. Flores, and W.A.Z. Cuba. 2018. Antibiotics and their environmental implications. Revista de Investigaciones Altoandinas 20:215–224. doi:10.18271/ria.2018.365

Wang, J.M., H. Lin, W.C. Sun, Y. Xia, J.W. Ma, J.R. Fu, Z.L. Zhang, H.Z. Wu, and M.R. Qian. 2016. Variations in the fate and biological effects of sulfamethoxazole, norfloxacin and doxycycline in different vegetable-soil systems following manure application. J. Hazard. Mater. 304:49–57. doi:10.1016/j.jhazmat.2015.10.038

Wang, N., X.Y. Guo, J. Xu, L.J. Hao, D.Y. Kong, and S.X. Gao. 2015. Sorption and transport of five sulfonamide antibiotics in agricultural soil and soil-manure systems. J. Environ. Sci. Health B. 50:23–33. doi:10.1080/03601234.2015.965612

Wang, Q.Q., S.A. Bradford, W. Zheng, and S.R. Yates. 2006. Sulfadimethoxine degradation kinetics in manure as affected by initial concentration, moisture, and temperature. J. Environ. Qual. 35:2162–2169. doi:10.2134/jeq2006.0178

Wegst-Uhrich, S.R., D.A.G. Navarro, L. Zimmerman, and D.S. Aga. 2014. Assessing antibiotic sorption in soil: A literature review and new case studies on sulfonamides and macrolides. Chem. Cent. J. 8:5. doi:10.1186/1752-153X-8-5

Wehrhan, A., T. Streck, J. Groeneweg, H. Vereecken, and R. Kasteel. 2010. Long-term sorption and desorption of sulfadiazine in soil: Experiments and modeling. J. Environ. Qual. 39:654–666. doi:10.2134/jeq2009.0001

Wei, R.C., F. Ge, L.L. Zhang, X. Hou, Y.N. Cao, L. Gong, M. Chen, R. Wang, and E.D. Bao. 2016. Occurrence of 13 veterinary drugs in animal manure-amended soils in Eastern China. Chemosphere 144:2377–2383. doi:10.1016/j.chemosphere.2015.10.126

Werner, J.J., K. McNeill, and W.A. Arnold. 2009. Photolysis of Chlortetracycline on a clay surface. J. Agric. Food Chem. 57:6932–6937. doi:10.1021/jf900797a

Winckler, C., and A. Grafe. 2001. Use of veterinary drugs in intensive animal production. J. Soils Sediments 1:66. doi:10.1007/BF02987711

Witte, W. 2000. Selective pressure by antibiotic use in livestock. Int. J. Antimicrob. Agents 16:19–24. doi:10.1016/S0924-8579(00)00301-0

Wu, Q.F., Z.H. Li, and H.L. Hong. 2012a. Influence of types and charges of exchangeable cations on ciprofloxacin sorption by montmorillonite. J. Wuhan Univ. Technol., Mater. Sci. Ed. 27:516–522. doi:10.1007/s11595-012-0495-2

Wu, X.Q., J.L. Conkle, and J. Gan. 2012b. Multi-residue determination of pharmaceutical and personal care products in vegetables. J. Chromatogr. A 1254:78–86. doi:10.1016/j.chroma.2012.07.041

Wu, Y., M. Williams, L. Smith, D.H. Chen, and R. Kookana. 2012c. Dissipation of sulfamethoxazole and trimethoprim antibiotics from manure-amended soils. J. Environ. Sci. Health B. 47:240–249. doi:10.1080/03601234.2012.636580

Xie, W.Y., Q. Shen, and F.J. Zhao. 2018. Antibiotics and antibiotic resistance from animal manures to soil: A review. Eur. J. Soil Sci. 69:181–195. doi:10.1111/ejss.12494

Xuan, R.C., L. Arisi, Q.Q. Wang, S.R. Yates, and K.C. Biswas. 2010. Hydrolysis and photolysis of oxytetracycline in aqueous solution. J. Environ. Sci. Health B. 45:73–81. doi:10.1080/03601230903404556

Yan, W., S. Hu, and C.Y. Jing. 2012. Enrofloxacin sorption on smectite clays: Effects of pH, cations, and humic acid. J. Colloid Interface Sci. 372:141–147. doi:10.1016/j.jcis.2012.01.016

Yang, J.F., G.G. Ying, L.J. Zhou, S. Liu, and J.L. Zhao. 2009. Dissipation of oxytetracycline in soils under different redox conditions. Environ. Pollut. 157:2704–2709. doi:10.1016/j.envpol.2009.04.031

Zhang, D.Q., R.M. Gersberg, T. Hua, J. Zhu, M.K. Goyal, W.J. Ng, and S.K. Tan. 2013. Fate of pharmaceutical compounds in hydroponic mesocosms planted with Scirpus validus. Environ. Pollut. 181:98–106. doi:10.1016/j.envpol.2013.06.016

Zhang, H.M., M.K. Zhang, and G.P. Gu. 2008. Residues of tetracyclines in livestock and poultry manures and agricultural soils from north Zhejiang Province. J. Ecol. Rural Environ. 24:69–73.

Zhang, H.Q., H.B. Xie, J.W. Chen, and S.S. Zhang. 2015. Prediction of hydrolysis pathways and kinetics for antibiotics under environmental pH conditions: A quantum chemical study on cephradine. Environ. Sci. Technol. 49:1552–1558. doi:10.1021/es505383b

Zhang, Y., S.Q. Hu, H.C. Zhang, G.X. Shen, Z.J. Yuan, and W. Zhang. 2017. Degradation kinetics and mechanism of sulfadiazine and sulfamethoxazole in an agricultural soil system with manure application. Sci. Total Environ. 607–608:1348–1356. doi:10.1016/j.scitotenv.2017.07.083

Zhang, Y.L., S.S. Lin, C.M. Dai, L. Shi, and X.F. Zhou. 2014. Sorption-desorption and transport of trimethoprim and sulfonamide antibiotics in agricultural soil: Effect of soil type, dissolved organic matter, and pH. Environ. Sci. Pollut. Res. 21:5827–5835. doi:10.1007/s11356-014-2493-8

Zheng, W., K.N. Wiles, N. Holm, N.A. Deppe, and C.R. Shipley. 2014. Uptake, translocation, and accumulation of pharmaceutical and hormone contaminants in vegetables. In: K. Myung, N.M. Satchivi, and C.K. Kingston, editors, Retention. Uptake, and translocation of agrochemicals in plants, Amer Chemical Soc., Washington. p. 167–181. doi:10.1021/bk-2014-1171.ch009

Zou, Y.H., and W. Zheng. 2013. Modeling manure colloid-facilitated transport of the weakly hydrophobic antibiotic florfenicol in saturated soil columns. Environ. Sci. Technol. 47:5185–5192. doi:10.1021/es400624w